INTRODUCTION TO ENVIRONMENTAL ENGINEERING AND SCIENCE

Introduction to Environmental Engineering and Science

Second Edition

Gilbert M. Masters
Stanford University

Prentice Hall
Upper Saddle River, New Jersey 07458

Masters, Gilbert M.
    Introductionto environmental engineering and science / Gilbert M. Masters. -- 2nd ed.
           p. sm.
    Includes index.
    ISBN 0-13-155384
    1. Environmental engineering. 2. Environmental sciences.
I. Title.
TD 145.M33 1996   97-27372
628--dc21   CIP

Acquisitions editor: Bill Stenquist
Editor-in-chief: Marcia Horton
Production editor: AnnMarie Longobardo
Copy editor: Patricia M. Daly
Director of production and manufacturing: David W. Riccardi
Art director: Jayne Conte
Managing editor: Bayani Mendoza de Leon
Cover designer: Karen Salzbach
Manufacturing buyer: Julia Meehan
Editorial assistant: Meg Weist

Cover photo: "Dawn at Port Augusta, South Australia" Bill Van Aken, CSIRO Land and Water Australia.

Prentice Hall

© 1998 by Prentice-Hall, Inc.
A Division of Pearson Education
Upper Saddle River, New Jersey 07458

The author and publisher of this book have used their best efforts in preparing this book. These efforts include the development, research, and testing of the theories and programs to determine their effectiveness. The author and publisher make no warranty of any kind, expressed or implied, with regard to these programs or the documentation contained in this book. The author and publisher shall not be liable in any event for incidental or consequential damages in connection with, or arising out of, the furnishing, performance, or use of these programs.

Printed in the United States of America

10   9

ISBN 0-13-155384-4

Prentice-Hall International (UK) Limited, London
Prentice-Hall of Australia Pty. Limited, Sydney
Prentice-Hall Canada, Inc., Toronto
Prentice-Hall Hispanoamericana, S.A., Mexico
Prentice-Hall of India Private Limited, New Delhi
Prentice-Hall of Japan, Inc., Tokyo
Prentice-Hall Asia Pte. Ltd., Singapore
Editora Prentice-Hall do Brasil, Ltda., Rio de Janeiro

To my parents,

Gil and Ruth,

who taught me the important things

Contents

Preface

The scope of environmental engineering and science continues to expand both in terms of the number of cities and countries of the world where water and air quality problems are in urgent need of attention, and in terms of the pollutants themselves, which now so often seem to have international and global impacts.

Due to diligent efforts of environmental engineers and scientists, great progress has been made in our understanding of the fate and transport of substances that contaminate our air, surface water, soil, and subsurface water systems. That progress has led to better technologies for controlling emissions and for cleaning up contaminated sites. With increased understanding and better technologies, it has been possible to craft more sophisticated legislation to address these problems. And, with a better sense of the enormous costs of cleaning up problems after they are created, we are beginning to focus on pollution prevention. In some parts of the developed world, the result has been air that is getting cleaner, greater areas of surface waters that allow beneficial uses such as fishing and swimming, some improvement in subsurface water quality, and, very importantly, continued access to safe drinking water. Unfortunately, the numbers of people globally who do not enjoy these environmental benefits continues to grow, some traditional environmental problems still seem intractable, the global implications of greenhouse gases and ozone-depleting substances seem even more threatening, and, in spite of the importance of the work to be done, our public will to face these challenges seems no longer assured.

It is hoped that some of the science, technology, and policy instruments that have enabled parts of the United States to approach clean air and water goals can be applied to the enormous pollution problems that are coming to light as the former Soviet Union and other eastern bloc countries transform their economies. Similarly, the continued rapid population growth and urbanization occurring in the developing countries of the world is causing unparalleled environmental health risks as people flock to cities that lack basic sanitation services and other infrastructure to control air and water contamination. Environmental engineers will play an increasingly important role as these countries attempt to improve their lot.

Since publication of the first edition of this book, there have been a number of significant studies and actions that are changing the way we think about and deal with our environmental challenges. The Clean Air Act Amendments of 1990, for example, are shifting the approach taken to emission controls from the traditional "command and control" methods, in which government dictates the use of technology, to a more market-based approach that allows major sources to buy and sell emission allowances. In the area of groundwater cleanup, which has received much of the research funding and offered a significant fraction of the professional environmental engineering work in the last decade, a 1994 national Research Council report concluded that the most commonly used remediation technologies have little hope of ever restoring many of our contaminated aquifers to drinking water quality. Our perceptions of "how clean is clean?" are changing. The Comprehensive Environmental Response, Compensation, and Liabilities Act that created Superfund, which deals with such sites, has been only moderately successful and has been severely criticized as providing too little risk reduction for too many dollars. Partly in response, the concept of "brownfields," with more flexible cleanup goals and much more limited liability to owners who redevelop abandoned urban industrial sites, is emerging as a way to stimulate local economies while speeding cleanup of modestly polluted sites.

In the area of global atmospheric contamination, the impact of the Montreal Protocol on Substances that Deplete the Ozone Layer is beginning to be felt as the ban on production and use of ozone-depleting substances takes effect. The atmospheric concentrations of the most important chlorofluorocarbons are no longer increasing, and models suggest that the ozone layer will begin to repair itself in the early twenty-first century. The apparent success of the Montreal Protocol is serving as a model for negotiations on the other global atmospheric problem, global warming. The 1996 assessments by the Intergovernmental Panel on Climate Change (IPCC) have concluded that the current global warming trend is not entirely natural in origin. Our understanding of the implications of continued warming, how fast it will occur, and how extensive it will be is fraught with uncertainties. How we make decisions about adaptation or mitigation for such a potentially critical environmental problem, given those uncertainties, is sure to be a contentious issue in the coming years.

This second edition has been completely updated, slightly reorganized, and modestly expanded. The goal continues to be to provide an environmental textbook that can bridge the gap between the qualitative descriptions so well presented in many environmental science books and the quantitative analysis and design techniques that are the essence of most advanced environmental engineering texts. A new chapter has been added on solid waste management, which fits in well with the emerging themes of pollution prevention and product stewardship. The material on risk assessment has been expanded and separated from hazardous substance legislation and is now a complete chapter in itself. By moving risk closer to the front of the book, it now stands along with the other three introductory chapters on mass and energy balances, environmental chemistry, and mathematics of growth as a topic that is fundamental to many environmental problems. The material on regulation of hazardous substances has been added to the water quality control chapter, which has been reorganized and renamed.

This book is not the only one designed for use in entry-level technical courses on the environment, but it does offer some features that are not commonly found in other texts. This book covers the basic, traditional materials on air and water pollution that have been the backbone of many introductory environmental engineering and science courses. It provides the necessary fundamental science and engineering principles that are generally assumed to be common knowledge for an advanced undergraduate, but that may be new, or may need to be reviewed, for many students in a mixed class of upper and lower division students. What it adds to the usual engineering introduction, though, is a basis for analyzing and understanding the newer environmental issues that have become the focus of much of the environmental attention in more recent years. Special attention is given to such topics as hazardous waste, risk assessment, groundwater contamination, indoor air quality, acid deposition, global climate change, and stratospheric ozone depletion.

The book has been organized and presented with a variety of possible courses in mind. First, it could be used, from start to finish, in a standard sophomore or junior-level environmental engineering class. In a more advanced undergraduate course, the first three chapters on fundamentals might be skipped or reviewed only lightly. Emphasis could them be placed on some of the more technical details provided in later chapters, supplemented with current environmental literature. I also envision the book being used as a text for a second course on the environment for less technical students, which would come after the usual general environmental science course that most colleges and universities now offer. Many of the more detailed quantitative aspects of the book can be covered lightly in such courses if necessary, while keeping the basic modeling and problem-solving techniques intact. Actually, I use the book at Stanford for a course that attracts both engineering and nontechnical students ranging from freshman to seniors. By carefully adjusting the level of presentation for the mixed classroom, and offering alternative assignments from the book for students with more intense and stronger backgrounds, it is possible to meet the needs of students with quite a range of abilities.

For most topics covered in this book, pertinent environmental legislation is described, simple engineering models are generated, and qualitative descriptions of treatment technologies are presented. The book has been designed to encourage self-teaching by providing numerous completely worked examples throughout. Virtually every topic that lends itself to quantitative analysis is illustrated with such examples. Each chapter ends with a relatively long list of problems that provide added practice for the student and should facilitate the preparation of homework assignments by the instructor.

The first four chapters use environmental problems to illustrate the application of certain key principles of engineering and science that are required for any quantitative treatment of environmental problems. The first chapter presents mass and energy transfer, the second reviews some of the essential chemistry with new material on enthalpy, and the third introduces certain mathematical functions used to model growth that are especially useful in developing future scenarios and projections. The fourth chapter introduces risk assessment and includes new material on risk perception. These four chapters provide the basic foundation needed for the more specialized topics that are the focus of subsequent chapters.

The remaining five chapters are much longer presentations of some of the major environmental problems of the day. these chapters are relatively modular and could be covered in virtually any order. My course at Stanford, for example, takes these chapters in almost reverse order, beginning with global atmospheric change. Chapters 5 and 6 cover topics that traditionally have been the essence of undergraduate civil engineering courses on the environment. In Chapter 5, an expanded introduction to water resources is followed by major sections on surface water contamination and groundwater. The description of groundwater remediation now includes material on the challenges posed by nonaqueous-phase liquids that contaminate the subsurface along with a summary of the National Resource Council's assessment of the ability of conventional pump-and-treat remediation technologies to restore aquifer quality. Chapter 6 focuses on water and wastewater treatment systems for conventional pollutants supplemented with sections on the treatment of hazardous substances: the Toxic Substances Control Act, the Resource Conservation and Recovery Act, and the Comprehensive Environmental Response, Compensation, and Liabilities Act.

Chapter 7 presents a rather thorough introduction to traditional air pollution problems involving criteria pollutants, local meteorology, simple dispersion models, and emission controls. The material on mobile source controls has been expanded to include a description of reformulated gasoline and other alternative fuels, and new material on two-stroke engines and hybrid electric vehicles has been added. The chapter has much more coverage of the often overlooked, but serious problems of indoor air quality, with special attention paid to environmental tobacco smoke and radon.

Chapter 8, on global atmospheric change, has been completely revised to match the perspectives, terminology, and approaches that have become the standard methods of addressing the problem of global warming as presented in the IPCC Scientific Assessments. The IPCC reports are lengthy, detailed documents that are periodically updated to present the latest summary of the technical literature on climate change. They assume a working knowledge of the basic attributes of radiatively active gases and how they affect the global energy balance, which Chapter 8 has been specifically rewritten to provide. The chapter has also been updated to include more extensive coverage of the status of ozone-depleting substances.

Chapter 9, on solid waste management and resource recovery, is a new addition to this book. The conventional topics of collection and transfer operations and municipal solid-waste landfills are briefly introduced, but the emphasis here is on the hierarchy of solid-waste management priorities, including source reduction, recycling, composting, and waste-to-energy combustion.

Obviously, for a single author to write a text that covers the range of topics introduced here is a challenging task. I have found it to be remarkably rewarding to continue to learn new things as I struggle to keep abreast of these fields, but I am always concerned about getting it right and rely heavily on the reviews and criticisms of my colleagues, who are much more focused in their work. Fortunately, the experts who have helped by providing suggestions, corrections, and criticisms have done so with generosity and gentle humor, and I am greatly in their debt. In particular, I would like to thank Professors Lynn Hildemann of Stanford; Susan Masten, Michigan State University; Brian J. Savilonis, Worcester Polytechnic Institute; and John T. Pfeffer, University of Illinois for their many helpful suggestions. Special thanks go to Kiersten West for her extensive help in gathering material for the new chapter on solid waste management, and to Professor John Ferguson from the University of Washington for offering his special spot on Orcas Island as a quiet place to write. Finally, I wish to acknowledge my wife, Mary, whose writing skills and expertise in hazardous waste management have made the book better, and whose encouragement and patience have been essential to its completion.

Gilbert M. Masters

INTRODUCTION TO ENVIRONMENTAL ENGINEERING AND SCIENCE

C H A P T E R   1

Mass and Energy Transfer

1.1 INTRODUCTION
1.2 UNITS OF MEASUREMENT
1.3 MATERIALS BALANCE
1.4 ENERGY FUNDAMENTALS
PROBLEMS

When you can measure what you are speaking about, and express it in numbers, you know something about it; but when you cannot measure it, when you cannot express it in numbers, your knowledge is of a meagre and unsatisfactory kind; it may be the beginning of knowledge, but you have scarcely, in your thoughts, advanced to the stage of science. —William Thomson, Lord Kelvin (1891)

1.1 INTRODUCTION

While most of the chapters in this book focus on specific environmental problems, such as pollution in surface waters or degradation of air quality, there are a number of important concepts that find application throughout the study of environmental engineering and science.

This chapter begins with a section on units of measurement. Engineers need to be familiar with both the American units of feet, pounds, hours, and degrees Fahrenheit as well as the more recommended International System of units. Both are used in the practice of environmental engineering and both will be used throughout this book.

Next, two fundamental topics, which should be familiar from the study of elementary physics, are presented: the *law of conservation of mass* and the *law of conservation of energy*. These laws tell us that within any environmental system we theoretically should be able to account for the flow of energy and materials into, and out of, that system. The law of conservation of mass, besides giving us an important tool for quantitatively tracking pollutants as they disperse in the environment, reminds us that pollutants have to go somewhere, and that we should be wary of approaches that merely transport them from one medium to another.

In a similar way, the law of conservation of energy is also an essential accounting tool with special environmental implications. When coupled with other thermodynamic principles, it will be useful in a number of applications, including the study of global climate change, thermal pollution, and the dispersion of air pollutants.

1.2 UNITS OF MEASUREMENT

In the Unites States, environmental quantities are measured and reported in both the *U.S. Customary System* (USCS) and the *International System of Units* (SI) and so it is important to be familiar with both. In this book, preference is given to SI units, though the American system will be used in some circumstances. Table 1.1 gives conversion factors between the SI and USCS systems for some of the most basic units that will be encountered. A more extensive table of conversions is given in the appendix at the end of the book.

In the study of environmental engineering, it is quite common to encounter both extremely large quantities and extremely small ones. The concentration of some toxic substance may be measured in parts per billion (ppb), for example, while a country's rate of energy use may be measured in thousands of billions of watts (terawatts). To describe quantities that may take on such extreme values, it is useful to have a system of prefixes that accompany the units. Some of the most important prefixes are presented in Table 1.2.

Quite often, it is the concentration of some substance in air or water that is of interest. In either medium, concentrations may be based on mass, volume, or a combination of the two, which can lead to some confusion.

Liquids

Concentrations of substances dissolved in water are usually expressed in terms of mass of substance per unit volume of mixture. Most often the units are milligrams (mg) or micrograms (μg) of substance per liter (L) of mixture. At times they may be expressed in grams per cubic meter (g/m^3).

Alternatively, concentrations in liquids are expressed as mass of substance per mass of mixture, with the most common units being parts per million (ppm), or parts

TABLE 1.1   Some Basic Units and Conversion Factors[a]

Quantity	SI Units	SI Symbol	× Conversion Factor	= USCS Units
Length	Meter	m	3.2808	ft
Mass	Kilogram	kg	2.2046	lb
Temperature	Celsius	°C	1.8 (°C) + 32	°F
Area	Square meter	m^2	10.7639	ft^2
Volume	Cubic meter	m^3	35.3147	ft^3
Energy	Kilojoule	kJ	0.9478	Btu
Power	Watt	W	3.4121	Btu/hr
Velocity	Meter/sec	m/s	2.2369	mi/hr
Flow rate	Meter^3/sec	m^3/s	35.3147	ft^3/s
Density	Kilogram/meter^3	kg/m^3	0.06243	lb/ft^3

[a]See the appendix for a more complete list.

___

TABLE 1.2   Common Prefixes

Quantity	Prefix	Symbol
10^{-15}	femto	f
10^{-12}	pico	p
10^{-9}	nano	n
10^{-6}	micro	μ
10^{-3}	milli	m
10^{-2}	centi	c
10^{-1}	deci	d
10	deka	da
10^2	hecto	h
10^3	kilo	k
10^6	mega	M
10^9	giga	G
10^{12}	tera	T
10^{15}	peta	P
10^{18}	exa	E
10^{21}	zetta	Z
10^{24}	yotta	Y

___

per billion (ppb). To help put these units in perspective, 1 ppm is about the same as 1 drop of vermouth added to 15 gallons of gin, while 1 ppb is about the same as one drop of pollutant in a fairly large (70 m^3) backyard swimming pool. Since most concentrations of pollutants are very small, 1 L of mixture has a mass that is essentially 1000 g, so for all practical purposes we can write

$$1 \text{ mg/L} = 1 \text{ g/m}^3 = 1 \text{ ppm (by weight)} \tag{1.1}$$

$$1 \text{ } \mu\text{g/L} = 1 \text{ mg/m}^3 = 1 \text{ ppb (by weight)} \tag{1.2}$$

In unusual circumstances, the concentration of liquid wastes may be so high that the specific gravity of the mixture is affected, in which case a correction to (1.1) and (1.2) may be required:

$$\text{mg/L} = \text{ppm (by weight)} \times \text{specific gravity of mixture} \tag{1.3}$$

Gases

For most air pollution work, it is customary to express pollutant concentrations in volumetric terms. For example, the concentration of a gaseous pollutant in parts per million is the volume of pollutant per million volumes of the air mixture:

$$\frac{1 \text{ volume of gaseous pollutant}}{10^6 \text{ volumes of air}} = 1 \text{ ppm (by volume)} = 1 \text{ ppmv} \tag{1.4}$$

To help remind us that this fraction is based on volume, it is common to add a "v" to the ppm; giving ppmv, as suggested in (1.4).

At times, concentrations are expressed as mass per unit volume, such as μg/m^3 or mg/m^3. The relationship between ppmv and mg/m^3 depends on the pressure, temperature, and molecular weight of the pollutant. The ideal gas law helps us establish that relationship:

$$PV = nRT \tag{1.5}$$

where

P = absolute pressure (atm)
V = volume (m^3)
n = mass (mol)
R = ideal gas constant = 0.082056 L·atm·K^{-1}·mol^{-1}
T = absolute temperature (K)

The mass in (1.5) is expressed as moles of gas. Recall from chemistry that one mole of any gas has Avogadro's number of molecules in it (6.02×10^{23} molecules/mol) and has a mass equal to its molecular weight. Also note the temperature is expressed in kelvins (K), where

$$K = °C + 273.15 \tag{1.6}$$

There are a number of ways to express pressure; in (1.5) we have used atmospheres. One atmosphere of pressure equals 101.325 kPa (Pa is the abbreviation for Pascals). One atmosphere is also equal to 14.7 pounds per square inch (psi), so 1 psi = 6.89 kPa. Finally, 100 kPa is called a bar and 100 Pa is a millibar, which is the unit of pressure often used in meteorology.

EXAMPLE 1.1   Volume of an Ideal Gas

Find the volume that 1 mol of an ideal gas would occupy at standard temperature and pressure (STP) conditions of 1 atm of pressure and 0°C temperature. Repeat the calculation for 1 atm and 25°C.

Solution   Using (1.5) at a temperature of 0°C (273.15 K) gives

$$V = \frac{1 \text{ mol} \times 0.082056 \text{ L·atm·K}^{-1}\text{·mol}^{-1} \times 273.15 \text{ K}}{1 \text{ atm}} = 22.414 \text{ L}$$

and at 25°C (298.15 K)

$$V = \frac{1 \text{ mol} \times 0.082056 \text{ L·atm·K}^{-1}\text{·mol}^{-1} \times 298.15 \text{ K}}{1 \text{ atm}} = 24.465 \text{ L}$$

■

From Example 1.1, 1 mol of an ideal gas at 0°C and 1 atm occupies a volume of 22.414 L (22.414×10^{-3} m^3). Thus we can write

$$\text{mg/m}^3 = \text{ppm} \times \frac{1 \text{ m}^3 \text{ pollutant}/10^6 \text{ m}^3 \text{ air}}{\text{ppm}} \times \frac{\text{mol wt (g/mol)}}{22.414 \times 10^{-3} \text{ m}^3/\text{mol}} \times 10^3 \text{ mg/g}$$

or, more simply,

$$\text{mg/m}^3 = \frac{\text{ppm} \times \text{mol wt}}{22.414} \quad (\text{at } 0 °C \text{ and 1 atm}) \tag{1.7}$$

Similarly, at 25 °C and 1 atm, which are the conditions that are assumed when air quality standards are specified in the United States,

$$\text{mg/m}^3 = \frac{\text{ppm} \times \text{mol wt}}{24.465} \quad \left(\text{at } 25\,°\text{C and } 1 \text{ atm}\right) \tag{1.8}$$

In general, the conversion from ppm to mg/m³ is given by

$$\text{mg/m}^3 = \frac{\text{ppm} \times \text{mol wt}}{22.414} \times \frac{273.15 \text{ K}}{T(\text{K})} \times \frac{P(\text{atm})}{1 \text{ atm}} \tag{1.9}$$

EXAMPLE 1.2   Converting ppm to μg/m³

The federal Air Quality Standard for carbon monoxide (based on an 8-hour measurement) is 9.0 ppm. Express this standard as a percent by volume as well as in mg/m³ at 1 atm and 25 °C.

Solution   Within a million volumes of this air there are 9.0 volumes of CO, no matter what the temperature or pressure (this is the advantage of the ppm units). Hence, the percentage by volume is simply

$$\text{percent CO} = \frac{9.0}{1 \times 10^6} \times 100 = 0.0009 \text{ percent}$$

To find the concentration in mg/m³, we need the molecular weight of CO, which is 28 (the atomic weights of C and O are 12 and 16, respectively). Using (1.8) gives

$$\text{CO} = \frac{9.0 \times 28}{24.465} = 10.3 \text{ mg/m}^3$$

Actually, it is usually rounded and listed as 10 mg/m³.   ∎

The fact that 1 mol of every ideal gas occupies the same volume (under the same temperature and pressure condition) provides several other interpretations of volumetric concentrations expressed as ppmv. For example, 1 ppmv is one volume of pollutant per million volumes of air, which is equivalent to saying 1 mol of pollutant per million moles of air. Similarly, since each mole contains the same number of molecules, 1 ppmv also corresponds to 1 molecule of pollutant per million molecules of air.

$$1 \text{ ppmv} = \frac{1 \text{ mol of pollutant}}{10^6 \text{ mol of air}} = \frac{1 \text{ molecule of pollutant}}{10^6 \text{ molecules of air}} \tag{1.10}$$

1.3 MATERIALS BALANCE

"Everything has to go somewhere" is a simple way to express one of the most fundamental engineering principles. More precisely, the *law of conservation of mass* says that when chemical reactions take place, matter is neither created nor destroyed (though in nuclear reactions mass can be converted to energy). What this concept allows us to do is track materials (e.g., pollutants) from one place to another with mass balance equations.

The first step in a mass balance analysis is to define the particular region in space that is to be analyzed. As examples, the region might include anything from a simple chemical mixing tank to an entire coal-fired power plant, a lake, a stretch of stream, an air basin above a city, or the globe itself. By picturing an imaginary boundary around the region, as is suggested in Figure 1.1, we can then begin to identify the flow of materials across the boundary as well as the accumulation of materials within the region.

A substance that enters the region has three possible fates. Some of it may leave the region unchanged; some of it may accumulate within the boundary; and some of it may be converted to some other substance (e.g., entering CO may be oxidized to CO_2 within the region). Thus, using Figure 1.1 as a guide, the following materials balance equation can be written for each substance of interest:

$$\begin{pmatrix} \text{Input} \\ \text{rate} \end{pmatrix} = \begin{pmatrix} \text{Output} \\ \text{rate} \end{pmatrix} + \begin{pmatrix} \text{Decay} \\ \text{rate} \end{pmatrix} + \begin{pmatrix} \text{Accumulation} \\ \text{rate} \end{pmatrix} \qquad (1.11)$$

Notice that the "decay" term in (1.11) does not imply a violation of the law of conservation of mass. Atoms are conserved, but there is no similar constraint on the chemical reactions, which may change one substance into another.

Frequently, (1.11) can be simplified. The most common simplification results when *steady-state* or *equilibrium* conditions can be assumed. Equilibrium simply means that nothing is changing with time; the system has had its inputs held constant for a long enough time that any transients have had a chance to die out. Pollutant concentrations are constant. Hence the "accumulation rate" term in (1.11) is set equal to zero and problems can usually be solved using just simple algebra.

A second simplification of (1.11) results when a substance is *conserved* within the region in question, meaning that there is no radioactive decay, bacterial decomposition, or chemical reaction occurring. For such conservative substances, the "decay rate" term in (1.11) is zero. Examples of substances that are typically modeled as conservative include total dissolved solids in a body of water, heavy metals in soils, and carbon dioxide in air. Nonconservative substances would include radioactive radon gas in a home or decomposing organic wastes in a lake. Many times problems involving nonconservative substances can be simplified when the reaction rate term is small enough to be ignored.

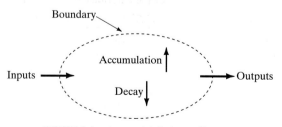

FIGURE 1.1 A materials balance diagram.

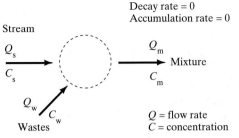

Decay rate = 0
Accumulation rate = 0

Stream

Q_s

C_s

Q_w
C_w
Wastes

Q_m
Mixture
C_m

Q = flow rate
C = concentration

FIGURE 1.2   A steady-state conservative system.
Pollutants enter and leave the region at the same rate.

Steady-state Conservative Systems

The simplest systems to analyze are those in which steady state is assumed and the substance in question is conservative. In these cases, (1.11) simplifies to the following:

$$\text{Input rate} = \text{Output rate} \tag{1.12}$$

Consider the steady-state conservative system shown in Figure 1.2.

The system contained within the boundaries might be a lake, a section of a free-flowing stream, or the mass of air above a city. One input to the system is a stream (of water or air, for instance) with a flow rate Q_s (volume/time) and pollutant concentration C_s (mass/volume). The other input is assumed to be a waste stream with flow rate Q_w and pollutant concentration C_w. The output is a mixture with flow rate Q_m and pollutant concentration C_m. If the pollutant is conservative, and if we assume steady-state conditions, then a mass balance based on (1.12) allows us to write the following:

$$C_s Q_s + C_w Q_w = C_m Q_m \tag{1.13}$$

The following example illustrates the use of this equation.

EXAMPLE 1.3   Two Polluted Streams

A stream flowing at $10.0 \text{ m}^3/\text{s}$ has a tributary feeding into it with a flow $5.0 \text{ m}^3/\text{s}$. The stream's concentration of chlorides upstream of the junction is 20.0 mg/L and the tributary chloride concentration is 40.0 mg/L. Treating chlorides as a conservative substance, and assuming complete mixing of the two streams, find the downstream chloride concentration.

Solution   We begin by sketching the problem and identifying the region that we want to analyze, as has been done in Figure 1.3.

Rearranging (1.13) to solve for the chloride concentration downstream gives us

$$C_m = \frac{C_s Q_s + C_w Q_w}{Q_m} = \frac{C_s Q_s + C_w Q_w}{Q_s + Q_w}$$

Note that since the mixture flow is the sum of the two stream flows, $Q_s + Q_w$ has been substituted for Q_m in this expression. All that remains is to substitute the appropriate values for the known quantities into the expression, which brings us to a question of units. The units given for C are mg/L and for Q they are m^3/s . Taking the product of concentrations and flow rates yields mixed units of $\text{mg/L} \cdot \text{m}^3/\text{s}$, which we could simplify by applying the conversion factor of $10^3 \text{ L} = 1 \text{ m}^3$. However, if we did so, we should have to reapply that same conversion factor to

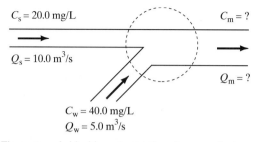

FIGURE 1.3 Flow rate and chloride concentrations for example stream and tributary.

get the mixture concentration back into the desired units of mg/L. In problems of this sort, it is much easier simply to leave the mixed units in the expression, even though they may look awkward at first, and let them work themselves out in the calculation. The downstream concentration of chlorides is thus

$$C_m = \frac{(20.0 \times 10.0 + 40.0 \times 5.0)\ \text{mg/L} \cdot \text{m}^3/\text{s}}{(10.0 + 5.0)\ \text{m}^3/\text{s}}$$

∎

Steady-state Systems with Nonconservative Pollutants

Many contaminants undergo chemical, biological, or nuclear reactions at a rate sufficient to necessitate treating them as nonconservative substances. If we continue to assume that steady-state conditions prevail so that the rate of accumulation is zero, but now treat the pollutants as nonconservative, then (1.11) becomes

$$\text{Input rate} = \text{Output rate} + \text{Decay rate} \qquad (1.14)$$

The decay of nonconservative substances is frequently modeled as a first-order reaction; that is, it is assumed that the rate of loss of the substance is proportional to the amount of the substance that is present. That is,

$$\frac{dC}{dt} = -KC \qquad (1.15)$$

where K is a reaction rate coefficient with dimensions of (time^{-1}), the negative sign implies a loss of substance with time, and C is the pollutant concentration. To solve this differential equation, we can rearrange the terms and integrate

$$\int_{C_0}^{C} \frac{dC}{C} = \int_{0}^{t} (-K)\, dt$$

which yields

$$\ln(C) - \ln(C_0) = \ln\left(\frac{C}{C_0}\right) = -Kt$$

Solving for concentration gives us

$$C = C_0 e^{-Kt} \tag{1.16}$$

where C_0 is the initial concentration. That is, assuming a first-order reaction, the concentration of the substance in question decays exponentially. This exponential function will appear so often in this text that it will be reintroduced and explored more fully in Chapter 3.

Equation (1.15) indicates the rate of change of *concentration* of the substance. If we assume that the substance is uniformly distributed throughout a volume V, then the total *amount* of substance is CV. The total rate of decay of the amount of a nonconservative substance is thus $d(CV)/dt = V\, dC/dt$, so using (1.15) we can write for a nonconservative substance:

$$\text{Decay rate} = KCV \tag{1.17}$$

Substituting (1.17) into (1.14) gives us our final simple, yet useful, expression for the mass balance involving a nonconservative pollutant in a steady-state system:

$$\text{Input rate} = \text{Output rate} + KCV \tag{1.18}$$

Implicit in (1.18) is the assumption that the concentration C is uniform throughout the volume V. This complete mixing assumption is common in the analysis of chemical tanks, called *reactors,* and in such cases the idealization is referred to as a *continuously stirred tank reactor* (CSTR) model. In other contexts, such as modeling air pollution, the assumption is referred to as a *complete mix box model.*

EXAMPLE 1.4   A Polluted Lake

Consider a $10.0 \times 10^6\ \text{m}^3$ lake fed by a polluted stream having a flow rate of $5.0\ \text{m}^3/\text{s}$ and pollution concentration equal to $10.0\ \text{mg/L}$ (Figure 1.4). There is also a sewage outfall that discharges $0.5\ \text{m}^3/\text{s}$ of wastewater having a pollutant concentration of $100\ \text{mg/L}$. The stream and sewage wastes have a reaction rate coefficient of $0.20/\text{day}$. Assuming the pollution is completely mixed in the lake, and assuming no evaporation or other water losses or gains, find the steady-state concentration.

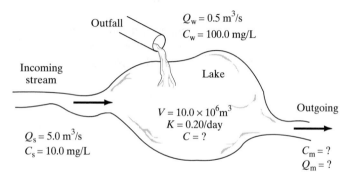

FIGURE 1.4   A lake with a nonconservative pollutant.

Solution   Assuming that complete and instantaneous mixing occurs in the lake implies that the concentration in the lake C is the same as the concentration of the mix leaving the lake, C_m. Using (1.18),

$$\text{Input rate} = \text{Output rate} + KCV \tag{1.18}$$

We can find each term as follows:

$$\text{Input rate} = Q_s\,C_s + Q_w\,C_w$$
$$= (5.0\ \text{m}^3/\text{s} \times 10.0\ \text{mg/L} + 0.5\ \text{m}^3/\text{s} \times 100.0\ \text{mg/L}) \times 10^3\ \text{L/m}^3$$
$$= 1.0 \times 10^5\ \text{mg/s}$$

$$\text{Output rate} = Q_m\,C_m = (Q_s + Q_w)C$$
$$= (5.0 + 0.5)\ \text{m}^3/\text{s} \times C\ \text{mg/L} \times 10^3\ \text{L/m}^3 = 5.5 \times 10^3 C\ \text{mg/s}$$

$$\text{Decay rate}\ = KCV = \frac{0.20/\text{d} \times C\ \text{mg/L} \times 10.0 \times 10^6\ \text{m}^3 \times 10^3\ \text{L/m}^3}{24\ \text{hr/d} \times 3600\ \text{s/hr}}$$
$$= 23.1 \times 10^3 C\ \text{mg/s}$$

So from (1.14)

$$1.0 \times 10^5 = 5.5 \times 10^3 C + 23.1 \times 10^3 C = 28.6 \times 10^3 C$$

$$C = \frac{1 \times 10^5}{28.6 \times 10^3} = 3.5\ \text{mg/L}$$

∎

Idealized models involving nonconservative pollutants in completely mixed, steady-state systems are used to analyze a variety of commonly encountered water pollution problems such as the one shown in the previous example. The same simple models can be applied to certain problems involving air quality, as the following example demonstrates.

EXAMPLE 1.5   A Smoky Bar

A bar with volume 500 m³ has 50 smokers in it, each smoking two cigarettes per hour (see Figure 1.5). An individual cigarette emits, among other things, about 1.4 mg of formaldehyde (HCHO). Formaldehyde converts to carbon dioxide with a reaction rate coefficient $K = 0.40/\text{hr}$. Fresh air enters the bar at the rate of 1000 m³/hr, and stale air leaves at the same rate. Assuming complete mixing, estimate the steady-state concentration of formaldehyde in the

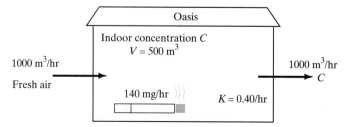

Oasis

Indoor concentration C
$V = 500\ \text{m}^3$

1000 m³/hr
Fresh air

1000 m³/hr
C

140 mg/hr
$K = 0.40/\text{hr}$

FIGURE 1.5   Tobacco smoke in a bar.

air. At 25 °C and 1 atm of pressure, how does the result compare with the threshold for eye irritation of about 0.05 ppm?

Solution   The rate at which formaldehyde enters the bar is

$$\text{Input rate} = 50 \text{ smokers} \times 2 \text{ cigs/hr} \times 1.4 \text{ mg/cig} = 140 \text{ mg/hr}$$

Since complete mixing is assumed, the concentration of formaldehyde C in the bar is the same as the concentration in the air leaving the bar, so

$$\text{Output rate} = 1000 \text{ m}^3/\text{hr} \times C(\text{mg/m}^3) = 1000C \text{ mg/hr}$$

and the decay rate is

$$\text{Decay rate} = KCV = (0.40/\text{hr}) \times (C \text{ mg/m}^3) \times (500 \text{ m}^3) = 200C \text{ mg/hr}$$

So, from (1.14),

$$\text{Input rate} = \text{Output rate} + \text{Decay rate}$$

$$140 = 1000C + 200C$$

$$C = \frac{140}{1200} = 0.117 \text{ mg/m}^3$$

We will use (1.8) to convert mg/m^3 to ppm. The molecular weight of formaldehyde is 30, so

$$\text{HCHO} = \frac{C (\text{mg/m}^3) \times 24/465}{\text{mol wt}} = \frac{0.117 \times 24.465}{30} = 0.095 \text{ ppm}$$

This is nearly double the 0.05 ppm threshold for eye irritation.   ■

Step Function Response

So far we have computed steady state concentrations in environmental systems that are contaminated with either conservative or nonconservative pollutants. Let us now extend the analysis to include conditions that are not steady state. Quite often, we will be interested in how the concentration will change with time when there is a sudden change in the amount of pollution entering the system. This is known as the *step function response* of the system.

In Figure 1.6 the environmental system to be modeled has been drawn as if it were a box of volume V that has flow rate Q out of the box. Again, let us assume the contents of the box are at all times completely mixed (a CSTR model) so that the pollutant concentration C in the box is the same as the concentration leaving the box. The total mass of pollutant in the box is therefore VC and the rate of increase of pollutant in the box is $V \, dC/dt$. Let us designate the total rate at which pollution enters

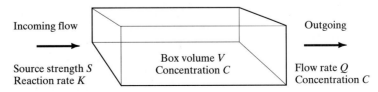

Incoming flow

Source strength S
Reaction rate K

Box volume V
Concentration C

Outgoing

Flow rate Q
Concentration C

FIGURE 1.6   A box model for a transient analysis.

the box as S, the source strength, with units of mass per unit time. If the pollutants are nonconservative, they will be modeled with a first-order reaction rate coefficient K. From (1.11) we can write

$$\begin{pmatrix} \text{Accumulation} \\ \text{rate} \end{pmatrix} = \begin{pmatrix} \text{Input} \\ \text{rate} \end{pmatrix} - \begin{pmatrix} \text{Output} \\ \text{rate} \end{pmatrix} - \begin{pmatrix} \text{Decay} \\ \text{rate} \end{pmatrix}$$

$$V\frac{dC}{dt} = S - QC - KCV \tag{1.19}$$

where

V = box volume (m³)
C = concentration in the box and exiting waste stream (g/m³)
S = total rate at which pollutants enter the box (g/hr)
Q = the total flow rate out of the box (m³/hr)
K = reaction rate coefficient (hr⁻¹)

The preceding units are representative of those that might be encountered; any consistent set will do.

An easy way to find the steady-state solution to (1.19) is simply to set $dC/dt = 0$, which yields

$$C_\infty = \frac{S}{Q + KV} \tag{1.20}$$

where C_∞ is the concentration in the box at time $t = \infty$. Our concern now, however, is with the concentration before it reaches steady state, so we must solve (1.19). Rearranging (1.19) gives

$$\frac{dC}{dt} = -\left(K + \frac{Q}{V}\right) \cdot \left[C - \frac{S}{Q + KV}\right] \tag{1.21}$$

which, using (1.20), can be rewritten as

$$\frac{dC}{dt} = -\left(K + \frac{Q}{V}\right) \cdot (C - C_\infty) \tag{1.22}$$

One way to solve this differential equation is to make a change of variable. If we let

$$y = C - C_\infty \tag{1.23}$$

then

$$\frac{dy}{dt} = \frac{dC}{dt} \tag{1.24}$$

so (1.22) becomes

$$\frac{dy}{dt} = -\left(K + \frac{Q}{V}\right) y \tag{1.25}$$

which is a differential equation just like (1.15), which we solved before. The solution is

$$y = y_0 e^{-(K+Q/V)t} \tag{1.26}$$

where y_0 is the value of y at $t = 0$. If C_0 is the concentration in the box at time $t = 0$, then from (1.23) we get

$$y_0 = C_0 - C_\infty \tag{1.27}$$

and substituting (1.23) and (1.27) into (1.26) yields

$$C - C_\infty = (C_0 - C_\infty)e^{-(K+Q/V)t} \tag{1.28}$$

Solving for the concentration in the box, writing it as a function of time $C(t)$, and expressing the exponential as exp( ) gives

$$C(t) = C_\infty + (C_0 - C_\infty)\exp\left[-(K + Q/V)t\right] \tag{1.29}$$

Equation (1.29) should make some sense. At time $t = 0$, the exponential function equals 1 and $C = C_0$. At $t = \infty$, the exponential term equals zero, and $C = C_\infty$. Equation (1.29) is plotted in Figure 1.7.

EXAMPLE 1.6   The Smoky Bar Revisited

The bar in Example 1.5 had volume 500 m^3 with fresh air entering at the rate of $1000 \text{ m}^3/\text{hr}$. Suppose the air in the bar is clean when it opens at 5 pm. If formaldehyde, with reaction rate $K = 0.40/\text{hr}$, is emitted from cigarette smoke at the constant rate of 140 mg/hr starting at 5 pm, what would the concentration be at 6 pm?

Solution   In this case, $Q = 1000 \text{ m}^3/\text{hr}$, $V = 500 \text{ m}^3$, $S = 140 \text{ mg/hr}$, and $K = 0.40/\text{hr}$. The steady-state concentration is found using (1.20):

$$C_\infty = \frac{S}{Q + KV} = \frac{140 \text{ mg/hr}}{1000 \text{ m}^3/\text{hr} + 0.40/\text{hr} \times 500\text{m}^3}$$

$$= 0.117 \text{ mg/m}^3$$

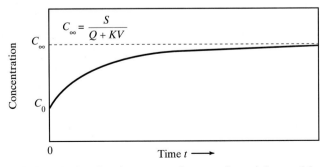

FIGURE 1.7   Step function response for a complete-mix box model.

which agrees with the result obtained in Example 1.5. To find the concentration at any time after 5 pm, we can apply (1.29) with $C_0 = 0$.

$$C(t) = C_\infty \{1 - \exp[-(K + Q/V)t]\}$$
$$= 0.117\{1 - \exp[-(0.40 + 1000/500)t]\}$$

at 6 pm, $t = 1$ hr, so

$$C(1 \text{ hr}) = 0.117[1 - \exp(-2.4 \times 1)] = 0.106 \text{ mg/m}^3$$ ∎

To demonstrate further the use of (1.29), let us reconsider the lake analyzed in Example 1.4. This time we will assume that the outfall suddenly stops draining into the lake so its contribution to the lake's pollution stops.

EXAMPLE 1.7   A Sudden Decrease in Pollutants Discharged into the Lake

Consider the $10 \times 10^6 \text{ m}^3$ lake analyzed in Example 1.4, which, under the conditions given, was found to have a steady-state pollution concentration of 3.5 mg/L. The pollution is nonconservative with reaction rate constant $K = 0.20/\text{day}$. Suppose the condition of the lake is deemed unacceptable, and to solve the problem it is decided to divert completely the sewage outfall around the lake, eliminating it as a source of pollution. The incoming stream still has flow $Q_s = 5.0 \text{ m}^3/\text{s}$ and concentration $C_s = 10.0 \text{ mg/L}$. With the sewage outfall removed, the outgoing flow Q is also $5.0 \text{ m}^3/\text{s}$. Assuming complete-mix conditions, find the concentration of pollution in the lake one week after the diversion and find the new final steady-state concentration.

Solution   For this situation,

$$C_0 = 3.5 \text{ mg/L}$$
$$V = 10 \times 10^6 \text{m}^3$$
$$Q = Q_s = 5.0 \text{ m}^3/\text{s} \times 3600 \text{ s/hr} \times 24 \text{ hr/day} = 43.2 \times 10^4 \text{m}^3/\text{day}$$
$$C_s = 10.0 \text{ mg/L} = 10.0 \times 10^3 \text{ mg/m}^3$$
$$K = 0.20/\text{day}$$

The total rate at which pollution is entering the lake from the incoming stream is

$$S = Q_s C_s = 43.2 \times 10^4 \text{ m}^3/\text{day} \times 10.0 \times 10^3 \text{ mg/m}^3 = 43.2 \times 10^8 \text{ mg/day}$$

The steady-state concentration can be obtained from (1.20):

$$C_\infty = \frac{S}{Q + KV} = \frac{43.2 \times 10^8 \text{ mg/day}}{43.2 \times 10^4 \text{ m}^3/\text{day} + 0.20/\text{day} \times 10^7 \text{ m}^3}$$
$$= 1.8 \times 10^3 \text{ mg/m}^3 = 1.8 \text{ mg/L}$$

Using (1.29), we can find the concentration in the lake one week after the drop in pollution from the outfall:

$$C(t) = C_\infty + (C_0 - C_\infty)\exp[-(K + Q/V)t]$$
$$C(7 \text{ days}) = 1.8 + (3.5 - 1.8)\exp\left[-\left(0.20/\text{day} + \frac{43.2 \times 10^4 \text{ m}^3/\text{day}}{10 \times 10^6 \text{ m}^3}\right) \times 7 \text{ days}\right]$$
$$C(7 \text{ days}) = 2.1 \text{ mg/L}$$

Figure 1.8 shows the decrease in contaminant concentration for this example.   ∎

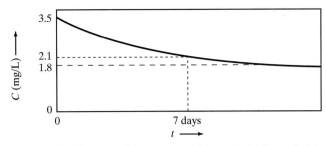

FIGURE 1.8   The contaminant concentration profile for Example 1.7.

1.4 ENERGY FUNDAMENTALS

Just as we are able to use the law of conservation of mass to write mass balance equations that are fundamental to understanding and analyzing the flow of materials, we can in a similar fashion use the *first law of thermodynamics* to write energy balance equations that will help us analyze energy flows.

One definition of energy is that it is the capacity for doing work, where work can be described by the product of force and the displacement of an object caused by that force. A simple interpretation of the *second law of thermodynamics* suggests that when work is done there will always be some inefficiency; that is, some portion of the energy put into the process will end up as waste heat. How that waste heat affects the environment is an important consideration in the study of environmental engineering and science.

Another important term to be familiar with is *power*. Power is the *rate* of doing work, so it has units of energy per unit of time. In SI units power is given in joules per second (J/s) or kilojoules per second (kJ/s). In honor of the Scottish engineer James Watt, who developed the reciprocating steam engine, the joule per second has been named the watt (1 J/s = 1 W = 3.412 Btu/hr).

The First Law of Thermodynamics

The first law of thermodynamics says, simply, that energy can be neither created nor destroyed. Energy may change forms in any given process, as when chemical energy in a fuel is converted to heat and electricity in a power plant, or when the potential energy of water behind a dam is converted to mechanical energy as it spins a turbine in a hydroelectric plant. No matter what is happening, the first law says we should be able to account for every bit of energy as it takes part in the process under study, so that in the end we have just as much as we had in the beginning. With proper accounting, even nuclear reactions involving conversion of mass to energy can be treated.

To apply the first law it is necessary to define the system being studied, much as was done in the analysis of mass flows. Realize that the system can be anything that we want to draw an imaginary boundary around—it can be an automobile engine, or a nuclear power plant, or a volume of gas emitted from a smokestack. Later, when we explore the topic of global temperature equilibrium, the system will be the earth itself. Once a boundary has been defined, the rest of the universe becomes the *surroundings*. Just because a boundary has been defined, however, does not mean that energy and/or materials cannot flow across that boundary. Systems in which both energy and matter can flow across the boundary are referred to as *open systems,* while those in which energy is allowed to flow across the boundary, but matter is not, are called *closed systems.*

Since energy is conserved, we can write the following for whatever system we have defined:

$$\begin{pmatrix} \text{Total energy} \\ \text{crossing boundary} \\ \text{as heat and work} \end{pmatrix} + \begin{pmatrix} \text{Total energy} \\ \text{of mass} \\ \text{entering system} \end{pmatrix} - \begin{pmatrix} \text{Total energy} \\ \text{of mass} \\ \text{leaving system} \end{pmatrix} = \begin{pmatrix} \text{Net change} \\ \text{of energy in} \\ \text{the system} \end{pmatrix} \quad (1.30)$$

For closed systems, there is no movement of mass across the boundary so the second and third term drop out of the equation. The accumulation of energy represented by the right side of (1.30) may cause changes in the observable, macroscopic forms of energy, such as kinetic and potential energies, or microscopic forms related to the atomic and molecular structure of the system. Those microscopic forms of energy include the kinetic energies of molecules and the energies associated with the forces acting between molecules, between atoms within molecules, and within atoms. The sum of those microscopic forms of energy is called the system's *internal energy* and is represented by the symbol U. The *total energy E* that a substance possesses can be described then as the sum of its internal energy U, its kinetic energy *KE,* and its potential energy *PE:*

$$E = U + KE + PE \quad (1.31)$$

In many applications of (1.30) the net energy added to a system will cause an increase in temperature. Waste heat from a power plant, for example, will raise the temperature of cooling water drawn into its condenser. The amount of energy needed to raise the temperature of a unit mass of a substance by 1 degree is called the *specific heat.* The specific heat of water is the basis for two important units of energy, namely the *British thermal unit,* or Btu, which is defined to be the energy required to raise 1 lb of water by $1\,°F$, and the *kilocalorie,* which is the energy required to raise 1 kg of water by $1\,°C$. In the definitions just given, the assumed temperature of the water is $15\,°C$ ($59\,°F$). Since kilocalories are no longer a preferred energy unit, values of specific heat in the SI system are given in kJ/kg $°C$, where 1 kcal/kg $°C$ = 1 Btu/lb $°F$ = 4.184 kJ/kg $°C$.

For most applications, the specific heat of a liquid or solid can be treated as a simple quantity that varies slightly with temperature. For gases on the other hand, the concept of specific heat is complicated by the fact that some of the heat absorbed by a gas may cause an increase in temperature and some may cause the gas to expand, doing work on its environment. That means it takes more energy to raise the temperature of a gas that is allowed to expand than the amount needed if the gas is kept at constant volume. The *specific heat at constant volume* c_v is used when a gas does not change volume as it is heated or cooled, or if the volume is allowed to vary but is brought back to its starting value at the end of the process. Similarly, the *specific heat at constant pressure* c_p applies for systems that do not change pressure. For incompressible substances, that is liquids and solids under the usual circumstances, c_v and c_p are identical and we will just use the symbol c. For gases, c_p is greater than c_v.

The added complications associated with gases that change pressure and volume are most easily handled by introducing another thermodynamic property of a substance called *enthalpy*. The enthalpy H of a substance is defined as

$$H = U + PV \qquad (1.32)$$

where U is its internal energy, P is its pressure, and V is its volume. The enthalpy of a unit mass of a substance depends only on its temperature. It has energy units (kJ or Btu) and historically it was referred to as a system's "heat content." Since heat is correctly defined only in terms of energy transfer across a system's boundaries, heat content is a somewhat misleading descriptor and is not used much anymore.

When a process occurs without a change of volume, the relationship between internal energy and temperature change is given by

$$\Delta U = m\, c_v\, \Delta T \qquad (1.33)$$

The analogous equation for changes that occur under constant pressure involves enthalpy

$$\Delta H = m\, c_p\, \Delta T \qquad (1.34)$$

For many environmental systems the substances being heated are solids or liquids for which $c_v = c_p = c$ and $\Delta U = \Delta H$. We can then write the following equation for the change in energy stored in a system when the temperature of mass m changes by an amount ΔT:

$$\text{Change in stored energy} = m\, c\, \Delta T \qquad (1.35)$$

Table 1.3 provides some examples of specific heat for several selected substances. It is worth noting that water has by far the highest specific heat of the substances listed; in fact, it is higher than almost all common substances. As will be noted in Chapter 5, this is one of water's very unusual properties and is in large part responsible for the major effect the oceans have on moderating temperature variations of coastal areas.

TABLE 1.3   Specific Heat Capacity c of Selected Substances

	(kJ/kg °C)	(kcal/kg °C, Btu/lb °F)
Water (15 °C)	4.18	1.00
Air (20 °C)	1.01	0.24
Aluminum	0.92	0.22
Copper	0.39	0.09
Dry soil	0.84	0.20
Ice	2.09	0.50
Steam (100 °C)[a]	2.01	0.48
Water vapor (20 °C)[a]	1.88	0.45

[a]Constant pressure values.

EXAMPLE 1.8   A Water Heater

How long would it take to heat the water in a 40-gallon electric water heater from 50 °F to 140 °F if the heating element delivers 5 kW? Assume all of the electrical energy is converted to heat in the water, neglect the energy required to raise the temperature of the tank itself, and neglect any heat losses from the tank to the environment.

Solution   The first thing to note is that the electric input is expressed in kilowatts, which is a measure of the *rate* of energy input (i.e., power). To get total energy delivered to the water, we must multiply rate × time. Letting Δt be the number of hours that the heating element is on gives

$$\text{Energy input} = 5 \text{ kW} \times \Delta t \text{ hrs} = 5 \, \Delta t \text{ kWhr}$$

Assuming no losses from the tank and no water withdrawn from the tank during the heating period, there is no energy output:

$$\text{Energy output} = 0$$

The change in energy stored corresponds to the water warming from 50 °F to 140 °F. Using (1.35) along with the fact that water weighs 8.34 lb/gal gives

$$
\begin{aligned}
\text{Change in energy stored} &= mc\Delta t \\
&= 40 \text{ gal} \times 8.34 \text{ lb/gal} \times 1 \text{ Btu/lb}°\text{F} \times (140 - 50)°\text{F} \\
&= 30 \times 10^3 \text{ Btu}
\end{aligned}
$$

Setting the energy input equal to the change in stored energy and converting units using Table 1.1 yields

$$5\Delta t \text{ kWhr} \times 3412 \text{ Btu/kWhr} = 30 \times 10^3 \text{ Btu}$$

$$\Delta t = 1.76 \text{ hr}$$

■

There are two key assumptions implicit in (1.35). First, the specific heat is assumed to be constant over the temperature range in question, though in actuality it does vary slightly. Second, (1.35) assumes that there is no change of *phase* as would occur if the substance were to freeze or melt (liquid-solid phase change) or evaporate or condense (liquid-gas phase change).

When a substance changes phase, energy is absorbed or released without a change in temperature. The energy required to cause a phase change of a unit mass from a solid to liquid (melting) at the same pressure is called the *latent heat of fusion* or, more correctly, the *enthalpy of fusion*. Similarly, the energy required to change phase from liquid to vapor at constant pressure is called the *latent heat of vaporization* or the *enthalpy of vaporization*. For example, 333 kJ will melt 1 kg of ice (144 Btu/lb) while 2257 kJ are required to convert 1 kg of water at 100 °C to steam at the same temperature (970 Btu/lb). When steam condenses or when water freezes, those same amounts of energy are released. To account for the latent heat stored in a substance we can include the following in our energy balance:

$$\text{Energy released or absorbed in phase change} = mL \qquad (1.36)$$

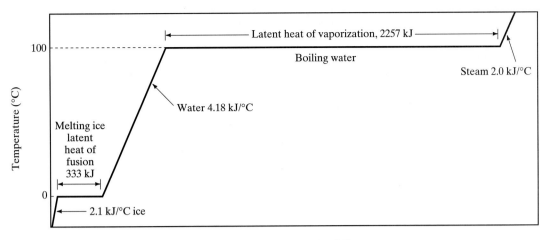

FIGURE 1.9   Heat needed to convert 1 kg of ice to steam. To change the temperature of 1 kg of ice, 2.1 kJ/ °C are needed. To completely melt that ice requires another 333 kJ (latent heat of fusion). Raising the temperature of that liquid water requires 4.184 kJ/ °C, and converting it to steam requires another 2257 kJ (latent heat of vaporization). To raise the temperature of 1 kg of steam (at atmospheric pressure) requires another 2.0 kJ/ °C.

where m is the mass and L is the latent heat of fusion or vaporization.

Figure 1.9 illustrates the concepts of latent heat and specific heat for water as it passes through its three phases from ice, to water, to steam.

Values of specific heat, heats of vaporization and fusion, and density for water are given in Table 1.4 for both SI and USCS units. An additional entry has been included in the table that shows the heat of vaporization for water at 15 °C. This is a useful number that can be used to estimate the amount of energy required to cause surface water on the earth to evaporate. The value of 15 °C has been picked as the starting temperature since that is approximately the current average surface temperature of the globe.

One way to demonstrate the concept of the heat of vaporization while at the same time introducing an important component of the global energy balance that will be encountered in Chapter 8 is to estimate the energy required to power the global hydrologic cycle.

TABLE 1.4   Important Physical Properties of Water

Property	SI Units	USCS Units
Specific heat (15 °C)	4.184 kJ/kg°C	1.00 Btu/lb°F
Heat of vaporization (100 °C)	2257 kJ/kg	972 Btu/lb
Heat of vaporization (15 °C)	2465 kJ/kg	1060 Btu/lb
Heat of fusion	333 kJ/kg	144 Btu/lb
Density (at 4 °C)	1000.00 kg/m³	62.4 lb/ft³

EXAMPLE 1.9 Power for the Hydrologic Cycle

Global rainfall has been estimated to average about 1 m of water per year across the entire 5.10×10^{14} m^2 of the earth's surface. Find the energy required to cause that much water to evaporate each year. Compare this to the estimated 1987 world energy consumption of 3.3×10^{17} kJ and compare it to the average rate at which sunlight is absorbed at the surface of the earth, which is about 168 W/m^2.

Solution In Table 1.4 the energy required to vaporize 1 kg of 15 °C water (roughly the average global temperature) is given as 2465 kJ. The total energy required to vaporize all of that water is

$$\text{Energy needed} = 1 \text{ m/yr} \times 5.10 \times 10^{14} \text{ m}^2 \times 10^3 \text{ kg/m}^3 \times 2465 \text{ kJ/kg}$$

$$= 1.25 \times 10^{21} \text{ kJ/yr}$$

This is roughly 4000 times the 3.3×10^{17} kJ/yr of energy we use to power our society. Averaged over the globe, the energy required to power the hydrologic cycle is

$$\frac{1.25 \times 10^{24} \text{ J/yr} \times 1 \dfrac{\text{W}}{\text{J/s}}}{365 \text{ day/yr} \times 24 \text{ hr/day} \times 3600 \text{ s/hr} \times 5.10 \times 10^{14} \text{ m}^2} = 78 \text{ W/m}^2$$

which is equivalent to almost half of the 168 W/m^2 of incoming sunlight absorbed by the earth's surface (see Figure 8.13). It might also be noted that the energy required to raise the water vapor high into the atmosphere once it has evaporated is negligible compared to the heat of vaporization. (See Problem 1.20 at the end of this chapter.) ∎

Many practical environmental engineering problems involve the flow of both matter and energy across system boundaries (open systems). For example, it is quite common for a hot liquid, usually water, to be used to deliver heat to a pollution control process or, the opposite, for water to be used as a coolant to remove heat from a process. In such cases, there are energy flow rates and fluid flow rates and Equation (1.35) needs to be modified as follows:

$$\text{Rate of change of stored energy} = \dot{m}c\Delta T \qquad (1.37)$$

where $\dot{m}$ is the mass flow rate across the system boundary, given by the product of fluid flow rate and density, and ΔT is the change in temperature of the fluid that is carrying the heat to, or away from, the process. For example, if water is being used to cool a steam power plant, then $\dot{m}$ would be the mass flow rate of coolant and ΔT would be the increase in temperature of the cooling water as it passes through the steam plant's condenser. Typical units for energy rates include watts, kJ/s, or Btu/hr, while mass flow rates might typically be in kg/s or lb/hr.

The use of a local river for power plant cooling is common, and the following example illustrates the approach that can be taken to compute the increase in river temperature that results. In Chapter 5, some of the environmental impacts of this thermal pollution will be explored.

EXAMPLE 1.10   Thermal Pollution of a River

A coal-fired power plant converts one-third of the coal's energy into electrical energy. The electrical power output of the plant is 1000 MW. The other two-thirds of the energy content of the fuel is rejected to the environment as waste heat. About 15 percent of the waste heat goes up the smokestack and the other 85 percent is taken away by cooling water that is drawn from a nearby river. The river has an upstream flow of 100.0 m³/s and a temperature of 20.0 °C.

  a. If the cooling water is only allowed to rise in temperature by 10.0 °C, what flow rate from the stream would be required?

  b. What would be the river temperature just after it receives the heated cooling water?

Solution   Since 1000 MW represents one-third of the power delivered to the plant by fuel, the total rate at which energy enters the power plant is

$$\text{Input power} = \frac{\text{Output power}}{\text{Efficiency}} = \frac{1000 \text{ MW}_e}{1/3} = 3000 \text{ MW}_t$$

Notice the subscript on the input and output power in the preceding equation. To help keep track of the various forms of energy, it is common to use MW_t for thermal power and MW_e for electrical power.

Total losses to the cooling water and stack are therefore $3000 \text{ MW} - 1000 \text{ MW} = 2000 \text{ MW}$. Of that 2000 MW,

$$\text{Stack losses} = 0.15 \times 2000 \text{ MW}_t = 300 \text{ MW}_t$$

and

$$\text{Coolant losses} = 0.85 \times 2000 \text{ MW}_t = 1700 \text{ MW}_t$$

  a. Finding the cooling water needed to remove 1700 MW_t with a temperature increase ΔT of 10.0 °C will require the use of (1.37) along with the specific heat of water, 4184 J/kg °C, given in Table 1.4:

$$\text{Rate of change in stored energy} = \dot{m} c \Delta T$$

$$1700 \text{ MW}_t = \dot{m} \text{ kg/s} \times 4184 \text{ J/kg °C} \times 10.0 °C \times 1 \text{ MW}/(10^6 \text{ J/s})$$

$$\dot{m} = \frac{1700}{4184 \times 10.0 \times 10^{-6}} = 40.6 \times 10^3 \text{ kg/s}$$

  or, since 1000 kg equals 1 m³ of water, the flow rate is 40.6 m³/s.

  b. To find the new temperature of the river, we can use (1.37) with 1700 MW_t being released into the river, which again has a flow rate of 100.0 m³/s.

$$\text{Rate of change in stored energy} = \dot{m} c \Delta T$$

$$\Delta T = \frac{1700 \text{ } MW \times \left(\dfrac{1 \times 10^6 \text{ J/s}}{\text{MW}} \right)}{100.00 \text{ m}^3/\text{s} \times 10^3 \text{ kg/m}^3 \times 4184 \text{ J/kg °C}} = 4.1 °C$$

  so the temperature of the river will be elevated by 4.1 °C, making it 24.1 °C.

The results of the calculations just performed are shown in Figure 1.10. Similar calculations for a nuclear plant are asked for in Problem 1.18.   ∎

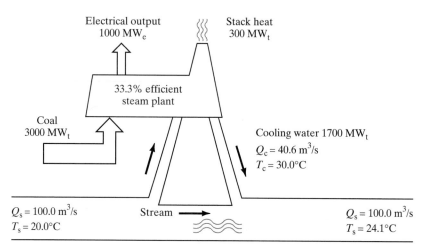

FIGURE 1.10   Cooling water energy balance for the 33.3 percent efficient, 1000 MW$_e$ power plant of Example 1.10.

The Second Law of Thermodynamics

In Example 1.10, you will notice that a relatively modest fraction of the fuel energy contained in the coal actually was converted to the desired output, electrical power, and a rather large amount of the fuel energy ended up as waste heat rejected to the environment. The second law of thermodynamics says that there will always be some waste heat; that is, it is impossible to devise a machine that can convert heat to work with 100 percent efficiency. There will always be "losses" (though, by the first law, the energy is not lost; it is merely converted into the lower-quality, less useful form of low-temperature heat).

The steam-electric plant just described is an example of a *heat engine,* a device studied at some length in thermodynamics. One way to view the steam plant is that it is a machine that takes heat from a high-temperature source (the burning fuel), converts some of it into work (the electrical output), and rejects the remainder into a low-temperature reservoir (the river and the atmosphere). It turns out that the maximum efficiency that our steam plant can possibly have depends on how high the source temperature is and how low the temperature is of the reservoir accepting the rejected heat. It is analogous to trying to run a turbine using water that flows from a higher elevation to a lower one. The greater the difference in elevation, the more power can be extracted.

Figure 1.11 shows a theoretical heat engine operating between two heat reservoirs, one at temperature T_h and one at T_c. An amount of heat energy Q_h is transferred from the hot reservoir to the heat engine. The engine does work W and rejects an amount of waste heat Q_c to the cold reservoir. The efficiency of this engine is the ratio of the work delivered by the engine to the amount of heat energy taken from the hot reservoir:

$$\text{Efficiency } \eta = \frac{W}{Q_h} \tag{1.38}$$

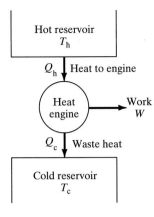

FIGURE 1.11   Definition of terms for a Carnot engine.

The most efficient heat engine that could possibly operate between the two heat reservoirs is called a *Carnot* engine after the French engineer Sadi Carnot, who first developed the explanation in the 1820s. Analysis of Carnot engines shows that the most efficient engine possible, operating between two temperatures, T_h and T_c, has an efficiency of

$$\eta_{max} = 1 - \frac{T_c}{T_h} \tag{1.39}$$

where these are absolute temperatures measured using either the Kelvin scale or Rankine scale. Conversions from Celsius to Kelvin and Fahrenheit to Rankine are

$$K = {}^{\circ}C + 273.15 \tag{1.40}$$

$$R = {}^{\circ}F + 459.67 \tag{1.41}$$

One immediate observation that can be made from (1.39) is that the maximum possible heat engine efficiency increases as the temperature of the hot reservoir increases or the temperature of the cold reservoir decreases. In fact, since neither infinitely hot temperatures nor absolute zero temperatures are possible, we must conclude that no real engine has 100 percent efficiency, which is just a restatement of the second law.

Equation (1.39) can help us understand the seemingly low efficiency of thermal power plants such as the one diagrammed in Figure 1.12. In this plant, fuel is burned in a firing chamber surrounded by metal tubing. Water circulating through this boiler tubing is converted to high-pressure, high-temperature steam. During this conversion of chemical to thermal energy, losses on the order of 10 percent occur due to incomplete combustion and loss of heat up the smokestack. Later, we shall consider local and regional air pollution effects caused by these emissions as well as their possible role in global warming.

The steam produced in the boiler then enters a steam turbine, which is in some ways similar to a child's pinwheel. The high-pressure steam expands as it passes through the turbine blades, causing a shaft that is connected to the generator to spin.

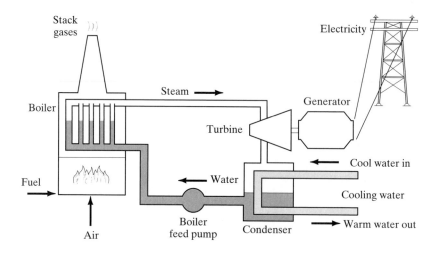

FIGURE 1.12 A fuel-fired, steam-electric power plant.

While the turbine in Figure 1.12 is shown as a single unit, in actuality turbines have many stages, with steam exiting one stage and entering another, gradually expanding and cooling as it goes. The generator converts the rotational energy of a spinning shaft into electrical power that goes out onto transmission lines for distribution. A well-designed turbine may have an efficiency that approaches 90 percent, while the generator may have a conversion efficiency even higher than that.

The spent steam from the turbine undergoes a phase change back to the liquid state as it is cooled in the condenser. The condenser pressure is below atmospheric pressure, which helps pull steam through the turbine, thereby increasing the turbine efficiency. The condensed steam is then pumped back to the boiler to be reheated.

The heat released when the steam condenses is transferred to cooling water that circulates through the condenser. Usually, cooling water is drawn from a lake or river, heated in the condenser, and returned to that body of water, in which case the process is called *once-through cooling*. A more expensive approach, which has the advantage of requiring less water, involves use of cooling towers that transfer the heat directly into the atmosphere rather than into a receiving body of water. In either case, the rejected heat is released into the environment. In terms of the heat engine concept shown in Figure 1.11, the cold reservoir temperature is thus determined by the temperature of the environment.

Let us estimate the maximum possible efficiency that a thermal power plant such as that diagrammed in Figure 1.12 can have. A reasonable estimate of T_h might be the temperature of the steam from the boiler, which is typically around 600 °C. For T_c, we might use an ambient temperature of about 20 °C. Using these values in (1.39), and remembering to convert temperatures to the absolute scale, gives

$$\eta_{max} = 1 - \frac{(20 + 273)}{(600 + 273)} = 0.66 = 66 \text{ percent}$$

New fossil-fuel-fired power plants have efficiencies around 40 percent. Nuclear plants have materials constraints that force them to operate at somewhat lower temperatures than fossil plants, which results in efficiencies of around 33 percent. The average efficiency of all thermal plants actually in use in the United States, including new and old (less efficient) plants, fossil and nuclear, is close to 33 percent. That suggests the following convenient rule of thumb:

> For every 3 units of energy entering the average thermal power plant, approximately 1 unit is converted to electricity and 2 units are rejected to the environment as waste heat.

The following example uses this rule of thumb for power plant efficiency combined with other emission factors to develop a mass and energy balance for a typical coal-fired power plant.

EXAMPLE 1.11   Mass and Energy Balance for a Coal-fired Power Plant

Typical coal burned in power plants in the United States has an energy content of approximately 24 kJ/g and an average carbon content of about 62 percent. For almost all new coal plants, Clean Air Act emission standards limit sulfur emissions to 260 g of sulfur dioxide (SO_2) per million kJ of heat input to the plant (130 g of elemental sulfur per 10^6 kJ). They also restrict particulate emissions to 13 g/10^6 kJ. Suppose the average plant burns fuel with 2 percent sulfur content and 10 percent unburnable minerals called *ash*. About 70 percent of the ash is released as *fly ash* and about 30 percent settles out of the firing chamber and is collected as *bottom ash*. Assume this is a typical coal plant with 3 units of heat energy required to deliver 1 unit of electrical energy.

a. Per kilowatt-hour of electrical energy produced, find the emissions of SO_2, particulates, and carbon (assume all of the carbon in the coal is released to the atmosphere).

b. How efficient must the sulfur emission control system be to meet the sulfur emission limitations?

c. How efficient must the particulate control system be to meet the particulate emission limits?

Solution

a. We first need the heat input to the plant. Since 3 kWhr of heat are required for each 1 kWhr of electricity delivered,

$$\frac{\text{Heat input}}{\text{kWhr electricity}} = 3 \text{ kWhr heat} \times \frac{1 \text{ kJ/s}}{\text{kW}} \times 3600 \text{ s/hr} = 10{,}800 \text{ kJ}$$

The sulfur emissions are thus restricted to

$$\text{S emissions} = \frac{130 \text{ g S}}{10^6 \text{ 6kJ}} \times 10{,}800 \text{ kJ/kWhr} = 1.40 \text{ g S/kWhr}$$

The molecular weight of SO_2 is $32 + 2 \times 16 = 64$, half of which is sulfur. Thus, 1.4 g of S corresponds to 2.8 g of SO_2, so 2.8 g SO_2/kWhr would be emitted.

Particulate emissions need to be limited to

$$\text{Particulate emissions} = \frac{13 \text{ g}}{10^6 \text{ kJ}} \times 10{,}800 \text{ kJ/kWhr} = 0.14 \text{ g/kWhr}$$

To find carbon emissions, let us first find the amount of coal burned per kWhr:

$$\text{Coal input} = \frac{10,800 \text{ kJ/kWhr}}{24 \text{ kJ/g coal}} = 450 \text{ g coal/kWhr}$$

Therefore, since the coal is 62 percent carbon,

$$\text{Carbon emissions} = \frac{0.62 \text{ g C}}{\text{g coal}} \times \frac{450 \text{ g coal}}{\text{kWhr}} = 280 \text{ g C/kWhr}$$

b. Burning 450 g coal containing 2 percent sulfur will release $0.02 \times 450 = 9.0$ g of S. Since the allowable emissions are 1.4 g, the removal efficiency must be

$$\text{S removal efficiency} = 1 - \frac{1.4}{9.0} = 0.85 = 85 \text{ percent}$$

c. Since 10 percent of the coal is ash and 70 percent of that is fly ash, the total fly ash generated will be

$$\text{Fly ash generated} = 0.70 \times 0.10 \times 450 \text{ g coal/kWhr} = 31.5 \text{ g fly ash/kWhr}$$

The allowable particulate matter is restricted to 0.14 g/ kWhr, so controls must be installed that have the following removal efficiency:

$$\text{Particulate removal efficiency} = 1 - \frac{0.14}{31.5} = 0.995 = 99.5 \text{ percent}$$

In Chapter 7 we will see how these emission control systems work. ■

The complete mass and energy balance for this coal plant is diagrammed in Figure 1.13. In this diagram it has been assumed that 85 percent of the waste heat is removed by cooling water and the remaining 15 percent is lost in stack gases (corresponding to the conditions given in Example 1.10).

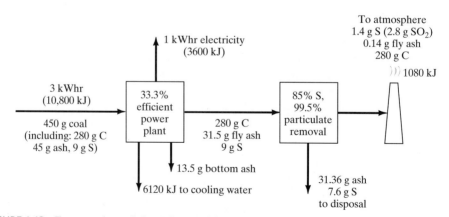

FIGURE 1.13 Energy and mass balance for a coal-fired power plant generating 1 kWhr of electricity (see Example 1.11).

The Carnot efficiency limitation provides insight into the likely performance of other types of thermal power plants in addition to the steam plants just described. For example, there have been many proposals to build power plants that would take advantage of the temperature difference between the relatively warm surface waters of the ocean and the rather frigid waters found below. In some locations, the sun heats the ocean's top layer to as much as 30 °C, while several hundred meters down, the temperature is a constant 4 or 5 °C. Power plants, called *ocean thermal energy conversion* (OTEC) systems, could be designed to operate on these small temperature differences in the ocean. However, as the following example shows, they would be quite inefficient.

EXAMPLE 1.12   OTEC System Efficiency

Consider an OTEC system operating between 30 °C and 5 °C. What would be the maximum possible efficiency for an electric generating station operating with these temperatures?

Solution   Using (1.39), we find

$$\eta_{max} = 1 - \frac{(5 + 273)}{(30 + 273)} = 0.08 = 8 \text{ percent}$$

An even lower efficiency, estimated at 2 to 3 percent for a real plant, would be expected. ■

Conductive and Convective Heat Transfer

When two objects are at different temperatures, heat will be transferred from the hotter object to the colder one. That heat transfer can be by *conduction* when there is direct physical contact between the objects; by *convection* when there is a liquid or gas between them; or by *radiation,* which can take place even in the absence of any physical medium between the objects.

Conductive heat transfer is usually associated with solids, as one molecule vibrates the next in the lattice. The rate of heat transfer in a solid is proportional to the thermal conductivity of the material. Metals tend to be good thermal conductors, which makes them very useful when high heat-transfer rates are desired. Other materials are much less so, with some being particularly poor thermal conductors (which makes them potentially useful as thermal insulation).

Convective heat transfer occurs when a fluid at one temperature comes in contact with a substance at another temperature. For example, warm air in a house in the winter that comes in contact with a cool wall surface will transfer heat to the wall. As that warm air loses some of its heat, it becomes cooler and denser, and it will sink, to be replaced by more warm air from the interior of the room. Thus there is a continuous movement of air around the room and with it a transference of heat from the warm room air to the cool wall. The cool wall, in turn, conducts heat to the cold exterior surface of the house, where outside air removes the heat by convection.

Figure 1.14 illustrates the two processes of convection and conduction through a hypothetical wall. In addition, there is radiative heat transfer from objects in the room

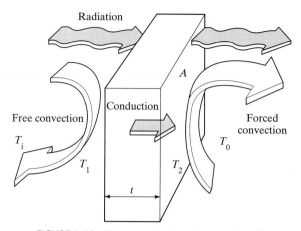

FIGURE 1.14   Heat transfer through a simple wall.

to the wall, and from the wall to the ambient outside. It is conventional practice in the building industry to combine all three processes into a single, overall heat-transfer process that is characterized by the following simple equation:

$$q = \frac{A(T_i - T_o)}{R}$$ (1.42)

where

q = heat transfer rate through the wall (W) or (Btu/hr)
A = wall area (m^2) or (ft^2)
T_i = air temperature on one side of the wall (°C) or (°F)
T_o = ambient air temperature (°C) or (°F)
R = overall thermal resistance (m^2-°C/W) or (hr-ft^2-°F/Btu)

The overall thermal resistance R is called the *R value*. If you buy insulation at the hardware store, it will be designated as having an R value in the American unit system (hr-ft^2- °F/Btu). For example, $3\frac{1}{2}$-inch-thick fiberglass insulation is usually marked R-11, while 6 inches of the same material is R-19.

As the following example illustrates, improving the efficiency with which we use energy can save money as well as reduce emissions of pollutants associated with energy consumption. This important connection between energy efficiency and pollution control has in the past been overlooked and underappreciated. However, as will be described in Chapter 7, that situation has been changing. The 1990 amendments to the Clean Air Act, for example, provide SO_2 emission credits for energy efficiency projects.

EXAMPLE 1.13   Reducing Pollution by Adding Ceiling Insulation

A home with 1500 ft^2 of poorly insulated ceiling is located in an area with an 8-month heating season during which time the outdoor temperature averages 40°F while the inside temperature is kept at 70°F (this could be Chicago, for example). It has been proposed to the owner that $1000 be spent to add more insulation to the ceiling, raising its total R value from 11 to 40 (ft²-°F-hr/Btu). The house is heated with electricity that costs 8 cents/kWhr.

a. How much money would the owner expect to save each year and how long would it take for the energy savings to pay for the cost of insulation?

b. Suppose 1 million homes served by coal plants like the one analyzed in Example 1.11 could achieve similar energy savings. Estimate the annual reduction in SO_2, particulate, and carbon emissions that would be realized.

Solution

a. Using (1.42) to find the heat loss rate with the existing insulation gives

$$q = \frac{A(T_i - T_o)}{R} = \frac{1500 \text{ ft}^2 \times (70 - 40)°F}{11 \text{ (ft}^2 \text{ °F hr/Btu)}} = 4090 \text{ Btu/hr}$$

After adding the insulation, the new heat loss rate will be

$$q = \frac{A(T_i - T_o)}{R} = \frac{1500 \text{ ft}^2 \times (70 - 40)°F}{40 \text{ (ft}^2 \text{ °F hr/Btu)}} = 1125 \text{ Btu/hr}$$

The annual energy savings can be found by multiplying the rate at which energy is being saved by the number of hours in the heating season. If we assume the electric heating system in the house is 100 percent efficient at converting electricity to heat (reasonable) and that it delivers all of that heat to the spaces that need heat (less reasonable, especially if there are heating ducts, which tend to leak), then we can use the conversion 3412 Btu = 1 kWhr.

$$\text{Energy saved} = \frac{(4090 - 1125) \text{ Btu/hr}}{3412 \text{ Btu/kWhr}} \times 24 \text{ hr/day} \times 30 \text{ day/mo} \times 8 \text{ mo/yr}$$

$$= 5005 \text{ kWhr/yr}$$

The annual savings in dollars would be

$$\text{Dollar savings} = 5005 \text{ kWhr/yr} \times \$0.08/\text{kWhr} = \$400/\text{yr}$$

Since the estimated cost of adding extra insulation is $1000, the reduction in electricity bills would pay for this investment in about $2\frac{1}{2}$ heating seasons.

b. One million such houses would save a total of 5 billion kWhr/yr (nearly the entire annual output of a typical 1000 MW_e power plant). Using the emission factors derived in Example 1.11, the reduction in air emissions would be

$$\text{Carbon reduction} = 280 \text{ g C/kWhr} \times 5 \times 10^9 \text{ kWhr/yr} \times 10^{-3} \text{ kg/g} = 1400 \times 10^6 \text{ kg/yr}$$

$$SO_2 \text{ reduction} = 2.8 \text{ g } SO_2/\text{kWhr} \times 5 \times 10^9 \text{ kWhr/yr} \times 10^{-3} \text{ kg/g} = 14 \times 10^6 \text{ kg/yr}$$

$$\text{Particulate reduction} = 0.14 \text{ g/kWhr} \times 5 \times 10^9 \text{ kWhr/yr} \times 10^{-3} \text{ kg/g} = 0.7 \times 10^6 \text{ kg/yr} \quad \blacksquare$$

Radiant Heat Transfer

Heat transfer by thermal radiation is the third way that one object can warm another. Unlike conduction and convection, radiant energy is transported by electromagnetic waves and does not require a medium to carry the energy. As is the case for other forms of electromagnetic phenomena, such as radio waves, X-rays, and gamma rays, thermal radiation can be described either in terms of wavelengths or, using the particle nature of electromagnetic radiation, in terms of discrete photons of energy. All electromagnetic waves travel at the speed of light. They can be described by their wavelength or their frequency, and the two are related as follows:

$$c = \lambda \nu \qquad (1.43)$$

where

c = speed of light $(3 \times 10^8 \text{ m/s})$
λ = wavelength (m)
ν = frequency (hertz, i.e., cycles per second)

When radiant energy is described in terms of photons, the relationship between frequency and energy is given by

$$E = h\nu \qquad (1.44)$$

where

E = energy of a photon (J)
h = Planck's constant $(6.6 \times 10^{-34} \text{ J-s})$

Equation (1.44) points out that higher-frequency, shorter-wavelength photons have higher energy content, which makes them potentially more dangerous when living things are exposed to them.

In this book, the most important application of radiant heat transfer will come in Chapter 8, when the effects of various gases on global climate and depletion of the stratospheric ozone layer will be discussed. The wavelengths of importance in that context are roughly in the range of about 0.1 μm up to about 100 μm (1 μm is 10^{-6} m, also called 1 micron). For perspective, Figure 1.15 shows a portion of the electromagnetic spectrum.

Every object emits thermal radiation. The usual way to describe how much radiation a real object emits, as well as other characteristics of the wavelengths emitted, is to compare it to a theoretical abstraction called a *blackbody*. A blackbody is defined to be a perfect emitter as well as a perfect absorber. As a perfect emitter, it radiates more energy per unit of surface area than any real object at the same temperature. As a perfect absorber, it absorbs all radiation that impinges on it; that is, none is reflected and none is transmitted through it. Actual objects do not emit as much radiation as this hypothetical blackbody, but most are quite close to this theoretical limit. The ratio of the amount of radiation an actual object would emit to the amount that a blackbody would emit at the same temperature is known as the emissivity ϵ. The emissivity of desert sand, dry ground, and most woodlands is estimated to be approximately 0.90, while water, wet sand, and ice all have estimated emissivities of roughly 0.95. A human body, no matter what pigmentation, has an emissivity of around 0.96.

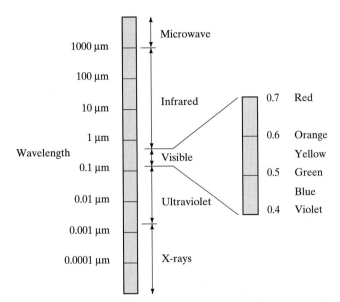

FIGURE 1.15 A portion of the electromagnetic spectrum. The wavelengths of greatest interest for this text are in the range of about 0.1 μm to 100 μm.

The wavelengths radiated by a blackbody depend on its temperature as described by *Planck's law:*

$$E_\lambda = \frac{C_1}{\lambda^5 \left(e^{C_2/\lambda T} - 1\right)} \tag{1.45}$$

where

E_λ = emissive power of a blackbody (W/m^2-μm)
T = absolute temperature of the body (K)
λ = wavelength (μm)
$C_1 = 3.74 \times 10^8$ W$-\mu$m^4/m^2
$C_2 = 1.44 \times 10^4$ μm-K

Figure 1.16 is a plot of the emissive power of radiation emitted from blackbodies at various temperatures. Curves such as these, which show the spectrum of wavelengths emitted, have as their vertical axis an amount of power per unit-wavelength. The way to interpret such spectral diagrams is to realize that the area under the curve between any two wavelengths is equal to the power radiated by the object within that band of wavelengths. Hence the total area under the curve is equal to the total power radiated. Objects at higher temperatures emit more power and, as the figure suggests, their peak intensity occurs at shorter wavelengths.

Extraterrestrial solar radiation (just outside the earth's atmosphere) shows spectral characteristics that are well approximated by blackbody radiation. While the

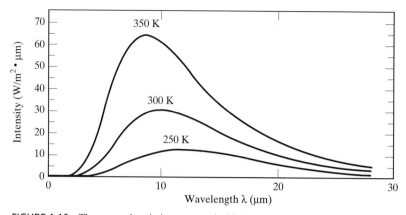

FIGURE 1.16 The spectral emissive power of a blackbody with various temperatures.

temperature deep within the sun is many millions of degrees, its effective surface temperature is about 5800 K. Figure 1.17 shows the close match between the actual solar spectrum and that of a 5800-K blackbody.

As has already been mentioned, the area under a spectral curve between any two wavelengths is the total power radiated by those wavelengths. For the solar spectrum of Figure 1.17, the area under the curve between 0.38 and 0.78 μm (the wavelengths visible to the human eye) is 47 percent of the total area. That is, 47 percent of the solar energy striking the outside of our atmosphere is in the visible portion of the spectrum. The ultraviolet range contains 7 percent of the energy, and the infrared wavelengths deliver the remaining 46 percent. The total area under the solar spectrum curve is called the *solar constant* and is estimated to be 1370 watts per m^2. As will be shown in Chapter 8, the solar constant plays an important role in determining the surface temperature of the earth.

The equation describing Planck's law (1.45) is somewhat tricky to manipulate—especially for calculations involving the area under a spectral curve. Two other radiation equations, however, are quite straightforward and are often all that is needed. The first, known as the *Stefan-Boltzmann law of radiation,* gives the total radiant energy emitted by a blackbody with surface area A and absolute temperature T:

$$E = \sigma A T^4 \tag{1.46}$$

where

E = total blackbody emission rate (W)
σ = the Stefan-Boltzmann constant = 5.67×10^{-8} W/m^2–K^4
T = absolute temperature (K)
A = surface area of the object (m^2)

The second is *Wien's displacement rule,* which tells us the wavelength at which the spectrum reaches its maximum point:

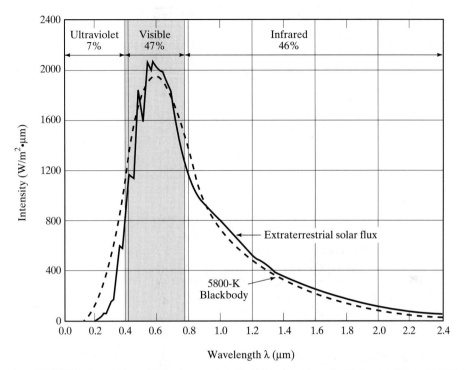

FIGURE 1.17   The extraterrestrial solar spectrum (solid line) compared with the spectrum of a 5800-K blackbody (dashed).

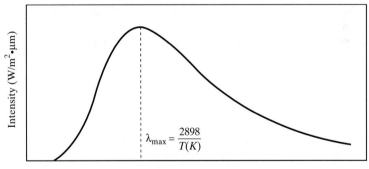

FIGURE 1.18   Wien's rule for finding the wavelength at which the spectral emissive power of a blackbody reaches its maximum value.

$$\lambda_{max}(\mu m) = \frac{2898}{T(\text{K})} \tag{1.47}$$

where wavelength is specified in micrometers and temperature is in kelvins. Figure 1.18 illustrates this concept and Example 1.14 shows how it can be used.

EXAMPLE 1.14   The Earth's Spectrum

Consider the earth to be a blackbody with average temperature $15.0\,°C$ and surface area equal to $5.1 \times 10^{14}\ m^2$. Find the rate at which energy is radiated by the earth and the wavelength at which maximum power is radiated. Compare this peak wavelength with that for a 5800-K blackbody (the sun).

Solution   Using (1.46), the earth radiates

$$E = \sigma A T^4$$

$$= 5.67 \times 10^{-8}\,W/m^2 - K^4 \times 5.1 \times 10^{14}\,m^2 \times (15.0 + 273.15\ K)^4$$

$$= 2.0 \times 10^{17}\ W$$

The wavelength at which the maximum point is reached in the earth's spectrum is

$$\lambda_{max}\,(\mu m) = \frac{2898}{T\,(K)} = \frac{2898}{288.15} = 10.1\ \mu m\ \text{(earth)}$$

For the 5800-K sun,

$$\lambda_{max}\,(\mu m) = \frac{2898}{5800} = 0.48\ \mu m\ \text{(sun)} \qquad\blacksquare$$

This tremendous rate of energy emission by the earth is balanced by the rate at which the earth absorbs energy from the sun. As shown in Example 1.14, however, the solar energy striking the earth has much shorter wavelengths than energy radiated back to space by the earth. This wavelength shift plays a crucial role in the greenhouse effect. As described in Chapter 8, carbon dioxide and other greenhouse gases are relatively transparent to the incoming short wavelengths from the sun, but they tend to absorb the outgoing, longer wavelengths radiated by the earth. As those greenhouse gases accumulate in our atmosphere, they act like a blanket that envelops the planet, upsets the radiation balance, and raises the earth's temperature.

PROBLEMS

1.1 The proposed air quality standard for ozone (O_3) is 0.08 ppm.

    (a) Express that standard in $\mu g/m^3$ at 1 atm of pressure and $25\,°C$.

    (b) At the elevation of Denver, the pressure is about 0.82 atm. Express the ozone standard at that pressure and at a temperature of $15\,°C$.

1.2 Suppose the exhaust gas from an automobile contains 1.0 percent by volume of carbon monoxide. Express this concentration in mg/m^3 at $25\,°C$ and 1 atm.

1.3 Suppose the average concentration of SO_2 is measured to be $400\ \mu g/m^3$ at $25\,°C$ and 1 atm. Does this exceed the (24-hr) air quality standard of 0.14 ppm? (See Table 2.1 for atomic weights.)

1.4 A typical motorcyle emits about 20 g of CO per mile.

    (a) What volume of CO would a 5-mile trip produce after the gas cools to $25\,°C$ (at 1 atm)?

    (b) Per meter of distance traveled, what volume of air could be polluted to the air quality standard of 9 ppm?

1.5 If we approximate the atmosphere to be 79 percent nitrogen (N_2) by volume and 21 percent oxygen (O_2), estimate the density of air (kg/m^3) at STP conditions (0°C, 1 atm).

1.6 Five million gallons per day (MGD) of a conservative substance, with concentration 10.0 mg/L , is released into a stream having an upstream flow of 10 MGD and substance concentration of 3.0 mg/L. Assume complete mixing.

   (a) What is the concentration in ppm just downstream?

   (b) How many pounds of substance per day pass a given spot downstream? (You may want the conversions 3.785 L/gal and 2.2 kg/lbm from the Appendix.)

1.7 A river with 400 ppm of salts (a conservative substance) and an upstream flow of 25.0 m^3/s receives an agricultural discharge of 5.0 m^3/s carrying 2000 mg/L of salts (see Figure P1.7). The salts quickly become uniformly distributed in the river. A municipality just downstream withdraws water and mixes it with enough pure water (no salt) from another source to deliver water having no more than 500 ppm salts to its customers. What should be the mixture ratio F of pure water to river water?

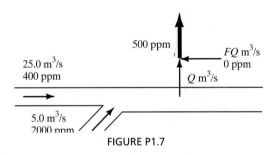

500 ppm

FQ m^3/s
0 ppm

25.0 m^3/s
400 ppm

Q m^3/s

5.0 m^3/s
2000 ppm

FIGURE P1.7

1.8 A lake with constant volume 10×10^6 m^3 is fed by a pollution-free stream with flow rate 50 m^3/s. A factory dumps 5 m^3/s of a nonconservative waste with concentration 100 mg/L into the lake. The pollution has a reaction rate coefficient K of 0.25/day. Assuming the pollution is well mixed in the lake, find the steady-state concentration of pollution in the lake.

1.9 The two-pond system shown in Figure P1.9 is fed by a stream with flow rate 1.0 MGD (million gallons per day) and BOD (a nonconservative pollutant) concentration 20.0 mg/L. The rate of decay of BOD is 0.30/day. The volume of the first pond is 5.0 million gallons and the second is 3.0 million.

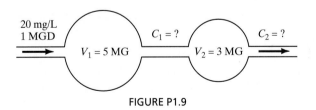

20 mg/L
1 MGD

$V_1 = 5$ MG

$C_1 = ?$

$V_2 = 3$ MG

$C_2 = ?$

FIGURE P1.9

Assuming complete mixing within each pond, find the BOD concentration leaving each pond.

1.10 A lagoon is to be designed to accomodate an input flow of 0.10 m^3/s of nonconservative pollutant with concentration 30.0 mg/L and reaction rate 0.20/day. The effluent from the lagoon must have pollutant concentration of less than 10.0 mg/L. Assuming complete mixing, how large must the lagoon be?

1.11 A simple way to model air pollution over a city is with a box model that assumes complete mixing and limited ability for the pollution to disperse horizontally or vertically except in the direction of the prevailing winds (for example, a town located in a valley with an inversion layer above it). Consider a town having an inversion at 250 m, a 20-km horizontal distance perpendicular to the wind, a windspeed of 2 m/s, and a carbon monoxide (CO) emission rate of 60 kg/s (see Figure P1.11). Assume the CO is conservative and completely mixed in the box.

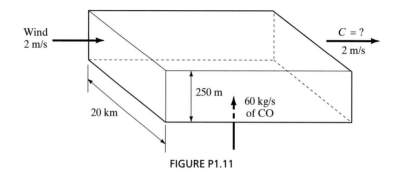

Wind
2 m/s

$C = ?$
2 m/s

250 m

20 km

60 kg/s
of CO

FIGURE P1.11

What would be the CO concentration in the box?

1.12 Consider the air over a city to be a box 100 km on a side that reaches up to an altitude of 1.0 km. Clean air is blowing into the box along one of its sides with a speed of 4 m/s. Suppose an air pollutant with reaction rate $K = 0.20/\text{hr}$ is emitted into the box at a total rate of 10.0 kg/s. Find the steady-state concentration if the air is assumed to be completely mixed.

1.13 If the windspeed in Problem 1.12 suddenly drops to 1 m/s, estimate the concentration of pollutants two hours later.

1.14 A lagoon with volume 1200 m³ has been receiving a steady flow of a conservative waste at a rate of 100 m³/day for a long enough time to assume that steady-state conditions apply. The waste entering the lagoon has a concentration of 10 mg/L. Assuming completely mixed conditions,

   (a) What would be the concentration of pollutant in the effluent leaving the lagoon?

   (b) If the input waste concentration suddenly increased to 100 mg/L, what would the concentration in the effluent be 7 days later?

1.15 Repeat Problem 1.14 for a nonconservative pollutant with reaction rate $K = 0.20/\text{d}$.

1.16 A 4 × 8-ft solar collector has water circulating through it at the rate of 1.0 gallon per minute (gpm) while exposed to sunlight with intensity 300 Btu/ft²-hr (see Figure P1.16). Fifty percent of that sunlight is captured by the collector and heats the water flowing through it. What would be the temperature rise of the water as it leaves the collector?

1.17 An uncovered swimming pool loses 1.0 in of water off of its 1000 ft² surface each week due to evaporation. The heat of vaporization for water at the pool temperature is 1050 Btu/lb. The cost of energy to heat the pool is $10.00 per million Btu. A salesman claims that a $500 pool cover that reduces evaporizative losses by two-thirds will pay for itself in one 15-week swimming season. Can it be true?

1.18 Two-thirds of the energy content of fuel entering a 1000-MW$_e$ nuclear power plant is removed by condenser cooling water that is withdrawn from a local river (there are no

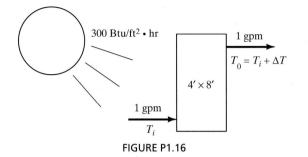

FIGURE P1.16

stack losses, as is the case for a fossil-fuel-fired plant). The river has an upstream flow of 100 m³/s and a temperature of 20 °C.

(a) If the cooling water is only allowed to rise in temperature by 10 °C, what flow rate from the river would be required? Compare it to the coal plant in Example 1.10.

(b) How much would the river temperature rise as it receives the heated cooling water? Again compare it to Example 1.10.

1.19 Consider a 354-ml can of soda cold enough to cause moisture from the air to condense on the outside of the can. If all of the heat released when 5 ml of vapor condenses on the can is transferred into the soda, how much would the soda temperature increase? Assume the density and specific heat of soda are the same as water, neglect the thermal capacitance of the can itself, and use 2500 kJ/kg as the latent heat of vaporization (condensation).

1.20 Compare the energy required to evaporate a kilogram of water at 15 °C to that required to raise it 3 km into the air. (Recall that 1 kg on earth weighs 9.8 N and 1 J = 1 N-m).

1.21 Compare the potential energy represented by 1 lb of water vapor at 59 °F (15 °C) and an elevation of 5000 ft to the energy that would be released when it condenses into rain (1 Btu = 778 ft-lb).

1.22 A 600-MW$_e$ power plant has an efficiency of 36 percent with 15 percent of the waste heat being released to the atmosphere as stack heat and the other 85 percent taken away in the cooling water (see Figure P1.22). Instead of drawing water from a river, heating it, and returning it to the river, this plant uses an evaporative cooling tower wherein heat is released to the atmosphere as cooling water is vaporized.

At what rate must 15 °C makeup water be provided from the river to offset the water lost in the cooling tower?

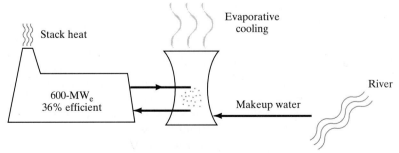

FIGURE P1.22

1.23 An electric water heater held at 140 °F is kept in a 70 °F room. When purchased, its insulation is equivalent to R 5. An owner puts a 25-ft^2 blanket on the water heater, raising its total R value to 15. Assuming 100 percent conversion of electriciy into heated water, how much energy (kWhr) will be saved each year? If electricity costs 8.0 cents/kWhr, how much money will be saved in energy each year?

1.24 A 15-W compact fluorescent lightbulb (CFL) produces the same amount of light as a 60-W incandescent while using only one-fourth the power. Over the 9000-hr lifetime of one CFL, compute carbon, SO_2 and particulate emissions that would be saved if one CFL replaces incandescents and the electricity comes from the coal-fired power plant described in Example 1.11.

1.25 Suppose a utility generates electricity with a 36 percent efficient coal-fired power plant emitting the legal limit of 0.6 lb of SO_2 per million BTUs of heat into the plant. Suppose the utility encourages its customers to replace their 75-W incandescents with 18-W compact fluorescent lamps (CFLs) that produce the same amount of light. Over the 10,000-hr lifetime of a single CFL,

  (a) How many kilowatt-hours of electricity would be saved?

  (b) How many 2000-lb tons of SO_2 would not be emitted?

  (c) If the utility can sell its rights to emit SO_2 at $800 per ton (these are called "allowances"; see Chapter 7), how much money could the utility earn by selling the SO_2 saved by a single CFL?

1.26 No. 6 fuel oil has a carbon content of 20 kg carbon per 10^9 J. If it is burned in a 40 percent efficient power plant, find the carbon emissions per kilowatt-hour of electricity produced, assuming all of the carbon in the fuel is released into the atmosphere. By law, new oil-fired power plant emissions are limited to 86 mg of SO_2 per million joules (MJ) of thermal input, and 130 mg NO_x/MJ. Estimate the maximum allowable SO_2 and NO_x emissions per kilowatt-hour.

1.27 Mars radiates energy with a peak wavelength of 13.2 μm.

  (a) Treating it as a blackbody, what would its temperature be?

  (b) What would be the frequency and energy content of a photon at that wavelength?

1.28 The rate at which sunlight reaches the outer edge of the atmosphere of earth is 1370 W/m^2 (the solar constant for earth). The earth's orbit has an average radius of 150×10^6 km. Since solar radiation decreases as the square of the distance from the sun, estimate the solar constants for

  (a) Mars, whose orbit has a radius of 228×10^6 km.

  (b) Venus, whose orbit has a radius of 108×10^6 km.

1.29 Objects not only radiate energy, but they absorb radiant energy as well. The net blackbody radiation for an object at temperature T_1 in an environment with temperature T_2 is given by

$$E_{net} = \sigma A \left[(T_1)^4 - (T_2)^4 \right]$$

Suppose an unclothed human body has a surface area of 1.35 m^2, an average skin temperature of 32 °C, and is in a room with surfaces at 15 °C. Treating this person as a blackbody (a very good approximation), find the net heat loss by radiation (watts).

1.30 A hot-water radiator has a surface temperature of 80 °C and a surface area of 2 m^2. Treating it as a blackbody, find the net rate at which it will radiate energy to a 20 °C room (see Problem 1.29).

C H A P T E R   2

Environmental Chemistry

> It often matters much how given atoms combine, in what arrange-
> ment, with what others, what impulse they receive, and what
> impart. The same ones make up earth, sky, sea, and stream; the
> same the sun, the animals, grain and trees, but mingling and mov-
> ing in ever different ways.   —Lucretius (95–52 B.C.)
>   in *The Nature of Things*

2.1 INTRODUCTION

Almost every pollution problem that we face has a chemical basis. Even the most qualitative descriptions of such problems as the greenhouse effect, ozone depletion, toxic
wastes, groundwater contamination, air pollution, and acid rain, to mention a few,
require at least a rudimentary understanding of some basic chemical concepts. And, of
course, an environmental engineer who must design an emission control system or a
waste treatment plant must be well grounded in chemical principles and the techniques of chemical engineering. In this brief chapter, the topics have been selected with
the goal of providing only the essential chemical principles required to understand the
nature of the pollution problems that we face and the engineering approaches to their
solutions.

2.2 STOICHIOMETRY

When a chemical reaction is written down, it provides both qualitative and quantitative information. Qualitatively, we can see what chemicals are interacting to produce
what end products. Quantitatively, the principle of conservation of mass can be applied
to give information about how much of each compound is involved to produce the

results shown. The balancing of equations so that the same number of each kind of atom appears on each side of the equation and the subsequent calculations, which can be used to determine amounts of each compound involved, is known as *stoichiometry*.

The first step is to balance the equation. For example, suppose we want to investigate the combustion of methane (CH_4), the principal component of natural gas and a major greenhouse gas. Methane combines with oxygen to produce carbon dioxide and water, as the following reaction suggests:

$$CH_4 + O_2 \rightarrow CO_2 + H_2O$$

The equation is not balanced. One atom of carbon appears on each side, which is fine, but there are four atoms of hydrogen on the left and only two on the right, and there are only two atoms of oxygen on the left while there are three on the right. We might try to double the water molecules on the right to balance the hydrogen on each side, but then there would be an imbalance of oxygen with two on the left and four on the right. So try doubling the oxygen on the left. This sort of trial-and-error approach to balancing simple reactions usually converges quickly. In this instance the following is a balanced equation with the same number of C, H, and O atoms on each side of the arrow:

$$CH_4 + 2O_2 \rightarrow CO_2 + 2\,H_2O \tag{2.1}$$

This balanced chemical equation can be read as follows: One molecule of methane reacts with two molecules of oxygen to produce one molecule of carbon dioxide and two molecules of water. It is of more use, however, to be able to describe this reaction in terms of the mass of each substance (that is, how many grams of oxygen are required to react with how many grams of methane, and so on). To do so requires that we know something about the mass of individual atoms and molecules.

The *atomic weight* of an atom is the mass of the atom measured in *atomic mass units* (amu), where one amu is defined to be exactly one-twelfth the mass of a carbon atom having six protons and six neutrons in its nucleus. While this might suggest that if we look up the atomic weight of carbon we would expect to find it to be exactly 12 amu, that is not the case. All carbon atoms do have six protons, but they do not all have six neutrons, so they do not all have the same atomic weight. Atoms having the same number of protons but differing numbers of neutrons are called *isotopes*. What is reported in tables of atomic weights, such as Table 2.1, is the average based on the relative abundance of different isotopes found in nature. Also shown in Table 2.1 is the *atomic number*, which is the number of protons in the nucleus. All isotopes of a given element have the same atomic number.

The *molecular* weight of a molecule is simply the sum of the atomic weights of all of the constituent atoms. If we divide the mass of a substance by its molecular weight, the result is the mass expressed in *moles* (mol). Usually the mass is expressed in grams, in which case the moles are *g-moles*; in like fashion, if the mass is expressed in pounds, the result would be *lb-moles*. In this text, all moles will be assumed to be g-moles. One g-mole contains 6.02×10^{23} molecules (Avogadro's number), while one lb-mole is made up of 2.7×10^{26} molecules.

$$\text{Moles} = \frac{\text{Mass}}{\text{Molecular weight}} \tag{2.2}$$

TABLE 2.1 Atomic Numbers and Atomic Weights

Element	Symbol	Atomic number	Atomic weight	Element	Symbol	Atomic number	Atomic weight
Actinium	Ac	89	227.03	Mercury	Hg	80	200.59
Aluminum	Al	13	26.98	Molybdenum	Mo	42	95.94
Americium	Am	95	243	Neodymium	Nd	60	144.24
Antimony	Sb	51	121.75	Neon	Ne	10	20.18
Argon	Ar	18	39.95	Neptunium	Np	93	237.05
Arsenic	As	33	74.92	Nickel	Ni	28	58.70
Astatine	At	85	210	Niobium	Nb	41	92.91
Barium	Ba	56	137.33	Nitrogen	N	7	14.01
Barkelium	Bk	97	247	Nobelium	No	102	259
Berylium	Be	4	9.01	Osmium	Os	76	190.2
Bismuth	Bi	83	208.98	Oxygen	O	8	16.00
Boron	B	5	10.81	Palladium	Pd	46	106.4
Bromine	Br	35	79.90	Phosphorus	P	15	30.97
Cadmium	Cd	48	112.41	Platinum	Pt	78	195.09
Calcium	Ca	20	40.08	Plutonium	Pu	94	244
Californium	Cf	98	251	Polonium	Po	84	209
Carbon	C	6	12.01	Potassium	K	19	39.09
Cerium	Ce	58	140.12	Praeseodymium	Pr	59	140.91
Cesium	Cs	55	132.90	Promethium	Pm	61	145
Chlorine	Cl	17	35.45	Protactinium	Pa	91	231.04
Chromium	Cr	24	51.99	Radium	Ra	88	226.03
Cobalt	Co	27	58.93	Radon	Rn	86	222
Copper	Cu	29	63.55	Rhenium	Re	75	186.2
Curium	Cm	96	247	Rhodium	Rh	45	102.91
Dysprosium	Dy	66	162.50	Robidium	Rb	37	85.45
Einsteinium	Es	99	254	Ruthenium	Ru	44	101.07
Erbium	Er	68	167.26	Samarium	Sm	62	150.4
Europium	Eu	63	151.96	Scandium	Sc	21	44.96
Fermium	Fm	100	257	Selenium	Se	34	78.96
Fluorine	F	9	19.00	Silicon	Si	14	28.09
Francium	Fr	87	223	Silver	Ag	47	107.89
Gadolinium	Gd	64	157.25	Sodium	Na	11	22.99
Gallium	Ga	31	69.72	Strontium	Sr	38	87.62
Germanium	Ge	32	72.59	Sulfur	S	16	32.06
Gold	Au	79	196.97	Tantalum	Ta	73	180.95
Hafnium	Hf	72	178.49	Technetium	Tc	43	97
Helium	He	2	4.00	Tellurium	Te	52	127.60
Holmium	Ho	67	164.93	Terbium	Tb	65	158.93
Hydrogen	H	1	1.01	Thallium	Tl	81	204.37
Indium	In	49	114.82	Thorium	Th	90	232.04
Iodine	I	53	126.90	Thulium	Tm	69	168.93
Iridium	Ir	77	192.22	Tin	Sn	50	118.69
Iron	Fe	26	55.85	Titanium	Tl	22	47.90
Krypton	Kr	36	83.80	Tungsten	W	74	183.85
Lanthanum	La	57	138.91	Uranium	U	92	238.03
Lawrencium	Lr	103	260	Vanadium	V	23	50.94
Lead	Pb	82	207.2	Xenon	Xe	54	131.30
Lithium	Li	3	6.94	Ytterbium	Yb	70	173.04
Lutetium	Lu	71	174.97	Yttrium	Y	39	88.91
Magnesium	Mg	12	24.31	Zinc	Zn	30	65.38
Manganese	Mn	25	54.94	Zirconium	Zr	40	91.22
Mendelevium	Md	101	258				

The special advantage of expressing amounts in moles is that one mole of any substance contains exactly the same number of molecules, which gives us another way to interpret a chemical equation. Consider (2.1), repeated here:

$$CH_4 + 2\,O_2 \rightarrow CO_2 + 2\,H_2O$$

On a molecular level we can say that one molecule of methane reacts with two molecules of oxygen to produce one molecule of carbon dioxide and two molecules of water. On a larger scale, we can say that one mole of methane reacts with two moles of oxygen to produce one mole of carbon dioxide and two moles of water. Since we know how many grams are contained in each mole, we can express our mass balance in those terms as well.

To express the preceding methane reaction in grams, we need first to find the number of grams per mole for each substance. Using Table 2.1, we find that the atomic weight of C is 12, H is 1, and O is 16. Notice that these values have been rounded slightly, which is common engineering practice. Thus, the molecular weights, and hence the number of grams per mole, are

$$
\begin{aligned}
CH_4 &= 12 + 4 \times 1 &&= 16 \ \ g/mol \\
O_2 &= 2 \times 16 &&= 32 \ \ g/mol \\
CO_2 &= 12 + 2 \times 16 &&= 44 \ \ g/mol \\
H_2O &= 2 \times 1 + 16 &&= 18 \ \ g/mol
\end{aligned}
$$

Summarizing these various ways to express the oxidation of methane, we can say

CH_4	$+$	$2\,O_2$	$\rightarrow$	CO_2	$+$	$2\,H_2O$
1 molecule of methane	$+$	2 molecules of oxygen	$\rightarrow$	1 molecule of carbon dioxide	$+$	2 molecules of water

or

1 mol of methane	$+$	2 mol of oxygen	$\rightarrow$	1 mol of carbon dioxide	$+$	2 mol of water

or

16 g of methane	$+$	64 g of oxygen	$\rightarrow$	44 g of carbon dioxide	$+$	36 g of water

Notice that mass is conserved in the last expression; that is, there are 80 grams on the left and 80 grams on the right.

EXAMPLE 2.1   Combustion of Butane

What mass of carbon dioxide would be produced if 100 g of butane (C_4H_{10}) is completely oxidized to carbon dioxide and water?

Solution   First write down the reaction:

$$C_4H_{10} + O_2 \rightarrow CO_2 + H_2O$$

and then balance it

$$2\,C_4H_{10} + 13\,O_2 \rightarrow 8\,CO_2 + 10\,H_2O$$

Find the grams per mole for butane:

$$C_4H_{10} = 4 \times 12 + 10 \times 1 = 58\text{ g/mol}$$

We already know that there are 44 grams per mole of CO_2, so we do not need to recalculate that. Two moles of butane $(2\text{ mol} \times 58\text{ g/mol} = 116\text{ g})$ yields 8 moles of carbon dioxide $(8\text{ mol} \times 44\text{ g/mol} = 352\text{ g CO}_2)$. So we can set up the following proportion:

$$\frac{116\text{ g }C_4H_{10}}{352\text{ g }CO_2} = \frac{100\text{ g }C_4H_{10}}{X\text{ g }CO_2}$$

Thus,

$$X = 100 \times 352/116 = 303\text{ g of }CO_2 \text{ produced.} \qquad \blacksquare$$

Many environmental problems involve concentrations of substances dissolved in water. In Chapter 1 we introduced two common sets of units, mg/L and ppm. However, it is also useful to express concentrations in terms of *molarity*, which is simply the number of moles of substance per liter of solution. A 1 molar (1 M) solution has 1 mol of substance dissolved into enough water to make the mixture have a volume of 1 L. Molarity is related to mg/L concentrations by the following:

$$\text{mg/L} = \text{Molarity (mol/L)} \times \text{ Molecular weight (g/mol)} \times 10^3 \text{ (mg/g).} \quad (2.3)$$

The following example illustrates the use of molarity and at the same time introduces another important concept having to do with the amount of oxygen required to oxidize a given substance.

EXAMPLE 2.2   Theoretical Oxygen Demand

Consider a 1.67×10^{-3} M glucose solution $(C_6H_{12}O_6)$ that is completely oxidized to CO_2 and H_2O. Find the amount of oxygen required to complete the reaction.

Solution   To find the oxygen required to oxidize this glucose completely, we first write a balanced equation, determine molecular weights, and find the mass of each constituent in the reaction:

$$\begin{array}{ccccccc}
C_6H_{12}O_6 & + & 6\,O_2 & \rightarrow & 6\,CO_2 & + & 6\,H_2O \\
6 \times 12 + 12 \times 1 + 6 \times 16 = 180 & & 6 \times 32 = 192 & & 6 \times 44 = 264 & & 6 \times 18 = 108
\end{array}$$

Thus it takes 192 g of oxygen to oxidize 180 g of glucose. From (2.3), the concentration of glucose is

$$\text{mg/L} = 1.67 \times 10^{-3} \text{ mol/L} \times 180 \text{ g/mol} \times 10^3 \text{ mg/g} = 300 \text{ mg/L}$$

so the oxygen requirement would be

$$300 \text{ mg/L glucose} \times \frac{192 \text{ g }O_2}{180 \text{ g glucose}} = 320 \text{ mg/L }O_2 \qquad \blacksquare$$

If the chemical composition of a substance is known, then the amount of oxygen required to oxidize it to carbon dioxide and water can be calculated using stoichiometry, as was done in the preceding example. That oxygen requirement is known as the *theoretical oxygen demand*. If that oxidation is carried out by bacteria using the substance for food, then the amount of oxygen required is known as the *biochemical oxygen demand, or BOD*. The BOD will be somewhat less than the theoretical oxygen since some of the original carbon is incorporated into bacterial cell tissue rather than being oxidized to carbon dioxide. Oxygen demand is an important measure of the likely impact that wastes will have on a receiving body of water, and much more will be said about it in Chapter 5.

The convenience of using moles to describe amounts of substances also helps when calculating atmospheric concentrations of pollutants. It was Avogadro's hypothesis, made back in 1811, that equal volumes of all gases, at a specified temperature and pressure, contain equal numbers of molecules. In fact, since 1 mol of any substance has Avogadro's number of molecules, it follows that 1 mol of gas, at a specified temperature and volume, will occupy a predictable volume. At standard temperature and pressure (STP), corresponding to $0\,°C$ and 1 atm (760 mm of mercury, 101.3 kPa), 1 g-mole of an ideal gas occupies 22.4 L, or $0.0224\,m^3$, and contains 6.02×10^{23} molecules. This fact was used in Chapter 1 to derive relationships between concentrations expressed in $\mu g/m^3$ and ppm (by volume).

Let us demonstrate the usefulness of Avogadro's hypothesis for gases by applying it to a very modern concern; that is, the rate at which we are pouring carbon dioxide into the atmosphere as we burn up our fossil fuels.

EXAMPLE 2.3   Carbon Emissions from Natural Gas

Worldwide combustion of methane CH_4 (natural gas) provides about 8.2×10^{16} kJ of energy per year. If methane has an energy content of 39×10^3 kJ/m^3 (at STP) what mass of CO_2 is emitted into the atmosphere each year? Also, express that emission rate as metric tons of carbon (not CO_2) per year. A metric ton, which is 1000 kg, is usually written as tonnes to distinguish it from the 2000-lb American, or short, tons.

Solution   We first need to express that consumption rate in moles. Converting kilojoules of energy into moles of methane is straightforward:

$$\text{moles } CH_4 = \frac{8.2 \times 10^{16}\text{kJ/yr}}{39 \times 10^3\text{kJ/m}^3} \times \frac{1}{22.4 \times 10^{-3}\text{m}^3\text{/mol}} = 9.4 \times 10^{13}\,\text{mol/yr}$$

We know from the balanced chemical reaction given in (2.1) that each mole of CH_4 yields one mole of CO_2, so there will be 9.4×10^{13} mol of CO_2 emitted. Since the molecular weight of CO_2 is 44, the mass of CO_2 emitted is

$$\text{Mass } CO_2 = 9.4 \times 10^{13}\,\text{mol/yr} \times 44\,\text{g/mol} = 4.1 \times 10^{15}\,\text{g/yr}$$

To express these emissions as tonnes of C per year, we must convert grams to tonnes and then sort out the fraction of CO_2 that is carbon. The fraction of CO_2 that is C is simply the ratio of the atomic weight of carbon (12) to the molecular weight of carbon dioxide (44):

$$\text{C emissions} = 4.1 \times 10^{15}\,\text{g } CO_2\text{/yr} \times \frac{1\text{ kg}}{1000\text{ g}} \times \frac{1\text{ tonne}}{1000\text{ kg}} \times \frac{12\text{ g C}}{44\text{ g } CO_2}$$

$$= 1.1 \times 10^9 \text{ tonnes/yr} = 1.1 \text{ gigatonne/yr} = 1.1 \text{ Gt/yr} \qquad\blacksquare$$

The 1.1×10^9 tonnes of carbon found in the preceding example is about 20 percent of the total, worldwide carbon emissions entering the atmosphere each year when fossil fuels (coal, oil, natural gas) are burned. As will be described in Chapter 8, the main worry about these emissions is their potential to increase the earth's greenhouse effect.

2.3 ENTHALPY IN CHEMICAL SYSTEMS

Just as we used conservation of mass to balance chemical equations, we can use the conservation of energy to learn something about heat absorbed or released during chemical reactions. Since energy must be conserved, we should be able to track it from beginning to end. The first law of thermodynamics lets us say that the energy in the reactants on the left side of the equation, plus any heat added to the system, should equal the energy contained in the reaction products on the right side plus any work done during the reaction.

$$U_1 + Q = U_2 + W \qquad (2.4)$$

where

U_1 = Internal energy of the chemical system at the beginning
U_2 = Internal energy at the end
Q = heat absorbed during the reaction
W = work done by the system during the reaction

While there are many forms of work that could be included in (2.4), our concern will be only with work that occurs when a system changes volume under constant pressure, which is typical of chemical reactions. Any possibility of electrical, magnetic, gravitational, or other forms work will be ignored. To analyze this work done by expansion, consider the cylinder in Figure 2.1, containing a volume of gas V exerting a constant pressure P against a piston with area A. The force exerted by the gas on the piston is $P \times A$. If the piston moves a distance d, then, since work is force $\times$ distance, we can write

$$W = Fd = PAd = P(V_2 - V_1) \qquad (2.5)$$

Substituting (2.5) into (2.4) and rearranging terms gives

$$(U_2 + PV_2) - (U_1 + PV_1) = Q \qquad (2.6)$$

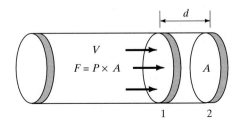

FIGURE 2.1   Work done when a substance expands at constant pressure is PΔV.

As described in Chapter 1, the *enthalpy*, H, of a system is

$$H = U + PV \tag{2.7}$$

then

$$H_2 - H_1 = \Delta H = Q \tag{2.8}$$

That is, the change in enthalpy during a constant pressure reaction is equal to the heat absorbed by the system. When ΔH is positive, heat is absorbed and the reaction is said to be *endothermic*. When ΔH is negative, heat is liberated and the reaction is called *exothermic*. The change in enthalpy, $H_2 - H_1$, is called the *heat of reaction*.

As is usually the case for discussions about energy, it is changes in energy, or in this case enthalpy, that are of interest. For example, we can talk about the potential energy of an object as being its weight times its height above some reference elevation. Our choice of reference elevation does not matter as long as we are only interested in the change in potential energy as an object is raised against gravity from one height to another. Similarly, since changes in enthalpy during a chemical reaction are of interest, it does not really matter what reference conditions are chosen. Tabulated values of enthalpy are usually based on 1 atm of pressure and 25 °C (298 K), in which case they are designated with the notation

$$H_{298}^0 = \text{Standard enthalpy} = \text{Enthalpy at 1 atm and 298 K (kJ/mol)} \tag{2.9}$$

It is also assumed that the reference condition for pure elements is the stable state of the substance at 1 atm and 25 °C. For example, the stable state of oxygen at 1 atm and 25 °C is gaseous O_2, so the standard enthalpy for O_2 (g), where (g) just means it is in the gaseous state, is defined to be zero. Similarly, mercury under those

TABLE 2.2 Standard enthalpies for selected species (kJ/mol)

Substance	State[a]	H_{298}^0	Substance	State[a]	H_{298}^0
Ca^{2+}	aq	−543.0	HCO_3^-	aq	−691.1
$CaCO_3$	s	−1207	H_2O	l	−285.8
$Ca(OH)_2$	s	−986.6	H_2O	g	−241.8
C	s	0	NO	g	90.4
CO	g	−110.5	NO_2	g	33.9
CO_2	g	−393.5	NO_2^-	aq	−106.3
CO_2	aq	−412.9	N	g	472.6
CH_4	g	−74.9	N_2	g	0
C_2H_4	g	52.3	N_2O	g	81.55
CH_3COOH	aq	−488.5	NH_3	aq	−80.8
$n-C_4H_{10}$	g	−124.7	O	g	247.5
$i-C_4H_{10}$	g	−131.6	O_2	g	0
H	g	217.9	O_3	g	142.9
H^+	aq	0	OH^-	aq	−229.9
H_2	g	0	SO_2	g	−296.9

[a]g = gas, aq = aqueous, s = solid, l = liquid.

conditions is a liquid, so the standard enthalpy for Hg (l) is zero, where (l) means the liquid state. A table of standard enthalpies for a number of substances is presented in Table 2.2. More extensive lists can be found in chemical handbooks as well as more advanced environmental engineering texts.

The sum of the enthalpies of the reaction products minus the sum of the enthalpies of the reactants is called the *heat of reaction*. When it is negative, heat is liberated during the reaction; when it is positive, heat is absorbed. Since standard enthalpies are expressed as energy per mole, we must first balance the chemical reaction and then for each species multiply the number of moles by the standard enthalpy to get its total enthalpy.

EXAMPLE 2.4   Gross Heat of Combustion for Methane

Find the heat of reaction when methane CH_4 is oxidized to CO_2 and liquid H_2O.

Solution   The reaction is written below, and beneath it are enthalpies taken from Table 2.2.

$$CH_4\,(g) + 2\,O_2\,(g) \rightarrow CO_2\,(g) + 2\,H_2O\,(l)$$
$$(-74.9) \qquad 2 \times (0) \qquad\quad (-393.5) \qquad 2 \times (-285.8)$$

Notice that we have used the enthalpy of liquid water for this calculation. The heat of reaction is the difference between the total enthalpy of the reaction products and the reactants:

$$[(-393.5) + 2 \times (-285.8)] - [(-74.9) + 2 \times (0)] = -890.2 \text{ kJ/mol of } CH_4$$

Since the heat of reaction is negative, heat is released during combustion (i.e., the reaction is exothermic).   ∎

When a fuel is burned, some of the energy released ends up as latent heat in the water vapor produced. Usually that water vapor, along with the latent heat it contains, exits the stack along with all the other combustion gases, and its heating value is, in essence, lost. That leads to two different values of what is called the *heat of combustion*. The *higher heating value* (HHV) includes the heat released when water condenses to liquid form, as was the case in Example 2.4. The HHV is also known as the *gross heat of combustion*. In the United States, the heat content of fossil fuels is usually expressed as this gross amount. The *lower heating value* (LHV), or *net heat of combustion*, is based on the heat of reaction when water is assumed to remain in the vapor state. The most fuel-efficient, modern furnaces used for space-heating buildings achieve their high efficiencies (above 90 percent) by causing the combustion gases to cool enough to condense the water vapor before it leaves the stack. Not unexpectedly, these are called condensing furnaces. We will encounter these HHV and LHV concepts again in Chapter 9 when incineration of solid waste is described.

EXAMPLE 2.5   The Net Heat of Combustion

Find the net heat of combustion when methane is burned.

Solution   We just repeat the procedure of Example 2.4, but this time the water will remain in the gaseous state. Again, using values from Table 2.2,

$$CH_4\,(g) \quad + \quad 2\,O_2\,(g) \quad \rightarrow \quad CO_2\,(g) \quad + \quad 2\,H_2O\,(g)$$
$$(-74.9) \qquad\qquad 2 \times (0) \qquad\qquad (-393.5) \qquad\quad 2 \times (-241.8)$$

The net heat of combustion is

$$[(-393.5) + 2 \times (-241.8)] - [(-74.9) + 2 \times (0)] = -802.2 \text{ kJ/mol of } CH_4$$

Again, the sign tells us this is an exothermic reaction. Notice that about 10 percent of the gross heating value is lost when water vapor is not condensed.   ■

There is another application of the concept of enthalpy that is useful in describing photochemical smog reactions in the atmosphere and for analyzing the chemical reactions that affect the stratospheric ozone layer. When a molecule is dissociated by absorbing a photon of energy, the process is called a *photochemical reaction, photochemical dissociation,* or, more simply, *photolysis.* We can use enthalpies to determine the amount of energy a photon must have to cause photolysis, and from the photon's energy we can determine the maximum wavelength that the photon can have.

For a photon to be able to cause photolysis it must have at least as much energy as the change in enthalpy for the reaction. It is important to realize that when a molecule absorbs a photon, the energy must be used almost immediately for photolysis or else the energy will be dissipated as waste heat as the molecule collides with neighboring molecules. Molecules cannot store up energy from a series of photon encounters, waiting until enough energy is accumulated to cause photolysis. Thus a *single* photon must have sufficient energy to cause photolysis all by itself.

In Chapter 1 the following relationship between energy contained in one photon and the wavelength and frequency associated with that photon was introduced:

$$E = h\nu = \frac{hc}{\lambda} \tag{2.10}$$

where

E = Energy of a photon (J)
h = Planck's constant $(6.6 \times 10^{-34}\,\text{J} - \text{s})$
ν = frequency (hertz, i.e., cycles per second)
c = speed of light $(3 \times 10^8\,\text{m/s})$
λ = wavelength (m)

We are interested in the maximum wavelength that a photon can have and still have enough energy for photolysis, so rearranging (2.10) gives

$$\lambda \leq \frac{hc}{E} \tag{2.11}$$

Before we can equate the energy in a photon with the enthalpy change, we must be sure the units are consistent. The reaction's enthalpy change ΔH^0 has units of kJ/mol and E is joules per photon. Using Avogadro's number, along with the fact that one photon dissociates one molecule, lets us write (Baird, 1995)

$$\lambda \leq \frac{6.6 \times 10^{-34}\,\text{J}\cdot\text{s} \times 3 \times 10^{8}\,\text{m/s} \times 6.02 \times 10^{23}\,\text{molecules/mol} \times 1\,\text{photon/molecule}}{\Delta H^{0}(\text{kJ/mol}) \times 10^{3}\text{J/kJ}}$$

$$\lambda(\text{m}) \leq \frac{1.19 \times 10^{-4}(\text{kJ}\cdot\text{m/mol})}{\Delta H^{0}(\text{kJ/mol})} \tag{2.12}$$

EXAMPLE 2.6   Photolysis of Ozone

What maximum wavelength of light would be capable of causing photolysis of ozone O_3 into O_2 and O?

Solution   First write the reaction, including enthalpies from Table 2.2. Even though those enthalpies are for standard conditions (1 atm, 298 K), they can be used under stratospheric pressure and temperature conditions with only modest error.

$$O_3 \quad + \quad h\nu \quad \rightarrow \quad O_2 \quad + \quad O$$
$$(142.9) \qquad\qquad\qquad (0) \qquad (247.5)$$

The enthalpy change is

$$\Delta H^{0} = 247.5 - 142.9 = 104.6\ \text{kJ/mol}$$

Since the sign of ΔH^{0} is positive, this is an endothermic reaction that needs to absorb energy in order to take place. That energy comes from the photon. From (2.12),

$$\lambda \leq \frac{1.19 \times 10^{-4}\,\text{kJ}\cdot\text{m/mol}}{104.6\ \text{kJ/mol}} = 1.13 \times 10^{-6}\ \text{m} = 1.13\ \mu\text{m}$$

Actually, the wavelength must be considerably shorter (see Problem 1.13). ∎

Absorption of incoming solar radiation by ozone, as described in Example 2.6, is part of the shielding that the stratospheric ozone layer provides for our planet. Destruction of stratospheric ozone by chlorofluorocarbons will be described in Chapter 8.

2.4 CHEMICAL EQUILIBRIA

In the reactions considered so far, the assumption has been that they proceed in one direction only. Most chemical reactions are, to some extent, reversible, proceeding in both directions at once. When the rates of reaction are the same (that is, products are being formed on the right at the same rate as they are being formed on the left), the reaction is said to have reached *equilibrium*.

The following represents a generalized reversible reaction:

$$a\text{A} \ + \ b\text{B} \ \rightleftharpoons \ c\text{C} \ + \ d\text{D} \tag{2.13}$$

in which the small letters $a, b, c,$ and d are coefficients corresponding to the number of molecules or ions of the respective substances that result in a balanced equation. The capital letters A, B, C, and D are the chemical species themselves. The double arrow designation indicates that the reaction proceeds in both directions at the same time.

In equilibrium, we can write that

$$\frac{[\text{C}]^c[\text{D}]^d}{[\text{A}]^a[\text{B}]^b} = K \tag{2.14}$$

where the [ ] designation represents concentrations of the substances in equilibrium, expressed in moles per liter. Do not use concentrations in mg/L! K is called the *equilibrium constant*. It should also be emphasized that (2.14) is valid only when chemical equilibrium is established, if ever. Natural systems are often subject to constantly changing inputs, and since some reactions occur very slowly, equilibrium may never be established. A practicing environmental engineer must therefore use this important equation with a certain degree of caution.

Many molecules, when dissolved in water, separate into positively charged ions, called *cations*, and negatively charged ions, called *anions*. Equation (2.14) can be applied to the dissociation of such molecules, in which case K is referred to as a *dissociation* constant or an *ionization constant*. In other circumstances, the quantity on the left is a solid and the problem is to determine the degree to which that solid enters solution. In such cases the equilibrium constant is called the *solubility product*.

Finally, often when dealing with very large and very small numbers, it is helpful to introduce the following logarithmic measure:

$$K = 10^{-pK} \tag{2.15}$$

or

$$pK = -\log K \tag{2.16}$$

Acid-Base Reactions

Water dissociates slightly into hydrogen ions (protons, H^+) and hydroxide ions (OH^-), as the following reaction suggests:

$$H_2O \ \rightleftharpoons \ H^+ \ + \ OH^- \tag{2.17}$$

The corresponding equilibrium expression for this reaction is

$$\frac{[H^+][OH^-]}{[H_2O]} = K \tag{2.18}$$

The molar concentration of water $[H_2O]$ is 1000 g/L divided by 18 g/mol, or 55.56 mol/L. Since water dissociates only slightly, the molar concentration after ionization is not changed enough to be of significance, so $[H_2O]$ is essentially a constant that can be included in the equilibrium constant. The result is the following:

$$[H^+][OH^-] = K_w = 1 \times 10^{-14} \quad \text{at } 25\,^\circ\text{C} \tag{2.19}$$

where K_w is the dissociation constant for water. For dilute aqueous solutions in general, $[H_2O]$ is considered constant and is included in the equilibrium constant. K_w is temperature dependent, but unless otherwise stated the value given in (2.19) at 25 °C will be the assumed value.

It is important to point out that (2.19) holds no matter what the source of hydrogen ions or hydroxide ions. That is, it is valid even if other substances dissolved in the water make their own contributions to the hydrogen and hydroxide supplies. It is always one of the equations that must be satisfied when chemical equilibria problems are analyzed.

It is customary to express $[H^+]$ and $[OH^-]$ concentrations using the logarithmic measure introduced in (2.15) and (2.16). To express hydrogen ion concentrations, the pH scale is used, where

$$pH = -\log[H^+] \tag{2.20}$$

or

$$[H^+] = 10^{-pH} \tag{2.21}$$

With the pH scale, it is easy to specify whether a solution is acidic, basic, or neutral. A *neutral* solution corresponds to the case where the concentration of hydrogen ions $[H^+]$ equals the concentration of hydroxide ions $[OH^-]$. From (2.19), for a neutral solution

$$[H^+][OH^-] = [H^+][H^+] = [H^+]^2 = 10^{-14}$$

so

$$[H^+] = 10^{-7}$$

That is, a neutral solution has a pH of 7 (written pH 7).

An *acidic* solution is one for which $[H^+]$ is greater than $[OH^-]$; that is, the hydrogen ion concentration is greater than 10^{-7} mol/L, and its pH is less than 7. A *basic* solution is the other way around, with more hydroxide ions than hydrogen ions and pH greater than 7. Notice that for every unit change in pH, the concentration of hydrogen ions changes by a factor of 10.

Figure 2.2 illustrates the pH scale, showing example values of pH for several common solutions. Notice in the figure that a distinction is made between distilled

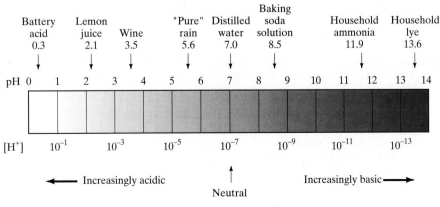

FIGURE 2.2   The pH scale.

water and "pure" rainfall. As will be seen in the next section on carbonates, as rainwater falls it absorbs carbon dioxide from the air and carbonic acid is formed. Unpolluted rainfall then has a natural pH of around 5.6. By the usual definition, acid rain, caused by industrial pollutants, has pH lower than 5.6. Actual acid rain and acid fog have been recorded with pH below 2.0.

EXAMPLE 2.7   pH of Tomato Juice

Find the hydrogen ion concentration and the hydroxide ion concentration in tomato juice having a pH of 4.1.

Solution   From (2.21), the hydrogen ion concentration is

$$[H^+] = 10^{-pH} = 10^{-4.1} = 7.94 \times 10^{-5} \, \text{mol/L}$$

and from (2.19) the hydroxide ion concentration is

$$[OH^-] = \frac{10^{-14}}{[H^+]} = \frac{10^{-14}}{7.94 \times 10^{-5}} = 1.25 \times 10^{-10} \, \text{mol/L} \qquad \blacksquare$$

Acid-base reactions are among the most important in environmental engineering. Most aquatic forms of life, for example, are very sensitive to the pH of their habitat. To protect local ecosystems, waste neutralization before release is common. In other circumstances, by manipulating pH, unwanted substances can be driven out of solution as precipitates or gases before the effluent is released.

As an example of the value of being able to control pH, consider the problem of removing nitrogen from wastewater. Nitrogen is a nutrient that can degrade water quality by stimulating excessive algal growth in receiving bodies of water. Nitrogen in the form of nitrate NO_3^- also poses another problem, especially in groundwater. When babies drink water with high nitrate content, a potentially lethal condition known as methemoglobinemia can result.

One way to remove nitrogen during wastewater treatment is with a process known as ammonia stripping. When organic matter decomposes, nitrogen is first released in the form of ammonia NH_3 or ammonium ion NH_4^+. By driving the equilibrum reaction

$$NH_3 + H_2O \rightleftharpoons NH_4^+ + OH^- \qquad (2.22)$$

toward the left, less soluble ammonia gas is formed, which can then be encouraged to leave solution and enter the air in a gas stripping tower. In such a tower, contaminated water is allowed to trickle downward over slats or corrugated surfaces while clean air is blown in from the bottom, aerating the dripping water. Gas stripping can be used in wastewater treatment facilities to remove such gases as ammonia and hydrogen sulfide, and it is also being used to remove *volatile organic chemicals* (VOCs) from contaminated groundwater.

To strip ammonia, the reaction given in (2.22) must be driven toward the left, which can be done by increasing the concentration of OH^-. That is, it can be accomplished by raising the pH, as the following example demonstrates.

EXAMPLE 2.8   Ammonia Stripping

Nitrogen in a wastewater treatment plant is in the form of ammonia and ammonium ion. Find the fraction of the nitrogen that is in the ammonia form (and hence strippable) as a function of pH (at 25°C) and draw a graph. The equilibrium constant is 1.82×10^{-5}.

Solution   The equilibrium equation for the reaction given in (2.22) is

$$\frac{[NH_4^+][OH^-]}{[NH_3]} = K_{NH_3} = 1.82 \times 10^{-5} \tag{2.23}$$

K_{NH_3} is like the equilibrium constant for water in that it treats the concentration of H_2O as a constant that is already accounted for in the equilibrium constant itself. What we want to find is the fraction of nitrogen in the form of ammonia, or

$$NH_3 \text{ fraction} = \frac{[NH_3]}{[NH_3] + [NH_4^+]} = \frac{1}{1 + [NH_4^+]/[NH_3]} \tag{2.24}$$

Equation (2.19) must also be satisfied:

$$[H^+][OH^-] = K_w = 10^{-14} \tag{2.19}$$

Rearranging (2.23) and substituting (2.19) into it gives

$$\frac{[NH_4^+]}{[NH_3]} = \frac{K_{NH_3}}{[OH^-]} = \frac{K_{NH_3}}{K_w/[H^+]} \tag{2.25}$$

and putting this into (2.24) gives

$$NH_3 \text{ fraction} = \frac{1}{1 + K_{NH_3}[H^+]/K_w}$$

$$= \frac{1}{1 + 1.82 \times 10^{-5} \times 10^{-pH}/10^{-14}}$$

$$= \frac{1}{1 + 1.82 \times 10^{(9-pH)}} \tag{2.26}$$

A table of values for this fraction can easily be generated from (2.26), and the results are plotted in Figure 2.3. ∎

As Example 2.5 suggests, to drive the reaction significantly toward the ammonia side requires a pH in excess of about 10. Since typical wastewaters seldom have a pH that high, it is necessary to add chemicals, the usual being lime (CaO), to raise the pH sufficiently. The lime, unfortunately, reacts with CO_2 in the air to form a calcium carbonate scale that can accumulate on stripping surfaces and that must be removed periodically, creating a sludge disposal problem of its own.

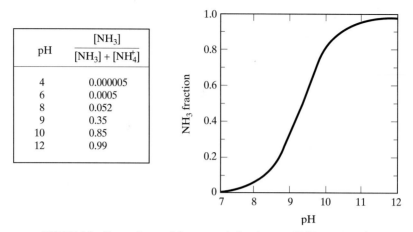

pH	$\dfrac{[NH_3]}{[NH_3] + [NH_4^+]}$
4	0.000005
6	0.0005
8	0.052
9	0.35
10	0.85
12	0.99

FIGURE 2.3　Dependence of the ammonia fraction on pH (Example 2.8).

The preceeding example was but one illustration of the value of being able to control pH in treatment processes. In other circumstances, pollution itself affects pH, driving reactions in directions that may be undesirable. A notable example of this phenomenon is the mobilization of trace metals, especially aluminum, when pH drops during acidification of a watershed. Aluminum is very toxic to fish, but its normal form is a quite insoluble solid and hence is relatively safe. However, when acid deposition acidifies a lake, the reduced pH drives equilibria reactions toward liberation of mobile Al^{3+} ions, which are very toxic. Mobilization of trace metals will be discussed further in Chapter 5.

Solubility Product

The preceeding comments about solids and gases in water suggest that we consider such phenomena a bit more carefully. We will deal with gases in the next section, after this brief introduction to solids.

All solids are to some degree soluble, some much more so than others. A generalized equation describing the equilibrium condition in which a solid is dissociating into its ionic components (*dissolution*) at the same rate that ionic components are recombining into the solid form (*precipitation*) is given as follows:

$$\text{Solid} \rightleftarrows a\text{A} + b\text{B} \tag{2.27}$$

where A and B are the ionic components that make up the solid. Applying (2.14) yields

$$\frac{[\text{A}]^a[\text{B}]^b}{[\text{Solid}]} = K \tag{2.28}$$

As long as there is still solid present in equilibrium, its effect can be incorporated into the equilibrium constant:

$$[\text{A}]^a[\text{B}]^b = K_{sp} \tag{2.29}$$

TABLE 2.3   Selected solubility-product constants at 25 °C

Equilibrium equation	K_{sp} at 25 °C	Significance in environmental engineering
$CaCO_3 \rightleftharpoons Ca^{2+} + CO_3^{2-}$	5×10^{-9}	Hardness removal, scaling
$CaSO_4 \rightleftharpoons Ca^{2+} + SO_4^{2-}$	2×10^{-5}	Flue gas desulfurization
$Cu(OH)_2 \rightleftharpoons Cu^{2+} + 2OH^-$	2×10^{-19}	Heavy metal removal
$Al(OH)_3 \rightleftharpoons Al^{3+} + 3OH^-$	1×10^{-32}	Coagulation
$Ca_3(PO_4)_2 \rightleftharpoons 3Ca^{2+} + 2PO_4^{3-}$	1×10^{-27}	Phosphate removal
$CaF_2 \rightleftharpoons Ca^{2+} + 2F^-$	3×10^{-11}	Fluoridation

Source: Sawyer et. al. (1994).

where K_{sp} is called the *solubility product*. Table 2.3 gives a short list of solubility products of particular importance in environmental engineering.

As an example of the use of (2.29), consider the fluoridation of water with calcium fluoride, CaF_2.

EXAMPLE 2.9   Fluoride Solubility

Find the equilibrium concentration of fluoride ions in pure water caused by the dissociation of CaF_2. Express the answer both in units of mol/L and mg/L.

Solution   From Table 2.3, the reaction and solubility product are

$$CaF_2 \rightleftharpoons Ca^{2+} + 2F^- \quad K_{sp} = 3 \times 10^{-11}$$

Remembering to square the term for the fluoride ion, the mass action equation becomes

$$[Ca^{2+}][F^-]^2 = 3 \times 10^{-11} \tag{2.30}$$

If we let s (mol/L) represent the concentration of Ca^{2+}, then the concentration of F^- will be $2s$ since 2 moles of F^- are formed for every mole of Ca^{2+}. Thus

$$[Ca^{2+}] = s$$
$$[F^-] = 2s$$

and from (2.30)

$$K_{sp} = s \times (2s)^2 = 4s^3 = 3 \times 10^{-11}$$
$$s = [Ca^{2+}] = 2 \times 10^{-4} \, \text{mol/L}$$
$$2s = [F^-] = 4 \times 10^{-4} \, \text{mol/L}$$

To find the concentration of fluoride ions in mg/L, we need the atomic weight of fluorine, which from Table 2.1 is 19.

$$(F^-) = 4 \times 10^{-4} \, \text{mol/L} \times 19 \, \text{g/mol} \times 10^3 \, \text{mg/g} = 7.6 \, \text{mg/L} \qquad \blacksquare$$

The fluoride concentration obtained in Example 2.9 is far above recommended drinking water levels of 1.8 mg/L. Fluoride concentrations of approximately 1 mg/L in drinking water help prevent cavities in children, but discoloration of teeth, called *mottling*, is relatively common when concentrations exceed 2.0 mg/L.

Solubility of Gases in Water

When air comes in contact with water, some of it dissolves into the water. Different constituents of air dissolve to different degrees and in amounts that vary with temperature and water purity. The behavior of gases in contact with water was reported by W. Henry in England in 1903, and the resulting relationship is known as *Henry's law*:

$$[\text{gas}] = K_H P_g \tag{2.31}$$

where

$$[\text{gas}] = \text{concentration of dissolved gas (mol/L)}$$
$$K_H = \text{Henry's law constant (mol/L} \cdot \text{atm)}$$
$$P_g = \text{the partial pressure of the gas in air (atm)}$$

The quantity P_g is the partial pressure of the gas in air, which is simply its volumetric concentration times the atmospheric pressure. The units suggested for pressure are atmospheres (atm), where 1 atm corresponds to 101,325 pascals (Pa), and 1 Pa equals one newton per square meter. For example, oxygen makes up about 21 percent of the atmosphere, so at 1 atm P_g would be 0.21×1 atm $= 0.21$ atm.

Each gas-liquid system has its own value for Henry's coefficient. The coefficient itself varies both with temperature (solubility decreases as temperature increases) and with concentration of other dissolved gases and solids (the solubility decreases as other dissolved material in the liquid increases). Henry's law is expressed in various ways with different units for the coefficient depending on the method of expression. The user must be careful, then, to check the units given for Henry's constant before applying the law. Table 2.4 gives some values of K_H for CO_2 and O_2, two of the gases that we will be most concerned with.

TABLE 2.4   Henry's Law Coefficients, K_H (mol/L $\cdot$ atm)

T (°C)	CO_2	O_2
0	0.076425	0.0021812
5	0.063532	0.0019126
10	0.053270	0.0016963
15	0.045463	0.0015236
20	0.039172	0.0013840
25	0.033363	0.0012630

Since atmospheric pressure changes with altitude, P_g will change as well. One estimate for atmospheric pressure as a function of altitude is the following (Thomann and Mueller, 1987):

$$P = P_0 - 1.15 \times 10^{-4}H \qquad (2.32)$$

where

P = atmospheric pressure at altitude H (atm)

H = altitude (m)

P_0 = atmospheric pressure at sea level (atm)

EXAMPLE 2.10   Solubility of Oxygen in Water

By volume, the concentration of oxygen in air is about 21 percent. Find the equilibrium concentration of O_2 in water (in mol/L and mg/L) at 25°C and 1 atm of pressure. Recalculate it for Denver at an altitude of 1525 m.

Solution   Air is 21 percent oxygen, so its partial pressure at 1 atm is

$$P_g = 0.21 \times 1 \text{ atm} = 0.21 \text{ atm}$$

From Table 2.4, at 25°C, $K_H = 0.0012630$ mol/L · atm; so, from (2.31),

$$[O_2] = K_H P_g = 0.0012630 \text{ mol/L} \cdot \text{atm} \times 0.21 \text{ atm}$$
$$= 2.65 \times 10^{-4} \text{ mol/L}$$
$$= 2.65 \times 10^{-4} \text{ mol/L} \times 32 \text{ g/mol} \times 10^3 \text{ mg/g} = 8.5 \text{ mg/L}$$

In Denver, at 1525 m, atmospheric pressure can estimated using (2.32):

$$P = P_0 - 1.15 \times 10^{-4}H = 1 - 1.15 \times 10^{-4} \times 1525 = 0.825 \text{ atm}$$

so

$$[O_2] = 0.0012630 \text{ mol/L} \cdot \text{atm} \times 0.21 \times 0.825 \text{ atm} = 2.19 \times 10^{-4} \text{ mol/L}$$
$$= 2.19 \times 10^{-4} \text{ mol/L} \times 32 \text{ g/mol} \times 10^3 \text{ mg/g} = 7.0 \text{ mg/L} \qquad \blacksquare$$

It should be emphasized that calculations based on Henry's law provide equilibrium concentrations of dissolved gases, or, as they are frequently called, *saturation* values. It is often the case that actual values differ considerably from those at equilibrium. There may be more than the saturation value of a dissolved gas, as when photosynthesis by plants pumps oxygen into the water at a faster rate than it can leave from the air/water interface. It is more common for dissolved gases to be less than the saturation value, as occurs when bacteria decompose large quantities of waste, drawing oxygen from the water (possibly leading to an oxygen deficiency that can kill fish). In either case, when an excess or a deficiency of a dissolved gas occurs, pressures act to try to bring the amount dissolved back to the saturation level.

The Carbonate System

The carbonate system about to be described is the most important acid-base system in natural waters because it controls pH. It is comprised of the following chemical species:

Aqueous carbon dioxide	$CO_2(aq)$
Carbonic acid	H_2CO_3
Bicarbonate ion	HCO_3^-
Carbonate ion	CO_3^{2-}

Aqueous CO_2 is formed when atmospheric CO_2 dissolves in water; we can find its concentration in fresh water using Henry's law (2.31):

$$[CO_2] = K_H P_g \tag{2.33}$$

where the concentration is in (mol/L) and P_{CO_2} is the partial pressure of gaseous CO_2 in the atmosphere (about 360 ppm or 360×10^{-6} atm). Aqueous CO_2 then forms carbonic acid (H_2CO_3), which, in turn, ionizes to form hydrogen ions (H^+) and bicarbonate (HCO_3^-):

$$CO_2(aq) + H_2O \rightleftharpoons H_2CO_3 \rightleftharpoons H^+ + HCO_3^- \tag{2.34}$$

The bicarbonate (HCO_3^-) ionizes to form more hydrogen ion (H^+) and carbonate (CO_3^{2-}):

$$HCO_3^- \rightleftharpoons H^+ + CO_3^{2-} \tag{2.35}$$

In addition, if there is a source of solid calcium carbonate ($CaCO_3$), such as occurs when water is in contact with limestone rocks, then the solubility reaction for $CaCO_3(s)$ applies in the system. Such waters are referred to as *calcareous* waters.

$$CaCO_3(s) \rightleftharpoons Ca^{2+} + CO_3^{2-} \tag{2.36}$$

If sufficient time is allowed for the system to reach equilibrium, then the equilibrium constants for reactions (2.34–2.36) can be used to analyze the system. Reaction (2.34) results in

$$\frac{[H^+][HCO_3^-]}{[CO_2(aq)]} = K_1 = 4.47 \times 10^{-7} \text{ mol/L} \tag{2.37}$$

and (2.35) yields

$$\frac{[H^+][CO_3^{2-}]}{[HCO_3^-]} = K_2 = 4.68 \times 10^{-11} \tag{2.38}$$

while (2.36) is governed by

$$[Ca^{2+}][CO_3^{2-}] = K_{sp} = 4.57 \times 10^{-9} \tag{2.39}$$

The values of K_1, K_2, and K_{sp} are temperature dependent. The values given in (2.37–2.39) are at 25 °C.

There are four possible conditions that determine the behavior of the carbonate system. There may or may not be a source of solid carbonate, as indicated in (2.36), and the solution may or may not be exposed to the atmosphere. If it is exposed to the atmosphere the system is said to be *open*; if it is not, the system is *closed*. We will analyze the common situation of an open system not in contact with a carbonate solid, such as occurs when rainwater passes through the atmosphere, absorbing CO_2 and forming carbonic acid.

Before proceeding to that calculation, it is useful to compare the relative concentrations of carbonate and bicarbonate as a function of pH. Dividing (2.38) by $[H^+]$ gives

$$\frac{[CO_3^{2-}]}{[HCO_3^-]} = \frac{K_2}{[H^+]} = \frac{K_2}{10^{-pH}} \tag{2.40}$$

and then incorporating the value of K_2 given in (2.38) yields:

$$\frac{[CO_3^{2-}]}{[HCO_3^-]} = \frac{4.68 \times 10^{-11}}{10^{-pH}} = 4.68 \times 10^{(pH-11)} \tag{2.41}$$

Equation (2.41) indicates that unless pH is extremely high, the carbonate concentration is usually negligible compared to the bicarbonate concentration, a fact that we will find useful in the following example, where we calculate the pH of pristine rainwater.

EXAMPLE 2.11   The pH of Natural Rainwater

Estimate the pH of natural rainwater, assuming that the only substance affecting it is the absorption of CO_2 from the atmosphere. Assume that the concentration of CO_2 is 360 ppm, and the temperature and pressure are 25 °C and 1 atm.

Solution   We have a number of equations to work with that are based on equilibrium constants, but there is another relationship that needs to be introduced, based on charge neutrality. The rainwater is assumed to have started without any electrical charge, and no net charge is added to it by the absorption of carbon dioxide. To have this neutrality maintained, the total positive charge contributed by the hydrogen ions (H^+) must equal the total negative charge of the bicarbonate (HCO_3^-), carbonate (CO_3^{2-}), and hydroxyl (OH^-) ions:

$$[H^+] = [HCO_3^-] + 2[CO_3^{2-}] + [OH^-] \tag{2.42}$$

Notice that since each carbonate ion (CO_3^{2-}) has two negative charges, its charge contribution is twice as much per mole, and hence its coefficient is 2.

Knowing that rainfall is likely to be slightly acidic will let us simplify (2.42). Equation (2.41) indicates that for an acidic solution, $[CO_3^{2-}] \ll [HCO_3^-]$, so (2.42) becomes

$$[H^+] \approx [HCO_3^-] + [OH^-] \tag{2.43}$$

Another equation that must be satisfied is (2.19):

$$[H^+][OH^-] = K_w = 10^{-14} \tag{2.19}$$

Substituting (2.19) into (2.43) gives

$$[H^+] = [HCO_3^-] + \frac{10^{-14}}{[H^+]} \qquad (2.44)$$

Putting $[HCO_3^-]$ from (2.37) into (2.44) yields

$$[H^+] = \frac{K_1[CO_2(aq)] + 10^{-14}}{[H^+]} \qquad (2.45)$$

So

$$[H^+]^2 = K_1[CO_2(aq)] + 10^{-14} \qquad (2.46)$$

We can find the saturation value of dissolved CO_2 from (2.33):

$$[CO_2] = K_H P_g \qquad (2.33)$$
$$= 0.033363 \, \text{mol/L} \cdot \text{atm} \times 360 \times 10^{-6} \, \text{atm} = 1.2 \times 10^{-5} \, \text{mol/L}$$

where K_H was taken from Table 2.4 and P_{CO_2} is given as 360 ppm or 360×10^{-6} atm. Equation (2.46) now gives us

$$[H^+]^2 = (4.47 \times 10^{-7} \times 1.2 \times 10^{-5} + 10^{-14})(\text{mol/L})^2$$

so

$$[H^+] = 2.31 \times 10^{-6} \, \text{mol/L}$$
$$pH = -\log[H^+] = 5.63$$

■

Example 2.11 indicates that the pH of pristine rainwater is not the same as the pH of pure water due to the presence of carbon dioxide in the atmosphere. There are other naturally occurring substances in the atmosphere that also affect pH. Sulfur dioxide (SO_2) lowers pH while ammonia and alkaline dust raise it. When all such natural substances are accounted for, rainfall is likely to have pH somewhere between 5 and 6 before it is influenced by human activities. As a result, some define acid rain caused by human activities to be rainfall with a pH of 5 or below, while others prefer to define it in terms of the carbonate calculation given previously; that is, as precipitation with pH less than 5.6. Acid deposition is discussed more fully in Chapters 5 and 7.

2.5 ORGANIC CHEMISTRY

The term *organic chemistry* has come to mean the chemistry of the compounds of carbon. This term is broader than the term *biochemistry*, which can be described as the chemistry of life. The need for a distinction between the two began with a discovery by Fredrich Wöhler in 1828. He accidentally converted an inorganic compound, ammonium cyanate, into urea, which until that time had been thought to be strictly associated with living things. That discovery demolished the *vital-force theory*, which held that organic and inorganic compounds were distinguished by some sort of "life force."

With carbon at the heart of the definition of organic chemistry, even DDT is just as organic as yogurt.

The science of organic chemistry is incredibly complex and varied. There are literally millions of different organic compounds known today, and 100,000 or so of these are products of synthetic chemistry, unknown in nature. About all that can be done here in a few pages is to provide the barest introduction to the origins of some of their names and structures, so that they will appear a bit less alien when we encounter them in the rest of the book.

One way to visualize the bonding of atoms is with electron-dot formulas that represent valence electrons in the outermost orbitals. These diagrams were developed by G. N. Lewis, in 1916 and are referred to as *Lewis structures*. According to Lewis's theory of covalence, atoms form bonds by losing, gaining, or sharing enough electrons to achieve the outer electronic configuration of a noble gas. For hydrogen, that means bonding with two shared electrons; for many other elements, including carbon, it means achieving a pattern with eight outermost electrons. For example, the following Lewis structure for the compound butane (C_4H_{10}) clearly shows each hydrogen atom sharing a pair of electrons and each carbon atom sharing four pairs of electrons:

$$\begin{array}{cccc} H & H & H & H \\ \cdot\cdot & \cdot\cdot & \cdot\cdot & \cdot\cdot \\ H\!:\!\overset{\cdot\cdot}{\underset{\cdot\cdot}{C}}\!:\!\overset{\cdot\cdot}{\underset{\cdot\cdot}{C}}\!:\!\overset{\cdot\cdot}{\underset{\cdot\cdot}{C}}\!:\!\overset{\cdot\cdot}{\underset{\cdot\cdot}{C}}\!:\!H \\ H & H & H & H \end{array} \qquad \textit{n-}\text{butane}$$

A common way to simplify Lewis structures is to use a single straight line to represent pairs of shared electrons, as shown in the following chemical formula. For convenience, the structural formula is often written linearly by combining hydrogens bonded to a given carbon and then showing those subunits connected by carbon-carbon bonds. Thus, for example, butane becomes $CH_3\!-\!CH_2\!-\!CH_2\!-\!CH_3$. Further simplification, referred to as a *condensed formula*, can be gained by not actually showing the C—C bonds, but merely grouping atoms connected to each carbon and writing the result. Butane becomes $CH_3(CH_2)_2CH_3$.

Another way to simplify the structural formula for some organic chemicals is with a symbolic "kinky" diagram. Each carbon in such a diagram is indicated with a kink, and the hydrogens are not shown at all. The number of hydrogens is easy to determine since the number of bonds to the carbon kink must be 4. A kinky diagram for *n*-butane is as follows:

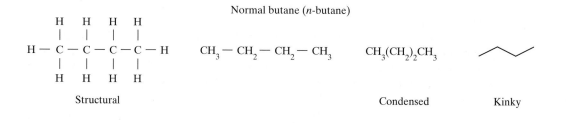

Normal butane (*n*-butane)

Structural   Condensed   Kinky

When butane has the structure shown—that is, with all of its carbons lined up in a straight chain (no side branches)—it is called *normal*-butane, or *n*-butane. Butane also occurs in another form known as isobutane. Isobutane (*i*-butane) has the same molecular formula, C_4H_{10}, but has a different structural formula, as shown:

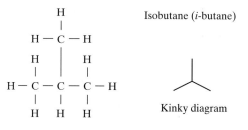

Isobutane (*i*-butane)

Kinky diagram

Compounds having the same molecular formula, but different structural formulas, are known as structural *isomers*. Isomers may have very different chemical and physical properties. As the number of carbon atoms per molecule increases, the number of possible isomers increases dramatically. Decane ($C_{10}H_{22}$), for example, has 75 possible isomers.

Butane and decane are examples of an important class of compounds called *hydrocarbons*, compounds containing only H and C atoms. Hydrocarbons in which each carbon atom forms four single bonds to other atoms are called *saturated hydrocarbons*, *paraffins*, or *alkanes*. Alkanes form a series beginning with methane (CH_4), ethane (C_2H_6), propane (C_3H_8), and butane (C_4H_{10}), which are all gases. The series continues with pentane (C_5H_{12}) through ($C_{20}H_{42}$), which are liquids and include gasoline and diesel fuel. Beyond that the alkanes are waxy solids. Notice that the name for each of these compounds ends in *-ane*. When one of the hydrogens is removed from an alkane, the result is called a *radical*, and the suffix becomes *-yl*. Thus, for example, removing one hydrogen from methane (CH_4) produces the methyl radical (CH_3-). Table 2.5 lists some of the methane-series radicals.

Hydrocarbons appear as building blocks in a great number of chemicals of environmental importance. By substituting other atoms or groups of atoms for some of the hydrogens, new compounds are formed. To get a feel for the naming of these compounds, consider the replacement of some of the hydrogen atoms with chlorine in an ethane molecule. Each carbon atom is given a number and the corresponding hydrogen replacements are identified by that number. For example, 1,1,1-trichloroethane, better known as TCA, has three chlorines all attached to the same carbon, while 1,2-dichloroethane (1,2-DCA) has one chlorine attached to each of the two carbon atoms. Similarly, 1,1,2-trichloro-1,2,2-trifluoroethane has chlorine and fluorine atoms attached to each carbon.

Other names for TCA include methyl chloroform and methyltrichloromethane, names that remind us that the molecule contains the methyl radical ($CH_3—$). TCA is commonly used in the electronics industry for flux cleaning and degreasing operations.

TABLE 2.5   Methane-series radicals

Parent compound	Radical	Formula
Methane	Methyl	$CH_3—$
Ethane	Ethyl	$C_2H_5—$
Propane	n-Propyl	$C_3H_7—$
Propane	Isopropyl	$(CH_3)_2CH—$
n-Butane	n-Butyl	$C_4H_9—$

EXAMPLE 2.12   Isomers of Trichloropropane

How many different isomers are there of trichloropropane, and how would they be numbered? Draw their kinky diagrams.

Solution   Propane has three carbons. The three chlorine atoms can be (1) all attached to an end carbon atom, (2) two attached to an end and one in the middle, (3) two attached to an end and one on the other end, (4) one on the end and two in the middle, or (5) one on each carbon atom. They would be numbered 1,1,1; 1,1,2; 1,1,3; 1,2,2; and 1,2,3 respectively.

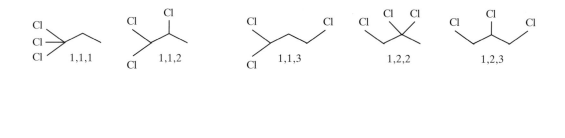

All of the examples so far have been of *saturated* hydrocarbons. *Unsaturated* compounds contain at least two carbon atoms joined by a double bond in which four electrons are shared. When there are double bonds the chemical name ends in *-ene*, and this category of hydrocarbons is called *alkenes*. The ethylene series of alkenes is analogous to the methane series of alkanes. It begins with ethene (ethylene), propene (propylene), butene (butylene), and so forth.

Ethene (ethylene)   Trichloroethylene (TCE)

A number of alkanes and alkenes of importance in environmental engineering have some of their hydrogen atoms replaced with chlorine atoms. For example, trichloroethylene, which is more commonly known as TCE, has been used as an industrial solvent for organic substances that are not soluble in water. Principal applications have included cleaning grease from metals, removing solder flux from printed circuit boards in the electronics industry, dry cleaning clothes, extracting caffeine from coffee, and stripping paint. It was quite widely used until it was shown to cause cancer in animals. It has the dubious distinction of being the most commonly found contaminant in groundwater at hazardous waste sites regulated under the Comprehensive Environmental Response, Compensation, and Liability Act (better known as Superfund). More will be said about this problem in Chapter 5.

Alkane and alkene hydrocarbons often have attachments of various functional groups of atoms. The attachment of the hydroxyl group —OH produces *alcohols*; the H—C=O group produces *aldehydes*; an oxygen —O— between carbons yields *ethers*; connecting an —OH group to a C=O group yields *carboxylic acids*; and attachments of the —NH₂ group result in *amines*. Examples of these hydrocarbon derivatives are as follows:

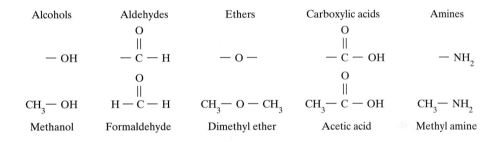

Hydrocarbons and their derivatives may have linear structures (e.g., *n*-butane), branch structures (e.g., isobutane), or ring structures. The compound benzene (C_6H_6), shown next, is one such ring structure.

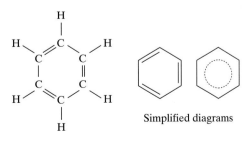

Benzene

In the figure for benzene, each carbon atom has been shown forming a single bond with one neighboring carbon and a double bond with the other. That representation is a gross simplification that should not be interpreted too literally; that is, the diagram is really a kind of average value of an actual benzene molecule. As shown, the benzene ring is usually simplified by leaving off the hydrogen atoms.

Molecules in which the hexagonal benzene ring occurs are called *aromatic compounds*. This category includes DDT, polychlorinated biphenyls (PCBs), and the defoliants used in Vietnam, 2,4,5-T and 2,4-D. When a benzene ring attaches to another molecule, the C_6H_5— that remains of the benzene ring is called a *phenyl* group, and so the name *phenyl* frequently appears in the compound name. Another physical characteristic of these compounds that is described by the nomenclature is the location of the attachments to the benzene ring. The terms *ortho, meta,* or *para,* or a numerical designation, specify the relative locations of the attachments, as suggested in Figure 2.4.

Figure 2.5 shows examples of the *o, m, p* nomenclature for two isomers of dichlorodiphenyltrichloroethane (DDT). Hopefully, this 30-letter name is somewhat less intimidating than it was before you read this section. Reading the name from right to left, the ethane on the end tells us it starts as a simple two-carbon ethane molecule; the trichloro refers to the three chlorines around one of the carbons; the diphenyl tells us there are two phenyl groups; and the dichloro refers to the chlorines on the phenyls. *p,p′*-DDT is an isomer with the chlorine and carbon attachments to each benzene located in the *para* position; the *o,p′*-DDT has one chlorine in the *ortho* position and one in the *para* position. Actually, dichlorodiphenyltrichloroethane as a name is insufficient to describe DDT unambiguously. An even more informative name is 1,1,1-trichloro-2-2-bis(*p*-chlorophenyl)-ethane, where *-bis* means "taken twice."

In Figure 2.4, the corners of a phenyl group are numbered, and that too provides a way to help describe the structure of a molecule. The herbicides 2,4-dichlorophenoxyacetic acid (2,4-D), and 2,4,5-trichlorophenoxyacetic acid (2,4,5-T) shown in Figure 2.6 provide examples of this numerical designation. Notice that the carboxylic acid functional group —COOH connected to a —CH$_2$— looks like acetic acid, which is contained in the names of these complex molecules.

Providing unambiguous names for complex organic molecules is extremely difficult, and what has been provided here is obviously just a very brief introduction. Even the structural formulas, which convey much more information than a name like DDT, are still highly simplified representations, and there are yet higher levels of descriptions based on quantum mechanics.

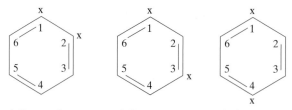

1, 2 - or *ortho*   1, 3 - or *meta*   1, 4 - or *para*

FIGURE 2.4   Designating the attachment points for phenyl.

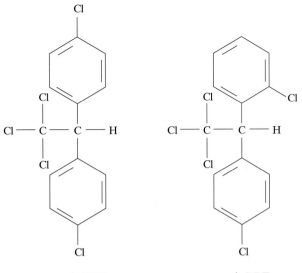

$$p, p' \text{ - DDT} \qquad\qquad o, p' \text{ - DDT}$$

FIGURE 2.5   Two isomers of dichlorodiphenyltrichloroethane (DDT).

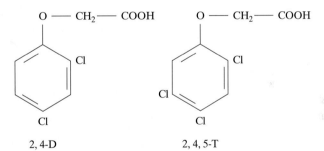

$$2, 4\text{-D} \qquad\qquad 2, 4, 5\text{-T}$$

FIGURE 2.6   2,4-Dichlorophenoxyacetic acid (2,4-D) and 2,4,5-trichlorophenoxyacetic acid (2,4,5-T).

2.6 NUCLEAR CHEMISTRY

As a simple but adequate model, we may consider an atom to consist of a dense nucleus containing a number of uncharged neutrons and positively charged protons, surrounded by a cloud of negatively charged electrons. The number of protons is the *atomic number*, and that is what defines a particular element. The sum of the number of protons and neutrons is the *mass number*. Elements with the same atomic number but differing mass numbers are called *isotopes*. The usual way to identify a given isotope is by giving its chemical symbol with the mass number written at the upper left and the atomic number at the lower left. For example, the two most important isotopes of uranium (which has 92 protons) are

$$^{235}_{92}\text{U} \qquad \text{and} \qquad ^{238}_{92}\text{U}$$

$$\text{Uranium-235} \qquad\qquad \text{Uranium-238}$$

When referring to a particular element, it is common to drop the atomic number subscript since it adds little information to the chemical symbol. Thus, U-238 is a common way to describe that isotope. When describing changes to an element as it decays, the full notation is usually helpful.

Some atomic nuclei are unstable (that is, *radioactive*), and during the spontaneous changes that take place within the nucleus, various forms of *radiation* are emitted. While all elements having more than 83 protons are naturally radioactive, it is possible artificially to produce unstable isotopes, or *radionuclides*, of virtually every element in the periodic table. Radioactivity is of interest in the study of environmental engineering both because it poses a hazard to exposed organisms and because it allows radionuclides to act as tracers to help measure the flow of materials in the environment.

There are three kinds of radiation: *alpha, beta,* and *gamma* radiation. *Alpha* radiation consists of a heavy two-proton, two-neutron particle emitted from the nucleus. When an unstable nucleus emits an alpha particle, its atomic number decreases by 2 units and its mass number decreases by 4 units. The following example shows the decay of plutonium into uranium:

$$^{239}_{94}\text{Pu} \;\rightarrow\; ^{235}_{92}\text{U} \;+\; ^{4}_{2}\alpha \;+\; \gamma$$

In this reaction, not only is an α particle ejected from the nucleus, but electromagnetic radiation is emitted as well, indicated by the γ.

As an alpha particle passes through an object, its energy is gradually dissipated as it interacts with other atoms. Its positive charge attracts electrons in its path, raising their energy levels and possibly removing them completely from their nuclei (*ionization*). Alpha particles are relatively massive and easy to stop. Our skin is sufficient protection for sources that are external to the body, but taken internally, such as by inhalation, α particles can be extremely dangerous.

Beta (β) particles are electrons that are emitted from an unstable nucleus as a result of the spontaneous transformation of a neutron into a proton plus an electron. Emission of a β particle results in an increase in the atomic number by one, while the mass number remains unchanged. A gamma ray may or may not accompany the transformation. The following example shows the decay of strontium-90 into yttrium:

$$^{90}_{38}\text{Sr} \;\rightarrow\; ^{90}_{39}\text{Y} \;+\; \beta$$

As negatively charged β particles pass through materials, they encounter swarms of electrons around individual atoms. The resulting Coulomb repulsive force between orbital electrons and β particles can raise the energy level of those electrons, possibly kicking them free of their corresponding atoms. While such ionizations occur less frequently than with α particles, β penetration is much deeper. Alpha particles travel less than 100 μm into tissue, while β radiation may travel several centimeters. They can be stopped with a modest amount of shielding, however. For example, to stop a β particle a 1-cm thickness of aluminum is sufficient.

Gamma (γ) rays have no charge or mass, being simply a form of electromagnetic radiation that travels at the speed of light. As such, they can be thought of either as waves with characteristic wavelengths, or as photon particles. Gamma rays have very short wavelengths in the range of 10^{-3} to 10^{-7} μm. Having such short wavelengths means that individual photons are highly energetic and easily cause biologically damaging ionizations. These rays are very difficult to contain and may require several centimeters of lead to provide adequate shielding.

All of these forms of radiation are dangerous to living things. The electron excitations and ionizations that are caused by such radiation cause molecules to become unstable, resulting in the breakage of chemical bonds and other molecular damage. The chain of chemical reactions that follows will result in the formation of new molecules that did not exist before the irradiation. Then, on a much slower time scale, the organism responds to the damage (so slow, in fact, that it may be years before the final effects become visible). Low-level exposures can cause *somatic* and/or *genetic* damage. Somatic effects may be observed in the organism that has been exposed and include higher risk of cancer, leukemia, sterility, cataracts, and a reduction in lifespan. Genetic damage, by increasing the mutation rate in chromosomes and genes, affects future generations.

As will be discussed in Chapter 7, one naturally occurring source of radiation exposure that is thought to be an important cause of lung cancer results from the inhalation of radon (Rn-222) and radon decay products (various isotopes of polonium). Radon is an α emitting, chemically inert gas that seeps out of the soil and that can sometimes accumulate in houses. It is an intermediate product in a naturally occurring decay chain that starts with uranium-238 and ends with a stable isotope of lead. Figure 2.7 diagrams this radioactive series. There are three other similar series that can

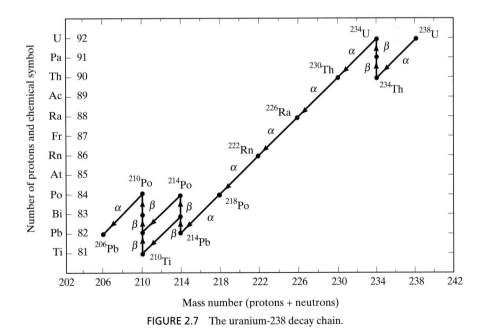

FIGURE 2.7   The uranium-238 decay chain.

be drawn. One begins with uranium-235 and ends with lead-207, another begins with thorium-232 and ends with lead-208, and the third begins with the artificial isotope, plutonium-241, and ends with bismuth-209.

An important parameter that characterizes a given radioactive isotope is its half-life, which is the time required for half of the atoms spontaneously to transform, or decay, into other elements. For example, if we start with 100 g of an isotope that has a half-life of 1 year, we would find 50 g remaining after 1 year, 25 g after 2 years, 12.5 g after 3 years, and so on. While the half-life for a given isotope is constant, half-lives of radionuclides in general vary from fractions of a second to billions of years. Half-lives for the radon portion of the U-238 decay chain are given in Table 2.6.

There are a number of commonly used radiation units, which unfortunately can easily be confused. The *curie* (Ci) is the basic unit of decay rate; 1 Ci corresponds to the disintegration of 3.7×10^{10} atoms per second, which is approximately the decay rate of 1 g of radium. A more intuitive unit is the *becquerel* (Bq), which corresponds to 1 radioactive decay per second; hence, $1 \text{ Ci} = 3.7 \times 10^{10}$ Bq. While the curie is a measure of the rate at which radiation is emitted by a source, it tells us nothing about the radiation dose that is actually absorbed by an object. The *roentgen* (R) is defined in terms of the number of ionizations produced in a given amount of air by X or γ rays. Of more interest is the amount of energy actually absorbed by tissue, be it bone, fat, muscle, or whatever. The *rad* (radiation *a*bsorbed *d*ose) corresponds to an absorption of 100 ergs of energy per gram of any substance. The rad has the further advantage that it may be used for any form of radiation, α, β, γ, or X. For water and soft tissue, the rad and roentgen are approximately equal.

Another unit, the *rem* (roentgen *e*quivalent *m*an) has been introduced to take into account the different biological effects that various forms of radiation have on humans. Thus, for example, if a 10-rad dose of β particles produces the same biological effect as a 1-rad dose of α particles, both doses would have the same value when expressed in rems. This unit is rather loosely defined, making it difficult to convert from rads to rems. However, in many situations involving X-rays, γ rays, and β radiation, rads and rems are approximately the same. The rem or millirem are units most often used when describing human radiation exposure.

TABLE 2.6 Half-lives for the radon decay chain

Isotope	Emission	Half-life
Rn-222	α	3.8 day
Po-218	α	3.05 min
Pb-214	β	26.8 min
Bi-214	β	19.7 min
Po-214	α	160 μsec
Pb-210	β	19.4 yr
Tl-210	β	1.32 min
Bi-210	β	4.85 day
Po-210	α	1.38 day
Pb-206		Stable

Nuclear Fission

Nuclear reactors obtain their energy from the heat that is produced when uranium atoms *fission*. In a fission reaction, uranium-235 captures a neutron and becomes unstable uranium-236, which splits apart, discharging two *fission fragments*, two or three neutrons, and γ rays (Figure 2.8). Most of the energy released is in the form of kinetic energy in the fission fragments. That energy is used to heat water, producing steam that powers a turbine and generator in much the same way as was diagrammed in Figure 1.12.

The fission fragments produced are always radioactive, and concerns for their proper disposal have created much of the controversy surrounding nuclear reactors. Typical fission fragments include cesium-137, which concentrates in muscles and has a half-life of 30 years; strontium-90, which concentrates in bone and has a half-life of 28 years; and iodine-131, which concentrates in the thyroid gland and has a half-life of 8.1 days. The half-lives of fission fragments tend to be no longer than a few tens of years, so after a period of several hundred years their radioactivity will decline to relatively insignificant levels.

Reactor wastes also include some radionuclides with very long half lives. Of major concern is plutonium, which has a half-life of 24,390 years. Only a few percent of the uranium atoms in reactor fuel is the fissile isotope ^{235}U, while essentially all of the rest is ^{238}U, which does not fission. Uranium-238 can, however, capture a neutron and be transformed into plutonium, as the following reactions suggest:

$$^{238}_{92}U + n \rightarrow {}^{239}_{92}U \xrightarrow{\beta} {}^{239}_{93}Np \xrightarrow{\beta} {}^{239}_{94}Pu \qquad (2.47)$$

This plutonium, along with several other long-lived radionuclides, makes nuclear wastes dangerously radioactive for tens of thousands of years, which greatly increases the difficulty of providing safe disposal. Removing the plutonium from nuclear wastes before disposal has been proposed as a way to shorten the decay period, but that introduces another problem. Plutonium not only is radioactive and highly toxic, it is also the critical ingredient in the manufacture of nuclear weapons. A single nuclear reactor produces enough plutonium each year to make dozens of small atomic bombs, and some

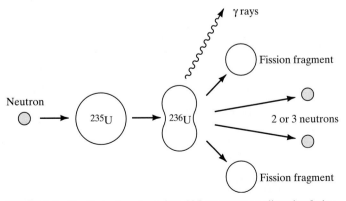

FIGURE 2.8 The fissioning of uranium-235 creates two radioactive fission fragments, neutrons, and gamma rays. (Masters, 1974)

have argued that if the plutonium is separated from nuclear wastes, the possibility of illicit diversions for such weapons would cause an unacceptable risk. Without reprocessing to reuse the plutonium and uranium in the spent fuel rods, the nuclear fuel cycle is as diagrammed in Figure 2.9.

If reprocessing is included, the nuclear fuel system more closely approaches a true cycle, as shown in Figure 2.9. Reprocessing high-level wastes recovers unused uranium and newly created plutonium that can be used to refuel reactors. The plutonium, however, also can be used for nuclear weapons. In a very controversial decision, Japan has begun to ship its nuclear wastes to France for reprocessing, and in 1995 the first shipments of recovered plutonium were transported back to Japan for reuse in their nuclear program.

As Figures 2.9 and 2.10 suggest, the power reactor itself is only one of several sources in the complete nuclear fuel cycle that creates waste disposal problems. Uranium ores must first be mined and the uranium extracted. Mine tailings typically contain toxic metals such as arsenic, cadmium, and mercury, as well as the radionuclides associated with the decay of ^{238}U shown in Figure 2.7. Only about 0.72 percent of naturally occurring uranium is the desired isotope ^{235}U, and an enrichment facility is needed to increase that concentration to 2 or 3 percent for use in reactors. The enriched uranium is then fabricated into fuel pellets and fuel rods that are shipped to reactors. Highly radioactive wastes from reactors are temporarily stored on site until, eventually, they will be transported to their ultimate disposal in a federal repository. Such a repository, however, is not expected to become operational in the United States until well into the twenty-first century. Low-level wastes can be disposed of in specially designed landfills, which are described in Chapter 6. When the reactor itself has reached the end of its lifetime, after roughly 30 years, it must be decommissioned and radioactive components eventually transported to a secure disposal site.

Providing proper disposal of radioactive wastes at every stage in the nuclear fuel cycle is a challenging engineering task, but there are many who argue that it is well within our capabilities.

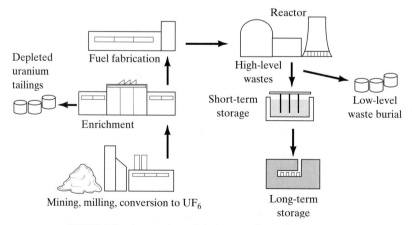

FIGURE 2.9 A once-through fuel system for nuclear reactors.

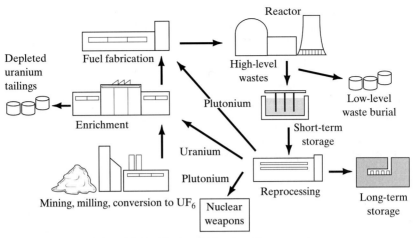

FIGURE 2.10   Nuclear fuel cycle with reprocessing.

PROBLEMS

2.1. Consider the following reaction representing the combustion of propane:

$$C_3H_8 + O_2 \rightarrow CO_2 + H_2O$$

(a) Balance the equation.

(b) How many moles of oxygen are required to burn 1 mol of propane?

(c) How many grams of oxygen are required to burn 100 g of propane?

(d) At standard temperature and pressure, what volume of oxygen would be required to burn 100 g of propane? If air is 21 percent oxygen, what volume of air at STP would be required?

(e) At STP what volume of CO_2 would be produced when 100 g of propane are burned?

2.2. Trinitrotoluene (TNT), $C_7H_5N_3O_6$, combines explosively with oxygen to produce CO_2, water, and N_2. Write a balanced chemical equation for the reaction and calculate the grams of oxygen requred for each 100 g of TNT.

2.3. An unknown substance is empirically determined to be 40.00 percent carbon by weight, 6.67 percent hydrogen, and 53.33 percent oxygen. Its molecular weight is roughly 55 g/mol. Determine the molecular formula and its correct molecular weight.

2.4. What is the molarity of 10 g of glucose ($C_6H_{12}O_6$) dissolved in 1 L of water?

2.5. Eighty-six-proof whiskey is 43 percent ethyl alcohol, CH_3CH_2OH, by volume. If the density of ethyl alcohol is 0.79 kg/L, what is its molarity in whiskey?

2.6. The earth's ozone layer is under attack in part by chlorine released when ultraviolet radiation breaks apart certain fluorocarbons. Consider three fluorocarbons known as CFCF-11 (CCl_3F), CFC-12 (CCl_2F_2), and HCFC-22 (CHF_2Cl).

(a) Calculate the percentage by weight of chlorine in each of these fluorocarbons.

(b) If each kilogram of CFC-11 could be replaced by 1 kg of HCFC-22, by what percentage would the mass of chlorine emissions be reduced?

2.7. Suppose total world energy consumption of fossil fuels, equal to 3×10^{17} kJ/yr, were to be obtained entirely by combustion of petroleum with the approximate chemical formula C_2H_3. Combustion of petroleum releases about 43×10^3 kJ/kg.

   (a) Estimate the emmisions of CO_2 per year.

   (b) What is the ratio of grams of C emitted per unit of energy for petroleum vs. methane? (Use the results of Example 2.3.)

2.8. The standard enthalpy for selected organic compounds is given in this problem. For each, compute the gross and net heats of combustion.

   (a) Propane (C_3H_8); -103.8 kJ/mol

   (b) n-Butane (C_4H_{10}); -124.7 kJ/mol

   (c) Isobutane (C_4H_{10}); -131.6 kJ/mol

   (d) Ethanol (l), (C_2H_5OH); -277.6 kJ/mol

   (e) Methanol (l), (CH_3OH); -238.6 kJ/mol

2.9. The gross heat of combustion for ethyl alcohol (C_2H_5OH) is -1370 kJ/mol. What is its standard enthalpy?

2.10. The standard enthalpy of propane (C_3H_8) is -103.8 kJ/mol. Find the gross heat released when 1 kg is burned.

2.11. For the following possible alternative automobile fuels, express their higher heating value (HHV) in Btu/gal:

   (a) Methanol (CH_3OH), density 6.7 lb/gal, $H^0 = -238.6$ kJ/mol

   (b) Ethanol (C_2H_5OH), density 6.6 lb/gal, $H^0 = -277.6$ kJ/mol

   (c) Propane (C_3H_8), density 4.1 lb/gal, $H^0 = -103.8$ kJ/mol

2.12. Repeat problem 2.11, but find the net heat of combustion for each fuel (Btu/gal).

2.13. In example 2.6, photolysis of ozone (O_3) was described as producing an atomic oxygen with enthalpy 247.5 kJ/mol. More correctly, photolysis yields an oxygen that is temporarily in an excited state (O*) with an enthalpy of 438 kJ/mol.

$$O_3 \quad + \quad h\nu \quad \rightarrow \quad O_2 \quad + \quad O*$$
$$H^0: (142.9) \qquad\qquad\qquad (0) \qquad (438) \text{ kJ/mol}$$

What maximum wavelength could cause this photolysis?

2.14. What maximum wavelength could a photon have to cause the dissociation of gaseous nitrogen dioxide?

$$NO_2 \quad + \quad h\nu \quad \rightarrow \quad NO \quad + \quad O$$

2.15. Hydrochloric acid, HCl, completely ionizes when dissolved in water. Calculate the pH of a solution containing 25 mg/L of HCl.

2.16. What is the pH of a solution containing 3×10^{-4} mg/L of OH^- ($25\,°C$)?

2.17. Find the hydrogen ion concentration and the hydroxide ion concentration in baking soda with pH 8.5.

2.18. Find the theoretical oxygen demand for the following solutions:

   (a) 200 mg/L of acetic acid, CH_3COOH

   (b) 30 mg/L of ethanol, C_2H_5OH

   (c) 50 mg/L of sucrose, $C_2H_{12}O_6$

2.19. Water is frequently disinfected with chlorine gas, forming hypochlorous acid (HOCl), which partially ionizes to hypoclorite and hydrogen ions as follows:

$$\text{HOCl} \rightleftharpoons \text{H}^+ + \text{OCl}^- \text{ with equilibrium constant } K = 2.9 \times 10^{-8}$$

The amount of [HOCl], which is the desired disinfectant, depends on the pH. Find the fraction that is hypochlorous acid—that is, $[\text{HOCl}]/\{[\text{HOCl}] + [\text{OCl}^-]\}$—as a function of pH. What would the hypochlorous fraction be for pH = 6, 8, and 10?

2.20. Hydrogen sulfide (H_2S) is an odorous gas that can be stripped from solution in a process similar to that described in Example 2.8 for ammonia. The reaction is

$$\text{H}_2\text{S} \rightleftharpoons \text{H}^+ + \text{HS}^- \text{ with equilibrium constant } K = 0.86 \times 10^{-7}$$

Find the fraction of hydrogen sulfide in H_2S form at pH 6 and pH 8.

2.21. Solid aluminum phosphate, AlPO_4, is in equilibrium with its ions in solution:

$$\text{AlPO}_4 \rightleftharpoons \text{Al}^{3+} + \text{PO}_4{}^{3-} \text{ with } K_{sp} = 10^{-22}$$

Find the equilibrium concentration of phosphate ions (in mg/L).

2.22. Calculate the equilibrium concentration of dissolved oxygen in 15 °C water at 1 atm, and again at 2000 m elevation.

2.23. Suppose the gas above the soda in a bottle of soft drink is pure CO_2 at a preassure of 2 atm. Calculate $[\text{CO}_2]$ at 25 °C.

2.24. Calculate the pH of the soft drink in Problem 2.23. Start with the following chemical reaction and realize the solution will be somewhat acidic (negligible carbonate):

$$\text{CO}_2 + \text{H}_2\text{O} \rightleftharpoons \text{H}^+ + \text{HCO}_3{}^-$$

2.25. It has been estimated that the concentration of CO_2 in the atmosphere before the Industrial Revolution was about 275 ppm. If the accumulation of CO_2 in the atmosphere continues, then by the middle of the next century it will probably be around 600 ppm. Calculate the pH of rainwater (neglecting the effect of any other gases)at 25 °C in each of these times.

2.26. One strategy for dealing with the acidifiction of lakes is periodically to add lime (CaCO_3) to them. Calculate the pH of a lake that has more than enough lime in it to saturate the water with its ions Ca^{2+} and $\text{CO}_3{}^{2-}$. *Suggestions:* Begin with the carbonate system equations (2.37), (2.38), and (2.39), and then add a charge balance equation:

$$[\text{H}^+] + 2[\text{Ca}^{2+}] = [\text{HCO}_3{}^-] + 2[\text{CO}_3{}^{2-}] + [\text{OH}^-] \approx [\text{HCO}_3{}^-] + [\text{OH}^-],$$

(The preceding holds for pH values that are less than about 10; when you work out the solution, check to see whether this was a valid assumption.) This calculation is the same as would be made to estimate the pH of the oceans, which are saturated with CaCO_3.

2.27. Draw the Lewis structure and the more common structural formulas for the following:

(a) Ethylene, C_2H_4

(b) 2-Chloropropane, $\text{CH}_3\text{CHClCH}_3$

(c) Methanol, CH_3OH

2.28. Draw kinky diagrams for the following molecules:

(a) $\text{CH}_3\text{CH}_2\text{CH}_2\text{CH}_2\text{CH}_3$

(b) $CH_3CH_2CHCH_2$

(c)

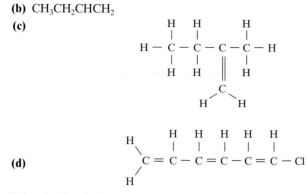

(d)

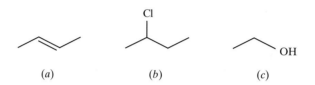

2.29. Write the chemical structures from the following kinky diagrams:

(a) (b) (c)

2.30. From the names alone, write the structures for the following organics:

(a) Dichloromethane

(b) Trichloromethane(chloroform)

(c) 1, 1-Dichloroethylene

(d) Trichlorofluoromethane(CFC-11)

(e) 1, 1, 2, 2-Tetrachlorethane

(f) *o*-Dichlorobenzene

(g) Tetrachloroethene(PCE)

(h) Dichlorofluoromethane (CFC-21)

2.31. What values of a and b would complete each of the following (X and Y are not meant to be any particular elements):

$$^{266}_{88}X \rightarrow \alpha + ^{a}_{b}Y$$

$$^{a}_{15}X \rightarrow \beta + ^{32}_{b}Y$$

2.32. The half-life of iodine-125 is about 60 days. If we were to start with 64 g of it, about how much would remain after 1 year?

REFERENCES

Baird, C., 1995, *Environmental Chemistry,* W. H. Freeman, New York.

Masters, G. M., 1974, *Introduction to Evironmental Science and Technology,* Wiley, New York.

Sawyer, C. N., P. L. McCarthy, and G. F. Parkin, 1994, *Chemistry for Environmental Engineering,* 5th ed., McGraw-Hill, New York.

Thomann, R. V., and J. A. Mueller, 1987, *Principles of Surface Water Quality and Control,* Harper & Row, New York.

CHAPTER 3

Mathematics of Growth

> *It is very difficult to make an accurate prediction, especially about the future.*
> —Niels Bohr

3.1 INTRODUCTION

A sense of the future is an essential component in the mix of factors that should be influencing the environmental decisions we make today. In some circumstances, only a few factors may need to be considered and the time horizon may be relatively short. For example, a wastewater treatment plant operator may need to predict the growth rate of bacteria in a digester over a period of hours or days. The designer of the plant, on the other hand, probably needed to estimate the local population growth rate over the next decade or two in order to size the facility. At the other extreme, to make even the crudest estimate of future carbon emissions and their affect on global warming, scientists need to forecast population and economic growth rates, anticipated improvements in energy efficiency, the global fossil fuel resource base and the consumption rate of individual fuels, the rate of deforestation or reforestation that could be anticipated, and so on. And these estimates need to be made for periods of time that are often measured in hundreds of years.

As the preceding quote of Niels Bohr suggests, we cannot expect to make accurate predictions of the future, especially when the required time horizon may be extremely long. We can, however, often make simple estimates that are robust enough that the insight they provide is most certainly valid. We can say, for example, with considerable certainty that world population growth at today's rates cannot continue for another hundred years. We can also use simple mathematical models to develop very useful "what if" scenarios: *If* population growth continues at a certain rate, and *if* energy demand is proportional to economic activity, and so forth, *then* the following would occur.

The purpose of this chapter is to develop some simple but powerful mathematical tools that can shed considerable light on the future of a number of environmental problems. We begin with what is probably the most useful and powerful mathematical function encountered in environmental studies: the *exponential function.* Other growth functions encountered often are also explored, including the *logistic* and the *Gaussian* functions. Applications that will be demonstrated using these functions include population growth, resource consumption, pollution accumulation, and radioactive decay.

3.2 EXPONENTIAL GROWTH

Exponential growth occurs in any situation where the increase in some quantity is proportional to the amount currently present. This type of growth is quite common, and the mathematics required to represent it is relatively simple, yet extremely important. We will approach this sort of growth first as discrete, year-by-year, increases, and then in the more usual way as a continuous growth function.

Suppose something grows by a fixed percentage each year. For example, if we imagine our savings at a bank earning 5 percent interest each year, compounded once a year, then the amount of increase in savings over any given year is 5 percent of the amount available at the beginning of that year. If we start now with \$1000, then at the end of one year we would have \$1050 ($1000 + 0.05 \times 1000$); at the end of two years we would have \$1102.50 ($1050 + 0.05 \times 1050$); and so on. We can represent this mathematically as follows:

$$N_0 = \text{initial amount}$$

$$N_t = \text{amount after } t \text{ years}$$

$$r = \text{growth rate (fraction per year)}$$

Then,

$$N_{t+1} = N_t + rN_t = N_t(1 + r).$$

For example,

$$N_1 = N_0(1 + r); N_2 = N_1(1 + r) = N_0(1 + r)^2;$$

and, in general,

$$N_t = N_0(1 + r)^t \tag{3.1}$$

EXAMPLE 3.1.   U.S. Electricity Growth (Annual Compounding)

In 1995, the United States produced 3.0×10^{12} kWhr of electricity. The average annual growth rate of U.S. electricity demand in the previous 15 years was about 1.8 percent. (For comparison, it had been about 7 percent per year for many decades before the 1973 oil embargo.) Estimate the electricity consumption in 2050 if the 1.8 percent per year growth rate were to remain constant over those 55 years.

Solution: From (3.1), we have

$$N_{100} = N_0(1 + r)^{55}$$
$$= 3.0 \times 10^{12} \times (1 + 0.018)^{55}$$
$$= 8.00 \times 10^{12} \text{ kWhr/yr}$$

■

Continuous Compounding

For most events of interest in the environment, it is usually assumed that the growth curve is a smooth, continuous function without the annual jumps that (3.1) is based on. In financial calculations this is referred to as continuous compounding. With the use of a little bit of calculus, the growth curve becomes the true exponential function that we will want to use most often.

One way to state the condition that leads to exponential growth is that the quantity grows in proportion to itself; that is, the rate of change of the quantity N is proportional to N. The proportionality constant r is called the rate of growth and has units of (time^{-1}).

$$\frac{dN}{dt} = rN \tag{3.2}$$

This is essentially the same as (1.15) solved in Chapter 1. The solution is

$$N = N_0 e^{rt} \tag{3.3}$$

which is plotted in Figure 3.1.

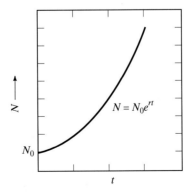

FIGURE 3.1   The exponential function.

EXAMPLE 3.2 U.S. Electricity Growth (Continuous Compounding)

Suppose we repeat Example 3.1, but now consider the 1.8 percent growth rate to be continuously compounded. Starting with the 1995 electricity consumption of 3.0×10^{12} kWhr/yr, what would consumption be in 2050 if the growth rate remains constant?

Solution Using (3.3),

$$N = N_0 e^{rt}$$
$$= 3.0 \times 10^{12} \times e^{0.018 \times 55}$$
$$= 8.07 \times 10^{12} \text{ kWhr/yr} \qquad \blacksquare$$

Example 3.1, in which increments were computed once each year, and Example 3.2, in which growth was continuously compounded, have produced nearly identical results. As either the period in time in question or the growth rate increases, the two approaches begin to diverge. At 12 percent growth, for example, those answers would differ by nearly 50 percent. In general, it is better to express growth rates as if they are continuously compounded so that (3.3) becomes the appropriate expression.

Doubling Time

Calculations involving exponential growth can sometimes be made without a calculator by taking advantage of the following special characteristic of the exponential function. A quantity that is growing exponentially requires a fixed amount of time to double in size, regardless of the starting point. That is, it takes the same amount of time to grow from N_0 to $2N_0$ as it does to grow from $2N_0$ to $4N_0$, and so on, as shown in Figure 3.2.

The *doubling time* (T_d) of a quantity that grows at a fixed exponential rate r is easily derived. From

$$N = N_0 e^{rt} \qquad (3.3)$$

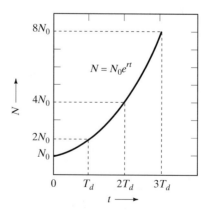

FIGURE 3.2 Illustrating the concept of doubling time.

the doubling time can be found by setting $N = 2N_0$ at $t = T_d$:

$$2N_0 = N_0 e^{rT_d}$$

Since N_0 appears on both sides of the equation, it can be canceled out, which is another way of saying the length of time required to double the quantity does not depend on how much you start with. So canceling N_0 and taking the natural log of both sides gives

$$\ln 2 = rT_d$$

or

$$T_d = \frac{\ln 2}{r} \cong \frac{0.693}{r} \tag{3.4}$$

If the growth rate r is expressed as a percentage instead of as a fraction, we get the following important result:

$$T_d \cong \frac{69.3}{r\,(\%)} \cong \frac{70}{r\,(\%)} \tag{3.5}$$

Equation (3.5) is well worth memorizing; that is,

The length of time required to double a quantity growing at r percent is about equal to 70 divided by r percent.

If your savings earns 7 percent interest, it will take about 10 years to double the amount you have in the account. If the population of a country grows continuously at 2 percent, then it will double in size in 35 years, and so on.

EXAMPLE 3.3   Historical World Population Growth Rate

It took about 300 years for the world's population to increase from 0.5 billion to 4.0 billion. If we assume exponential growth at a constant rate over that period of time, what would that growth rate be? Do this example first with a calculator and (3.3), and then with the rule of thumb suggested in (3.5).

Solution Rearranging (3.3) and taking the natural log of both sides gives

$$\ln(e^{rt}) = rt = \ln\left(\frac{N}{N_0}\right)$$

$$r = \frac{1}{t}\ln\left(\frac{N}{N_0}\right) \tag{3.6}$$

$$r = \frac{1}{300}\ln\left(\frac{4.0}{0.5}\right) = 0.00693 = 0.693 \text{ percent}$$

Using the rule of thumb approach given by (3.5), we can say that three doublings would be required to have the population grow from 0.5 billion to 4.0 billion. Three doublings in 300 years means each doubling took 100 years.

$$T_d = 100 \text{ yrs} \cong \frac{70}{r\%}$$

so

$$r(\%) \cong \frac{70}{100} = 0.7 \text{ percent}$$

Our answers would have been exactly the same if the rule of thumb in (3.5) had not been rounded off. ∎

Quantities that grow exponentially very quickly also increase in size at a deceptively fast rate. As Table 3.1 suggests, quantities growing at only a few percent per year increase in size with incredible speed after just a few doubling times. For example, at the 1995 rate of world population growth ($r = 1.5$ percent), the doubling time was about 46 years. If that rate were to continue for just 4 doubling times, or 184 years, world population would increase by a factor of 16, from 5.7 billion to 91 billion. In 20 doubling times, there would be 6 million billion (quadrillion) people, or more than one person for each square foot of surface area of the earth. The absurdity of these figures simply points out the impossibility of the underlying assumption that exponential growth, even at this relatively low sounding rate, could continue for such periods of time.

Half-Life

When the rate of *decrease* of a quantity is proportional to the amount present, exponential growth becomes exponential decay. We dealt with quantities that decay exponentially in Chapter 1, when nonconservative pollutants were discussed, and in Chapter 2, when radioactivity was introduced. Exponential decay can be described using either a *reaction rate coefficient* (K) or a *half-life* ($T_{1/2}$), and the two are easily related to each other. A reaction rate coefficient for an exponential decay plays the same role that r played for exponential growth.

Exponential decay can be expressed as

$$N = N_0 e^{-Kt} \tag{3.7}$$

where K is a reaction rate coefficient (time^{-1}), N_0 is an initial amount, and N is an amount at time t.

TABLE 3.1 Exponential Growth Factors for Various Numbers of Doubling Times

Number of Doublings (n)	Growth Factor (2^n)
1	2
2	4
3	8
4	16
5	32
10	1024
20	$\approx 1.05 \times 10^6$
30	$\approx 1.07 \times 10^9$

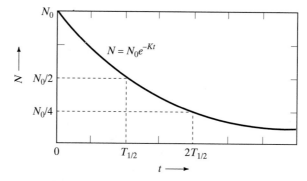

FIGURE 3.3   Exponential decay and half-life.

A plot of (3.7) is shown in Figure 3.3 along with an indication of *half-life*. To relate half-life to reaction rate, we set $N = N_0/2$ at $t = T_{1/2}$, giving

$$\frac{N_0}{2} = N_0 e^{-KT_{1/2}}$$

Canceling the N_0 term and taking the natural log of both sides gives us the desired expression

$$T_{1/2} = \frac{\ln 2}{K} \cong \frac{0.693}{K} \tag{3.8}$$

EXAMPLE 3.4   Radon Half-Life

If we start with a 1.0-Ci radon-222 source, what would its activity be after 5 days?

Solution   Table 2.6 indicates that the half-life for radon is 3.8 days. Rearranging (3.8) to give a reaction rate coefficient results in

$$K = \frac{\ln 2}{T_{1/2}} = \frac{0.693}{3.8 \text{ day}} = 0.182/\text{day}$$

Using (3.7) gives

$$N = N_0 e^{-Kt} = 1 \text{ Ci} \times e^{-0.182/\text{day} \times 5 \text{ day}} = 0.40 \text{ Ci} \qquad \blacksquare$$

Disaggregated Growth Rates

Quite often the quantity that we want to model can be considered to be the product of a number of individual factors. For example, to estimate future carbon emissions we might *disaggregate* demand into the following three factors:

$$\text{Carbon emissions} = (\text{Population}) \times \left(\frac{\text{Energy}}{\text{Person}}\right) \times \left(\frac{\text{Carbon}}{\text{Energy}}\right) \tag{3.9}$$

By considering demand this way, we would hope to increase the accuracy of our forecast by estimating separately the rates of change of population, per capita energy consumption, and carbon emissions per unit of energy. Equation (3.9) is an example of the following simple conceptual model of factors that drive the environmental impacts of human activity:

$$\text{Impacts} = (\text{Population}) \times (\text{Affluence}) \times (\text{Technology}) \tag{3.10}$$

In (3.9), affluence is indicated by per capita energy demand, while technology is represented by the carbon emissions per unit of energy.

Expressing a quantity of interest as the product of individual factors, such as has been done in (3.9), leads to a very simple method of estimating growth. Beginning with

$$P = P_1 P_2 \dots P_n \tag{3.11}$$

If each factor is itself growing exponentially

$$P_i = p_i e^{r_i t}$$

then

$$P = (p_1 e^{r_1 t})(p_2 e^{r_2 t}) \dots (p_n e^{r_n t})$$
$$= (p_1 p_2 \dots p_n) e^{(r_1 + r_2 + \dots r_n)t}$$

which can be written as

$$P = P_0 e^{rt}$$

where

$$P_0 = (p_1 p_2 \dots p_n) \tag{3.12}$$

and

$$r = r_1 + r_2 + \dots r_n \tag{3.13}$$

Equation (3.13) is a very simple, very useful result. That is,

If a quantity can be expressed as a product of factors, each growing exponentially, then the total rate of growth is just the sum of the individual growth rates.

EXAMPLE 3.5   Future U.S. Energy Demand

One way to disaggregate energy consumption is with the following product:

$$\text{Energy demand} = (\text{Population}) \times \left(\frac{\text{GDP}}{\text{person}} \right) \times \left(\frac{\text{Energy}}{\text{GDP}} \right)$$

where GDP is the gross domestic product. Suppose we project per capita GDP to grow at 2.3 percent and population at 0.6 percent. Suppose we assume that through energy efficiency efforts, we expect energy required per dollar of GDP to *decrease* exponentially at the rate that it did between 1978 and 1995. In 1978, (energy/GDP) was 17.38 kBtu/$, and in 1995 it was 13.45 kBtu/$

(in constant 1987 dollars). Total U.S. energy demand in 1995 was 90.6 quads (quadrillion Btu). If the aforementioned rates of change continue, what would energy demand be in the year 2010?

Solution   First, let us find the exponential rate of decrease of (energy/GDP) over the 17 years between 1978 and 1995. Using (3.6) gives

$$r_{\text{energy/GDP}} = \frac{1}{t} \ln \left(\frac{N}{N_0} \right) = \frac{1}{17} \ln \left(\frac{13.45}{17.38} \right) = -0.015 = -1.5 \text{ percent}$$

Notice that the equation automatically produces the negative sign. The overall energy growth rate is projected to be the sum of the three rates:

$$r = 0.6 + 2.3 - 1.5 = 1.4 \text{ percent}$$

(For comparison, energy growth rates in the United States before the 1973 oil embargo were typically 3 to 4 percent per year.) Projecting out 15 years to the year 2010, (3.3) gives

$$\text{Energy demand in 2010} = 90.6 \times 10^{15} \text{ Btu/yr} \times e^{0.014 \times 15}$$
$$= 112 \times 10^{15} \text{ Btu/yr} = 112 \text{ quads/yr} \qquad ■$$

3.3 RESOURCE CONSUMPTION

To maintain human life on earth, we depend on a steady flow of energy and materials to meet our needs. Some of the energy we use is *renewable* (e.g. hydroelectric power, windpower, solar heating systems, and even firewood if the trees are replanted), but most comes from *depletable* fossil fuels. The minerals we extract from the earth's crust, such as copper, iron, and aluminum, are limited as well. The future of our way of life to a large extent depends on the availability of abundant supplies of inexpensive, depletable energy and materials.

How long will those resources last? The answer, of course, depends on how much there is and how quickly we use it. We will explore two different ways to model the rate of consumption of a resource: One is based on the simple exponential function that we have been working with, and the other is based on a more realistic bell-shaped curve.

Exponential Resource Production Rates

When a mineral is extracted from the earth, geologists traditionally say the resource is being *produced* (rather than *consumed*). If we plot the rate of production of a resource versus time, as has been illustrated in Figure 3.4, the area under the curve between any two times will represent the total resource that has been produced during that time interval. That is, if P is the production rate (e.g., barrels of oil per day, tonnes of aluminum per year, etc.), and Q is the resource produced (total barrels, total tonnes) between times t_1 and t_2, we can write

$$Q = \int_{t_2}^{t_1} P \, dt \qquad (3.14)$$

If we assume that the production rate of a resource grows exponentially, we can easily determine the total amount produced during any time interval; or, conversely, if

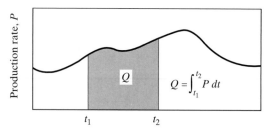

FIGURE 3.4 The total amount of a resource produced between times t_1 and t_2 is the cross-hatched area under the curve.

we know the total amount to be produced, we can estimate the length of time that it will take to produce it. The basic assumption of exponential growth is probably not a good one if the time frame is very long, but nonetheless it will give us some very useful insights. In the next section we will work with a more reasonable production rate curve.

Figure 3.5 shows an exponential rate of growth in production. If the time interval of interest begins with $t = 0$, we can write

$$Q = \int_0^t P_0 e^{rt}\, dt = \left.\frac{P_0}{r} e^{rt}\right|_0^t$$

which has as a solution

$$Q = \frac{P_0}{r}\left(e^{rt} - 1\right) \tag{3.15}$$

where

$$Q = \text{the total resource produced from time 0 to time } t$$
$$P_0 = \text{the initial production rate}$$
$$r = \text{the exponential rate of growth in production}$$

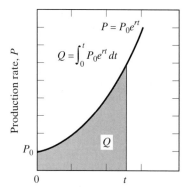

FIGURE 3.5 Total consumption of a resource experiencing exponential growth.

Equation (3.15) tells us how much of the resource is produced in a given period of time if the production rate grows exponentially. If we want to know how long it will take to produce a given amount of the resource, we can rearrange (3.15) to give

$$T = \frac{1}{r} \ln \left(\frac{rQ}{P_0} + 1 \right) \tag{3.16}$$

where T = the length of time required to produce an amount Q.

Applications of (3.16) often lead to startling results, as should be expected anytime we assume that exponential rates of growth will continue for a prolonged period of time.

EXAMPLE 3.6   World Coal Production

World coal production in 1995 was estimated to be 5.0 billion (short) tons, and the estimated total recoverable reserves of coal were estimated at 1.1 trillion tons. Growth in world coal production in the previous decade averaged 1.5 percent per year. How long would it take to use up those reserves at current production rates, and how long would it take if production continues to grow at 1.5 percent?

Solution   At those rates coal reserves would last

$$\frac{\text{Reserves}}{\text{Production}} = \frac{1.1 \times 10^{12} \text{ tons}}{5.0 \times 10^9 \text{ tons/yr}} = 220 \text{ years}$$

If production grows exponentially, we need to use (3.16):

$$T = \frac{1}{r} \ln \left(\frac{rQ}{P_0} + 1 \right)$$

$$= \frac{1}{0.015} \ln \left(\frac{0.015 \times 1.1 \times 10^{12}}{5.0 \times 10^9} + 1 \right) = 97 \text{ years}$$

Even though the growth rate of 1.5 percent might seem modest, compared to a constant production rate it cuts the length of time to deplete those reserves by more than half. ∎

Example 3.6 makes an important point: If we simply divide the remaining amount of a resource by the current rate of production, we can get a misleading estimate of the remaining lifetime for that resource. If exponential growth is assumed, what would seem to be an abundant resource may actually be consumed surprisingly quickly. Continuing this example for world coal, let us perform a sensitivity analysis on the assumptions used. Table 3.2 presents the remaining lifetime of coal reserves, assuming an initial production rate of 5.0 billion tons per year, for various estimates of total available supply and for differing rates of production growth.

Notice how the lifetime becomes less and less sensitive to estimates of the total available resource as exponential growth rates increase. At a constant 3 percent growth rate, for example, a total supply of 500 billion tons of coal would last 46 years; if

TABLE 3.2 Years required to consume all of the world's recoverable coal reserves, assuming various reserve estimates and differing production growth rates.[a]

Growth rate (%)	500 Billion tons (yrs)	1000 Billion tons (yrs)	2000 Billion tons (yrs)
0	100	200	400
1	69	110	160
2	55	80	109
3	46	65	85
4	40	55	71
5	35	48	61

[a]Initial production rate 5.0 billion tons/year (1995); actual reserves estimated at 1100 billion tons.

our supply is four times as large, 2000 billion tons, the resource only lasts another 39 years. Having four times as much coal does not even double the projected lifetime.

This is a good time to mention the important distinction between the *reserves* of a mineral and the ultimately producible *resources*. As shown in Figure 3.6, *reserves* are quantities that can reasonably be assumed to exist and are producible with existing technology under present economic conditions. *Resources* include present reserves as well as deposits not yet discovered, or deposits that have been identified but are not recoverable under present technological and economic conditions.

As existing reserves are consumed, further exploration, advances in extraction technology, and higher acceptable prices constantly shift mineral resources into the reserves category. Estimating the lifetime of a mineral based on the available reserves rather than on the ultimately producible resources can therefore be quite misleading. World oil reserves in 1970, for example, were estimated at 550 billion barrels (1 barrel equals 42 gallons) while production was 17 billion barrels per year. The reserves-to-production ratio was about 32 years; that is, at 1970 production rates those reserves would be depleted in 32 years. However, 25 years later (1995), instead of almost being out of reserves, they had grown to 1000 billion barrels. Production in 1995 was 22.5 billion barrels per year, giving a reserves-to-production ratio of 45 years (Hobbs, 1995).

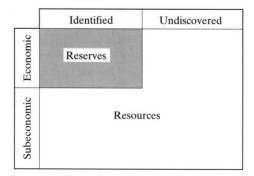

FIGURE 3.6 Classification of minerals. Reserves are a subcategory of resources.

A Symmetrical Production Curve

There are many ways that we could imagine a complete production cycle to occur. The model of exponential growth until the resource is totally consumed that we just explored is just one example. It might seem an unlikely one since the day before the resource collapses, the industry is at full production, and the day after, it is totally finished. It is a useful model, however, to dispel any myths about the possibility of long-term exponential growth in the consumption rate of any finite resource.

A more reasonable approach to estimating resource cycles was suggested by M. King Hubbert (1969). Hubbert argued that consumption would more likely follow a course that might begin with exponential growth while the resource is abundant and relatively cheap. But as new sources get harder to find, prices go up and substitutions would begin to take some of the market. Eventually, consumption rates would peak and then begin a downward trend as the combination of high prices and resource substitutions would prevail. The decline could be precipitous when the energy needed to extract and process the resource exceeds the energy derived from the resource itself. A graph of resource consumption versus time would therefore start at zero, rise, peak, and then decrease back to zero, with the area under the curve equaling the total resource consumed.

Hubbert suggested that a symmetrical production cycle that resembles a bell-shaped curve be used. One such curve is very common in probability theory, where it is called the *normal* or *Gaussian* function. Figure 3.7 shows a graph of the function and identifies some of the key parameters used to define it.

The equation for the complete production cycle of a resource corresponding to Figure 3.7 is

$$P = P_{\mathrm{m}} \exp\left[-\frac{1}{2}\left(\frac{t - t_{\mathrm{m}}}{\sigma}\right)^2 \right] \tag{3.17}$$

where

P = the production rate of the resource

P_{m} = the maximum production rate

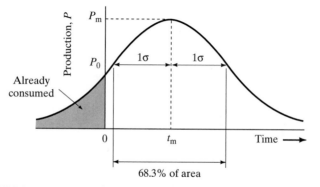

FIGURE 3.7   Resource production following a Gaussian distribution function.

t_m = time at which the maximum production rate occurs

σ = standard deviation, a measure of the width of the bell-shaped curve

exp [ ] = the exponential function

The parameter σ is the standard deviation of a normal density function, and in this application it has units of time. Within $\pm 1\ \sigma$ away from the time of maximum production, 68.3 percent of the production occurs; within $\pm 2\ \sigma$, 95 percent of the production occurs. Notice that with this bell-shaped curve, it is not possible to talk about the resource ever being totally exhausted. It is more appropriate to specify the length of time required for some major fraction of it to be used up. Hubbert uses 80 percent as his criterion, which corresponds to $\pm 1.3\ \sigma$ away from the year of maximum production.

We can find the total amount of the resource ever produced, Q_∞, by integrating (3.17):

$$Q_\infty = \int_{-\infty}^{\infty} P\, dt = \int_{-\infty}^{\infty} P_m \exp\left[-\frac{1}{2}\left(\frac{t - t_m}{\sigma}\right)^2 \right] dt$$

which works out to

$$Q_\infty = \sigma P_m \sqrt{2\pi} \tag{3.18}$$

Equation (3.18) can be used to find σ if the ultimate amount of the resource ever to be produced Q_∞ and the maximum production rate P_m can be estimated.

It is also interesting to find the length of time required to reach the maximum production rate. If we set $t = 0$ in (3.17), we find the following expression for the initial production rate, P_0:

$$P_0 = P_m \exp\left[-\frac{1}{2}\left(\frac{t_m}{\sigma}\right)^2 \right] \tag{3.19}$$

which leads to the following expression for the time required to reach maximum production:

$$t_m = \sigma\sqrt{2 \ln \frac{P_m}{P_0}} \tag{3.20}$$

Let us demonstrate the use of these equations by fitting a Gaussian curve to data for U.S. coal resources.

EXAMPLE 3.7   U.S. Coal Production

Suppose ultimate total production of U.S. coal is double the 1995 recoverable reserves, which were estimated at 268×10^9 (short) tons. The U.S. coal production rate in 1995 was 1.0×10^9 tons/year. How long would it take to reach a peak production rate equal to four times the 1995 rate if a Gaussian production curve is followed?

Solution   Equation (3.18) gives us the relationship we need to find an appropriate σ.

$$\sigma = \frac{Q_{\infty}}{P_{m}\sqrt{2\pi}} = \frac{2 \times 268 \times 10^{9} \text{ ton}}{4 \times 1.0 \times 10^{9} \text{ ton/yr } \sqrt{2\pi}} = 53.5 \text{ yr}$$

A standard deviation of 53.5 years says that in a period of 107 years ($2\,\sigma$) about 68 percent of the coal would be consumed.

To find the time required to reach peak production, use (3.20):

$$t_{m} = \sigma\sqrt{2 \ln \frac{P_{m}}{P_{0}}} = 53.5\sqrt{2 \ln 4} = 89 \text{ yr}$$

The complete production curve is given by (3.17):

$$P = P_{m} \exp\left[-\frac{1}{2}\left(\frac{t - t_{m}}{\sigma}\right)^{2}\right] = 4 \times 1.0 \times 10^{9} \exp\left[-\frac{1}{2}\left(\frac{t - 89}{53.5}\right)^{2}\right] \text{tons/yr}$$

which is plotted in Figure 3.8 along with the production curves that result from the maximum production rate reaching two and eight times the current production rates. The tradeoffs between achieving high production rates versus making the resource last beyond the next century are readily apparent from this figure.   ■

Hubbert's analysis of world oil production is presented in Figure 3.9 for two scenarios. One scenario is based on fitting a Gaussian curve to the historical production record before 1975; the other is based on a conservation scenario in which oil production follows the Gaussian curve until 1975, but then remains constant at the 1975 rate for as long as possible before a sudden decline takes place. With the bell-shaped curve, world oil consumption would peak just before the turn of the century; with the conservation scenario, supplies would last until about 2040 before the decline would begin. The actual production curve between 1975 and 1995 fell between these two scenarios.

The curves drawn in Figure 3.9 were based on Hubbert's 1977 estimate of 2000 billion barrels of oil ultimately produced. For perspective, by 1995 approximately

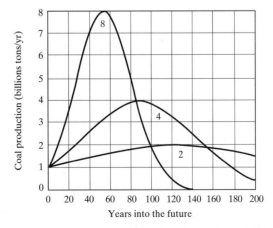

FIGURE 3.8   Gaussian curves fitted to U.S. production of coal. Ultimate production set equal to twice the 1995 reserves. Parameter is ratio of peak production rate to initial (1995) production rate.

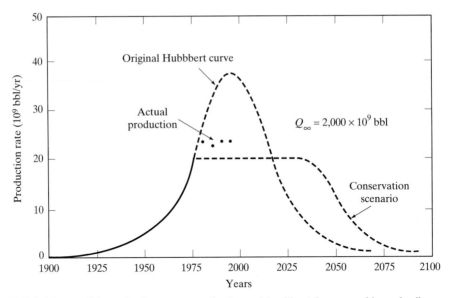

FIGURE 3.9 Two possible production rate curves for the world's ultimately recoverable crude oil resources, estimated by Hubbert in 1977 to be 2 trillion barrels. The original Hubbert curve indicates that peak production would be reached before the year 2000, while the "conservation" scenario, which holds production to the 1975 level as long as possible, extends the resource for only a few decades more. (After Hubbert, 1977, with added data from Energy Information Administration, 1996)

750 billion barrels had been produced (and consumed) and proved reserves stood at 1000 billion barrels (two-thirds of which were in the Middle East). Amounts already produced plus proved reserves total 1750 billion barrels, not much below Hubbert's estimate for the total amount ever to be produced.

At current production rates of about 22.5 billion barrels per year, the reserves to production ratio is 45 years. In other words, at current rates proved reserves would last until 2040, which pretty much agrees with the conservation scenario in Figure 3.9. Will more oil ultimately be produced than the current 1 trillion barrels of proved reserves? Finding new sources, using enhanced oil recovery techniques to remove some of the oil that is now left behind, and eventually tapping into unconventional hydrocarbon resources such as extra heavy oil, bitumen, and shale oil could stretch petroleum supplies much further into the future, but at significantly higher environmental and economic costs.

3.4 POPULATION GROWTH

The simple exponential growth model can serve us well for short time horizons, but obviously, even small growth rates cannot continue for long before environmental constraints limit further increases.

The typical growth curve for bacteria, shown in Figure 3.10, illustrates some of the complexities that natural biological systems often exhibit. The growth curve is divided into phases designated as lag, exponential, stationary, and death. The *lag phase,* characterized by little or no growth, corresponds to an initial period of time when bacteria are first inoculated into a fresh medium. After the bacteria have adjusted to their new environment, a period of rapid growth, the *exponential phase,* follows. During this time, conditions are optimal and the population doubles with great regularity. (Notice that the vertical scale in Figure 3.10 is logarithmic so that exponential growth produces a straight line.) As the bacterial food supply begins to be depleted, or as toxic metabolic products accumulate, the population enters the no-growth, or *stationary phase.* Finally, as the environment becomes more and more hostile, the *death phase* is reached and the population declines.

Logistic Growth

Population projections are quite often mathematically modeled with a *logistic* or S-shaped (*sigmoidal*) growth curve such as the one shown in Figure 3.11. Such a curve

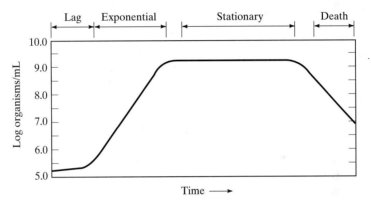

FIGURE 3.10   Typical growth curve for a bacterial population. (*Source:* Brock, *Biology of Microorganisms,* 2nd ed., © 1974. Reprinted by permission of Prentice Hall, Inc., Englewood Cliffs, New Jersey.)

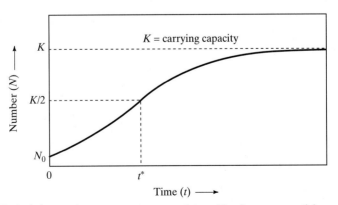

FIGURE 3.11   The logistic growth curve suggests a smooth transition from exponential growth to a steady-state population.

has great intuitive appeal. It suggests an early exponential growth phase while conditions for growth are optimal, followed by slower and slower growth as the population nears the carrying capacity of its environment. Biologists have successfully used logistic curves to model populations of many organisms, including protozoa, yeast cells, water fleas, fruit flies, pond snails, worker ants, and sheep (Southwick, 1976).

Mathematically, the logistic curve is derived from the following differential equation:

$$\frac{dN}{dt} = rN\left(1 - \frac{N}{K}\right) \tag{3.21}$$

where N is population size, K is called the *carrying capacity* of the environment, and r is the exponential growth rate that would apply if the population size is far below the carrying capacity. Notice that when the population N is much less than the carrying capacity K, the rate of change of population is proportional to population size. That is, population grows exponentially with a growth rate r. As N increases, the rate of growth slows down, and eventually, as N approaches K, the growth stops and the population stabilizes at a level equal to the carrying capacity. The factor $(1 - N/K)$ is often called the *environmental resistance*. As population grows, the resistance to further population growth continuously increases.

The solution to (3.21) is

$$N = \frac{K}{1 + e^{-r(t-t^*)}} \tag{3.22}$$

Note that t^* corresponds to the time at which the population is half of the carrying capacity, $N = K/2$. Substituting $t = 0$ into (3.22) lets us solve for t^*:

$$t^* = \frac{1}{r} \ln\left(\frac{K}{N_0} - 1\right) \tag{3.23}$$

where N_0 is the population at time $t = 0$. In the usual application of (3.22), the population growth rate is known at $t = 0$, but this is not the same as the growth rate r. To find r, let us introduce another factor, R_0. Let R_0 = instantaneous rate of growth at $t = 0$. If we characterize the growth at $t = 0$ as exponential, then

$$\frac{dN}{dt}\Big|_{t=0} = R_0 N_0 \tag{3.24}$$

But from (3.21)

$$\frac{dN}{dt}\Big|_{t=0} = rN_0\left(1 - \frac{N_0}{K}\right) \tag{3.25}$$

so that equating (3.24) with (3.25) yields

$$r = \frac{R_0}{(1 - N_0/K)} \tag{3.26}$$

Equation (3.26) lets us use quantities that are known at $t = 0$; namely, the population size N_0 and the population growth rate R_0, to find the appropriate growth factor r for (3.22). The following example demonstrates this process.

EXAMPLE 3.8   Logistic Human Population Curve

Suppose the human population follows a logistic curve until it stabilizes at 15.0 billion. In 1995 the world's population was 5.7 billion and its growth rate was 1.5 percent. When would the population reach 7.5 billion? 14 billion?

Solution   We need to find r using (3.26):

$$r = \frac{R_0}{(1 - N_0/K)} = \frac{0.015}{(1 - 5.7 \times 10^9/15 \times 10^9)} = 0.024$$

The time required to reach 7.5 billion, or half of the final population size, can be found from (3.23):

$$t^* = \frac{1}{r} \ln\left(\frac{K}{N_0} - 1\right) = \frac{1}{0.024} \ln\left(\frac{15 \times 10^9}{5.7 \times 10^9} - 1\right) = 20 \text{ yrs} \quad (\text{that is, 2015})$$

To determine the number of years that it will take to reach 14.0 billion, we need to solve (3.22) for t:

$$t = t^* - \frac{1}{r} \ln\left(\frac{K}{N} - 1\right) \tag{3.27}$$

$$= 20 - \frac{1}{0.024} \ln\left(\frac{15}{14} - 1\right) = 130 \text{ yrs} \qquad ■$$

Maximum Sustainable Yield

The logistic curve can also been used to introduce another useful concept in population biology called the *maximum sustainable yield* of an ecosystem. The maximum sustainable yield is the maximum rate at which individuals can be harvested (removed) without reducing the population size. Imagine, for example, harvesting fish from a pond. If the pond is at its carrying capacity, there will be no population growth, so that any fish removed will reduce the population. Therefore, the maximum sustainable yield will correspond to some population size less than the carrying capacity. In fact, since yield is the same as dN/dt, the maximum yield will correspond to the point on the logistic curve where the slope is a maximum. If we set the derivative of the slope equal to zero, we can find that point. The slope of the logistic curve is given by (3.21):

$$\text{Yield} = \text{Slope} = \frac{dN}{dt} = rN\left(1 - \frac{N}{K}\right) \tag{3.21}$$

Setting the derivative of the slope equal to zero gives

$$\frac{d}{dt}\left(\frac{dN}{dt}\right) = r\frac{dN}{dt} - \frac{r}{K}\left(2N\frac{dN}{dt}\right) = 0$$

Letting N* be the population at the maximum yield point gives

$$1 - \frac{2N^*}{K} = 0$$

so that

$$N^* = \frac{K}{2} \qquad \text{(for maximum sustainable yield)} \qquad (3.28)$$

That is, if population growth is logistic, then the maximum sustainable yield will be obtained when the population is half the carrying capacity. The yield at that point can be determined by substituting (3.28) into (3.21):

$$\text{Maximum yield} = \left(\frac{dN}{dt}\right)_{max} = r\frac{K}{2}\left(1 - \frac{K/2}{K}\right) = \frac{rK}{4} \qquad (3.29)$$

Using (3.26) lets us express the maximum yield in terms of the current growth rate R_0 and current size N_0:

$$\left(\frac{dN}{dt}\right)_{max} = \left(\frac{R_0}{1 - N_0/K}\right) \times \frac{K}{4} = \frac{R_0 K^2}{4(K - N_0)} \qquad (3.30)$$

EXAMPLE 3.9 Harvesting Fish

Observations of a pond newly stocked with 100 fish shows their population doubling every year for the first year or two, but after many years their population stabilizes at 4000 fish. Assuming a logistic growth curve, what would be the maximum sustainable yield from this pond?

Solution In the early years, with no harvesting, the population doubles every year. We can find the initial rate of growth R_0 from the doubling-time equation (3.5):

$$R_0 = \frac{\ln 2}{T_d} = \frac{\ln 2}{1 \text{ yr}} = 0.693/\text{yr}$$

The initial population N_0 is 100 fish, so from (3.30) the maximum yield would be

$$\left(\frac{dN}{dt}\right)_{max} = \frac{R_0 K^2}{4(K - N_0)} = \frac{0.693 \times (4000)^2}{4(4000 - 100)} = 710 \text{ fish/yr}$$

We could have gone about this in a more intuitive way by realizing that, since the initial population N_0 is so much lower than the carrying capacity K, the growth rate R_0 is approximately equal to what it would be when there are no growth constraints. That is, $r \approx R_0 = 0.693/\text{yr}$. Using this approximation along with the fact that population size when yield is highest is $K/2 = 2000$. From (3.21) we have

$$\text{Maximum yield} = rN\left(1 - \frac{N}{K}\right) \approx 0.693 \times 2000 \, (1 - 2000/4000) = 693 \text{ fish/yr} \qquad \blacksquare$$

3.5 HUMAN POPULATION GROWTH

The logistic growth equations just demonstrated are frequently used with some degree of accuracy and predictive capability for density-dependent populations, but they are not often used to predict human population growth. More detailed information on fertility and mortality rates and population age composition are usually available for humans, increasing our ability to make population projections. Use of such data enables us to ask such questions as What if every couple had just two children? What if such replacement-level fertility were to take 40 years to achieve? What would be the effect of reduced fertility rates on the ratios of retired people to workers in the population? The study of human population dynamics is known as *demography* and what follows is but a brief glimpse at some of the key definitions and most useful mathematics involved.

The simplest measure of fertility is the *crude birth rate,* which is the number of live births per 1000 population in a given year. It is called crude because it does not take into account the fraction of the population that is physically capable of giving birth in any given year. The crude birth rate is typically in the range of 30 to 40 per 1000 per year for the poorest, least developed countries of the world, while it is typically around 10 for the more developed ones. For the world as a whole, in 1995, the crude birth rate was 24 per thousand. That means that the 5.7 billion people alive then would have had about $5.7 \times 10^9 \times (24/1000) = 137$ million live births in that year. (These statistics, and all others in this section, are taken from data supplied by the Population Reference Bureau in a useful annual publication called the *World Population Data Sheet.*)

The *total fertility rate* (TFR) is the average number of children that would be born alive to a woman, assuming that current age-specific birth rates remain constant through the woman's reproductive years. It is a good measure of the average number of children each woman is likely to have during her lifetime. In the developed countries of the world in 1995, the total fertility rate was 1.6 children per female, while in the less developed countries the average was 3.5. In some of the poorest countries, the TFR is over 7 children per woman. Figure 3.12 shows the history of the total fertility rate for the United States. During the depression years of the 1930s, TFR was barely above 2; during the baby-boom era, from 1946 to 1964, it was well over 3 children per woman; in the mid-1970s it dropped back to a level close to 2.

The number of children that a woman must have, on the average, to replace herself with one daughter in the next generation is called *replacement level fertility.* Replacement-level fertility accounts for differences in the ratio of male to female births as well as child mortality rates. For example, in the United States, replacement-level fertility is 2.11 children per female. At that fertility level, 100 women would bear 211 children on the average. Statistically, 108 of the children would be boys and 103 would be girls. About 3 percent of the girls would be expected to die before bearing children, leaving a net of 100 women in the next generation. Notice that differing mortality rates in other countries cause the replacement level of fertility to vary somewhat compared to the 2.11 just described. In many developing countries, for example, higher infant mortality rates raise the level of fertility needed for replacement to approximately 2.7 children.

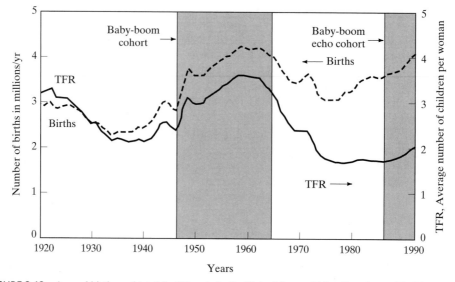

FIGURE 3.12 Annual births and total fertility rate in the United States. (After Bouvier and DeVita, 1991)

As will be demonstrated later, having achieved replacement-level fertility does not necessarily mean that a population has stopped growing. When a large fraction of the population is young people, as they pass through their reproductive years, the total births will continue to grow despite replacement fertility. Figure 3.12 illustrates this concept for the United States. Even though replacement-level fertility had been achieved by the mid-1970s, the total number of births rose as the baby boomers had their children. The cohorts born between 1985 and 1995 are often referred to as the baby-boom echo. The continuation of population growth, despite achievement of replacement-level fertility, is a phenomenon known as *population momentum,* and it will be described more carefully a bit later in this chapter.

The simplest measure of mortality is the *crude death rate,* which is the number of deaths per 1000 population per year. Again, caution should be exercised in interpreting *crude* death rates since the age composition of the population is not accounted for. The United States, for example, has a higher crude death rate than Guatemala, but that in no way indicates equivalent risks of mortality. In Guatemala only 3 percent of the population is over 65 years of age (and hence at greater risk of dying), while in the United States a much larger fraction of the population, 12 percent, is over 65.

An especially important measure of mortality is the *infant mortality rate,* which is the number of deaths to infants (under one year of age) per 1000 live births in a given year. The infant mortality rate is one of the best indicators of poverty in a country. In some of the poorest countries of the world, infant mortality rates are over 150, which means that one child in six will not live to see its first birthday. In the more developed countries, infant mortality rates are typically around 10 per 1000 per year.

EXAMPLE 3.10   Birth and Death Statistics

In 1995, 80 percent of the world's population, some 4.5 billion people, lived in the less developed countries of the world. In those countries the average crude birth rate was 28, the crude death rate was 9, and the infant mortality rate was 67. What fraction of the total deaths is due to infant mortality? If the less developed countries were able to care for their infants as well as they are cared for in most developed countries, resulting in an infant mortality rate of 10, how many infant deaths would be avoided each year?

Solution   To find the number of infant deaths each year, we must first find the number of live births and then multiply that by the fraction of those births that die within the first year:
Now:

$$\text{Infant deaths} = \text{Population} \times \text{Crude birth rate} \times \text{Infant mortality rate}$$
$$= 4.5 \times 10^9 \times (28/1000) \times (67/1000) = 8.4 \times 10^6 \text{ per year}$$
$$\text{Total deaths} = \text{Population} \times \text{Crude death rate}$$
$$= 4.5 \times 10^9 \times (9/1000) = 40.5 \times 10^6 \text{ per year}$$
$$\text{Fraction infants} = 8.4/40.5 = 0.21 = 21 \text{ percent}$$

With lowered infant mortality rate:

$$\text{Infant deaths} = 4.5 \times 10^9 \times (28/1000) \times (10/1000) = 1.3 \times 10^6 \text{ per year}$$
$$\text{Avoided deaths} = (8.4 - 1.3) \times 10^6 = 7.1 \text{ million per year}$$

It is often argued that reductions in the infant mortality rate would eventually result in reduced birth rates as people began to gain confidence that their offspring would be more likely to survive. Hence, the reduction of infant mortality rates through such measures as better nutrition, cleaner water, and better medical care is thought to be a key to population stabilization.   ∎

The difference between crude birth rate b and crude death rate d is called the *rate of natural increase* of the population. While it can be expressed as a rate per 1000 of population, it is more common to express it either as a decimal fraction or as a percentage rate; as a simple, but important, equation

$$r = b - d \tag{3.31}$$

where r is the rate of natural increase. If r is treated as a constant, then the exponential relationships developed earlier can be used. For example, in 1995 the crude birth rate for the world was 24 and the crude death rate was 9 per thousand, so the rate of natural increase was $(24 - 9)/1000 = 15/1000 = 0.015 = 1.5$ percent. If this rate continued, then the world would double in population in about $70/1.5 = 47$ years.

While it is reasonable to talk in terms of a rate of natural increase for the world, it is often necessary to include effects of migration when calculating growth rates for an individual country. Letting m be the *net migration rate*, which is the difference between immigration (in-migration) and emigration (out-migration), we can rewrite the preceding relationship as follows:

$$r = b - d + m \tag{3.32}$$

At the beginning of the Industrial Revolution, it is thought that crude birth rates were around 40 per 1000 while death rates were around 35 per 1000, yielding a rate of natural increase of about 0.5 percent for the world. As parts of the world began to achieve the benefits of a better and more assured food supply, improved sanitation, and modern medicines, death rates began to drop. As economic and social development has proceeded in the currently more developed countries, birth rates have fallen to about 12 per thousand and the crude death rate to about 10 per thousand. These countries have undergone what is referred to as the *demographic transition,* a transition from high birth and death rates to low birth and death rates, as shown in Figure 3.13.

The less developed countries of the world have also experienced a sizable drop in death rates, especially during the last half century. Imported medicines and better control of disease vectors have contributed to a rather sudden drop death rates to their

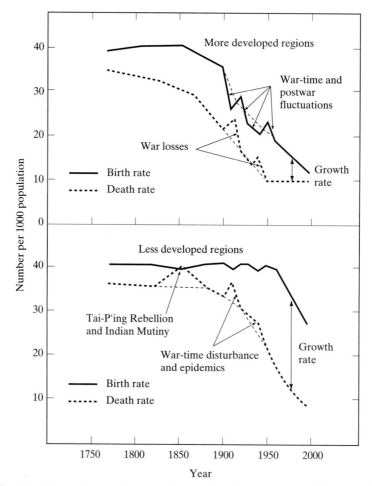

FIGURE 3.13 The demographic transition to low birth and death rates took over 100 years in the more developed nations. The less developed regions are in the middle of the process. (United Nations, 1971)

1995 level of 9 per thousand. In general, birth rates have not fallen nearly as fast, however. They are still up around 28 per thousand, which yields a growth rate of 1.9 percent for 80 percent of the world's population. The rapid (and, historically speaking, extraordinary) population growth that the world is experiencing now is almost entirely due to the drop in death rates in developing countries without a significant corresponding drop in birth rates. Many argue that decreases in fertility depend on economic growth, and many countries today face the danger that economic growth may not be able to exceed population growth. Such countries may be temporarily stuck in the middle of the demographic transition and are facing the risk that population equilibrium may ultimately occur along the least desirable path, through rising death rates.

One developing country that stands out from the others is China. After experiencing a disastrous famine from 1958–1961, China instituted aggressive family planning programs coupled with improved health care and living standards, which cut birth rates in half in less than two decades. Figure 3.14 shows China's rapid passage through the demographic transition, which was unprecedented in speed (the developed countries took more than 100 years to make the transition).

Some of these data on fertility and mortality are presented in Table 3.3 for the world, for more developed countries, for China, and for the less developed countries excluding China. China has been separated out because, in terms of population control, it is so different from the rest of the less developed countries and because its size so greatly influences the data.

Age Structure

A great deal of insight and predictive power can be obtained from a table or diagram representing the age composition of a population. A graphical presentation of the data, indicating numbers of people (or percentages of the population) in each age

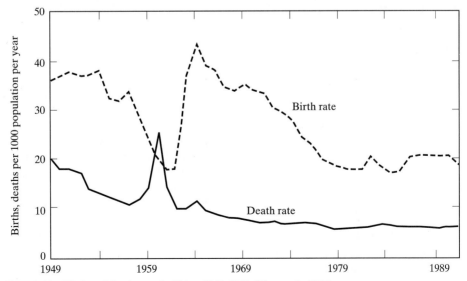

FIGURE 3.14   Birth and death rates in China, 1949–1991. (Tien et al., 1992)

TABLE 3.3 Some Important Population Statistics (1995)

	World	More developed countries	Less developed countries (non-China)	China	United States
Population (millions)	5702	1169	3314	1219	263
% of world population	100	20	58	22	4.6
Crude birth rate b	24	12	31	18	15
Crude death rate d	9	10	9	6	9
Natural increase r %	1.5	0.2	2.2	1.1	0.7
% Population under age 15	32	20	38	27	22
Total fertility rate	3.1	1.6	4.0	1.9	2.0
Infant mortality rate	62	10	72	44	8
% of total added 1995–2025	100	4	85	11	3
Per capita GNP (US$)	4500	17270	1250	490	24750
% urban	43	74	38	28	75
Est. population 2025 (millions)	8312	1271	5518	1523	338
Added pop. 1995–2025 (millions)	2610	102	2204	304	75

Source: Population Reference Bureau, 1996.

category, is called an *age structure* or a *population pyramid*. An age structure diagram is a snapshot of a country's population trends that tells us a lot about both the recent past as well as the near future. Developing countries with rapidly expanding populations, for example, have pyramids that are triangular in shape, with each cohort larger than the cohort born before it. An example pyramid is shown in Figure 3.15 for Morocco. It is not uncommon in such countries to have nearly half of the population younger than 15 years old. The age structure shows us that even if replacement fertility is achieved in the near future, there are so many young women already born who will be having their children that population size will continue to grow for many decades.

The second pyramid in Figure 3.15 is an example of a population that is not growing very rapidly, if at all. Notice that the sides of the structure are much more vertical. There is also a pinching down of the age structure in the younger years, suggesting a fertility rate below replacement level, so that if these trends continue, the population will eventually begin to decrease.

There are two important terms used by demographers to describe the shape of an age structure. A *stable population* is one that has had constant age-specific birth and death rates for such a long time that the percentage of the population in any age category does not change. That is, the shape of the age structure is unchanging. A population that is stable does not have to be fixed in size; it can be growing or shrinking at a constant rate. When a population is both stable and unchanging in size it is called a *stationary population*. Thus all stationary populations are stable, but not all stable populations are stationary.

The population pyramid for the United States in 1990 is shown in Figure 3.16. The nearly vertical sides on the lower part of the pyramid suggest that this is a population that is well underway toward a stationary condition. The dominant feature, however, is

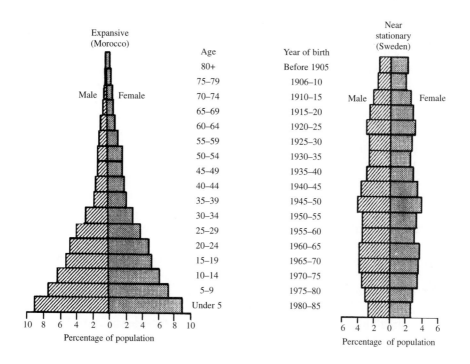

Age	Year of birth
80+	Before 1905
75–79	1906–10
70–74	1910–15
65–69	1915–20
60–64	1920–25
55–59	1925–30
50–54	1930–35
45–49	1935–40
40–44	1940–45
35–39	1945–50
30–34	1950–55
25–29	1955–60
20–24	1960–65
15–19	1965–70
10–14	1970–75
5–9	1975–80
Under 5	1980–85

FIGURE 3.15   A rapidly growing, expansive population (Morocco), compared with a population that is nearing stationary condition (Sweden). *Source:* Haupt and Kane (1985).

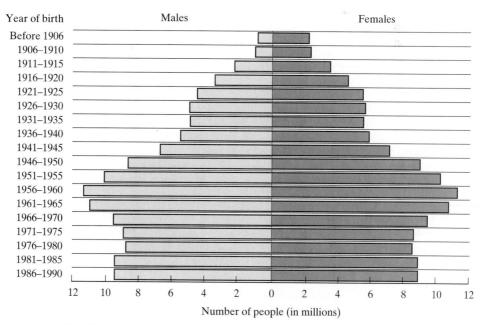

FIGURE 3.16   The age structure for the United States in 1990. *Source:* Weeks (1994).

the bulge in the middle corresponding to the baby-boom cohorts born between 1946 and 1964. Thinking of the age structure as a dynamic phenomenon, the bulge moving upward year after year suggests an image of a python that has swallowed a pig, which slowly works its way through the snake. As the boomers pass through the age structure, resources are stretched to the limit. When they entered school there were not enough classrooms and teachers, so new schools were built and teacher training programs expanded. When they left, schools closed and there was an oversupply of teachers. Universities found themselves competing for smaller and smaller numbers of high school graduates. During their late teens and early twenties, arrest rates increased since those are the ages that are, statistically speaking, responsible for the majority of crimes. As they entered the workforce, competition was fierce and unemployment rates were persistently high, housing was scarce, and the price of real estate jumped. The baby-boom generation is now nearing retirement, and there is a boom in construction of retirement communities coupled with growing concern about the ability of the social security system to accommodate so many retirees.

By contrast, children born in the 1930s, just ahead of the baby-boom generation, are sometimes referred to as the "good times cohorts." Since there are so few of them, as they sought their educations, entered the job market, bought homes, and are now retiring, the competition for those resources has been relatively modest. Children born in the 1970s also benefit by being part of a narrow portion of the age structure. However, their future is less rosy since they will be part of the working population that will face the responsibility of supporting the more elderly baby-boom generation. At the bottom of the age structure are the baby-boom echo cohorts born in the 1980s and 1990s.

Figure 3.17 shows age structures for the world in 1990 and projections for 2025. The developed countries essentially have stationary populations in both years, but the developing countries are still growing rapidly.

Human Population Projections

If the age structure for a population is combined with data on age-specific birth and death rates, it is possible to make realistic projections of future population size and composition. The techniques that we will explore now are especially useful for predicting the effects and implications of various changes in fertility patterns that might be imagined or advocated. For example, suppose we want to determine the effects of replacement-level fertility. If replacement fertility were achieved today, how long would it take to achieve a stationary population? What would the age composition look like in the interim? What fraction of the population would be young and looking for work, and what fraction would be retired and expecting support? Or, suppose a policy is established that sets a target for maximum size of the population (as is the case for China). How many children would each couple need to have to achieve that goal, and how would the number change with time?

Not long ago, the calculations required to develop these scenarios were tedious and best left to professional demographers. Now, however, with the widespread use of simple spreadsheet programs on personal computers, it is possible to work them out with relative ease.

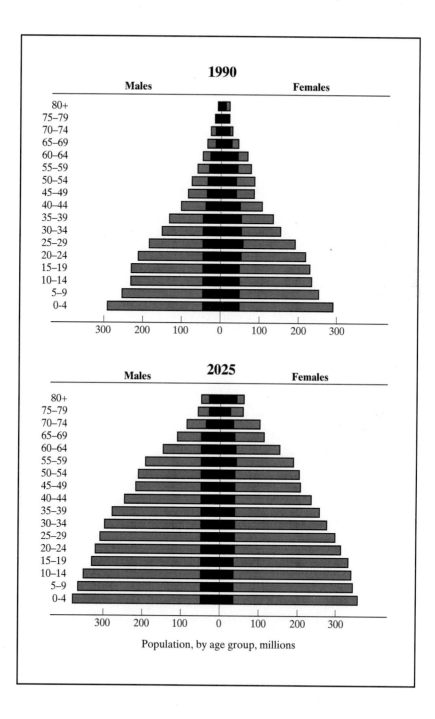

FIGURE 3.17   Age structures for the world in 1990 and projections for 2025, showing the developed countries in darker tones than the developing countries. *Source: The Economist,* January 27, 1996.

The starting point in a population projection is the current age structure combined with mortality data obtained from *life tables*. Life tables are the mainstay of insurance companies for predicting the average number of years of life remaining as a function of the age of their clients. A life table is developed by applying a real population's age-specific death rates (the fraction of the people in a given age category who will die each year) to *hypothetical* stable and stationary populations having 100,000 live births per year, evenly distributed through the year, with no migration. As the 100,000 people added each year get older, their ranks are thinned in accordance with the age-specific death rates. It is then possible to calculate the numbers of people who would be alive within each age category in the following year.

Table 3.4 presents a life table for the United States that has been simplified by broadening the age intervals to five-year increments rather than the one-year categories used by insurance companies and demographers. These data, remember, are for a hypothetical population with 100,000 live births each year (and 100,000 deaths each year as well since this is a stationary population). The first column shows the age interval (e.g., 10–14 means people who have had their tenth birthday but not their fifteenth). The second column is the number of people who would be alive at any given time in the corresponding age interval, and is designated L_x, where x is the age at the beginning of the interval. The third column, L_{x+5}/L_x, is the ratio of number of people

TABLE 3.4   Age distribution for a hypothetical, stationary population with 100,000 live births per year in the United States[a]

Age interval x to $x + 5$	Number in interval L_x	$\dfrac{L_{x+5}}{L_x}$
0–4	494,285	0.9979
5–9	493,250	0.9989
10–14	492,699	0.9973
15–19	491,382	0.9951
20–24	488,972	0.9944
25–29	486,219	0.9940
30–34	483,310	0.9927
35–39	479,794	0.9898
40–44	474,877	0.9841
45–49	467,346	0.9745
50–54	455,448	0.9597
55–59	437,078	0.9381
60–64	410,020	0.9082
65–69	372,397	0.8658
70–74	322,425	0.8050
75–79	259,548	0.7163
80–84	185,922	0.9660
85 and over	179,601	0.0000

[a]Age-specific death rates are U.S. 1984 values.
Source: Abstracted from data in U.S. Dept. of Health and Human Services (1987).

in the next interval to the number in the current interval; it is the probability that individuals will live five more years (except in the case of those 80 and older, where the catch-all category of 85+ years old modifies the interpretation).

If we assume that the age-specific death rates that were used to produce Table 3.4 remain constant, we can use the table to make future population projections at five-year intervals for all but the 0–4 age category. The 0–4 age category will depend on fertility data, to be discussed later.

If we let $P_x(0)$ be the number of people in age category x to $x + 5$ at time $t = 0$, and $P_{x+5}(5)$ be the number in the next age category five years later, then

$$P_{x+5}(5) = P_x(0) \frac{L_{x+5}}{L_x} \tag{3.33}$$

That is, five years from now, the number of people in the next five-year age interval will be equal to the number in the interval now times the probability of surviving for the next five years (L_{x+5}/L_x).

For example, in the United States in 1985, there were 18.0 million people in the age group 0–4 years; that is, $P_0(1985) = 18.0$ million. We would expect that in 1990 the number of people alive in the 5–9 age category would be

$$P_5(1990) = P_0(1985)(L_5/L_0)$$
$$= 18.0 \times 10^6 \times 0.9979 = 17.98 \text{ million}$$

Let's take what we have and apply it to the age composition of the United States in 1985 to predict as much of the structure as we can for 1990. This involves application of (3.33) to all of the categories, giving us a complete 1990 age distribution except for the 0–4 year olds. The result is presented in Table 3.5.

To find the number of children ages 0–4 to enter into the age structure, we need to know something about fertility patterns. Demographers use *age-specific fertility rates*, which are the number of live births per woman in each age category, to make these estimates. To do this calculation carefully, we would need to know the number of people in each age category who are women as well statistical data on child mortality during the first five years of life. Since our purpose here is to develop a tool for asking questions of the "what if" sort, rather than making actual population projections, we can use the following simple approach to estimate the number of children ages 0–4 to put into our age composition table:

$$P_0(n + 5) = b_{15}P_{15}(n) + b_{20}P_{20}(n) + \ldots b_{45}P_{45}(n) \tag{3.34}$$

where b_x is the number of the surviving children born per person in age category x to $x + 5$. Equation (3.34) has been written with the assumption that no children are born to individuals under 15 or over 49 years of age, although it could obviously have been written with more terms to include them. Notice that we can interpret the sum of the b_x's (Σb_x) to be the number of children each person would be likely to have. The total fertility rate, which is the expected number of children per woman, is therefore close to $2\Sigma b_x$.

For example, these fertility factors and the numbers of people in their reproductive years for the United States in 1985 are shown in Table 3.6. The total fertility rate can be estimated as $2\Sigma b_x$, which is $2 \times 1.004 = 2.01$ children per woman.

TABLE 3.5   1990 U.S. Population Projection Based on the 1985 Age Structure (Ignoring Immigration)

Age interval x to $x + 5$	$\dfrac{L_{x+5}}{L_x}$	P_x (thousands)	
		1985	1990
0–4	0.9979	18,020	$P_0(1990)$
5–9	0.9989	17,000	17,982
10–14	0.9973	16,068	16,981
15–19	0.9951	18,245	16,025
20–24	0.9944	20,491	18,156
25–29	0.9940	21,896	20,376
30–34	0.9927	20,178	21,765
35–39	0.9898	18,756	20,031
40–44	0.9841	14,362	18,564
45–49	0.9745	11,912	14,134
50–54	0.9597	10,748	11,609
55–59	0.9381	11,132	10,314
60–64	0.9082	10,948	10,443
65–69	0.8658	9,420	9,943
70–74	0.8050	7,616	8,156
75–79	0.7163	5,410	6,131
80–84	0.9660	3,312	3,875
85 and over	0.0000	2,113	3,199
		Total = 237,627	227,684 + $P_0(1990)$

Source: 1985 data from Vu (1985).

TABLE 3.6   Calculating births in five-year increments for the United States, 1985

Age category	$P_x(1985)$ (thousands)	b_x (births per person during five-year period)	Births (thousands in five years)
15–19	18,245	0.146	2,664
20–24	20,491	0.289	5,922
25–29	21,896	0.291	6,372
30–34	20,178	0.190	3,834
35–39	18,756	0.075	1,407
40–45	14,362	0.013	187
Totals		1.004	20,385

With this 20,385,000 added to the 227,684,000 found in Table 3.5, the total population in the United States in 1990 would have been 248,069,000. Notice that, in spite of the fact that fertility was essentially at the replacement level in 1985, the population would still be growing by about 2 million people per year. This is an example of *population momentum,* a term used to describe the fact that a youthful age structure causes a population to continue growing for several decades after achieving replacement fertility. To finish off the aforementioned estimate, we would have to add in net migration, which is currently 650,000 legal immigrants per year plus probably hundreds of thousands of immigrants entering the country illegally.

Population Momentum

Let us begin with an extremely simplified example, which does not model reality very well, but which does keep the arithmetic manageable. Then we will see the results of more carefully done population scenarios.

Suppose we have an age structure with only three categories: 0–24 years, 25–49 years, and 50–74 years, subject to the following fertility and mortality conditions:

1. All births occur to women as they leave the 0–24 age category.
2. All deaths occur at age 75.
3. The total fertility rate (TFR) is 4.0 for 25 years; then it drops instantaneously to 2.0.

Suppose a population of 5.0 billion people has an age structure at time t = 0, as shown in Figure 3.18*a.* Equal numbers of men and women will be assumed so that the

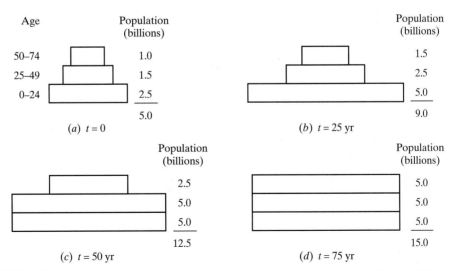

FIGURE 3.18   Age structure diagrams illustrating population momentum. For the first 25 years, TFR = 4; thereafter, it is at the replacement level of 2.0 yet population continues to grow for another 50 years.

number of children per person is half the total fertility rate. During the first 25 years, the 2.5 billion people who start in the 0–24 age category will all pass their twenty-fifth birthday and all of their children will be born. Since TFR is 4.0, these 2.5 billion individuals will bear 5.0 billion children (two per person). Those 2.5 billion people will now be in the 25–49 age category. Similarly, the 1.5 billion people who were 25–49 years old will now be ages 50–74. Finally, the 1.0 billion 50–74 year olds will have all passed their seventy-fifth birthdays and, by the rules, they will all be dead. The total population will have grown to 9.0 billion, and the age structure will be as shown in Figure 3.18b.

After the first 25 years have passed, TFR drops to the replacement level of 2.0 (one child per person). During the ensuing 25 years, the 5.0 billion 0–24 year olds will have 5.0 billion children and, following the logic given previously, the age structure at $t = 50$ years will be as shown in Figure 3.18c. With the replacement level continuing, at $t = 75$ years the population stabilizes at 15 billion. A plot of population versus time is shown in Figure 3.19. Notice that the population stabilizes 50 years after replacement-level fertility is achieved, during which time it grows from 9 billion to 15 billion.

A more carefully done set of population scenarios is presented in Figure 3.20. In the constant-path scenario, population is assumed to continue growing at the 1980 growth rate into the indefinite future. The slow fertility-reduction path assumes that the world's fertility will decline to reach replacement level by 2065, which results in a stabilized population of about 14 billion by the end of the twenty-first century. The moderate fertility-reduction path assumes that replacement-level fertility will be reached by 2035, with the population stabilizing at about 10 billion. Finally, the rapid fertility-reduction path assumes that the world's fertility will decline to replacement level by 2010. Even under this most optimistic scenario, the world's population would still grow to approximately 7.5 billion. The implications of population momentum are obviously of tremendous importance.

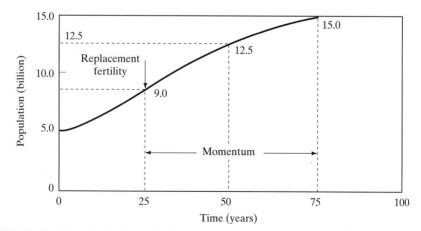

FIGURE 3.19 In the hypothetical example, it takes 50 years of replacement-level fertility before population stops growing.

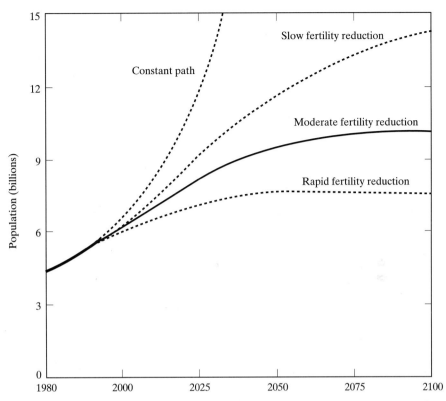

FIGURE 3.20 Four scenarios for the future of the world's population. The moderate path has fertility declining to replacement level by 2035, with momentum carrying the population to approximately 10 billion by the end of the twenty-first century. *Source:* Haupt and Kane (1985).

PROBLEMS

3.1. World population in 1850 has been estimated to have been about 1 billion. It reached 4 billion in 1975.

    (a) Use the doubling time approximation (3.5) to estimate the exponential rate of growth that would produce those numbers.

    (b) Use the exponential growth equation (3.6) to find the growth rate.

3.2. Tuition at a major university rose from $1500/yr in 1962 to $20,000/yr in 1995.

    (a) What exponential rate of growth characterized that period of time?

    (b) If that growth rate were to continue for another 25 years (enough time for current students to have children in college!) what would the tuition be?

3.3. The world's population 10,000 years ago has been estimated at about 5 million. What exponential rate of growth would have resulted in the population in 1850, which is estimated to have been 1 billion? Had that rate continued, what would the population be in the year 2000?

3.4. Suppose we express the amount of land under cultivation as the product of four factors:

$$\text{Land} = (\text{land/food}) \times (\text{food/kcal}) \times (\text{kcal/person}) \times (\text{population})$$

The annual growth rates for each factor are (1) the land required to grow a unit of food, -1 percent (due to greater productivity per unit of land); (2) the amount of food grown per calorie of food eaten by a human, $+0.5$ percent (with affluence, people consume more animal products, which greatly reduces the efficiency of land use); (3) the per capita calorie consumption, $+0.1$ percent; and (4) the size of the population, $+1.5$ percent. At these rates, how long would it take to double the amount of cultivated land needed? At that time, how much less land would be required to grow a unit of food?

3.5. Suppose world carbon emissions are expressed as the following product:

$$\text{Carbon emissions} = (\text{Population}) \times (\text{Energy/person}) \times (\text{Carbon/energy})$$

If per capita energy demand increases at 1.5 percent per year, fossil fuel emissions of carbon per unit of energy increase at 1 percent per year, and world population grows at 1.5 percent per year,

(a) How long would it take before we are emitting carbon at twice the current rate?

(b) At that point, by what fraction would per capita energy demand have increased?

(c) At that point, by what fraction would total energy demand have increased?

3.6. Under the assumptions stated in Problem 3.5, if our initial rate of carbon emission is 5×10^9 tonnes C/yr and if there are 700×10^9 tonnes of carbon in the atmosphere now,

(a) How long would it take to emit a total amount of carbon equal to the amount now contained in the atmosphere?

(b) If half of the carbon that we emit stays in the atmosphere (the other half being "absorbed" in other parts of the biosphere), how long would it take for fossil fuel combustion to double the atmospheric carbon concentration?

3.7. Consider the following disaggregation of carbon emissions:

$$\text{Carbon emissions (kg C/yr)} = \text{Population} \times \frac{\text{Energy (kJ/yr)}}{\text{Person}} \times \frac{\text{Carbon (kgC)}}{\text{Energy (kJ)}}$$

Using the following estimates for the United States and assuming that growth rates remain constant,

	Population	(kJ/yr)/Person	kg C/kJ
1990 amounts	250×10^6	320×10^6	15×10^{-6}
Growth, r (%/yr)	0.6	0.5	-0.3

(a) Find the carbon emission rate in 2020.

(b) Find the carbon emitted in those 30 years.

(c) Find total energy demand in 2020.

(d) Find per capita carbon emission rate in 2020.

3.8. World reserves of chromium are about 800 million tons, and current usage is about 2 million tons per year. If growth in demand for chromium increases exponentially at a constant rate of 2.6 percent per year, how long would it take to use up current reserves? Suppose the total resource is five times current reserves, if the use rate continues to grow at 2.6 percent, how long would it take to use up the resource?

3.9. Suppose a Gaussian curve is used to approximate the production of chromium. If production peaks at six times its current rate of 2 million tons per year, how long would it take to reach that maximum if the total chromium ever mined is 4 billion tons (five times the current reserves)? How long would it take to consume about 80 percent of the total resource?

3.10. Suppose we assume the following:

(a) Any chlorofluorocarbons (CFCs) released into the atmosphere remain in the atmosphere indefinitely.

(b) At current rates of release, the atmospheric concentration of fluorocarbons would double in 100 years (what does that say about Q/P_0?).

(c) Atmospheric release rates are, however, not constant but growing at 2 percent per year.

How long would it take to double atmospheric CFC concentrations?

3.11. Bismuth-210 has a half-life of 4.85 days. If we start with 10 g of it now, how much would we have left in 7 days?

3.12. Suppose some sewage drifting down a stream decomposes with a reaction rate coefficient K equal to 0.2/day. What would be the half-life of this sewage? How much would be left after 5 days?

3.13. Suppose human population grows from 6.3 billion in 2000 to an ultimate population of 10.3 billion following the logistic curve. Assuming a growth rate of 1.5 percent in 2000, when would the population reach 9 billion? What would the population be in 2050? Compare this to the moderate fertility reduction scenario of Figure 3.20.

3.14. Suppose a logistic growth curve had been used to project the world's population back in 1970, when there were 3.65 billion people and the growth rate was 2.0 percent per year. If a steady-state population of 10.3 billion had been used (the moderate path in Figure 3.20), what would the projected population have been for 1995 (when it was actually 5.7 billion) and 2025?

3.15. Suppose we stock a pond with 100 fish and note that the population doubles every year for the first couple of years (with no harvesting), but after quite a number of years, the population stabilizes at what we think must be the carrying capacity of the pond, 2000 fish. Growth seems to have followed a logistic curve.

(a) What population size should be maintained to achieve maximum yield, and what would be the maximum sustainable fish yield?

(b) If the population is maintained at 1500 fish, what would be the sustainable yield?

3.16. What would be the sustainable yield from the pond in Example 3.9 if the population is maintained at 3000 fish?

3.17. A lake has a carrying capacity of 10,000 fish. At the current level of fishing, 2000 fish per year are taken and the fish population seems to hold fairly steady at about 4000. If you wanted to maximize the sustainable yield, what would you suggest in terms of population size and yield?

3.18. The following statistics are for India in 1985: population, 762 million; crude birth rate, 34; crude death rate, 13; infant mortality rate, 118 (rates are per thousand per year). Find (a) the fraction of the total deaths that are to infants less than one year old; (b) the avoidable deaths, assuming that any infant mortality above 10 could be avoided with better sanitation, food, and health care; and (c) the annual increase in the number of people in India.

3.19. The following statistics are for India in 1995: population, 931 million; crude birth rate, 29; crude death rate, 9; infant mortality rate, 74 (rates are per thousand per year). Find (a) the fraction of the total deaths that are to infants less than one year old; (b) the avoidable deaths, assuming that any infant mortality above 10 could be avoided with better sanitation, food, and health care; and (c) the annual increase in the number of people in India. (For comparison, in 1985 the population was growing at 16 million per year and the avoidable deaths were 2.8 million per year.)

3.20. Consider a simplified age structure that divides a population into three groups: ages 0–24, with 3.0 million; 25–49, with 2.0 million; and 50–74, with 1.0 million. Suppose we impose the following simplified fertility and mortality constraints: All births occur just as the woman leaves the 0–24 age category, and no one dies until their seventy-fifth birthday, at which time they all die. Suppose we have replacement-level fertility starting now. Draw the age structure in 25 years, 50 years, and 75 years. What is the total population size at each of these times?

3.21. Using the same initial population structure given in Problem 3.20, with all births on the twenty-fifth birthday, draw the age structure in 25, 50, and 75 years under the following conditions: No deaths occur until the fiftieth birthday, at which time 20 percent die; the rest die on their seventy-fifth birthday. For the first 25 years, the total fertility rate is 4 (2 children/person), and thereafter is it 2.

3.22. Consider the following simplified age structure: All births are on the twentieth birthday and all deaths are on the sixtieth birthday. Total population starts at 290,000 (half males, half females) and is growing at a constant rate of 3.5 percent per year (see Figure P3.22). (*Hint:* what does that say about the doubling time?)

Draw the age structure in 20 years. If the total fertility rate is a single, constant value during those 20 years, what is it?

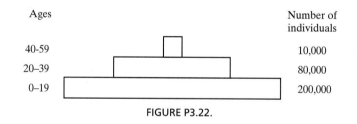

Ages — Number of individuals

Ages	Number of individuals
40-59	10,000
20–39	80,000
0–19	200,000

FIGURE P3.22.

3.23. The following age structure and survival data are for China in 1980. Suppose the birth factors (corresponding to a total fertility rate of 1.0) are as shown. Estimate the population of China in 1990.

Age	Population (millions)	L_{x+10}/L_x	b_x
0–9	235	0.957	0
10–19	224	0.987	0.25
20–29	182	0.980	0.25
30–39	124	0.964	0
40–49	95	0.924	0
50–59	69	0.826	0
60–69	42	0.633	0
70–79	24	0.316	0
80–89	6	0	0
Total 1001			

3.24. Use a spreadsheet to project the China population data given in Problem 3.23 out to the year 2030. What is the population at that time?

3.25. Use a spreadsheet to project the China population data given in Problem 3.23 out to the year 2030, but delay the births by one 10-year interval (that is, $b_{10} = 0$, $b_{20} = 0.25$, and $b_{30} = 0.25$). Compare the peak population in Problem 3.23 to that obtained by postponing births by 10 years.

3.26. The birth-rate data for China in Problem 3.23 were for the very optimistic one-child per family scenario (total fertility rate = 1). In reality, the total fertility rate has been closer to 2. Assuming that each woman had 2 children while she was in the 20 to 30 age group (and none at any other age),

   (a) Repeat the population projection for China for 1990.

   (b) Continuing that birth rate, find the population in 2000.

REFERENCES

Brock, T. D., 1974, *Biology of Microorganisms,* 2nd ed., Prentice Hall, Englewood Cliffs, NJ.

Bouvier, L. F., and C. J. De Vita, 1991, *The Baby Boom—Entering Midlife,* Population Reference Bureau, Inc., Washington, DC.

Energy Information Administration, 1996, *Annual Energy Review, 1996,* U.S. Department of Energy, Washington, DC.

Haupt, A., and T. T. Kane, 1985, *Population Handbook,* 2nd ed., Population Reference Bureau, Washington, DC.

Hubbert, M. K., 1969, *Resources and Man,* National Academy of Sciences, Freeman, San Francisco.

Hubbert, M. K., 1977, *Project Independence: U. S. and World Energy Outlook through 1990,* Congressional Research Service, Washington, DC.

Hobbs, G. W., 1995, Oil, gas, coal, uranium, tar sand resource trends on rise, *Oil and Gas Journal,* September 4, 104–108.

Population Reference Bureau, 1996, *1995 World Population Data Sheet,* Washington, DC.

Southwick, C. H., 1976, *Ecology and the Quality of our Environment,* 2nd ed., Van Nostrand, New York.

Tien, H. Y., Z. Tianhu, P. Yu, L. Jingnent, and L. Zhongtang, 1992, *China's Demographic Dilemas,* Population Reference Bureau, Inc., Washington, DC.

United Nations, 1971, *A Concise Summary of the World Population Situation in 1970,* Population Studies No. 48, New York.

U. S. Dept of Health and Human Services, 1987, *Vital Statistics of the United States, 1984, Life Tables,* National Center for Health Statistics, Hyattsville, MD.

Vu, M. T., 1985, *World Population Projections, 1985, Short- and Long-term Estimates by Age and Sex with Related Demographic Statistics,* World Bank, Johns Hopkins Press, Baltimore.

Weeks, J. R., 1994, *Population, An Introduction to Concepts and Issues,* Wadsworth, Belmont, CA.

Risk Assessment

> *All substances are poisons; there is none which is not a poison. The right dose differentiates a poison and a remedy.* —Paracelsus
> (1493–1541)

4.1 INTRODUCTION

One of the most important changes in environmental policy in the 1980s was the acceptance of the role of risk assessment and risk management in environmental decision making. In early environmental legislation, such as the Clean Air and Clean Water Acts, the concept of risk is hardly mentioned; instead, these acts required that pollution standards be set that would allow adequate margins of safety to protect public health. Intrinsic to these standards was the assumption that pollutants have thresholds, and that exposure to concentrations below these thresholds would produce no harm. All of that changed when the problems of toxic waste were finally recognized and addressed. Many toxic substances are suspected carcinogens; that is, they may cause cancer, and for carcinogens the usual assumption is that even the smallest exposure creates some risk.

If any exposure to a substance causes some risk, how can air quality and water quality standards be set? When cleaning up a hazardous waste site, at what point is the project completed; that is, how clean is clean? At some point in the cleanup, the remaining health and environmental risks may not justify the continued costs and, from a risk perspective, society might be better off spending the money elsewhere. Almost by definition, achieving zero risk would cost an infinite amount of money, so policy makers have had to grapple with the tradeoff between acceptable risk and acceptable cost. Complicating those decisions is our very limited understanding of diseases such as cancer coupled with a paucity of data on the tens of thousands of synthetic chemicals that are in widespread use today. Unfortunately, those who have responsibility for creating and administering environmental regulations have to take action even if definitive answers from the scientific community on the relationship between exposure and risk are not available.

The result has been the emergence of the controversial field of environmental risk assessment. Hardly anyone is comfortable with it. Scientists often deplore the notion of condensing masses of frequently conflicting, highly uncertain, often ambiguous data that have been extrapolated well beyond anything actually measured down to a single number or two. Regulatory officials are battered by the public when they propose a level of risk that they think a community living next to a toxic waste site should tolerate. Critics of government spending think risks are being systematically overestimated, resulting in too much money being spent for too little real improvement in public health. Others think risks are underestimated since risk assessments are based on data obtained for exposure to individual chemicals, ignoring the synergistic effects that are likely to occur when we are exposed to thousands of them in our daily lives.

Some of the aforementioned conflicts can best be dealt with if we make the distinction between risk assessment and risk management. *Risk assessment* is the scientific side of the story. It is the gathering of data that are used to relate response to dose. Such dose-response data can then be combined with estimates of likely human exposure to produce overall assessments of risk. *Risk management,* on the other hand, is the process of deciding what to do. It is decision making, under extreme uncertainty, about how to allocate national resources to protect public health and the environment. Enormous political and social judgment is required to make those decisions. Is a one-in-a-million lifetime risk of getting cancer acceptable and, if it is, how do we go about trying to achieve it?

4.2 PERSPECTIVES ON RISKS

The usual starting point for an explanation of risk is to point out that there is some risk in everything we do, and since we will all die someday, our lifetime risk of death from all causes is 1.0, or 100 percent. It is easy to gather good data on the causes of death, such as are shown in Table 4.1, which gives us one approach to talking about risk. For example, of the total 2.177 million deaths in the United States in 1992, there were roughly 521 thousand cancer deaths. Neglecting age structure complications, we could say that, on the average, the risk, or probability, of an American dying of cancer is therefore about 24 percent (521,000/2,177,000 = 0.24).

Notice that there are no units associated with risk, although other clarifiers may be needed, such as whether the risk is a lifetime risk or an annual risk, whether it is an average risk to the general public or a risk faced by individuals who engage in some activity, or whether it is being expressed as a percentage or as a decimal fraction. For example, in the United States, smoking is thought to be causing approximately 400,000 deaths per year. On the average, the probability of death caused by smoking is therefore about 18 percent (400,000/2,177,000 = 0.18). Obviously, however, the risk that an individual faces from smoking depends on how much that person smokes and how

TABLE 4.1 Leading Causes of Death in the United States, 1992

Cause	Annual deaths (thousands)	Percent
Cardiovascular (heart) disease	720	33
Cancer (malignant neoplasms)	521	24
Cerebrovascular diseases (strokes)	144	7
Pulmonary diseases (bronchitis, emphysema, asthma)	91	4
Pneumonia and influenza	76	3
Diabetes mellitus	50	2
Non-motor-vehicle accidents	48	2
Motor vehicle accidents	42	2
HIV/AIDS	34	1.6
Suicides	30	1.4
Homicides	27	1.2
All other causes	394	18
Total annual deaths (rounded)	2177	100

Source: Kolluru (1996).

much exposure is caused by smoke from other people's cigarettes. The probability that a pack-a-day smoker will die of lung cancer, heart disease, or emphysema brought on by smoking is about 0.25, or 25 percent (Wilson and Crouch, 1987). Statistically, smokers shorten their life expectancy by about 5 minutes for each cigarette smoked, which is roughly the time it takes to smoke that cigarette.

Environmental risk assessments deal with incremental probabilities of some harm occurring. For example, the Environmental Protection Agency (EPA) attempts to control our exposure to toxics to levels that will pose incremental lifetime cancer risks to the most exposed members of the public of roughly 10^{-6} (one additional cancer in one million people) to 10^{-4} (100 additional cancers per million people). For perspective, suppose all 260 million Americans faced a 10^{-6} lifetime risk of cancer from exposure to a particular toxic chemical. That would mean 260 extra cancers during their lifetimes. Suppose we assume a typical lifetime of 70 years. Then spreading those 260 cancers out over 70 years suggests roughly four extra cancers per year in the United States. Table 4.1 tells us there are 521,000 cancer deaths per year, so the four extra cancers caused by toxic exposure would be less than 0.001 percent of the nominal rate.

Presenting risks as an annual probability of death to individuals who engage in some activity is a much more specific way to express risks than simply looking at the population as a whole. Table 4.2 shows risk data for some common activities. For example, the probability of dying in a motorcycle accident among those who ride motorcycles is 2000 deaths per year per 100,000 motorcyclists. Another example is the risk associated with consuming 4 tablespoons of peanut butter per day. It turns out that mold on peanuts creates a group of chemicals called *aflatoxins* that are known to cause malignant tumors in a number of animals, such as rats, mice, guinea pigs, and monkeys. The Food and Drug Administration (FDA) restricts the aflatoxin concentration in peanut products to 20 ppb (a risk-based decision), and at that level, eating 4 tablespoons of peanut butter per day may cause an estimated 0.8 cancer deaths per year per

TABLE 4.2 Annual risks of death associated with certain activities.

Activity/exposure	Annual risk (Deaths per 100,000 persons at risk)
Motorcycling	2000
Smoking, all causes	300
Smoking (cancer)	120
Hang gliding	80
Coal mining	63
Farming	36
Motor vehicles	24
Chlorinated drinking water (chloroform)	0.8
4 tbsp peanut butter per day (aflatoxin)	0.8
3 oz charcoal broiled steak per day (PAHs)	0.5
1-in-a-million lifetime risk	0.0014

Source: Based on Wilson and Crouch, 1987.

100,000. Another interesting item on the list is associated with the polycyclic aromatic hydrocarbons (PAHs) created when food is charbroiled. As will be described in Chapter 7, PAHs are formed when carbon-containing materials are not completely oxidized during combustion, so consumption of burned steak poses some cancer risk. In Table 4.2 that risk is estimated at 0.5 deaths per year per 100,000 people who eat 3 ounces of broiled steak per day. Notice, for comparison, that the annual risk associated with a "one-in-a-million" lifetime risk is about 0.0014 per 100,000.

The data in Table 4.1 are based on actuarial data, so they may be considered accurate, but the data in Tables 4.2 and 4.3 are a mix of actuarial values and estimates based on various risk models. It should be always be kept in mind that when risks are based on models, there are generally very large uncertainties in the estimates.

Wilson (1979) provides some perspectives on risk by comparing various activities on the basis of equal, one-in-one-million (10^{-6}) risks. For example, aircraft statistics indicate that 100 billion passenger miles of travel each year in the United States result in about 100 deaths per year; that is, one death per billion passenger miles. A trip of 1000 miles would therefore result in a risk of about 10^{-6}. As another example, Wilson cites statistics on death rates caused by sulfur emissions from coal plants east of the Mississippi. At 20,000 deaths per year among 100 million people exposed to this dirty air, the average risk would be 20,000/100,000,000, or 0.0002 per year of exposure. Two days of breathing this polluted air would pose a risk of $2/365 \times 0.0002 = 10^{-6}$. Other examples of one-in-one-million risks are given in Table 4.3. As suggested there, for example, smoking 1.4 cigarettes is equivalent in risk terms to living 50 years within 5 miles of a nuclear power plant. Again, bear in mind that these values are at best rough approximations.

One of the purposes of risk assessment is to provide a starting point in balancing the tradeoffs between an acceptable incremental risk and the cost of controlling risk to that level. Table 4.4 shows some estimated expenditures to prevent a life from being shortened by one year. Immunizations and phasing out leaded gasoline are indicated to have no cost to them because the direct savings in health care far exceed their cost.

TABLE 4.3   Activities that increase mortality risk by one in a million

Activity	Type of risk
Smoking 1.4 cigarettes	Cancer, heart disease
Drinking 1/2 liter of wine	Cirrhosis of the liver
Spending 1 hour in a coal mine	Black lung disease
Living 2 days in New York or Boston	Air pollution
Traveling 300 miles by car	Accident
Flying 1000 miles by jet	Accident
Flying 6000 miles by jet	Cancer by cosmic radiation
Traveling 10 miles by bicycle	Accident
Traveling 6 minutes by canoe	Accident
Living 2 summer months in Denver (vs. sea level)	Cancer by cosmic radiation
Living 2 months with a cigarette smoker	Cancer, heart disease
Eating 40 tablespoons of peanut butter	Liver cancer caused by aflatoxin
Eating 100 charcoal-broiled steaks	Cancer from benzopyrene
Living 50 years within 5 miles of a nuclear reactor	Accident releasing radiation

Source: Wilson (1979).

TABLE 4.4   Estimated expenditures per life-year saved for selected programs

Program	1990 U.S. $
Childhood immunizations	Direct savings
Eliminating lead in gasoline	Direct savings
Safety rules at underground construction sites	52,000
Hemodialysis at a dialysis center	56,000
Coronary artery bypass surgery	68,000
Front seat air bags in new cars	109,000
Dioxin effluent controls at paper mills	5,570,000

Source: Kolluru (1996) based on data from the Harvard School of Public Health.

Pollution control in the case of lead emissions is very cost-effective, but the table suggests that saving lives by controlling dioxin at a paper mill is very costly indeed.

4.3 PERCEPTION OF RISK

Data such as those given in the preceding tables are often used to try to put the health risk associated with pollution into perspective. It usually turns out, however, that the perceptions of risk as seen by an engineer or scientist familiar with the numbers are very different from those of an individual who lives next to a toxic waste site. Social scientists have studied this phenomenon and conclude that there are a number of attributes of risk that can increase the anxiety level of someone evaluating his or her own personal exposures.

For example, people are more likely to be outraged when they have no control of the risks they are exposed to, and they are more fearful of unknown risks than ones they are familiar with. We quite readily take on the risk of crashing into a tree when skiing, because it is a voluntary activity and we are familiar with and understand the risks. We put ourselves at great risk by driving cars, but we feel somewhat in control and believe the risk is worth the benefits. We also accept natural risks such as earthquakes and hurricanes much more readily than unnatural ones, such as the 1984 explosion of a methyl isocyanate storage tank in Bhopal, India, which killed 3400 people. And we are probably more comfortable living next to a gas station, despite the exposure to the carcinogen benzene, than living anywhere near a nuclear power plant, with its perceived unknown and uncertain risks. Table 4.5 illustrates this notion by comparing attributes that seem to be associated with elevated perceptions of risk.

That these risk attributes can be so important to the public can be a source of frustration in the technical community, but they are real and must be acknowledged by anyone who needs to communicate risk concepts to the public. Finding a way to help people act on the real risks in life and worry less about the minor ones is a difficult challenge.

4.4 RISK ASSESSMENT

Our concern is with the probability that exposure of some number of people to some combination of chemicals will cause some amount of response, such as cancer, reproductive failure, neurological damage, developmental problems, or birth defects. That is, we want to begin to develop the notions of risk assessment. The National Academy of Sciences (1983) suggests that risk assessment be divided into the following four steps: hazard identification, dose-response assessment, exposure assessment, and risk charac-

TABLE 4.5   Some characteristics that elevate the perception of risk.

Attributes that elevate the perception of risk	Attributes that lower perception
Involuntary	Voluntary
Exotic	Familiar
Uncontrollable	Controllable
Controlled by others	Controlled by self
Dread	Accept
Catastrophic	Chronic
Caused by humans	Natural
Inequitable	Equitable
Permanent effect	Temporary effect
No apparent benefits	Visible benefits
Unknown	Known
Uncertainty	Certainty
Untrusted source	Trusted source

Source: based on Slovic (1987) and Slovic et al. (1980).

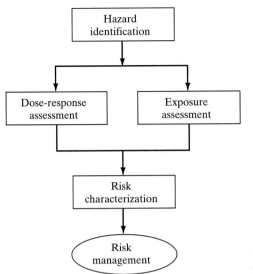

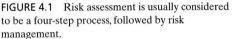

FIGURE 4.1   Risk assessment is usually considered
to be a four-step process, followed by risk
management.

terization. After a risk assessment has been completed, the important stage of risk
management follows, as shown in Figure 4.1.

- *Hazard identification* is the process of determining whether or not a particular
  chemical is causally linked to particular health effects, such as cancer or birth
  defects. Since human data are so often difficult to obtain, this step usually focuses
  on whether a chemical is toxic in animals or other test organisms.

- *Dose-response assessment* is the process of characterizing the relationship
  between the dose of an agent administered or received and the incidence of an
  adverse health effect. Many different dose-response relationships are possible
  for any given agent depending on such conditions as whether the response is carcinogenic (cancer causing) or noncarcinogenic and whether the experiment is a
  one-time acute test or a long-term chronic test. Since most tests are performed
  with high doses, the dose-response assessment must include a consideration for
  the proper method of extrapolating data to low exposure rates that humans are
  likely to experience. Part of the assessment must also include a method of extrapolating animal data to humans.

- *Exposure assessment* involves determining the size and nature of the population
  that has been exposed to the toxicant under consideration, and the length of time
  and toxicant concentration to which they have been exposed. Consideration
  must be given to such factors as the age and health of the exposed population,
  smoking history, the likelihood that members of the population might be pregnant, and whether or not synergistic effects might occur due to exposure to multiple toxicants.

- *Risk characterization* is the integration of the foregoing three steps, which results
  in an estimate of the magnitude of the public-health problem.

4.5 HAZARD IDENTIFICATION

The first step in a risk analysis is to determine whether or not the chemicals that a population has been exposed to are likely to have any adverse health effects. This is the work of toxicologists, who study both the nature of the adverse effects caused by toxic agents as well as the probability of their occurrence. We shall start our description of this hazard identification process by summarizing the pathways that a chemical may take as it passes through a human body and the kinds of damage that may result. A simple diagram of the human circulatory system is shown in Figure 4.2 that identifies some of the principal organs and the nomenclature for toxic effects.

A toxicant can enter the body using any of three pathways: by ingestion with food or drink, through inhalation, or by contact with the skin (dermal) or other exterior surfaces, such as the eyes. Once in the body it can be absorbed by the blood and distributed to various organs and systems. The toxicant may then be stored (for example, in fat, as in the case of DDT), or it may be eliminated from the body by excretion

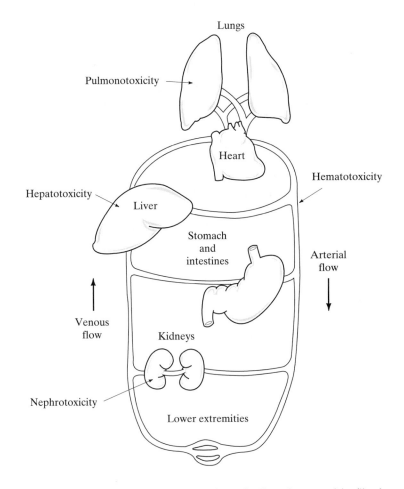

FIGURE 4.2 The circulatory system and nomenclature for toxic effects: hepatotoxicity (liver), nephrotoxicity (kidneys), pulmonotoxicity (lungs), hematotoxicity (blood). (*Source:* Based on James, 1985)

or by transformation into something else. The biotransformation process usually yields metabolites that can be more readily eliminated from the body than the original chemicals; however, metabolism can also convert chemicals to more toxic forms. Figure 4.3 presents the most important movements of chemical toxicants in the body, showing absorption, distribution, storage, and excretion. Although these are shown as separate operations, they all occur simultaneously.

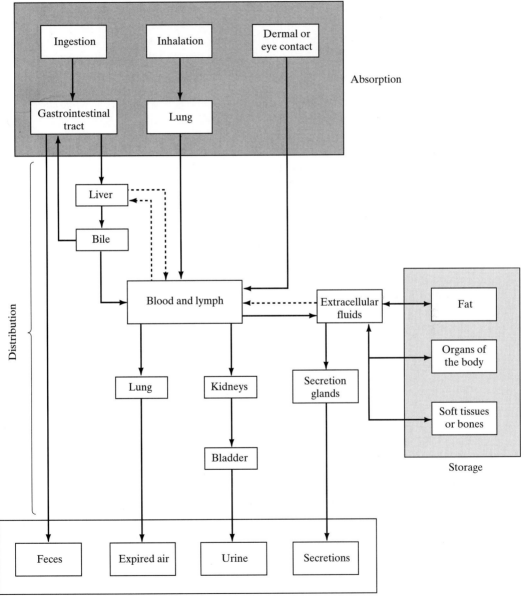

FIGURE 4.3   Fate of chemical toxicants in the body. (*Source:* Environ, 1988)

There are several organs that are especially vulnerable to toxicants. The liver, for example, which filters the blood before it is pumped through the lungs, is often the target. Since toxics are transported by the bloodstream, and since the liver is exposed to so much of the blood supply, it can be directly damaged by toxics. Moreover, since a major function of the liver is to metabolize substances, converting them into forms that can be excreted more easily from the body, it is also susceptible to chemical attack by toxic chemicals formed during the biotransformation process itself. Chemicals that can cause liver damage are called *hepatotoxins.* Examples of hepatotoxic agents include a number of synthetic organic compounds, such as carbon tetrachloride (CCl_4), chloroform ($CHCl_3$), and trichloroethylene (C_2HCl_3); pesticides, such as DDT and paraquat; heavy metals, such as arsenic, iron, and manganese; and drugs, such as acetaminophen and anabolic steroids. The kidneys also filter the blood, and they too are frequently susceptible to damage.

Toxic chemicals often injure other organs and organ systems as well. The function of the kidneys is to filter blood to remove wastes that will be excreted in the form of urine. Toxicants that damage the kidneys, called *nephrotoxics,* include metals such as cadmium, mercury, and lead, as well as a number of chlorinated hydrocarbons. Excessive kidney damage can decrease or stop the flow of urine, causing death by poisoning from the body's own waste products. *Hematotoxicity* is the term used to describe the toxic effects of substances on the blood. Some hematotoxins, such as carbon monoxide in polluted air and nitrates in groundwater, affect the ability of blood to transport oxygen to the tissues. Other toxicants, such as benzene, affect the formation of platelets, which are necessary for blood clotting. The lungs and skin, due to their proximity to pollutants, are also often affected by chemical toxicants. Lung function can be impaired by such substances as cigarette smoke, ozone, asbestos, and quartz rock dust. The skin reacts in a variety of ways to chemical toxicants, but the most common and serious environmentally related skin problem is cancer induced by excessive ultraviolet radiation, as will be described in Chapter 8.

Acute Toxicity

This chapter began with a quote by Philippus Aureolus Theophrastus Bombastus von Hohenheim-Paracelsus to the effect that it is the dose that makes the poison. One measure of the toxicity of something is the amount needed to cause some acute response, such as organ injury, coma, or even death. Acute toxicity refers to effects that are caused within a short period of time after a single exposure to the chemical; later we will discuss chronic toxicity effects that take place after prolonged exposure periods.

One way to describe the toxicity of chemicals is by the amount that is required to kill the organism. Table 4.6 shows a conventional toxicity rating scheme that expresses the dose in terms of milligrams of chemical ingested per kilogram of body weight. That is, ingestion of a given amount of toxin will be more dangerous for a small person, such as a child, than a larger adult. Normalizing the dose using body weight is the first step in trying to relate a lethal dose to a laboratory animal to what might be expected in a human. For example, it takes on the order of 20,000 mg of ordinary sucrose per kilogram to kill a rat. Using the rating system in Table 4.6, sucrose would be considered

TABLE 4.6 A conventional rating system for the acute toxicity of chemicals in humans

	Probable lethal oral dose for humans	
Toxicity rating	Dose (mg/kg of body weight)	For average adult
1. Practically nontoxic	more than 15,000	More than 1 quart
2. Slightly toxic	5,000–15,000	1 pint to 1 quart
3. Moderately toxic	500–5,000	1 ounce to 1 pint
4. Very toxic	50–500	1 teaspoon to 1 ounce
5. Extremely toxic	5–50	7 drops to 1 teaspoon
6. Supertoxic	Less than 5	Less than 7 drops

practically nontoxic. If we scale up that dose to a 70-kg human (without any other adjustments), it might take something like 1.4 kg of sucrose (3 pounds) ingested all at one time to be lethal. At the other extreme, the bacteria *Clostridium botulinum* responsible for botulism (food poisoning), is lethal with a single dose of only 0.00001 mg/kg, so it is supertoxic (Rodricks, 1992).

Not every member of an exposed population will react the same way to a toxin, so one way to illustrate the variation is with a dose-response curve that shows the percentage of a population that is affected as a function of the dose received. In the dose-response curves of Figure 4.4, a logarithmic scale for dose is shown, which tends to yield the familiar S-shaped curve. Also notice that the dose is expressed as milligram of chemical ingested per kilogram of body weight. Normalizing with body weight allows the dose to be extrapolated to individuals of different sizes, such as a child versus an adult. Also, it provides a first cut at extrapolating the likely effects on a human when the dose-response curve has been generated using animal tests.

The curves of Figure 4.4 show the response to chemical exposure as a mortality rate. The dose that will kill 50 percent of a population is designated LD_{50}, where LD stands for lethal dose. In Figure 4.4a, the dose-response curves for two chemicals are shown. Chemical A has a lower LD_{50} than Chemical B and it is always more toxic. Figure 4.4b warns us to be aware that just because one chemical has a lower LD_{50} than

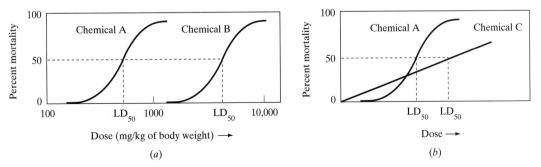

FIGURE 4.4 Dose-response mortality curves for acute toxicity: (a) Chemical A is always more toxic than B; (b) but Chemical A is less toxic than C at low doses even though it has a lower LD_{50}.

another does not necessarily mean it is always more toxic. Chemical A has a lower LD_{50}, which would normally suggest it is more toxic than C, but notice it is not as toxic as C at low doses. So the dose-response curve does provide more information than a simple table of LD_{50} doses.

Mutagenesis

In contrast to the short-term responses associated with acute toxicity, most risk assessments are focused on responses that may take years to develop. Measuring the ability of specific chemicals to cause cancer, reproductive failure, and birth defects is much more difficult than the acute toxicity testing just described.

Deoxyribonucleic acid (DNA) is an essential component of all living things and a basic material in the chromosomes of the cell nucleus. It contains the genetic code that determines the overall character and appearance of every organism. Each molecule of DNA has the ability to replicate itself exactly, transmitting that genetic information to new cells. Our interest here in DNA results from the fact that certain chemical agents, as well as ionizing radiation, are *genotoxic;* that is, they are capable of altering DNA. Such changes, or *mutations,* in the genetic material of an organism can cause cells to misfunction, leading in some cases to cell death, cancer, reproductive failure, or abnormal offspring. Chemicals that are capable of causing cancer are called *carcinogens;* chemicals that can cause birth defects are *teratogens.*

Mutations may affect somatic cells, which are the cells that make up the tissues and organs of the body itself, or they may cause changes in germ cells (sperm or ovum) that may be transmitted to future offspring. As is suggested in Figure 4.5, one possible outcome of a mutagenic event is the death of the cell itself. If the mutation is in a

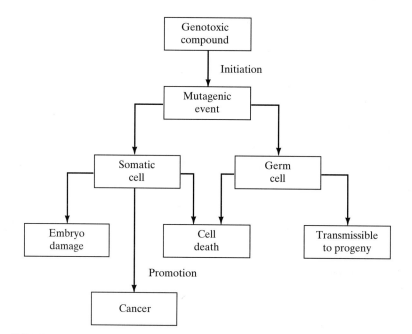

FIGURE 4.5 Possible consequences of a mutagenic event in somatic and germinal cells.

somatic cell and it survives, the change may be such that the cell no longer responds to signals that normally control cell reproduction. If that occurs, the cell may undergo rapid and uncontrolled cellular division, forming a tumor. Mutations in somatic cells may damage or kill the affected individual, and if the individual is a pregnant female, the embryo may be damaged, leading to a birth defect. Germ cell mutations, on the other hand, have the potential to become established in the gene pool and be transmitted to future generations.

Carcinogenesis

Cancer is second only to heart disease in terms of the number of Americans killed every year. Every year close to 1 million people are diagnosed with cancer, and over one-half million die each year. Cancer is truly one of the most dreaded diseases.

Chemically induced carcinogenesis is thought to involve two distinct stages, referred to as *initiation* and *promotion*. In the initiation stage, a mutation alters a cell's genetic material in a way that may or may not result in the uncontrolled growth of cells that characterizes cancer. In the second, or promotion, stage of development, affected cells no longer recognize growth constraints that normally apply and a tumor develops. Promoters can increase the incidence rate of tumors among cells that have already undergone initiation, or they can shorten the latency period between initiation and the full carcinogenic response. The model of initiation followed by promotion suggests that some carcinogens may be initiators, others may be promoters, and some may be complete carcinogens capable of causing both stages to occur. Current regulations do not make this distinction, however, and any substance capable of increasing the incidence of tumors is considered a carcinogen, subject to the same risk assessment techniques. Tumors, in turn, may be *benign* or *malignant* depending on whether or not the tumor is contained within its own boundaries. If a tumor undergoes *metastasis*—that is, it breaks apart and portions of it enter other areas of the body—it is said to be malignant. Once a tumor has metastasized, it is obviously much harder to treat or remove.

The theoretical possibility that a single genotoxic event can lead to a tumor is referred to as the *one-hit hypothesis*. Based on this hypothesis, exposure to even the smallest amount of a carcinogen leads to some nonzero probability that a malignancy will result. That is, in a conservative, worst-case risk assessment for carcinogens, it is assumed that there is no threshold dose below which the risk is zero.

A brief glossary of carcinogenesis terminology is presented in Table 4.7.

Toxicity Testing in Animals

With several thousand new chemicals coming onto the market each year, a backlog of tens of thousands of relatively untested chemicals already in commerce, and a limited number of facilities capable of providing the complex testing that might be desired, it is just not possible to test each and every chemical for its toxicity. As a result, a hierarchy of testing procedures has been developed that can be used to help select those chemicals that are most likely to pose serious risks.

The starting point is the relatively straightforward acute toxicity testing already described. The next step may be to compare the structure of the chemical in question with other chemicals that are known or suspected to be human carcinogens, such as

TABLE 4.7 Glossary of Carcinogenesis Terminology

Acute toxicity	Adverse effects caused by a toxic agent occurring within a short period of time following exposure
Benign tumor	A new tumor composed of cells that, though proliferating in an abnormal manner, do not spread to surrounding, normal tissue
Cancer	An abnormal process in which cells begin a phase of uncontrolled growth and spread
Carcinogen	Any cancer-producing substance
Carcinoma	A malignant tumor in the tissue that covers internal or external surfaces of the body such as the stomach, liver, or skin
Chronic toxicity	Adverse effects caused by a toxic agent after a long period of exposure
Initiator	A chemical that initiates the change in a cell that irreversibly converts the cell into a cancerous or precancerous state
Malignant tumor	Relatively autonomous growth of cells or tissue that invade surrounding tissue and have the ability to metastasize
Mutagenesis	Alteration of DNA in either somatic or germinal cells not associated with the normal process of recombination
Mutation	A permanent, transmissible change in DNA that changes the function or behavior of the cell
Neoplasm	Literally, new growth, usually of an abnormally fast-growing tissue
Oncogenic	Giving rise to tumors or causing tumor formation
Pharmacokinetics	The study of how a chemical is absorbed, distributed, metabolized, and excreted
Promoter	A chemical that can increase the incidence of response to a carcinogen previously administered
Sarcoma	A cancer that arises from mesodermal tissue (e.g., fat, muscle, bone)
Teratogen	Any substance capable of causing malformation during development of the fetus
Toxicity	A relative term generally used in comparing the harmful effect of one chemical on some biological mechanism with the effect of another chemical

Source: Based on Williams and Burson, 1985.

those shown in Figure 4.6. New chemicals that are similar to these, and other suspected carcinogens, would be potential candidates for further testing.

The prevailing carcinogenesis theory, that human cancers are initiated by gene mutations, has led to the development of short-term, in vitro (in glassware) screening procedures, which are one of the first steps taken to determine whether a chemical is carcinogenic. It is thought that if a chemical can be shown to be mutagenic, then it *may* be carcinogenic, and further testing may be called for. The most widely used short-term test, called the *Ames mutagenicity assay,* subjects special tester strains of bacteria to the chemical in question. These tester strains have previously been rendered incapable of normal bacterial division so, unless they mutate back to a form that is capable of division, they will die. Bacteria that survive and form colonies do so through mutation; therefore, the greater the survival rate of these special bacteria, the more mutagenic is the chemical.

Intermediate testing procedures involve relatively short-term (several months duration) carcinogenesis bioassays in which specific organs in mice and rats are subjected to known mutagens to determine whether tumors develop.

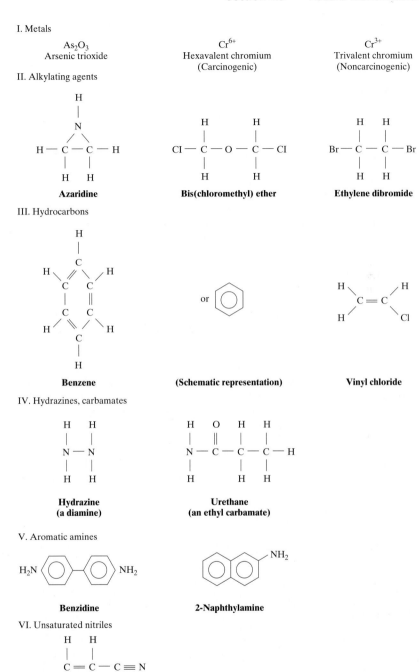

I. Metals

As_2O_3
Arsenic trioxide

Cr^{6+}
Hexavalent chromium
(Carcinogenic)

Cr^{3+}
Trivalent chromium
(Noncarcinogenic)

II. Alkylating agents

Azaridine

Bis(chloromethyl) ether

Ethylene dibromide

III. Hydrocarbons

Benzene

or

(Schematic representation)

Vinyl chloride

IV. Hydrazines, carbamates

Hydrazine
(a diamine)

Urethane
(an ethyl carbamate)

V. Aromatic amines

Benzidine

2-Naphthylamine

VI. Unsaturated nitriles

Acrylonitrile

FIGURE 4.6   Selected structural formulae for some classes of carcinogenic chemicals. (Williams and Burson, 1985)

Finally, the most costly, complex, and long-lasting test, called a *chronic carcinogenesis bioassay,* involves hundreds or thousands of animals over a time period of several years. To assure comparable test results and verifiable data, the National Toxicology Program in the United States has established minimum test requirements for an acceptable chronic bioassay, which include the following:

- *Two species of rodents must be tested.* Mice and rats, using specially inbred strains for consistency, are most often used. They have relatively short lifetimes, and their small size makes them easier to test in large numbers.
- *At least 50 males and 50 females of each species for each dose must be tested.* Many more animals are required if the test is to be sensitive enough to detect risks of less than a few percent.
- *At least two doses must be administered (plus a no-dose control).* One dose is traditionally set at the maximum tolerated dose (MTD), a level that can be administered for a major portion of an animal's lifetime without significantly impairing growth or shortening the lifetime. The second dose is usually one-half or one-fourth the MTD.

Exposure begins at 6 weeks of age and ends when the animal reaches 24 months of age. At the end of the test, all animals are killed and their remains are subjected to detailed pathological examinations. These tests are expensive as well as time consuming. Testing a typical new chemical costs between $500,000 and $1.5 million, takes up to two or three years, and may entail the sacrifice of thousands of animals (Goldberg and Frazier, 1989).

Notice that, following the aforementioned protocol, the minimum number of animals required for a bioassay is 600 (2 species $\times$ 100 animals $\times$ 3 doses), and at that number it is still only relatively high risks that can be detected. With this number of animals, for the test to show a statistically significant effect, the exposed animals must have at least 5 or 10 percent more tumors than the controls in order to conclude that the extra tumors were caused by the chemical being tested. That is, the risk associated with this chemical can be measured only down to roughly 0.05 or 0.10 unless we test a lot more animals.

A simple example may help to clarify this statistical phenomenon. Suppose we test 100 rats at a particular dose and find one tumor. To keep it easy, let's say the control group never gets tumors. Can the actual probability (risk) of tumors caused by this chemical at this dose be 1 percent? Yes, definitely. If the risk is 1 percent we would expect to get one tumor, and that is what we got. Could the actual probability be 2 percent? Well, if the actual risk is 2 percent, and *if we were able to run the test over and over again* on sets of 100 rats each, some of those groups would have no tumors, some would certainly have one tumor, and some would have more. So our actual test of only one group of 100, which found one tumor, is not at all inconsistent with an actual risk of 2 percent. Could the actual risk be 3 percent? Running many sets of 100 rats through the test would likely result in at least one of those groups having only one tumor. So it would not be out of the question to find one tumor in a single group of 100 rats even if the actual risk is 3 percent. Getting back to the original test of 100 rats and finding one tumor, we have just argued that the actual risk could be anything from 0 percent to, say,

2 or 3 percent, maybe even more, and still be consistent with finding just one tumor. We certainly cannot conclude that the risk is only 1 percent. In other words, with 100 animals we cannot perform a statistically significant test and be justified in concluding that the risk is anything less than a few percent. Bioassays designed to detect lower risks require many thousands of animals. In fact, the largest experiment ever performed involved over 24,000 mice and yet was still insufficiently sensitive to measure a risk of less than 1 percent (Environ, 1988).

The inability of a bioassay to detect small risks presents one of the greatest difficulties in applying the data so obtained to human risk assessment. Regulators try to restrict human risks due to exposure to carcinogens to levels of about 10^{-6} (one in a million), yet animal studies are only capable of detecting risks of down to 0.01 to 0.1. It is necessary, therefore, to find some way to extrapolate the data taken for animals exposed to high doses to humans who will be exposed to doses that are several orders of magnitude lower.

Human Studies

Another shortcoming in the animal testing methods just described, besides the necessity to extrapolate the data toward zero risk, is the obvious difficulty in interpreting the data for humans. How does the fact that some substance causes tumors in mice relate to the likelihood that it will cause cancer in humans as well? Animal testing can always be criticized in this way, but since we are not inclined to perform the same sorts of tests directly on humans, other methods must be used to gather evidence of human toxicity.

Sometimes human data can be obtained by studying victims of tragedies, such as the chemical plant explosion that killed and injured thousands in Bhopal, India, and the atomic bombing of Hiroshima and Nagasaki, Japan. The most important source of human risk information, however, comes from epidemiologic studies. Epidemiology is the study of the incidence rate of diseases in real populations. By attempting to find correlations between disease rates and various environmental factors, an epidemiologist attempts to show in a quantitative way the relationship between exposure and risk. Such data can be used to complement animal data, clinical data, and scientific analyses of the characteristics of the substances in question.

Epidemiologists have a number of strategies for gathering useful information, but they share the common feature of trying to identify two populations of people having different exposures to the risk factor being studied. Preliminary data analysis usually involves setting up a simple 2×2 matrix such as the one shown in Figure 4.7. The rows divide the populations according to those who have, and those who have not,

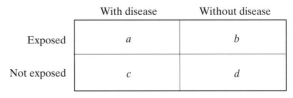

FIGURE 4.7   A 2×2 matrix for an epidemiologic rate comparison. Rows divide people by exposure; columns divide them by disease.

been exposed to the risk factor. The columns are based on the numbers of individuals who have acquired the disease being studied and those who have not.

Various measures can be applied to the data given in Figure 4.7 to see whether they *suggest* an association between exposure and disease.

- The *relative risk* is defined as

$$\text{Relative risk} = \frac{a/(a + b)}{c/(c + d)} \qquad (4.1)$$

Notice that the numerator is the fraction of those exposed who have the disease, and the denominator is the fraction of those exposed who do not have the disease. If those two ratios are the same, the odds of having the disease would not depend on whether an individual had been exposed to the risk factor, and the relative risk would be 1.0. Above 1.0, the higher the relative risk the more the data suggests an association between exposure and risk.

- The *attributable risk* is defined as

$$\text{Attributable risk} = \frac{a}{a + b} - \frac{c}{c + d} \qquad (4.2)$$

The attributable risk is the difference between the odds of having the disease with exposure and the odds of having the disease without exposure. An attributable risk of 0.0 suggests no relationship between exposure and risk.

- The *odds ratio* is defined as the cross product of the entries in the matrix:

$$\text{Odds ratio} = \frac{ad}{bc} \qquad (4.3)$$

The odds ratio is similar to the relative risk. Numbers above 1.0 suggest a relationship between exposure and risk.

EXAMPLE 4.1   Epidemiologic Data Analysis

An evaluation of personnel records for employees of a plant that manufactures vinyl chloride finds that out of 200 workers, 15 developed liver cancer. A control group consisting of individuals with smoking histories similar to the exposed workers, and who were unlikely to have encountered vinyl chloride, had 24 with liver cancers and 450 who did not develop liver cancer. Find the relative risk, attributable risk, and odds ratio for these data.

Solution   Putting the data into the 2 x 2 matrix gives

	D	$\overline{D}$
E	15	185
$\overline{E}$	24	450

Solving for each measure yields:

$$\text{Relative risk} = \frac{15/(15 + 185)}{24/(24 + 450)} = \frac{.075}{0.05} = 1.48$$

$$\text{Attributable risk} = \frac{15}{200} - \frac{24}{474} = 0.024$$

$$\text{Odds ratio} = \frac{15 \times 450}{185 \times 24} = 1.52$$

The relative risk and the odds ratio both are above 1.0, so they suggest a relationship between exposure and risk. For those who were exposed, the risk of cancer has increased by 0.024 (the attributable risk) over that of their cohorts who were not exposed. All three measures indicate that further study of the relationship between vinyl chloride and liver cancer would be warranted. ∎

Caution must be exercised in interpreting every epidemiologic study since any number of confounding variables may lead to invalid conclusions. For example, the study may be biased because workers are compared with nonworkers (workers are generally healthier), or because relative rates of smoking have not been accounted for, or there may be other variables that are not even hypothesized in the study that may be the actual causal agent. As an example of the latter, consider an attempt to compare lung cancer rates in a city having high ambient air pollution levels with rates in a city having less pollution. Suppose the rates are higher in the more polluted city, even after accounting for smoking history, age distribution, and working background. To conclude that ambient air pollution is causing those differences may be totally invalid. Instead, it might well be different levels of radon in homes, or differences in other indoor air pollutants associated with the type of fuel used for cooking and heating, that are causing the cancer variations.

Weight-of-Evidence Categories for Potential Carcinogens

Based on the accumulated evidence from clinical studies, epidemiologic evidence, in vitro studies, and animal data, the EPA uses the following categories to describe the likelihood that a chemical substance is carcinogenic (U.S. EPA, 1986a). Using both human and animal data, five categories, A through E, have been established as follows:

Group A: Human carcinogen. A substance is put into this category only when there is sufficient epidemiologic evidence to support a causal association between exposure to the agent and cancer.

Group B: Probable human carcinogen. This group is actually made up of two subgroups. An agent is categorized as B1 if there is limited epidemiologic evidence; and it is put into B2 if there is inadequate human data but sufficient evidence of carcinogenicity in animals.

Group C: Possible human carcinogen. This group is used for agents with limited evidence of carcinogenicity in animals and an absence of human data.

Group D: Not classified. This group is for agents with inadequate human and animal evidence or for which no data are available.

Group E: Evidence of noncarcinogenicity. This group is used for agents that show no evidence for carcinogenicity in at least two adequate animal tests in different species or in both adequate epidemiologic and animal studies.

Table 4.8 summarizes this categorization scheme.

4.6 DOSE-RESPONSE ASSESSMENT

As the name suggests, the fundamental goal of a dose-response assessment is to obtain a mathematical relationship between the amount of a toxicant that a human is exposed to and the risk that there will be an unhealthy response to that dose. We have seen dose-response curves for acute toxicity, in which the dose is measured in milligram per kilogram of body weight. The dose-response curves that we are interested in here are the result of chronic toxicity; that is, the organism is subjected to a prolonged exposure over a considerable fraction of its life. For these curves the abscissa is dose, which is usually expressed as the average milligrams of substance per kilogram of body weight per day (mg/kg-day). The dose is an exposure averaged over an entire lifetime (for humans, assumed to be 70 years). The ordinate is the response, which is the risk that there will be some adverse health effect. As usual, response (risk) has no units; it is a probability that there will be some adverse health effect. For example, if prolonged exposure to some chemical would be expected to produce 700 cancers in a population of 1 million, the response could be expressed as 0.0007, 7×10^{-4}, or 0.07 percent. The annual risk would be obtained by spreading that risk over an assumed 70-year lifetime, giving a risk of 0.00001 or 1×10^{-5} per year.

For substances that induce a carcinogenic response, it is always conventional practice to assume that exposure to any amount of the carcinogen will create some likelihood of cancer. That is, a plot of response versus dose is required to go through the origin. For noncarcinogenic responses, it is usually assumed that there is some

---

TABLE 4.8   Weight-of-Evidence Categories for Human Carcinogenicity

Human Evidence	Animal Evidence				
	Sufficient	Limited	Inadequate	No Data	No Evidence
Sufficient	A	A	A	A	A
Limited	B1	B1	B1	B1	B1
Inadequate	B2	C	D	D	D
No data	B2	C	D	D	E
No evidence	B2	C	D	D	E

Source: USEPA (1986a).

threshold dose, below which there will be no response. As a result of these two assumptions, the dose-response curves and the methods used to apply them are quite different for carcinogenic and noncarcinogenic effects, as suggested in Figure 4.8. The same chemical, by the way, may be capable of causing both kinds of response.

To apply dose-response data obtained from animal bioassays to humans, a *scaling factor* must be introduced. Sometimes the scaling factor is based on the assumption that doses are equivalent if the dose per unit of body weight in the animal and human is the same. Sometimes, especially if the exposure is dermal, equivalent doses are normalized to body surface area rather than body weight when scaling up from animal to human. In either case, the resulting human dose-response curve is specified with the standard mg/kg-day units for dose. Adjustments between animal response and human response may also have to be made to account for differences in the rates of chemical absorption. If enough is known about the differences between the absorption rates in test animals and in humans for the particular substance in question, it is possible to account for those differences later in the risk assessment. Usually, though, there is insufficient data and it is simply assumed that the absorption rates are the same.

Extrapolations from High Doses to Low Doses

The most controversial aspect of dose-response curves for carcinogens is the method chosen to extrapolate from the high doses actually administered to test animals to the low doses to which humans are likely to be exposed. Recall that even with extremely large numbers of animals in a bioassay, the lowest risks that can be measured are usually a few percent. Since regulators attempt to control human risk to several orders of magnitude less than that, there will be no actual animal data anywhere near the range of most interest.

Many mathematical models have been proposed for the extrapolation to low doses. Unfortunately, no model can be proved or disproved from the data, so there is no way to know which model is the most accurate. That means the choice of models is strictly a policy decision. One commonly used model is called the *one-hit model,* in

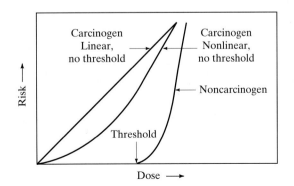

FIGURE 4.8   Dose-response curves for carcinogens are assumed to have no threshold; that is, any exposure produces some chance of causing cancer.

which the relationship between dose (d) and lifetime risk (probability) of cancer, $P(d)$, is given in the following form (Crump, 1984):

$$P(d) = 1 - e^{-(q_0 + q_1 d)} \tag{4.4}$$

where q_0 and q_1 are parameters picked to fit the data. The one-hit model corresponds to the simplest mechanistic model of carcinogenesis, in which it is assumed that a single chemical hit is capable of inducing a tumor.

If we substitute $d = 0$ into (4.1), the result will be an expression for the background rate of cancer incidence, $P(0)$. Using the mathematical expansion for an exponential

$$e^x = 1 + x + \frac{x^2}{2!} + \cdots + \frac{x^n}{n!} \cong 1 + x \quad \text{(for small } x) \tag{4.5}$$

and assuming that the background cancer rate is small allows us to write

$$P(0) = 1 - e^{-q_0} \cong 1 - [1 + (-q_0)] = q_0 \tag{4.6}$$

That is, the background rate for cancer incidence corresponds to the parameter q_0. Using the exponential expansion again, the one-hit model suggests that the lifetime probability of cancer for small dose rates can be expressed as

$$P(d) \cong 1 - [1 - (q_0 + q_1 d)] = q_0 + q_1 d = P(0) + q_1 d \tag{4.7}$$

For low doses, the additional risk of cancer above the background rate would be

$$\text{Additional risk} = A(d) = P(d) - P(0) \tag{4.8}$$

Substituting (4.7) into (4.8) yields the following equation for the additional cancer risk incurred when the organism in question is exposed to a dose d:

$$\text{Additional risk} = A(d) \cong q_1 d \tag{4.9}$$

That is, the one-hit model predicts that for low doses the extra lifetime probability of cancer is linearly related to dose.

The one-hit model relating risk to dose is not the only one possible. Another mathematical model that has been proposed has its roots in the multistage model of tumor formation; that is, that tumors are the result of a sequence of biological events (Crump, 1984). The *multistage* model expresses the relationship between risk and dose as

$$P(d) = 1 - e^{-(q_0 + q_1 d + q_2 d^2 + \cdots q_n d^n)} \tag{4.10}$$

where the individual parameters q_i are positive constants picked to best fit the dose-response data. Again, it is easy to show that for small values of dose d, the multistage model also has the simplifying feature of producing a linear relationship between additional risk and dose. Figure 4.9 illustrates the use of a one-hit model and a multistage model to fit experimental data. The multistage model will always fit the data better since it includes the one-hit model as a special case.

Since the choice of an appropriate low-dose model is not based on experimental data, there is no model that can be proved to be more correct than another. To protect public health, EPA chooses to err on the side of safety and overemphasize risk. The EPA's model of choice is a modified multistage model, called the *linearized multistage model*. It is linear at low doses with the constant of proportionality picked in a way that the probability of overestimating the risk is 95 percent.

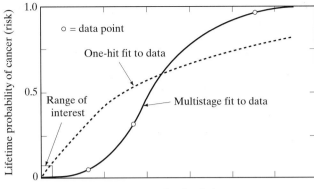

FIGURE 4.9   Dose-response curves showing two methods of fitting an equation to the data. The range of interest is well below the point where any data actually exist. (Based on Crump, 1984)

Potency Factor for Carcinogens

For chronic toxicity studies, a low dose is administered over a significant portion of the animal's lifetime. The resulting dose-response curve has the incremental risk of cancer (above the background rate) on the y-axis, and the lifetime average daily dose of toxicant along the x-axis. At low doses, where the dose-response curve is assumed to be linear, the slope of the dose-response curve is called the *potency factor* (PF), or *slope factor*.

$$\text{Potency factor} = \frac{\text{Incremental lifetime cancer risk}}{\text{Chronic daily intake (mg/kg-day)}} \qquad (4.11)$$

The denominator in (4.11) is the dose averaged over an entire lifetime; it has units of average milligrams of toxicant absorbed per kilogram of body weight per day, which is usually written as (mg/kg-day) or (mg/kg/day). Since risk has no units, the units for potency factor are therefore $(\text{mg/kg-day})^{-1}$.

 If we have a dose-response curve, we can find the potency factor from the slope. In fact, one interpretation of the potency factor is that it is the risk produced by a chronic daily intake of 1 mg/kg-day, as shown in Figure 4.10.

 Rearranging (4.11) shows us where we are headed. If we know the chronic daily intake CDI (based on exposure data) and the potency factor (from EPA), we can find the lifetime, incremental cancer risk from

$$\text{Incremental lifetime cancer risk} = \text{CDI} \times \text{Potency factor} \qquad (4.12)$$

The linearized multistage risk-response model assumptions built into (4.12) should make this value an upper-bound estimate of the actual risk. Moreover, (4.12) estimates the risk of getting cancer, which is not necessarily the same as the risk of dying of cancer, so it should be even more conservative as an upper-bound estimate of cancer death rates.

 Potency factors needed for (4.12) can be found in an EPA database on toxic substances called the Integrated Risk Information System (IRIS). Included in the rather extensive background information on each potential carcinogen in IRIS is the potency factor and the weight-of-evidence category (recall Table 4.8). A short list of some of

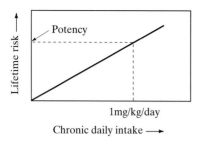

FIGURE 4.10 The potency factor is the slope of the dose-response curve. It can also be thought of as the risk that corresponds to a chronic daily intake of 1 mg/kg-day.

these chemicals, potency factors (for both oral and inhalation exposure routes), and cancer categories is given in Table 4.9.

The other factor we need to develop more fully in order to use the basic risk equation (4.12) is the concept of chronic daily intake. The CDI is, by definition,

$$\text{CDI (mg/kg-day)} = \frac{\text{Average daily dose (mg/day)}}{\text{Body weight (kg)}} \qquad (4.13)$$

The numerator in (4.13) is the total lifetime dose averaged over an assumed 70-year lifetime. The next example shows how to combine CDI and potency to find risk.

TABLE 4.9 Toxicity data for selected potential carcinogens

Chemical	Category	Potency factor oral route $(\text{mg/kg-day})^{-1}$	Potency factor inhalation route $(\text{mg/kg-day})^{-1}$
Arsenic	A	1.75	50
Benzene	A	2.9×10^{-2}	2.9×10^{-2}
Benzol(a)pyrene	B2	11.5	6.11
Cadmium	B1	—	6.1
Carbon tetrachloride	B2	0.13	—
Chloroform	B2	6.1×10^{-3}	8.1×10^{-2}
Chromium VI	A	—	41
DDT	B2	0.34	—
1,1-Dichloroethylene	C	0.58	1.16
Dieldrin	B2	30	—
Heptachlor	B2	3.4	—
Hexachloroethane	C	1.4×10^{-2}	—
Methylene chloride	B2	7.5×10^{-3}	1.4×10^{-2}
Nickel and compounds	A	—	1.19
Polychlorinated biphenyls (PCBs)	B2	7.7	—
2,3,7,8-TCDD (dioxin)	B2	1.56×10^{5}	—
Tetrachloroethylene	B2	5.1×10^{-2}	$1.0 - 3.3 \times 10^{-3}$
1,1,1-Trichloroethane (1,1,1-TCA)	D	—	—
Trichloroethylene (TCE)	B2	1.1×10^{-2}	1.3×10^{-2}
Vinyl chloride	A	2.3	0.295

Source: U.S. EPA http://www.epa.gov/iris.

EXAMPLE 4.2   Risk Assessment for Chloroform in Drinking Water

When drinking water is disinfected with chlorine an undesired byproduct, chloroform ($CHCl_3$), is formed. Suppose a 70-kg person drinks 2 L of water every day for 70 years with a chloroform concentration of 0.10 mg/L (the drinking water standard).

a. Find the upper-bound cancer risk for this individual.

b. If a city with 500,000 people in it also drinks the same amount of this water, how many extra cancers per year would be expected? Assume the standard 70-year lifetime.

c. Compare the extra cancers per year caused by chloroform in the drinking water with the expected number of cancer deaths from all causes. The cancer death rate in the United States is 193 per 100,000 per year.

Solution

a. From Table 4.9 we see that chloroform is a Class B2 probable human carcinogen with a potency factor of 6.1×10^{-3} (mg/kg-day)$^{-1}$. Using (4.13), the chronic daily intake is

$$\text{CDI (mg/kg-day)} = \frac{\text{Average daily dose (mg/day)}}{\text{Body weight (kg)}}$$

$$= \frac{0.10 \text{ mg/L} \times 2 \text{ L/day}}{70 \text{ kg}} = 0.00286 \text{ mg/kg-day}$$

From (4.12), the incremental lifetime cancer risk is

$$\text{Risk} = \text{CDI} \times \text{Potency factor}$$

$$= 0.00286 \text{ (mg/kg-day)} \times 6.1 \times 10^{-3} \text{ (mg/kg-day)}^{-1} = 17.4 \times 10^{-6}$$

So over a 70-year period the upper-bound estimate of the probability that a person will get cancer from this drinking water is about 17 in one million.

b. If there are 17.4 cancers per million people over a 70-year period, then in any given year in a population of one-half million, the number of cancers caused by chloroform would be

$$500{,}000 \text{ people} \times \frac{17.4 \text{ cancer}}{10^6 \text{ people}} \times \frac{1}{70 \text{ yr}} = 0.12 \text{ cancers/yr}$$

c. The total number of cancer deaths that would be expected in a city of 500,000 would be

$$500{,}000 \text{ people} \times \frac{193 \text{ cancer/yr}}{100{,}000 \text{ people}} = 965 \text{ cancer deaths/yr}$$

It would seem that an additional 0.12 new cancers per year would not be detectable. ■

Once again, it is necessary to emphasize that the science behind a risk assessment calculation of the sort demonstrated in Example 4.2 is primitive, and there are enormous uncertainties associated with any particular answer so computed. There is still great value, however, to this sort of procedure since it does organize a mass of data into a format that can be communicated to a much wider audience, and it can greatly help that audience find legitimate perspectives based on that data. For example, it matters little whether the annual extra cancers associated with chloroform in the preceding example are found to be 0.12 or even ten times as much, 1.2; the conclusion that the extra cancers would be undetectable would not change.

Another use for these risk calculations is to estimate the concentration of a contaminant in drinking water that would result in a politically acceptable risk level. Often that risk goal is 10^{-6} and the concentration that will produce that risk is called the *drinking water equivalent level* (DWEL). To find the DWEL, it is usually assumed that a 70-kg adult consumes 2 L of water per day. As Example 4.3 shows, we can find the DWEL from the potency factor using (4.12).

EXAMPLE 4.3   Drinking water concentration of chloroform for a 10^{-6} risk

Find the concentration of chloroform in drinking water that would result in a 10^{-6} risk for a 70-kg person who drinks 2 L/day throughout his or her entire lifetime.

Solution   Rearranging (4.12) and using the potency factor from Table 4.9 gives

$$CDI = \frac{Risk}{Potency\ factor} = \frac{10^{-6}}{6.1 \times 10^{-3}\ (kg\text{-}day/mg)} = 1.64 \times 10^{-4}\ (mg/kg\text{-}day)$$

Since CDI is just the average daily intake divided by body mass, we can write

$$CDI = \frac{C(mg/L) \times 2\ L/day}{70\ kg} = 1.64 \times 10^{-4}\ (mg/kg\text{-}day)$$

where C (mg/L) is the allowable concentration of chloroform. Solving for C gives

$$C = 70 \times 1.64 \times 10^{-4}/2 = 0.0057 \cong 6 \times 10^{-3}\ mg/L = 6\mu g/L$$

So a DWEL of 6 μg/L for chloroform would result in an upper-bound risk of 10^{-6}.   ∎

In Examples 4.2 and 4.3 it was assumed that everyone drinks 2 L of contaminated water every day for 70 years. When a risk assessment is made for exposures that do not last the entire lifetime, we need to develop the chronic daily intake a little more carefully.

If the contaminant is in drinking water, the CDI can be expressed as

$$CDI = \frac{Concentration\ (mg/L) \times Intake\ rate\ (L/day) \times Exposure\ (days/life)}{Body\ weight\ (kg) \times 70\ (yr/life) \times 365\ (days/yr)} \quad (4.14)$$

where *Concentration* refers to the contaminant concentration, *Intake rate* is the amount of water ingested each day, and *Exposure* is the number of days in a lifetime that the person drinks contaminated water.

If the exposure route is inhalation of a contaminant, the chronic daily intake can be expressed as

$$CDI = \frac{Concentration\ (mg/m^3) \times Intake\ rate\ (m^3/day) \times Exposure\ (days/life)}{Body\ weight\ (kg) \times 70\ (yr/life) \times 365\ (days/yr)} \quad (4.15)$$

where *Concentration* is the contaminant concentration in air, and the *Intake rate* is the amount of air inhaled during each day that the person is exposed to the contamination. Similar expressions can be used for consumption of contaminated food or soil and for dermal contact with contaminated soil. For some of these circumstances, the CDI needs to include an absorption factor if it is thought that the absorption rate by a human is different from the absorption rate of the test animals that were used to establish the dose-response curve.

EXAMPLE 4.4   An Occupational Exposure

Estimate the incremental cancer risk for a 60-kg worker exposed to a particular carcinogen under the following circumstances. Exposure time is 5 days per week, 50 weeks per year, over a 25-year period of time. The worker is assumed to breathe 20 m^3 of air per day. The carcinogen has a potency factor of 0.02 $(mg/kg\text{-}day)^{-1}$ and its average concentration is 0.05 mg/m^3.

Solution   Since this is an inhalation exposure, we will use (4.15).

$$CDI = \frac{0.05 \text{ mg/m}^3 \times 20 \text{ m}^3/\text{day} \times 5 \text{ days/wk} \times 50 \text{ wk/yr} \times 25 \text{ yr}}{60 \text{ kg} \times 70 \text{ yr/life} \times 365 \text{ days/yr}}$$

$$= 0.0041 \text{ mg/kg-day}$$

Using (4.12), the upper-bound, incremental cancer risk is CDI × potency:

$$\text{Incremental risk} = 0.0041 \text{ mg/kg-day} \times 0.02 \text{ (mg/kg-day)}^{-1} = 81 \times 10^{-6}$$

which is considerably higher than the usual goal of 10^{-6} risk.   ■

The EPA has developed a set of recommended default values for daily intakes, exposures, and body weights to be used in risk calculations when more site-specific information is not available. Table 4.10 shows some of these default factors, and the next example illustrates their use.

EXAMPLE 4.5   A Proposed Source of Benzene in Your Neighborhood

Suppose an industrial facility that emits benzene into the atmosphere is being proposed for a site near a residential neighborhood. Air quality models predict that 60 percent of the time, prevailing winds will blow the benzene away from the neighborhood, but 40 percent of the time the benzene concentration will be 0.01 mg/m^3. Use standard exposure factors from Table 4.10 to assess the incremental risk to adults in the neighborhood if the facility is allowed to be built. If the acceptable risk is 10^{-6}, should this plant be allowed to be built?

Solution   Using factors from Table 4.10, the chronic daily intake will be

$$CDI = \frac{0.01 \text{ mg/m}^3 \times 20 \text{ m}^3/\text{day} \times 350 \text{ day/yr} \times 30 \text{ yr}}{70 \text{ kg} \times 365 \text{ day/yr} \times 70 \text{ yr}} \times 0.40$$

$$= 0.00047 \text{ mg/kg-day}$$

The potency factor from Table 4.9 for benzene is 2.9×10^{-2} $(mg/kg\text{-}day)^{-1}$, so the incremental risk would be

$$\text{Incremental risk} = 0.00047 \text{ mg/kg-day} \times 2.9 \times 10^{-2} \text{ (mg/kg-day)}^{-1} = 1.3 \times 10^{-5}$$

The risk is higher than the acceptable level, so the facility should not be built as it is being proposed.   ■

TABLE 4.10   Example EPA Exposure Factors Recommended for Risk Assessments

Land use	Exposure pathway	Daily intake	Exposure frequency, days/year	Exposure duration, years	Body weight, kg
Residential	Ingestion of potable water	2 L (adult) 1 L (child)	350	30	70 (adult) 15 (child)
	Ingestion of soil and dust	200 mg (child) 100 mg (adult)	350	6 24	15 (child) 70 (adult)
	Inhalation of contaminants	20 m^3 (adult) 12 m^3 (child)	350	30	70
Industrial and commercial	Ingestion of potable water	1 L	250	25	70
	Ingestion of soil and dust	50 mg	250	25	70
	Inhalation of contaminants	20 m^3 (workday)	250	25	70
Agricultural	Consumption of homegrown produce	42 g (fruit) 80 g (veg.)	350	30	70
Recreational	Consumption of locally caught fish	54 g	350	30	70

Source: U.S. EPA (1991).

The Reference Dose for Noncarcinogenic Effects

The key assumption for noncarcinogens is that there is an exposure threshold; that is, any exposure less than the threshold would be expected to show no increase in adverse effects above natural background rates. One of the principal goals of toxicant testing is therefore to identify and quantify such thresholds. Unfortunately, for the usual case, inadequate data are available to establish such thresholds with any degree of certainty and, as a result, it has been necessary to introduce a number of special assumptions and definitions.

Suppose there exists a precise threshold for some particular toxicant for some particular animal species. To determine the threshold experimentally, we might imagine a testing program in which animals would be exposed to a range of doses. Doses below the threshold would elicit no response; doses above the threshold would produce responses. The lowest dose administered that results in a response is given a special name: the *lowest-observed-effect level* (LOEL). Conversely, the highest dose administered that does not create a response is called the *no-observed-effect level* (NOEL). NOELs and LOELs are often further refined by noting a distinction between effects that are *adverse* to health and effects that are not. Thus, there are also *no-observed-adverse-effect levels* (NOAELs) and *lowest-observed-adverse-effect levels* (LOAELs).

Figure 4.11 illustrates these levels and introduces another exposure called the *reference dose,* or RfD. The RfD used to be called the *acceptable daily intake* (ADI), and as that name implies, it is intended to give an indication of a level of human exposure that is likely to be without appreciable risk. The units of RfD are mg/kg-day averaged over a lifetime, just as they were for the chronic daily intake CDI. The RfD is obtained by dividing the NOAEL by an appropriate *uncertainty factor* (sometimes called a safety factor). A 10-fold uncertainty factor is used to account for differences in sensitivity between the most sensitive individuals in an exposed human population, such as pregnant women, babies, and the elderly, and "normal, healthy" people. Another factor of 10 is introduced when the NOAEL is based on animal data that is to be extrapolated to humans. And finally, another factor of 10 is sometimes applied when there are no good human data and the animal data available are limited. Thus, depending on the strength of the available data, human RfD levels are established at doses that are anywhere from one-tenth to one-thousandth of the NOAEL, which is itself somewhat below the actual threshold. Table 4.11 gives a short list of some commonly encountered toxicants and their RfDs.

The Hazard Index for Noncarcinogenic Effects

Since the reference dose RfD is established at what is intended to be a safe level, well below the level at which any adverse health effects have been observed, it makes sense to compare the actual exposure to the RfD to see whether the actual dose is supposedly safe. The hazard quotient is based on that concept:

$$\text{Hazard quotient} = \frac{\text{Average daily dose during exposure period (mg/kg-day)}}{\text{RfD}} \quad (4.16)$$

Notice that the daily dose is averaged only over the period of exposure, which is different from the average daily dose used in risk calculations for carcinogens. For noncarcinogens, the toxicity is important only during the time of exposure. Recall that for a cancer risk calculation (e.g., Eq. 4.13) the dose is averaged over an assumed 70-year lifetime.

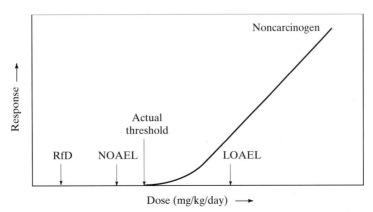

FIGURE 4.11   The reference dose RfD is the no-observed-adverse-effects-level (NOAEL) divided by an uncertainty factor typically between 10 and 1000.

TABLE 4.11 Oral RfDs for chronic noncarcinogenic effects of selected chemicals.

Chemical	RfD (mg/kg-day)
Acetone	0.100
Arsenic	0.0003
Cadmium	0.0005
Chloroform	0.010
1,1-dichloroethylene	0.009
cis-1,2-Dichloroethylene	0.010
Fluoride	0.120
Mercury (inorganic)	0.0003
Methylene chloride	0.060
Phenol	0.600
Tetrachloroethylene	0.010
Toluene	0.200
1,1,1-Trichloroethane	0.035
Xylene	2.000

Source: U.S. EPA. http://www.epa.gov/iris

The hazard quotient has been defined so that if it is less than 1.0, there should be no significant risk of systemic toxicity. Ratios above 1.0 could represent a potential risk, but there is no way to establish that risk with any certainty.

When exposure involves more than one chemical, the sum of the individual hazard quotients for each chemical is used as a measure of the potential for harm. This sum is called the *hazard index:*

$$\text{Hazard index} = \text{Sum of the hazard quotients} \qquad (4.17)$$

EXAMPLE 4.6 Hazard Index

Suppose drinking water contains 1.0 mg/L of toluene and 0.01 mg/L of tetrachloroethylene (C_2Cl_4). A 70-kg adult drinks 2 L per day of this water for 10 years.

    a. Would the hazard index suggest that this was a safe level of exposure?

    b. Tetrachloroethylene is a B2 carcinogen. What would be the carcinogenic risk faced by someone drinking this water? Would it be less than a goal of 10^{-6}?

Solution

    a. First we need to find the average daily doses (ADDs) for each of the chemicals and then their individual hazard quotients.

For toluene, the RfD is given in Table 4.11 as 0.200 mg/kg-day, so

$$\text{ADD (toluene)} = \frac{1.0 \text{ mg/L} \times 2 \text{ L/day}}{70 \text{ kg}} = 0.029 \text{ mg/kg-day}$$

$$\text{Hazard quotient (toluene)} = \frac{0.029 \text{ mg/kg-day}}{0.200 \text{ mg/kg-day}} = 0.14$$

The RfD for tetrachloroethylene is 0.01 mg/kg-day, so

$$\text{ADD (C}_2\text{Cl}_4) = \frac{0.01 \text{ mg/L} \times 2 \text{ L/day}}{70 \text{ kg}} = 0.00029 \text{ mg/kg-day}$$

$$\text{Hazard quotient (C}_2\text{Cl}_4) = \frac{0.00029 \text{ mg/kg-day}}{0.01 \text{ mg/kg-day}} = 0.029$$

So

$$\text{Hazard index} = 0.14 + 0.029 = 0.17 < 1.0$$

The hazard index suggests that this water is safe. By the way, notice that we did not need to know that the person drank this water for 10 years.

b. The incremental carcinogenic risk associated with the C_2Cl_4 is

$$\text{Risk} = \text{CDI} \times \text{Potency factor}$$

$$\text{CDI} = \frac{0.01 \text{ mg/L} \times 2 \text{ L/day} \times 365 \text{ days/yr} \times 10 \text{ yrs}}{70 \text{ kg} \times 365 \text{ days/yr} \times 70 \text{ yrs}} = 4.0 \times 10^{-5} \text{ mg/kg-day}$$

From Table 4.9 the oral potency is 5.1×10^{-2} $(\text{mg/kg-day})^{-1}$, so the risk is

$$\text{Risk} = \text{CDI} \times \text{Potency factor}$$

$$= 4.0 \times 10^{-5} \text{ mg/kg-day} \times 5.1 \times 10^{-2} (\text{mg/kg-day})^{-1} = 2 \times 10^{-6}$$

So, from a cancer risk standpoint, this water does not meet the 10^{-6} risk goal. Notice how the tetrachloroethylene was way below the RfD but was above the desired risk goal. This is not uncommon when the hazard index is computed for carcinogens.   ■

4.7 HUMAN EXPOSURE ASSESSMENT

One of the most elementary concepts of risk assessment is one that is all too often overlooked in public discussions: that risk has two components—the toxicity of the substance involved, and the amount of exposure to that substance. Unless individuals are exposed to the toxicants, there is no human risk.

A human exposure assessment is itself a two-part process. First, pathways that allow toxic agents to be transported from the source to the point of contact with people must be evaluated. Second, an estimate must be made of the amount of contact that is likely to occur between people and those contaminants. Figure 4.12 suggests some of the transport mechanisms that are common at a toxic waste site. Substances that are exposed to the atmosphere may volatilize and be transported with the prevailing winds (in which case, plume models such as the ones introduced in Chapter 7 are often used). Substances in contact with soil may leach into groundwater and eventually be transported to local drinking water wells (groundwater flows will be analyzed in Chapter 5). As pollutants are transported from one place to another, they may undergo various transformations that can change their toxicity and/or concentration. Many of these fate and transport pathways for pollutants will be covered later in this book. A useful summary of exposure pathway models that the EPA uses is given in the Superfund Exposure Assessment Manual (U.S. EPA, 1988).

Once the exposure pathways have been analyzed, an estimate of the concentrations of toxicants in the air, water, soil, and food at a particular exposure point can be

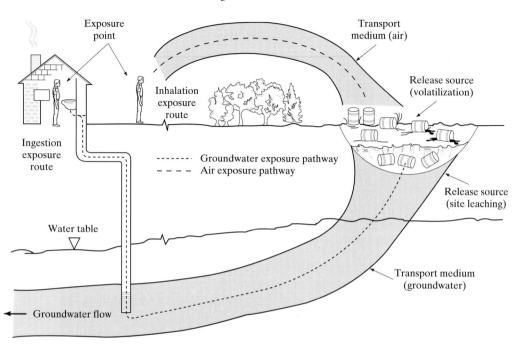

FIGURE 4.12   Illustration of exposure pathways. (Source: U.S. EPA, 1986b)

made. With the concentrations of various toxic agents established, the second half of an exposure assessment begins. Human contact with those contaminants must be estimated. Necessary information includes numbers of people exposed, duration of exposure, and amounts of contaminated air, water, food, and soil that find their way into each exposed person's body. Often, the human intake estimates are based on a lifetime of exposure, assuming standard, recommended daily values of amounts of air breathed, water consumed, and body weight, such as are given in Table 4.10. In some circumstances, the exposure may be intermittent and adjustments might need to be made for various body weights, rates of absorption, and exposure periods, as was illustrated in Example 4.4.

Bioconcentration

One potentially important exposure route is human consumption of contaminated fish. It is relatively straightforward to estimate concentrations of contaminants in water, and it also is reasonable to make estimates of consumption of fish that individuals may consume (for example, the standard intake values given in Table 4.10). What is more difficult is the task of estimating the concentration of a contaminant in fish, given only the chemical concentration in water. The *bioconcentration factor* (BCF) provides the key link. It is a measure of the tendency for a substance to accumulate in fish tissue. The equilibrium concentration of a chemical in fish can be estimated by multiplying the chemical concentration in water by the bioconcentration factor:

Concentration in fish = (concentration in water) × (bioconcentration factor)   (4.18)

The units of BCF (L/kg) are picked to allow the concentration of substance in water to be the usual mg/L and the concentration in fish to be milligrams of substance per kilogram of fish. Some example values of BCF are given in Table 4.12. Note the high bioconcentration factors for chlorinated hydrocarbon pesticides, such as chlordane, DDT, and heptachlor, and the especially high concentration factor for polychlorinated biphenyls (PCBs). These high bioconcentration factors played an important role in the decision to reduce or eliminate the use of these chemicals in the United States.

The following example illustrates the use of bioconcentration factors in a carcinogenic risk assessment.

TABLE 4.12   Bioconcentration Factors (BCFs) for a Selected List of Chemicals.

Chemical	Bioconcentration Factor (L/kg)
Aldrin	28
Arsenic and compounds	44
Benzene	5.2
Cadmium and compounds	81
Carbon tetrachloride	19
Chlordane	14,000
Chloroform	3.75
Chromium III, VI, and compounds	16
Copper	200
DDE	51,000
DDT	54,000
1,1-Dichloroethylene	5.6
Dieldrin	4760
Formaldehyde	0
Heptachlor	15,700
Hexachloroethane	87
Nickel and compounds	47
Polychlorinated biphenyls (PCBs)	100,000
2,3,7,8-TCDD (Dioxin)	5000
Tetrachloroethylene	31
1,1,1-Trichloroethane	5.6
Trichloroethylene (TCE)	10.6
Vinyl chloride	1.17

Source: U.S. EPA (1986b).

EXAMPLE 4.7   Bioconcentration of TCE

Using the standard exposure factors in Table 4.10 for a person eating locally caught fish, estimate the lifetime cancer risk from fish taken from waters containing a concentration of trichloroethylene (TCE) equal to 100 ppb (0.1 mg/L).

Solution   In Table 4.12 the bioconcentration factor for TCE is given as 10.6 L/kg. From (4.18) the expected concentration of TCE in fish is therefore

TCE concentration = 0.1 mg/L × 10.6 L/kg = 1.06 mg TCE/kg fish

From Table 4.10, standard exposure factors include a 70-kg person consuming 54g of fish, 350 days per year for 30 years. The chronic daily intake CDI is thus

$$CDI = \frac{0.054 \text{ kg/day} \times 1.06 \text{ mg TCE/kg} \times 350 \text{ days/yr} \times 30 \text{yrs}}{70 \text{ kg} \times 365 \text{ days/yr} \times 70 \text{ yrs}}$$
$$= 3.36 \times 10^{-4} \text{ mg/kg-day}$$

From Table 4.9 the cancer potency factor for an oral dose of TCE is $1.1 \times 10^{-2} (\text{mg/kg-day})^{-1}$. Using (4.12), the upper-bound, incremental lifetime risk of cancer is

$$\text{Risk} = \text{CDI} \times \text{potency factor}$$
$$= 3.36 \times 10^{-4} \text{ mg/kg-day} \times 1.1 \times 10^{-2} (\text{mg/kg-day})^{-1} = 3.6 \times 10^{-6}$$

or about 4 in 1 million. ∎

Contaminant Degradation

Many toxic chemicals of concern are nonconservative; that is, they degrade with time. Degradation may be the result of a number of processes that remove pollutants from the medium in which they reside. There may be phase transfer as a chemical volatilizes; chemical transformation if it reacts with other substances; or biological transformation if it is consumed by microorganisms. The persistence of a chemical as it moves through various environmental media may be affected by some combination of these mechanisms. A convenient way to deal with such complexity is simply to combine the degradation processes into a single, overall *half-life*. The half-life of a given substance will depend on whether it appears in soil, air, surface water, or groundwater. Some representative half-lives are given in Table 4.13.

TABLE 4.13 Range of Half-Lives (days) of Various Contaminants in Air and Surface Water

Chemical	Air		Surface Water	
	Low	High	Low	High
Benzene	6	—	1	6
Benzo(a)pyrene	1	6	0.4	—
Carbon Tetrachloride	8030	—	0.3	300
Chlordane	40	—	420	500
Chloroform	80	—	0.3	30
DDT	—	—	56	110
1,1-Dichloroethane	45	—	1	5
Formaldehyde	0.8	—	0.9	3.5
Heptachlor	40	—	0.96	—
Hexachloroethane	7900	—	1.1	9.5
Polychlorinated biphenyls (PCBs)	58	—	2	12.9
2,3,7,8-TCDD (Dioxin)	—	—	365	730
1,1,1-Trichloroethane	803	1752	0.14	7
Trichloroethylene	3.7	—	1	90
Vinyl chloride	1.2	—	1	5

Source: U.S. EPA (1986b).

The relationship between reaction rate coefficient K and half-life ($T_{1/2}$) was derived in Chapter 3. Recall that if the concentration of a substance is modeled with a simple exponential decay relationship,

$$C(t) = C(0)e^{-Kt} \tag{4.19}$$

then the time required for the concentration to be decreased by 50 percent is the half-life, given by

$$T_{1/2} = \frac{\ln 2}{K} = \frac{0.693}{K} \tag{4.20}$$

An example of how to use half lives is given in the following example.

EXAMPLE 4.8   A Leaking Underground Storage Tank Exposure Assessment

Suppose an underground storage tank has been leaking for many years, contaminating the groundwater and causing a contaminant concentration directly beneath the site of 0.30 mg/L. The contamination is flowing at the rate of 0.5 ft per day toward a public drinking water well 1 mile away. The half-life of the contaminant is 10 years.

  a. Estimate the steady-state pollutant concentration expected at the well.
  b. If the potency factor for the contaminant is 0.02 $(mg/kg\text{-}day)^{-1}$, estimate the cancer risk if a 70-kg person drank 2 L of this water per day for 10 years.

Solution

  a. The time required to travel to the well 1 mile away would be

$$\text{Time to well} = \frac{5280 \text{ ft}}{0.5 \text{ ft/day}} = 10{,}560 \text{ days}$$

The pollutant is assumed to degrade exponentially, so the reaction rate coefficient K can be found using (4.20):

$$K = \frac{0.693}{T_{1/2}} = \frac{0.693}{10 \text{ yr} \times 365 \text{ days/yr}} = 1.9 \times 10^{-4}/\text{day}$$

In the 10,560 days required to travel to the drinking water well, (4.19) suggests that the initial 0.30 mg/L would be reduced to

$$C(t) = C(0)e^{-Kt} = 0.30e^{-(1.9 \times 10^{-4}/d \times 10{,}560\, d)} = 0.040 \text{ mg/L}$$

  b. The chronic daily intake for someone drinking this water for 10 years out of a 70-year lifetime would be

$$\text{CDI} = \frac{0.040 \text{ mg/L} \times 2 \text{ L/day} \times 10 \text{ yr}}{70 \text{ kg} \times 70 \text{ yr}} = 1.6 \times 10^{-4} \text{ mg/kg-day}$$

so the lifetime cancer risk would be

$$\text{Risk} = \text{CDI} \times \text{potency factor}$$

$$= 1.6 \times 10^{-4} \text{ mg/kg-day} \times 0.020 \, (mg/kg\text{-}day)^{-1} = 3.2 \times 10^{-6}$$

This is probably an upper-bound estimate of the individual risk and is subject to all of the uncertainties that currently characterize all risk assessments. ∎

4.8 RISK CHARACTERIZATION

The final step in a risk assessment is to bring the various studies together into an overall risk characterization. In its most primitive sense, this step could be interpreted to mean simply multiplying the exposure (dose) by the potency to get individual risk, and then multiplying that by the number of people exposed to get an estimate of overall risk to some specific population.

While there are obvious advantages to presenting a simple, single number for extra cancers, or some other risk measure, a proper characterization of risk should be much more comprehensive. The final expressions of risk derived in this step will be used by regulatory decision makers in the process of weighing health risks against other societal costs and benefits, and the public will use them to help them decide on the adequacy of proposed measures to manage the risks. Both groups need to appreciate the extraordinary leaps of faith that, by necessity, have had to be used to determine these simple quantitative estimates. It must always be emphasized that these estimates are preliminary, subject to change, and extremely uncertain.

The National Academy of Sciences (1983) suggests a number of questions that should be addressed in a final characterization of risk, including the following:

- What are the statistical uncertainties in estimating the extent of health effects? How are these uncertainties to be computed and presented?
- What are the biological uncertainties? What are their origins? How will they be estimated? What effect do they have on quantitative estimates? How will the uncertainties be described to agency decision makers?
- Which dose-response assessments and exposure assessments should be used?
- Which population groups should be the primary targets for protection, and which provide the most meaningful expression of the health risk?

Rodricks (page 181, 1992) offers the following example of the sort of qualifying statement that ought to accompany all risk assessments (in this case for a hypothetical contaminant difluoromuckone, DFM):

> Difluoromuckone (DFM) has been found to increase the risk of cancer in several studies involving experimental animals. Investigations involving groups of individuals exposed in the past to relatively high levels of DFM have not revealed that the chemical increases cancer risk in humans. Because these human studies could not detect a small increase in risk, and because there is a scientific basis for assuming results from animal experiments are relevant to humans, exposure to low levels of DFM may create an increase in risk of cancer for people. The magnitude of this risk is unknown, but probably does not exceed one in 50,000. This figure is the lifetime chance of developing cancer from a daily exposure to the highest levels of DFM detected in the environment. Average levels, which are more likely to be experienced over the course of a lifetime, suggest a lifetime risk more like one in 200,000. These risk figures were derived using scientific assumptions that are not recognized as plausible by all scientists, but which are consistently used by regulatory scientists when attempting to portray the risks of environmental chemicals. It is quite plausible that

actual risks are lower than the ones cited above; higher risks are not likely but cannot be ruled out. Regulators typically seek to reduce risks that exceed a range of one in 100,000 to one in 1,000,000. Note that the lifetime cancer risk we face from all sources of these diseases is about 1 in 5 (1 in 10 for non-smokers), so that, even if correct, the DFM risk is a minor contributor to the overall cancer problem. Prudence may dictate the need for some small degree of risk reduction for DFM in the environment.

4.9 COMPARATIVE RISK ANALYSIS

In 1987, the EPA released a report entitled *Unfinished Business: A Comparative Assessment of Environmental Problems* (U.S. EPA, 1987), in which the concepts of risk assessment were applied to a variety of pressing environmental problems. The goal of the study was to attempt to use risk as a policy tool for ranking major environmental problems in order to help the agency establish broad, long-term priorities.

At the outset it was realized that direct comparisons of different environmental problems would be next to impossible. Not only are the data usually insufficient to quantify risks, but the kinds of risk associated with some problems, such as global warming, are virtually incomparable with risks of others, such as hazardous waste. In most cases, considerable professional judgment rather than hard data was required to finalize the EPA's rankings. In spite of difficulties such as these, the report is noteworthy both in terms of its methodology and its conclusions.

The study was organized around a list of 31 environmental problems, including topics as diverse as conventional (criteria) air pollutants, indoor radon, stratospheric ozone depletion, global warming, active hazardous waste sites regulated by the Resource Conservation and Recovery Act (RCRA), and inactive (Superfund) hazardous waste sites, damage to wetlands, mining wastes, and pesticide residues on foods. Each of these 31 problems was analyzed in terms of four different types of risk: cancer risks, noncancer health risks, ecological effects, and welfare effects (visibility impairment, materials damage, etc.). In each assessment, it was assumed that existing environmental control programs continue so that the results represent risks as they exist now, rather than what they would have been had abatement programs not already been in place.

The ranking of cancer risks was perhaps the most straightforward part of the study since the EPA has already established risk assessment procedures and there are considerable data already available from which to work. Rankings were based primarily on overall cancer risk to the entire U.S. population, although high risks to specific groups of individuals, such as farm workers, were noted. A number of caveats were emphasized in the final rankings on such issues as lack of complete data, uneven quality of data, and the usual uncertainties in any risk assessment that arise from such factors as interspecies comparisons, adequacy of the low-dose extrapolation model, and estimations of exposures. Ordinal rankings were given, but it was emphasized that these should not be interpreted as being precise, especially when similarly ranked problems are being compared.

Given all of the uncertainties, in the cancer working group's final judgment two problems were tied at the top of the list: (1) worker exposure to chemicals, which does not involve a large number of individuals but does result in high individual risks to those exposed; and (2) indoor radon exposure, which is causing significant risk to a large number of people. Inactive (Superfund) hazardous waste sites ranked eighth and active (RCRA) hazardous waste sites were thirteenth. Interestingly, it was noted that with the exception of pesticide residues on food, the major route of exposure for carcinogens is inhalation. Their final ranking of carcinogenic risks is reproduced in Table 4.14.

The other working groups had considerably greater difficulty ranking the 31 environmental problem areas since there are no accepted guidelines for quantitatively assessing relative risks. As noted in *Unfinished Business,* a perusal of the rankings of the 31 problem areas for each of the four types of risk (cancer, noncancer health effects, ecological, and welfare effects) produced the following general results:

TABLE 4.14 Consensus Ranking of Environmental Problem Areas on the Basis of Population Cancer Risk

Rank	Problem Area	Selected Comments
1 (tied)	Worker exposure to chemicals	About 250 cancer cases per year estimated based on exposure to 4 chemicals; but workers face potential exposures to over 20,000 substances. Very high individual risk possible.
1 (tied)	Indoor radon	Estimated 5000 to 20,000 lung cancers annually from exposure in homes.
3	Pesticide residues on foods	Estimated 6000 cancers annually, based on exposure to 200 potential carcinogens.
4 (tied)	Indoor air pollutants (nonradon)	Estimated 3500 to 6500 cancers annually, mostly due to tobacco smoke.
4 (tied)	Consumer exposure to chemicals	Risk from 4 chemicals investigated is about 100 to 135 cancers annually; an estimated 10,000 chemicals in consumer products. Cleaning fluids, pesticides, particleboard, and asbestos-containing products especially noted.
6	Hazardous/toxic air pollutants	Estimated 2000 cancers annually based on an assessment of 20 substances.
7	Depletion of stratospheric ozone	Ozone depletion projected to result in 10,000 additional annual deaths in the year 2100. Not ranked higher because of the uncertainties in future risk.
8	Hazardous waste sites, inactive	Cancer incidence of 1000 annually from 6 chemicals assessed. Considerable uncertainty since risk based on extrapolation from 35 sites to about 25,000 sites.
9	Drinking water	Estimated 400 to 1000 annual cancers, mostly from radon and trihalomethanes.
10	Application of pesticides	Approximately 100 cancers annually; small population exposed but high individual risks.
11	Radiation other than radon	Estimated 360 cancers per year. Mostly from building materials. Medical exposure and natural background levels not included.
12	Other pesticide risks	Consumer and professional exterminator uses estimated cancers of 150 annually. Poor data.
13	Hazardous waste sites, active	Probably fewer than 100 cancers annually; estimates sensitive to assumptions regarding proximity of future wells to waste sites.

TABLE 4.14   Continued

Rank	Problem Area	Selected Comments
14	Nonhazardous waste sites, industrial	No real analysis done, ranking based on consensus of professional opinion.
15	New toxic chemicals	Difficult to assess; done by consensus.
16	Nonhazardous waste sites, municipal	Estimated 40 cancers annually, not including municipal surface impoundments.
17	Contaminated sludge	Preliminary results estimate 40 cancers annually, mostly from incineration and landfilling.
18	Mining waste	Estimated 10 to 20 cancers annually, largely due to arsenic. Remote locations and small population exposure reduce overall risk though individual risk may be high.
19	Releases from storage tanks	Preliminary analysis, based on benzene, indicated low cancer incidence (< 1).
20	Nonpoint-source discharges to surface water	No quantitative analysis available; judgment.
21	Other groundwater contamination	Lack of information; individual risks considered less than 10^{-6}, with rough estimate of total population risk at < 1.
22	Criteria air pollutants	Excluding carcinogenic particles and volatile organic chemicals (VOCs) (included under Hazardous/Toxic Air Pollutants); ranked low because remaining criteria pollutants have not been shown to be carcinogens.
23	Direct point-source discharges to surface water	No quantitative assessment available. Only ingestion of contaminated seafood was considered.
24	Indirect, point-source discharges to surface water	Same as above.
25	Accidental releases—toxics	Short-duration exposure yields low cancer risk; noncancer health effects of much greater concern.
26	Accidental releases—oil spills	See above. Greater concern for welfare and ecological effects.

Not ranked: Biotechnology; global warming; other air pollutant; discharges to estuaries, costal waters and oceans; discharges to wetlands/

Source: Based on data from U.S. EPA (1987).

- No problems rank relatively high in all four types of risk, or relatively low in all four.
- Problems that rank relatively high in three of the four risk types, or at least medium in all four, include criteria air pollutants (see Chapter 7); stratospheric ozone depletion (Chapter 8); pesticide residues on food; and other pesticide risks (runoff and air deposition of pesticides).
- Problems that rank relatively high in cancer and noncancer health risks, but low in ecological and welfare risks, include hazardous air pollutants; indoor radon; indoor air pollution other than radon; pesticide application; exposure to consumer products; and worker exposures to chemicals.
- Problems that rank relatively high in ecological and welfare risks, but low in both health risks, include global warming; point and nonpoint sources of surface water

pollution; physical alteration of aquatic habitats (including estuaries and wet-lands), and mining wastes.
- Areas related to groundwater consistently rank medium or low.

In spite of the great uncertainties involved in making their assessments, the divergence between the EPA effort in the 1980s and relative risks is noteworthy. As concluded in the study, areas of relatively high risk but low EPA effort include indoor radon; indoor air pollution; stratospheric ozone depletion; global warming; nonpoint sources; discharges to estuaries, coastal waters, and oceans; other pesticide risks; accidental releases of toxics; consumer products; and worker exposures. Areas of high EPA effort but relatively medium or low risks include RCRA sites, Superfund sites, underground storage tanks, and municipal nonhazardous waste sites.

The *Unfinished Business* report was the first major example of what has come to be known as *comparative risk analysis*. Comparative risk analysis differs from conventional risk assessment since its purpose is not to establish absolute values of risk, but rather to provide a process for ranking environmental problems by their seriousness. A subsequent 1990 report, *Reducing Risks*, by EPA's Science Advisory Board, recommended that the EPA reorder its priorities on the basis of reducing the most serious risks. The combination of these two reports has had considerable influence on the way that the EPA perceives its role in environmental protection. EPA's Office of Research and Development (U.S. EPA, 1996) has incorporated these recommendations in setting forth its strategic principles, which include the following:

- Focus research and development on the greatest risks to people and the environment, taking into account their potential severity, magnitude, and uncertainty.
- Focus research on reducing uncertainty in risk assessment and on cost-effective approaches for preventing and managing risks.
- Balance human health and ecological research.

Based on those strategic principles, the EPA has recently defined its six highest-priority research topics for the next few years (U.S. EPA, 1996):

- *Drinking water disinfection.* Some microorganisms, especially the protozoan *Cryptosporidium,* are able to survive conventional disinfection processes, and some carcinogens, such as chloroform, are created during chlorination of drinking water. Questions to be addressed include the comparative risk between waterborne microbial disease and the disinfection byproducts formed during drinking water disinfection.
- *Particulate matter.* Inhalation of particulate matter in the atmosphere poses a high potential human health risk. The relationship between morbidity/mortality and low ambient levels of particulate matter, and cost-effective methods to reduce particulate matter emissions, are principle areas of interest.
- *Endocrine disruptors.* Declines in the quality and quantity of human sperm production and increased incidence of certain cancers that may have an endocrine-related basis form the basis of concern for this high-priority research topic.
- *Improved ecosystem risk assessment.* Understanding the impacts of human activities on ecosystems has not developed as rapidly as human health impacts. Topics

such as forest decline, toxic microorganisms in estuaries, reproductive failure of wildlife, and the reappearance of vectorborne epidemic diseases need to be addressed.

- *Improved health risk assessment.* Continued focus on reducing the uncertainty in source-exposure-dose relationships, including the impacts of mixtures of chemical insults, is needed.
- *Pollution prevention and new technologies.* Avoiding the creation of environmental problems is the most cost-effective risk-management strategy, but it is not clear how best to integrate pollution prevention into government and private-sector decision making.

PROBLEMS

4.1. Consider a carcinogenic VOC with the dose-response curve shown in Figure P4.1. If 70-kg people breath 20 m³/day of air containing 10^{-3} mg/m³ of this VOC throughout their entire 70-year lifetime, find the cancer risk (you first need to find the potency).

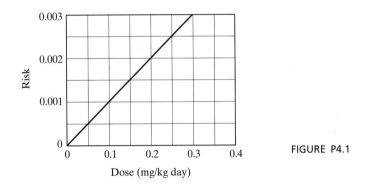

FIGURE P4.1

4.2. Suppose a city's water supply has 0.2 ppb (1 ppb = 10^{-3} mg/L) of polychlorinated biphenyls (PCBs) in it. Using the PCB oral potency factor (Table 4.9) and EPA recommended exposure factors given in Table 4.10,

  (a) What would be the chronic daily intake, CDI?

  (b) What would be the individual lifetime cancer risk for an adult residential consumer?

  (c) In a city of 1 million people, use this risk assessment to estimate the number of extra cancers per year caused by the PCBs in the water supply.

  (d) The average cancer death rate in the United States is 193 per 100,000 per year. How many cancer deaths would be expected in a city of 1 million? Do you think the incremental cancers caused by PCBs in the drinking water would be detectable?

4.3. Suppose 30 out of 500 rats exposed to a potential carcinogen develop tumors. A control group of 300 rats not exposed to the carcinogen develops only 10 tumors. Based on these data, compute (a) the relative risk, (b) the attributable risk, and (c) the odds ratio. Do these indicators suggest that there might be a relationship between exposure and tumor risk?

4.4. Suppose 5 percent of individuals exposed to a chemical get a tumor and 2 percent of those not exposed get the same kind of tumor. Find (a) the relative risk, (b) attributable risk, and (c) the odds ratio.

4.5. Suppose a 70-kg individual drinks 2 L/day of water containing 0.1 mg/L of 1,1-dichloroethylene for 20 years.

   (a) Find the hazard quotient for this exposure.

   (b) Find the cancer risk.

   (c) If the individual drinks this water for 30 years instead of just 20, recompute the hazard quotient and the cancer risk.

4.6. Compute the drinking water equivalent level (DWEL) for methylene chloride based on a 10^{-6} risk for a lifetime consumption of 2 L of water per day for a 70-kg individual.

4.7. Based on a 10^{-6} risk for a 70-kg individual consuming 2 L of water per day, the DWEL for a contaminant is 10 μg/L. What potency would produce this value?

4.8. The drinking water standard for 2,3,7,8-TCDD (dioxin) is 3×10^{-8} mg/L. Using EPA exposure factors for residential consumption, what lifetime risk would this pose?

4.9. The drinking water standard for tetrachloroethylene is 0.005 mg/L. Using EPA exposure factors for residential consumption, what lifetime risk would this pose?

4.10. Human exposure to radiation is often measured in rems (roentgen-equivalent man), or millirems (mrem). The cancer risk caused by exposure to radiation is thought to be approximately 1 fatal cancer per 8000 person-rems of exposure (e.g., 1 cancer death if 8000 people are exposed to 1 rem each, or 10,000 people exposed to 0.8 rems each, etc.).

   (a) Natural radioactivity in the environment is thought to expose us to roughly 130 mrem/yr. How many cancer deaths in the United States (population 260 million) would be expected per year from this exposure?

   (b) A single 3000-mile, cross-country jet flight exposes an individual to about 4 mrem. How many cross-country flights would be required to elevate your cancer risk by one in 1 million? How does this answer compare with the value given in Table 4.3?

4.11. Exposure to cosmic radiation increases with increasing altitude. At sea level it is about 40 mrem/yr, while at the elevation of Denver it is about 120 mrem/yr. Using the radiation potency factor given in Problem 4.10,

   (a) Compare the lifetime (70-yr) probability of dying of cancer induced by cosmic radiation for a person living at sea level with that of a person living in Denver.

   (b) Estimate the incremental cancer deaths per year caused by the elevated cosmic radiation exposure in Denver (population 0.5 million). Compare that incremental death rate with the expected cancer deaths per year at the typical U.S. rate of 193 per 100,000 per year.

   (c) How long would an individual have to live in Denver to cause an incremental cancer risk of 10^{-6} (compared with sea level)?

   (d) If all 260 million Americans lived at sea level, estimate the total cancer deaths per year caused by cosmic radiation.

4.12. Living in a home with 1.5 pCi/L of radon is thought to cause a cancer risk equivalent to that caused by approximately 400 mrem/yr of radiation. Using the radiation potency factor given in Problem 4.10,

   (a) Estimate the annual cancers in the United States (population 260 million) caused by radon in homes.

   (b) Estimate the lifetime risk for an individual living in a home with that amount of radon.

4.13. It has been estimated that about 75 million people in the Ukraine and Byelorussia were exposed to an average of 0.4 rem of radiation as a result of the Chernobyl nuclear accident. Using potency from Problem 4.10,

   (a) How many extra cancer deaths might eventually be expected from this exposure?

   (b) If the normal probability of dying of cancer from all causes is 0.22, how many cancer deaths would you normally expect among those 75 million people?

4.14. In Table 4.3 it was estimated that living 50 years within 5 miles of a nuclear reactor would increase risk by 10^{-6}. Using radiation potency given in Problem 4.10, what mrem/yr exposure rate would yield this risk?

4.15. One way to estimate maximum acceptable concentrations of toxicants in drinking water or air is to pick an acceptable lifetime risk and calculate the concentration that would give that risk assuming agreed-on exposures such as the residential factors given in Table 4.10. Find the acceptable concentrations of the following substances:

   (a) Benzene in drinking water (mg/L), at a lifetime acceptable risk of 1×10^{-5}

   (b) Trichloroethylene in air (mg/m^3), at a lifetime acceptable risk of 1×10^{-6}

   (c) Benzene in air (mg/m^3), at a lifetime acceptable risk of 1×10^{-5}

   (d) Vinyl chloride in drinking water (mg/L), at a lifetime acceptable risk of 1×10^{-4}

4.16. Using exposure factors in Table 4.10, what would be an acceptable concentration of trichloroethylene in the air of an industrial facility if worker risk is to be less than 10^{-4}? Express the answer in mg/m^3 and ppm (see Eq. 1.8).

4.17. Suppose an individual eats fish from a river contaminated by benzene. What concentration of benzene (mg/L) in the water would produce a lifetime risk of 1×10^{-6} to an individual who eats the amount of fish suggested by EPA exposure factors in Table 4.10? Use the oral potency factor for benzene and the bioconcentration factor given in Table 4.12.

4.18. Estimate the cancer risk for a 70-kg individual consuming 2 g of fish every day for 70 years from a stream with 20 ppb of DDT.

4.19. Suppose a 50-kg individual drinks 1 L/day of water containing 2 mg/L of 1,1,1-trichloroethane, 0.04 mg/L of tetrachloroethylene, and 0.1 mg/L of 1,1-dichloroethylene. What is the hazard index? Is there cause for concern?

4.20. Suppose 1.0 g/day of heptachlor leaks into a 30,000 m^3 pond. If heptachlor has a reaction rate coefficient of 0.35/day, and complete mixing occurs,

   (a) What would be the steady-state concentration in the pond?

   (b) Suppose a 70-kg individual drank 2 L/day of that water for 5 years. Estimate the maximum risk of cancer due to that exposure to heptachlor.

4.21. Mainstream smoke inhaled by a 70-kg smoker contains roughly 0.03 mg per cigarette of the class B2 carcinogen, benzo(a)pyrene. For an individual who smokes 20 cigarettes per day for 40 years, estimate the lifetime risk of cancer caused by that benzo(a)pyrene (there are other carcinogens in cigarettes as well).

4.22. Consider the problem of indoor air pollution caused by sidestream smoke (unfiltered smoke from an idling cigarette). Suppose the sidestream smoke from one cigarette in a small apartment will produce an airborne benzo(a)pyrene (BaP) concentration that will average 6×10^{-4} mg/m^3 for 1 hour. How many cigarettes would need to be smoked in the presence of a nonsmoking roommate to raise the nonsmoker's cancer risk by 10^{-6} just due to this BaP? Assume an inhalation rate of 0.83 m^3/hr (20 m^3/day). At eight cigarettes of exposure per day, how many days would it take to create this risk?

4.23. The sidestream smoke from one cigarette releases about 0.1 mg of benzo(a)pyrene (BaP). In an apartment with fresh air entering through holes and cracks (infiltration) at an average rate of 120 m^3/hr (see Example 1.5),

(a) What would be the steady-state indoor concentration of BaP if one cigarette per hour is smoked? (Assume that BaP is a conservative pollutant.)

(b) What would be the incremental cancer risk to a nonsmoking roommate who spends 8 hr/day for 1 year in this apartment? (Assume an inhalation rate of 20 m^3/day.)

4.24. For the following carcinogens, the U.S. drinking water standards are given. For each, find the lifetime individual cancer risk and the incremental cancers per year in a population of 260 million as computed using a standard risk assessment based on residential exposure factors recommended by the EPA.

(a) Trichloroethylene (TCE), 0.005 mg/L

(b) Benzene, 0.005 mg/L

(c) Arsenic, 0.05 mg/L

(d) Carbon tetrachloride, 0.005 mg/L

(e) Vinyl chloride, 0.002 mg/L

(f) Polychlorinated biphenyls (PCBs), 0.0005 mg/L

4.25. One way to express cancer potency for substances that are inhaled is in terms of risk caused by a lifetime (20 m^3/day for 70 years) of breathing air with a concentration of 1.0 μg/m^3 of carcinogen. The potency for formaldehyde in these terms is 1.3×10^{-5} cancer/μg/m^3. What is the cancer risk caused by a lifetime of breathing formaldehyde at the not unusual (in smoggy cities) concentration of 50 μg/m^3 (the threshold of eye irritation)?

4.26. Trichloroethylene (TCE) is a common groundwater contaminant. In terms of cancer risk, which would be better: (1) to drink unchlorinated groundwater with 10 ppb (0.010 mg/L) of TCE; or (2) switch to a surface water supply that, as a result of chlorination, has a chloroform concentration of 50 ppb?

4.27. Suppose a 70-kg man is exposed to 0.1 mg/m^3 of tetrachloroethylene in the air at his workplace. If he inhales 1 m^3/hr, 8 hours per day, 5 days per week, 50 weeks per year, for 30 years, and if tetrachloroethylene has an absorption factor of 90 percent, an inhalation potency of 2×10^{-3} (mg/kg-day)$^{-1}$, what would be his lifetime cancer risk? What would the risk be to a 50-kg woman similarly exposed?

4.28. Suppose a factory releases a continuous flow of wastewater into a local stream, resulting in an in-stream carcinogen concentration of 0.1 mg/L just below the outfall. Suppose this carcinogen has an oral potency factor of 0.30 (mg/kg-day)$^{-1}$ and that it is degradable with a reaction rate coefficient K of 0.10/day. To keep the problem simple, assume that the stream is uniform in cross section, flowing at the rate of 1 mph, and that there are no other sources or sinks for this carcinogen. At a distance of 100 miles downstream, a town uses this stream as its only source of water. Estimate the individual residential lifetime cancer risk caused by drinking this water.

4.29. The following tumor data were collected for rats exposed to ethylene-thiourea (ETU) (data from Crump, 1984):

Dietary concentration	Animals with tumors
125 ppm	3%
250 ppm	23%
500 ppm	88%

A one-hit model fitted to the data has coefficients: $q_0 = 0.01209$ and $q_1 = 0.001852/\text{ppm}$. A multistage model has coefficients

$$q_0 = 0.02077$$
$$q_1 = q_2 = 0.0$$
$$q_3 = 1.101 \times 10^{-8}/(\text{ppm})^3$$
$$q_4 = 1.276 \times 10^{-11}/(\text{ppm})^4$$

(a) For each of the three concentrations given, compare the measured data with the values derived from each of these two models.

(b) For a concentration of 1 ppm, compare the values that each of the two models would predict for percent tumors.

4.30. Suppose 10 million people are exposed to a carcinogen that poses an individual lifetime (70-yr) cancer risk of 10^{-4}.

(a) How many cancers per year might be caused by this carcinogen?

(b) If spending $1 per year per person (for an indefinitely long time) to reduce exposure to that carcinogen reduces that risk to 10^{-5}, what would be the cost of each cancer avoided?

REFERENCES

Crump, K. S., 1984, An improved procedure for low-dose carcinogenic risk assessment from animal data, *Journal of Environmental Pathology, Toxicology, and Oncology*, 5–4/5:339–349.

Environ, 1988, *Elements of Toxicology and Chemical Risk Assessment*, Environ Corporation, Washington, DC.

Goldberg, A. M., and J. M. Frazier, 1989, Alternatives to animals in toxicity testing, *Scientific American*, August, 24–30.

James, R. C., 1985, Hepatotoxicity: Toxic Effects in the Liver, in *Industrial Toxicology*, P. L.Williams and J. L. Burson (eds.), Van Nostrand Reinhold, New York.

Kolluru, R. V., S. M. Bartell, R. M. Pitblado, and R. S. Stricoff, 1996, *Risk Assessment and Management Handbook*, McGraw-Hill, New York.

National Academy of Sciences, 1983, *Risk Assessment in the Federal Government: Managing the Process*, National Academy Press, Washington, DC.

Rodricks, J. V., 1992, *Calculated Risks: The Toxicity and Human Health Risks of Chemicals in Our Environment*, Cambridge University Press, Cambridge, UK.

Slovic, P., B. Fischhoff, and S. Lichtenstein, 1980, Facts and Fears: Understanding Perceived Risk, in *Societal Risk Assessment: How Safe Is Safe Enough?*, R. Schwing and W. Albers (eds.), Plenum Press, New York.

Slovic, P., 1987, Perception of risk, *Science*, April 17, 236:280–285.

U.S. EPA, 1986a, *Guidelines for Carcinogen Risk Assessment*, Federal Register, Vol. 51, No. 185, pp. 33992–34003, September 24, 1986.

U.S. EPA, 1986b. *Superfund Public Health Evaluation Manual*, Office of Emergency and Remedial Response, Washington, DC.

U.S. EPA, 1987, *Unfinished Business: A Comparative Assessment of Environmental Problems,* Office of Policy, Planning and Evaluation, EPA/2302–87/025, Washington, DC.

U.S. EPA, 1988, *Superfund Exposure Assessment Manual,* Environmental Protection Agency, Office of Remedial Response, EPA/540/1-88/001, Washington, DC.

U.S. EPA, 1990, *Reducing Risks: Setting Priorities and Strategies for Environmental Protection,* Science Advisory Board, U.S. Environmental Protection Agency, Washington, DC.

U.S. EPA, 1991, *Human Health Evaluation Manual,* Supplemental Guidance: Standard Default Exposure Factors, OSWER Directive 9285.6-03, U.S. Environmental Protection Agency, Washington, DC.

U.S. EPA, 1996, *Strategic Plan for the Office of Research and Development,* EPA/600/R-96/059, U.S. Environmental Protection Agency, Washington, DC.

Williams, P. L., and J. L. Burson, 1985, *Industrial Toxicology, Safety and Health Applications in the Workplace,* Van Nostrand Reinhold, New York.

Wilson, R., 1979, Analyzing the daily risks of life, *Technology Review,* 81(4):41–46.

Wilson, R., and E. A. C. Crouch, 1987, Risk assessment and comparisons: An introduction, *Science,* April 17, 267–270.

C H A P T E R   5

Water Pollution

When the well's dry, we know the worth of water. —Ben Franklin,
Poor Richard's Almanac

5.1 INTRODUCTION

In the late 1960s, the Cuyahoga River in Ohio was clogged with debris and trash, floating in layers of black, heavy oil several inches thick. When it caught fire in 1969, it became a symbol of the environmental degradation that can result from uncontrolled industrialization. Along with images of thick mats of algae along the shores of Lake Erie and dead fish floating on its surface, visible manifestations of pollution quickly galvanized the public, created the modern environmental movement, and led to the enactment of some of our most important environmental legislation, including the Clean Water Act of 1972, which has jurisdiction over water quality in rivers, lakes, estuaries, and wetlands, and the Safe Drinking Water Act of 1974, which regulates tap water quality.

In many ways, the quality of our surface waters has improved significantly as a result of the Clean Water Act, but problems remain. The level of municipal sewage treatment and the fraction of the population served by these facilities has risen sharply as a result of almost 130 billion dollars of federal, state, and local expenditures on wastewater treatment plants. Municipal treatment plant upgrades, coupled with a pollution discharge permit system, have dramatically reduced point-source discharges of raw sewage and industrial wastes. But diffuse, nonpoint-source pollution, such as runoff from agricultural lands and urban streets, has proven to be much more difficult to regulate and control, resulting in surface water quality that still does not meet the Clean Water Act's goals of "fishable and swimmable" waters throughout the nation by 1983. In fact, by the late-1990s, roughly 40 percent of the surface water in the United States still was *not* suitable for fishing, swimming, or other designated uses.

Water pollution problems in many parts of the world are far worse. It has been estimated that in the year 2000, 2.2 billion people in the developing countries will lack access to safe drinking water, and 2.7 billion will lack access to sanitation services. As shown in Figure 5.1, growth in numbers lacking access to safe water and sanitation will be driven in large part by the growth rate of people living in urban areas (Gleick, 1993). Most urban centers in Africa and Asia have no sewage system at all, including many cities with populations over 1 million people. The result is a tragic rate of morbidity and mortality in the less developed parts of the world. Waterborne diseases, such as cholera and typhoid, cause more than 1.5 billion episodes of diarrhea each year, resulting in 4 million deaths annually (UNEP, 1993).

5.2 WATER RESOURCES

Water is so common that we take it for granted. After all, it covers nearly three-fourths of the surface of the earth. And we probably think it is much like any other liquid, but it is not. In fact, nearly every physical and chemical property of water is unusual when compared with other liquids, and these differences are essential to life as we know it.

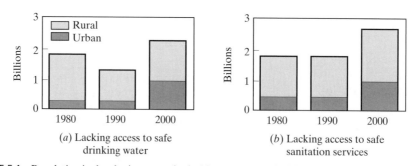

FIGURE 5.1   Population in developing countries lacking access to safe drinking water (*a*); and lacking access to sanitation services (*b*) (Based on Gleick, 1993).

Unusual Properties of Water

Consider a simple molecule of water, H_2O. As shown in Figure 5.2, the two hydrogen-to-oxygen chemical bonds form a 105 ° angle with each other, resulting in a molecule that has a slightly positive charge at one end and a slightly negative charge at the other. This *dipolar* character means water molecules are attracted to each other, which helps explain why water boils at such a high temperature and why it takes an unusual amount of energy to cause it to vaporize. It also helps explain water's high surface tension, which allows it to support relatively heavy objects, such as insects that skate along its surface.

The dipolar nature of water molecules also causes them to adhere to other surfaces easily, and the combination of surface tension and adhesion lets water crawl up the sides of objects. This *capillary* action is crucial to life since, for example, it causes sap to rise in trees, water to rise in soil, and food to move through organisms. The dipolar property also makes water a very effective solvent since water molecules tend to surround charged ions and effectively neutralize them.

Density.   Water is the only common liquid that expands when it freezes. In fact, a plot of density versus temperature shows a maximum density at 4 °C, which means that as temperatures move away from this point, water continuously becomes lighter and more buoyant. As a result, ice floats. If it did not, ice that would form on the surface of bodies of water would sink to the bottom, making it possible for rivers and lakes to freeze solid from the bottom up. The expansion of water as it freezes also contributes to the weathering of rocks by literally breaking them apart when water freezes in the cracks. When water is warmed beyond 4 °C, it becomes buoyant once again, so warm water floats on top of cold water in lakes. This thermal stratification affects aquatic life in unusual ways, as will be described later in this chapter.

Melting and Boiling Points.   Water has unusually high boiling and freezing temperatures for a compound having such a low molecular weight. If water were similar to other "H_2X" substances, such as H_2S, H_2Se, and H_2Te, it would boil at normal earth temperatures, so it would exist mostly as a gas rather than a liquid or solid. It also has an unusually high difference in temperature between the melting point and boiling point, thus remaining a liquid over most of the globe. It is the only substance that appears in all three states, gaseous, liquid, and ice, within the normal range of temperatures on earth. With only slightly different phase change temperatures, life on earth would be very different, if it could exist at all.

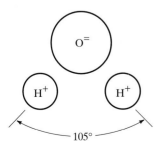

FIGURE 5.2   A water molecule is dipolar; that is, it appears to have a positive charge at one end and a negative charge at the other. This dipolar character helps explain a number of water's unusual properties.

Specific Heat.   Water has a higher heat capacity (4184 J/kg °C) than any other known liquid except ammonia. It is five times higher than the specific heat of most common heavy solids, such as rock and concrete. As a result, it takes longer to heat up and to cool down water than almost anything else. This high heat capacity helps make the oceans the major moderating factor in maintaining the temperature of the surface of the earth. It also serves the important function of protecting life from rapid thermal fluctuations, which are often lethal.

Heat of Vaporization.   The heat required to vaporize water (2258 kJ/kg) is one of the highest of all liquids. This high heat of vaporization means that water vapor stores an unusually large amount of energy, energy that is released when the water vapor condenses. This property is important in distributing heat from one place on the globe to another and is a major factor affecting the earth's climate.

Water as a Solvent.   Water dissolves more substances than any other common solvent. As a result, it serves as an effective medium for both transporting dissolved nutrients to tissues and organs in living things as well as eliminating their wastes. Water also transports dissolved substances throughout the biosphere.

Greenhouse Effect.   The 105° angle bond angle for water is actually an average value. The hydrogen atoms vibrate back and forth, causing the bond angle to oscillate. This H—O—H *bending vibration* resonates with certain wavelengths of electromagnetic radiation coming from the sun, allowing water vapor in the atmosphere to absorb solar energy. Other vibrations in water molecules also cause absorption of infrared radiation leaving the earth's surface. Earth's temperature depends to a significant degree on these absorptions of incoming solar and outgoing infrared radiation. As will be described in Chapter 8, water vapor is in fact the most important greenhouse gas in our atmosphere.

The Hydrologic Cycle

Almost all of the world's water (97 percent) is located in the oceans, but as might be expected, the high concentration of salts renders the oceans virtually unusable as a source of water for municipal, agricultural, or most industrial needs. We do use the oceans, however, for thermal cooling of power plants and as a sink for much of our pollution. While desalination technologies exist, the capital and energy requirements to produce significant quantities of fresh water are generally prohibitive, although some small regions of the world do rely quite heavily on this approach. Fortunately, the sun performs that desalination service for us when it provides the energy needed to evaporate water, leaving the salts behind. In fact, close to one-half of the sun's energy that is absorbed on the earth's surface is converted to latent heat, removing water from wet surfaces by evaporation and from the leaves of plants by transpiration. The combination of processes, called *evapotranspiration,* requires an enormous amount of energy, equivalent to roughly 4000 times the rate at which we use energy resources to power our societies.

Evapotranspiration removes an amount of water equivalent to a layer about 1 m thick around the globe each year. About 88 percent of that is evaporation from the

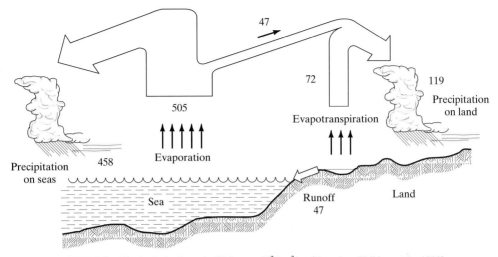

FIGURE 5.3   The hydrologic cycle. Units are 10^3 km^3/yr. (Based on Shiklomanov, 1993)

oceans, while the remaining 12 percent is evapotranspiration from the land. The result-ing water vapor is transported by moving air masses and eventually condenses and returns to the earth's surface as precipitation. Over the oceans, there is more evapora-tion than precipitation, and over the land it is the other way around, more precipitation than evapotranspiration. The difference between precipitation and evapotranspiration on land is water that is returned to the oceans, both by stream flow and groundwater flow, as *runoff.* Figure 5.3 illustrates this simple concept of evapotranspiration, precipi-tation, and runoff as components of the *hydrologic cycle.*

This representation of the hydrologic cycle is highly simplified, masking many of the complexities of timing and distribution. Snowfall may remain locked in polar ice for thousands of years; groundwater may emerge at the surface, contributing to surface water flow and vice versa; droughts and floods attest to the erratic rates of precipita-tion; and our own activities, most important those contributing to global climate change, are probably beginning to alter these balances.

As the data in Figure 5.3 indicate, 60 percent of the precipitation falling on the earth's land masses is eventually returned to the atmosphere as evapotranspiration. Of the 40 percent that does not evaporate, most collects on the surface, flowing into streams and rivers and emptying into the oceans, while some seeps into the soil to become under-ground water that slowly moves toward the seas. This combined groundwater and sur-face water runoff, 47,000 km^3/yr, is a renewable supply of freshwater that can potentially be used year after year without ever depleting the freshwater resources of the world.

While the rates of evaporation, precipitation, and runoff are obviously important, the amounts of water stored in various locations and forms are also critical. It has already been mentioned that almost all of the world's water is contained in the oceans. The remainder is distributed as shown in Table 5.1. Freshwater lakes and rivers, which are the main source of water for human use, account for just 0.007 percent of the world's stock of water, or about 93,000 km^3.

TABLE 5.1 Stocks of Water on Earth

Location	Amount (10^6 km^3)	Percentage of World Supply
Oceans	1338.0	96.5
Glaciers and permanent snow	24.1	1.74
Groundwater	23.4	1.7
Ground ice/permafrost	0.30	0.022
Freshwater lakes	0.091	0.007
Saline lakes	0.085	0.006
Swamp water	0.011	0.008
Atmosphere	0.013	0.001
Average in stream channels	0.002	0.0002
Water in living biomass	0.001	0.0001

Source: Shiklomanov (1993).

Water Usage

Roughly 10 percent of the world's annual runoff is withdrawn for human use each year. While that small figure may suggest ample supplies for the future, that is not at all the case. Some areas of the world are inundated with water, while others have so little rainfall that human existence is barely possible. Even areas with adequate average precipitation are vulnerable to chaotic variations from one year to the next. Unless major water storage and conveyance facilities are constructed, a region may have plenty of water on the average, but not enough to cover needs during dry spells. As population grows, the demands for water will rise and the amount available for each person will drop, as Figure 5.4 suggests.

The geographic distribution of water does not match well the distribution of people on the planet. Asia, with 60 percent of the world's population, has only 36 percent of global runoff, while South America, with only 5 percent of the world's population, has 25 percent of the runoff. Variations within regions or continents can be extreme.

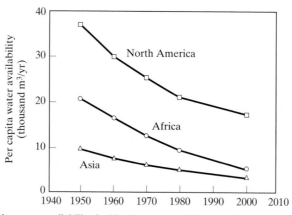

FIGURE 5.4 Per capita water availability for North America, Africa, and Asia, showing the implications of growing population. (*Source:* Shiklomanov, 1993, Reprinted by permission of Oxford University Press)

TABLE 5.2   Dependence on imported surface water for selected countries.

Country	Percent of total flow originating outside of border	Ratio of external water supply to internal supply
Egypt	97	32.3
Hungary	95	17.9
Mauritania	95	17.5
Gambia	86	6.4
Syria	79	3.7
Sudan	77	3.3
Iraq	66	1.9
Bangladesh	42	0.7

Source: Gleick (1993).

For example, the per capita water availability in north Africa is less than 7 percent of the African average, which is already low.

As populations grow and development proceeds, rising demands for water increase the potential for internal disruption within countries and external conflict with other countries. Many countries depend on local rivers for their water supply, but their upstream neighbors control the flow, examples of which are given in Table 5.2. Egypt, for example, gets 32.3 times as much surface water from precipitation that falls outside of Egypt as it does from its own rainfall. Moreover, Egypt depends on the Nile for 97 percent of its surface water supplies, while its neighbor, Ethiopia, controls almost all of the Nile's total flow. Similar circumstances exist all around the globe. Turkey, Syria, and Iraq, for example, share the Euphrates; Bangladesh relies on the Ganges, which is controlled by India; Israel and Jordan share an uneasy dependence on the Jordan River. As these and other such countries increase their demands for water, stresses between neighbors will no doubt intensify.

When describing how water is used, and especially if comparisons are to be made with available supplies, we have to be careful about the terminology. For example, what does it mean to say that a sector of society "uses" a certain amount of water? For example, almost all of the water that is withdrawn from a river for power plant cooling is returned to the river (at a slightly higher temperature) and can be used again. The rest is lost to the atmosphere by evaporation. Downriver, that same water may sprinkle someone's lawn. Again evaporation takes its toll. To keep track of water usage, it is important to distinguish between consumptive uses of water, in which water is made unavailable for future use (lost mostly to evaporation), and nonconsumptive uses, in which water serves its purpose for one user and is made available for the next. A simple equation relates the key terms: water *withdrawals,* water *returns,* and water *consumption.*

$$\text{Withdrawals} = \text{Consumption} + \text{Returns} \qquad (5.1)$$

Figure 5.5 provides data on freshwater withdrawals, consumption, and returns in the United States. Total withdrawals are about 500 km³/yr, which is about one-third of

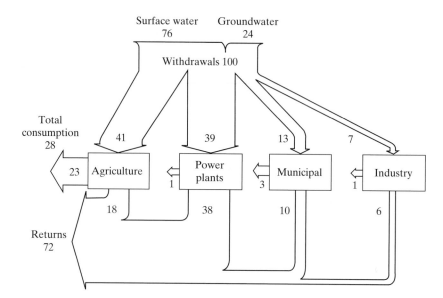

FIGURE 5.5 Freshwater use in the United States, 1990. Total annual withdrawals are about 500 km³. *Note: Industry* includes mining, *agriculture* includes irrigation and livestock, and *municipal* includes public, domestic, and commercial uses. (Data from Gleick, 1993, and U.S. Water Resources Council, 1978)

the annual runoff. For the whole country, groundwater provides one-fourth of all withdrawals and surface water provides the other three-fourths, but that distribution is very uneven across the country. In Kansas, for example, close to 90 percent of withdrawals are from groundwater, while in Montana the total is only 2 percent.

While water to cool power plants accounts for almost 40 percent of freshwater withdrawals, essentially all of that water is returned, and it is returned in a relatively unpolluted condition. If those withdrawals are removed from Figure 5.5, the importance of agriculture stands out. Excluding power plant cooling, agriculture accounts for 67 percent of freshwater withdrawals in the United States and 85 percent of consumption. In some areas, the predominant role of agriculture is even more dramatic. In drought-prone California, for example, irrigation accounts for over 80 percent of both withdrawals and consumption (excluding cooling water), while municipal withdrawals account for only about 15 percent of total withdrawals. As the state continues to grow, water transfers from agricultural uses to cities would be one way to meet increasing urban needs. A reduction of only 20 percent in agriculture, for example, could double the water supply in cities.

Average per-capita household water use in the United States is about 400 liters per day (110 gal/person-day). If water for landscaping is not included, it is not uncommon for one-third of the remaining water to be used for toilet flushings, another one-third for bathing, and the remaining third for cooking, laundry, and everything else. Table 5.3 shows some examples of water use for various household activities, giving standard and water-conserving estimates.

TABLE 5.3   Examples of water withdrawals to supply various end uses

Household water usage[a]	Liters	Gallons
Standard toilet, per flush	10–30	3–8
Ultralow volume toilet, per flush	6 or less	1.6 or less
Dishwasher, per load	50–120	13–30
Water-saver dishwasher, per load	40–100	10–25
Washing dishes in a filled sink	20–40	5–10
Shower head, per minute	20–30	5–8
Low-flow shower head, per minute	6–11	1.5–3
Slowly dripping faucet, per day	40–80	10–20
Washing car with running water, 20 min	400–800	100–200
Washing car with pistol-grip faucet, 20 min	60 or more	15 or more
Uncovered 60-m^2 pool, per day	100–400	25–100
Covered pool, per day	10–40	2.5–10
Agricultural items[b]		
One egg	150	40
Glass of milk	380	100
One pound of rice	2120	560
One pound of grain-fed beef	3030	800
One pound of cotton	7730	2040

[a]From Gleick, 1993.
[b]From U.S. GS, 1984.

5.3 WATER POLLUTANTS

Water that has been withdrawn, used for some purpose, and then returned will be polluted in one way or another. Agricultural return water contains pesticides, fertilizers, and salts; municipal return water carries human sewage; power plants discharge water that is elevated in temperature; industry contributes a wide range of chemical pollutants and organic wastes. The list of pollutants that contaminate return water is lengthy, so it helps to try to organize them into a smaller number of major categories, as has been done in this section.

Pathogens

It has long been known that contaminated water is responsible for the spread of many contagious diseases. In a famous study published in 1849, Dr. John Snow provided some of the earliest evidence of the relationship between human waste, drinking water, and disease. He noted that individuals who drank from a particular well on Broad Street in London were much more likely to become victims of a local cholera epidemic than those from the same neighborhood who drank from a different well. He not only found a likely source of the contamination—sewage from the home of a cholera patient—but he was able to end the epidemic by simply removing the handle from the pump on the Broad Street well. It was not until later in the nineteenth century, however, when Pasteur and others convincingly established the germ theory of disease, that the role of pathogenic microorganisms in such epidemics was understood.

Pathogens are disease-causing organisms that grow and multiply within the host. The resulting growth of microorganisms in a host is called an infection. Examples of pathogens associated with water include *bacteria,* which are responsible for cholera, bacillary dysentery (shigellosis), typhoid, and paratyphoid fever; *viruses,* which are responsible for infectious hepatitis and poliomyelitis; *protozoa,* which cause amebic dysentery, giardiasis, and cryptosporidiosis; and *helminths,* or parasitic worms, which cause diseases such as schistosomiasis and dracunculiasis (guinea-worm disease). Table 5.4 provides a more complete list of pathogens excreted in human feces. The intestinal discharges of an infected individual, a carrier, may contain billions of these pathogens, which, if allowed to enter the water supply, can cause epidemics of immense proportions. Carriers may not even necessarily exhibit symptoms of their disease, which makes it even more important to protect all water supplies carefully from any human waste contamination. In developing countries, where

TABLE 5.4 Typical Pathogens Excreted In Human Feces

Pathogen Group and Name	Associated Diseases
Virus	
Adenoviruses	Respiratory, eye infections
Enteroviruses	
Polioviruses	Aseptic meningitis, poliomyelitis
Echoviruses	Aseptic meningitis, diarrhea, respiratory infections
Coxsackie viruses	Aseptic meningitis, herpangina, myocarditis
Hepatitis A virus	Infectious hepatitis
Reoviruses	Not well known
Other viruses	Gastroenteritis, diarrhea
Bacterium	
Salmonella typhi	Typhoid fever
Salmonella paratyphi	Paratyphoid fever
Other salmonellae	Gastroenteritis
Shigella species	Bacillary dysentery
Vibrio cholerae	Cholera
Other vibrios	Diarrhea
Yersinia enterocolitica	Gastroenteritis
Protozoan	
Entamoeba histolytica	Amoebic dysentery
Giardia lamblia	Diarrhea
Cryptosporidium species	Diarrhea
Helminth	
Ancylostoma duodenale (hookworm)	Hookworm
Ascaris lumbricoides (roundworm)	Ascariasis
Hymenolepis nana (dwarf tapeworm)	Hymenolepiasis
Necator americanus (hookworm)	Hookworm
Strongyloides stercoralis (threadworm)	Strongyloidiasis
Trichuris trichiura (whipworm)	Trichuriasis

Source: Hammer and Hammer, 1996.

resources are scarce, even simple measures can be quite effective. Household water, for example, is often taken from open wells or streams that are easily contaminated. By enclosing a well and replacing the dirty rope and bucket with a less easily contaminated handpump, the incidence rate of these diseases can be greatly reduced.

Epidemics of infectious diseases periodically emerge in areas where crowded conditions and poor sanitation enable the microbes to reach new victims at a rapid rate. With international travel being so common these days, local epidemics can become global pandemics. A pandemic of cholera, for example, that began in 1961 in Indonesia has been tracked by the World Health Organization (WHO) as the disease spread around the globe (Figure 5.6). By 1970, it had reached Africa, where it spread quickly and remained for another two decades. It finally jumped to Peru in 1991, where it killed almost 4000 people. Within one year 400,000 new cases of cholera were reported in South America, where there had been almost none before.

There are many ways that contaminated water is associated with infectious diseases. The most commonly used system of classification to describe and track diseases that involve water for their transmission is described in Table 5.5. *Waterborne* diseases, such as cholera and typhoid, are spread by ingestion of contaminated water; *water-washed* diseases, such as trachoma and scabia, are associated with lack of sufficient water to maintain cleanliness; *water-based* diseases, such as schistosomiasis and dracunculiasis, involve water contact but do not require ingestion; and *water-related* diseases, such as malaria and dengue, involve a host that depends on water for its habitat (e.g., mosquitoes), but human contact with the water is not required. Table 5.6 shows estimates of global morbidity, mortality, and numbers of people living in areas where

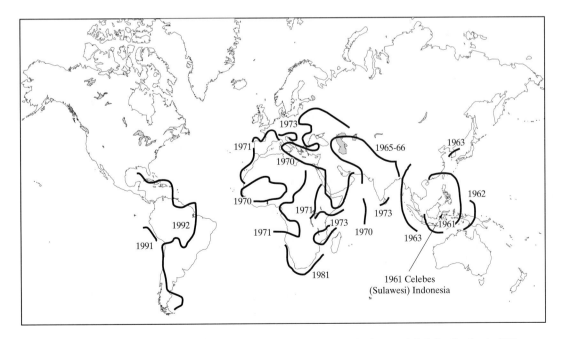

FIGURE 5.6   The progressive spread of cholera from Indonesia in 1961 to the known global distribution in 1992. (UNEP, 1993)

TABLE 5.5   Classification Of Infectious Diseases Associated With Water

Transmission mechanism	Description	Examples of diseases
Waterborne	Oral ingestion of pathogens in water contaminated by urine or feces	Cholera, typhoid, bacillary dysentery, infectious hepatitis
Water-washed	Disease spread enhanced by scarcity of water making cleanliness difficult	Trachoma, scabia, dysentery, louseborne fever
Water-based	Water provides the habitat for inter-mediate host organisms, transmission to humans through water contact	Schistosomiasis (bilharziasis), dracunculiasis (guinea worm)
Water-related	Insect vectors (e.g., mosquitoes) rely on water for habitat, but human water contact not needed	Malaria, filariasis, yellow fever onchocerciasis (river blindness), dengue

TABLE 5.6   Selected Examples of Global Morbidity, Mortality, and Populations at Risk, for Infectious Diseases Associated With Water

Disease	Vector	Morbidity	Mortality	Population at risk
Diarrheal diseases	Microorganisms	> 1.5 billion	4 million	> 2 billion
Schistosomiasis	Water snails	200 million	200,000	500–600 million
Malaria	Mosquitoes	267 million	1–2 million	2.1 billion
Onchocerciasis	Blackflies	18 million	20 to 50,000	90 million

Source: UNEP (1993).

these diseases are common, for a number of the most prevalent diseases associated with water.

One disease, schistosomiasis (bilharzia), is particularly insidious since mere contact with contaminated water is sufficient to cause infection and since its incidence rate is to a large extent the result of water-resource development projects. Schistosomiasis is one of the most common water-associated diseases in the world, affecting approximately 200 million people, 200,000 of whom die each year. It is spread by free-swimming larva in the water, called *cercaria,* that attach themselves to human skin, penetrate it, and enter the bloodstream. Cercaria mature in the liver into worms that lay masses of eggs on the walls of the intestine. When these eggs are excreted into water, they hatch and have only a few hours to find a snail host, in which they develop into new cercaria. Cercaria excreted by the snails then have a few days to find another human host, continuing the cycle. Continuation of the cycle requires continued contamination by schistosomiasis carriers in waters that are still enough to allow snails to thrive. Unfortunately, development projects such as dams and irrigation canals, built in countries with poor sanitation, often lead to an increase in schistosomiasis by creating the still-water conditions needed by the intermediate hosts, the snails.

Not that long ago, even developed countries such as the United States experienced numerous epidemics of waterborne diseases such as typhoid and cholera. At the turn of the century, typhoid, for example, was killing approximately 28,000 Americans each year. In one tragic incident in 1885, almost 90,000 people in Chicago died of typhoid or cholera when untreated sewage was drawn directly into the drinking water supply during a severe storm. It was only after the advent of chlorination, which began in the United States in 1908, that outbreaks of waterborne diseases such as these became rare. Even now, however, inadequate sanitation in developing countries continues to contribute to high rates of disease and death. The World Health Organization, for example, estimates that 80 to 100 percent of the illness caused by cholera, typhoid, and guinea-worm infection could be eliminated in developing countries with improved water supply and sanitary disposal of excreta (UNEP, 1993).

In the United States, most waterborne diseases are adequately controlled, and we usually do not need to worry about water coming from the tap. There are times, however, when even disinfection is not totally effective, as was the case in Milwaukee, in 1993, when an outbreak of the gastrointestinal illness, cryptosporidiosis, occurred. *Cryptosporidium parvum* is a protozoan parasite that causes diarrhea, abdominal pain, nausea, and vomiting in its host. For individuals whose immune systems are compromised, it can be life threatening. Most incidents of cryptosporidiosis have been traced to water supplies that were contaminated with agricultural runoff containing cattle feces. An infected calf sheds billions of oocysts (eggs) in feces per day, and these oocysts are resistant to the usual chlorine, or chlorine dioxide, disinfection processes used in water treatment plants. At present, the most suitable method of removing them at water treatment plants is with coagulation, flocculation, and filtration processes that will be described later in this chapter.

Another protozoan, *Giardia lamblia,* is less deadly than *Cryptosporidium,* but it shares a number of similar characteristics. *Giardia* cysts can be carried by wild animals as well as humans; they survive for months in the environment, and they are not easily destroyed by chlorination. As every backpacker knows these days, it is no longer safe to drink surface water from even the most sparkling stream because of the risk of giardiasis caused by the *Giardia lamblia.*

Oxygen-demanding Wastes

One of the most important measures of the quality of a water source is the amount of dissolved oxygen (DO) present. As mentioned in Chapter 2, the saturated value of dissolved oxygen in water is modest, on the order of 8 to 15 mg of oxygen per liter of water, depending on temperature and salinity. The minimum recommended amount of DO for a healthy fish population has often been set at 5 mg/L, but a more careful analysis of the needs of different species of fish at different times in their life cycle yields a range of oxygen requirements. For example, the EPA recommends at least 8 mg/L for coldwater species, such as trout and salmon, during their embryonic and larval stages and the first 30 days after hatching, but 5 mg/L is the recommendation for early life stages of warmwater species such as bluegill and bass. For older fish, EPA recommends a seven-day mean minimum of 5 mg/L for coldwater fish and 4 mg/L for warmwater fish.

Oxygen-demanding wastes are substances that oxidize in the receiving body of water. As bacteria decompose these wastes, they utilize oxygen dissolved in the water, which reduces the remaining amount of DO. As DO drops, fish and other aquatic life are threatened and, in the extreme case, killed. In addition, as dissolved oxygen levels fall, undesirable odors, tastes, and colors reduce the acceptability of that water as a domestic supply and reduce its attractiveness for recreational uses. Oxygen-demanding wastes are usually biodegradable organic substances contained in municipal wastewaters or in effluents from certain industries, such as food processing and paper production. In addition, the oxidation of certain inorganic compounds may also contribute to the oxygen demand. Even naturally occurring organic matter, such as leaves and animal droppings, that finds its way into surface water contributes to oxygen depletion.

There are several measures of oxygen demand commonly used. The *chemical oxygen demand,* or COD, is the amount of oxygen needed to oxidize the wastes chemically, while the *biochemical oxygen demand,* or BOD, is the amount of oxygen required by microorganisms to degrade the wastes biologically. BOD has traditionally been the most important measure of the strength of organic pollution, and the amount of BOD reduction in a wastewater treatment plant is a key indicator of process performance. The importance and use of these measures will be discussed more fully in later sections.

Nutrients

Nutrients are chemicals, such as nitrogen, phosphorus, carbon, sulfur, calcium, potassium, iron, manganese, boron, and cobalt, that are essential to the growth of living things. In terms of water quality, nutrients can be considered as pollutants when their concentrations are sufficient to allow excessive growth of aquatic plants, particularly algae. When nutrients stimulate the growth of algae, the attractiveness of the body of water for recreational uses, as a drinking water supply, and as a viable habitat for other living things can be adversely affected.

Nutrient enrichment can lead to blooms of algae, which eventually die and decompose. Their decomposition removes oxygen from the water, potentially leading to levels of DO that are insufficient to sustain normal life forms. Algae and decaying organic matter add color, turbidity, odors, and objectionable tastes to water that are difficult to remove and that may greatly reduce its acceptability as a domestic water source. The process of nutrient enrichment, called *eutrophication,* is an especially important one in lakes, and it is described more fully in a later section of this chapter.

Aquatic species require a long list of nutrients for growth and reproduction, but from a water quality perspective, the three most important ones are carbon, nitrogen, and phosphorus. Plants require relatively large amounts of each of these three nutrients, and unless all three are available, growth will be limited. The nutrient that is least available relative to the plant's needs is called the *limiting nutrient.* This suggests that algal growth can be controlled by identifying and reducing the supply of that particular nutrient. Carbon is usually available from a number of natural sources, including alkalinity, dissolved carbon dioxide from the atmosphere, and decaying organic matter, so it is not often the limiting nutrient. Rather, it is usually either nitrogen or phosphorus that controls algal growth rates. In general, seawater is most often limited by nitrogen, while freshwater lakes are most often limited by phosphorus.

Major sources of nitrogen include municipal wastewater discharges, runoff from animal feedlots, chemical fertilizers, and nitrogen deposition from the atmosphere, especially in the vicinity of power plants. In addition, certain bacteria and blue-green algae can obtain their nitrogen directly from the atmosphere. These life forms are usually abundant in lakes that have high rates of biological productivity, making the control of nitrogen in such places extremely difficult.

Not only is nitrogen capable of contributing to eutrophication problems, but when found in drinking water in a particular form it can pose a serious public health threat. Nitrogen in water is commonly found in the form of nitrate (NO_3), which is itself not particularly toxic. However, certain bacteria commonly found in the intestinal tract of infants can convert nitrates to highly toxic nitrites (NO_2). Hemoglobin in the bloodstream has a greater affinity for nitrites than it does for oxygen, and when oxygen is replaced by nitrites a condition known as *methemoglobinemia* can result. The oxygen starvation characteristic of methemoglobinemia causes a bluish discoloration of the infant; hence it is commonly referred to as the "blue baby" syndrome. In extreme cases the victim may die from suffocation. Usually after the age of about six months, the digestive system of a child is sufficiently developed that this syndrome does not occur.

While there are usually enough natural sources of nitrogen to allow algae and aquatic weeds to grow, there is not much phosphorus available from nature, so it tends to become the controlling nutrient in rivers and lakes. Human activities, however, often provide enough phosphorus to allow excessive growth of aquatic weeds and algae. Human sources of phosphorus include agricultural runoff in heavily fertilized areas and domestic sewage. In sewage, part of the phosphorus is from human feces and part is from detergents. Sewage treatment plants do not usually remove very much of that phosphorus, so the simplest way to control releases is to reduce the amount of phosphorus reaching the facility in the first place, and the best way to do that is to limit the use of phosphorus in detergents.

Detergents were developed just after World War II to replace soaps, which tended to form a scum of insoluble precipitates (the ring around the bathtub) in hard water. When they were first introduced, detergents were nonbiodegradable, which led to mountains of foam on rivers, lakes, and sewage treatment plants, and in some areas foamy water came out the tap. The problem was caused by the choice of surfactant, which is the ingredient in detergents that lowers the surface tension in water and allows dirt particles to be lifted or floated from the soiled material during washing. By 1965, the choice of surfactants had changed so that all detergents became biodegradable, but that caused a new problem. Detergents contained large amounts of phosphorus, and when that was released during degradation the phosphorus acted as a stimulant to algal growth, and enormous blooms resulted.

The phosphorus in detergents is usually in the form of sodium tripolyphosphate (STP), $Na_5P_3O_{10}$. When washwater containing this ingredient is discarded, the following reaction occurs and slowly releases the orthophosphate ion, PO_4^{3-}:

$$P_3O_{10}^{5-} + 2\,H_2O \longrightarrow 3\,PO_4^{3-} + 4H^+ \qquad (5.2)$$

Tripolyphosphate   Orthophosphate

Orthophosphate is the form of phosphorus that is directly usable by plants, so it immediately begins to act as a fertilizer once it is released. Concern for the environmental effects of phosphorus has led to reductions in its use in detergents. In some countries,

STP has been replaced by sodium nitrilotriacetate (NTA), but concerns for its safety are still unresolved and it is not used in the United States.

Salts

Water naturally accumulates a variety of dissolved solids, or *salts,* as it passes through soils and rocks on its way to the sea. These salts typically include such cations as sodium, calcium, magnesium, and potassium, and anions such as chloride, sulfate, and bicarbonate. While a careful analysis of salinity would result in a list of the concentrations of the primary cations and anions, a simpler, more commonly used measure of salinity is the concentration of *total dissolved solids* (TDS). As a rough approximation, *freshwater* can be considered to be water with less than 1500 mg/L TDS; *brackish* waters may have TDS values up to 5000 mg/L; and, *saline* waters are those with concentrations above 5000 mg/L (Tchobanoglous and Shroeder, 1985). Seawater contains 30,000 to 34,000 mg/L TDS.

The concentration of dissolved solids is an important indicator of the usefulness of water for various applications. Drinking water, for example, has a recommended maximum TDS concentration of 500 mg/L. Livestock can tolerate higher concentrations. Upper limits for stock water concentrations quoted by the U.S. Geological Survey (U.S. GS, 1985) include poultry at 2860 mg/L, pigs at 4290 mg/L, and beef cattle at 10,100 mg/L. Of greater importance, however, is the salt tolerance of crops. As the concentration of salts in irrigation water increases above 500 mg/L, the need for careful water management to maintain crop yields becomes increasingly important. With sufficient drainage to keep salts from accumulating in the soil, up to 1500 mg/L TDS can be tolerated by most crops with little loss of yield (Frederick and Hanson, 1982), but at concentrations above 2100 mg/L, water is generally unsuitable for irrigation except for the most salt tolerant of crops.

All naturally occurring water has some amount of salt in it. In addition, many industries discharge high concentrations of salts, and urban runoff may contain large amounts in areas where salt is used to keep ice from forming on roads in the winter. While such human activities may increase salinity by adding salts to a given volume of water, it is more often the opposite process, the removal of freshwater by evaporation, that causes salinity problems. When water evaporates, the salts are left behind, and since there is less remaining freshwater to dilute them, their concentration increases.

Irrigated agriculture, especially in arid areas, is always vulnerable to an accumulation of salts due to this evapotranspiration on the cropland itself. The salinity is enhanced by the increased evaporation in reservoirs that typically accompany irrigation projects. In addition, irrigation drainage water may pick up additional salt as it passes over and through soils. As a result, irrigation drainage water is always higher in salinity than the supply water and, with every reuse, its salt concentration increases even more. In rivers that are heavily used for irrigation, the salt concentration progressively increases downstream as the volume of water available to dilute salts decreases due to evaporation and as the salt load increases due to salty drainage water returning from irrigated lands. As an example, Table 5.7 shows decreasing flows and increasing TDS for the Rio Grande as it travels from New Mexico to Texas.

TABLE 5.7   Mean Annual Flow and TDS Levels in the Rio Grande as it Travels Through New Mexico and Texas

Station	Flow ($10^6 \, m^3$/yr)	Dissolved solids (mg/L)
Otowi Bridge, NM	1.33	221
San Marcial, NM	1.05	449
Elephant Butte Outlet, NM	0.97	478
Caballo Dam, NM	0.96	515
Leasburg Dam, NM	0.92	551
El Paso, TX	0.65	787
Fort Quitman, TX	0.25	1691

Source: Skogerboe and Law (1971).

It has been estimated that roughly one-third of the irrigated lands in the western part of the United States have a salinity problem that is increasing with time, including regions in the Lower Colorado River Basin and the west side of the San Joaquin Valley in California. Salinity problems are also having major impacts on irrigated lands in Iraq, Pakistan, India, Mexico, Argentina, Mali, and North Africa, among others. The collapse of ancient civilizations, such as those that once flourished in the Fertile Crescent in what is now Iraq, is now thought to have been precipitated by the demise of irrigated agriculture caused by accumulating salt (Reisner, 1986).

Salt accumulation in soils is quite often controlled by flushing the salts away with additional amounts of irrigation water. This increases costs and wastes water, which may not be abundantly available in the first place, and unless adequate drainage is available, it increases the likelihood that a rising water table will drown plant roots in salt-laden water. Providing adequate drainage can be an expensive and challenging task involving extensive on-farm subsurface drainage systems coupled with a central drain and disposal system. Since irrigation return water contains not only salts but fertilizers and pesticides as well, finding an acceptable method of disposal is difficult.

Thermal Pollution

A large steam-electric power plant requires an enormous amount of cooling water. A typical nuclear plant, for example, warms about 150,000 m^3/hr of cooling water by 10 °C as it passes through the plant's condenser. If that heat is released into a local river or lake, the resulting rise in temperature can adversely affect life in the vicinity of the thermal plume. For some species, such as trout and salmon, any increase in temperature is life threatening, although for others warmed water might be considered beneficial. Within certain limits, thermal additions can promote fish growth, and fishing may actually be improved in the vicinity of a power plant. On the other hand, sudden changes in temperature caused by periodic plant outages, both planned and unanticipated, can make it difficult for the local ecology to acclimate.

As water temperature increases, two factors combine to make it more difficult for aquatic life to get sufficient oxygen from the water. The first results from the fact that metabolic rates tend to increase with temperature, generally by a factor of 2 for each 10 °C rise in temperature. This causes an increase in the amount of oxygen required by organisms. At the same time, the available supplies of dissolved oxygen are reduced both because waste assimilation is quicker, drawing down DO at a faster rate, and because the amount of DO that the water can hold decreases with temperature. Thus as temperatures increase, the demand for oxygen goes up while the amount of DO available goes down.

Heavy Metals

In some contexts, the definition of a *metal* is based on physical properties. Metals are characterized by high thermal and electrical conductivity, high reflectivity and metallic luster, strength, and ductility. From a chemical perspective, however, it is more common to use a broader definition that says a metal is an element that will give up one or more electrons to form a cation in an aqueous solution. With this latter definition, there are about 80 elements that can be called metals. The term *heavy metal* is less precisely defined. It is often used to refer to metals with specific gravity greater than about 4 or 5. In terms of their environmental impacts, the most important heavy metals are mercury (Hg), lead (Pb), cadmium (Cd), and arsenic (As).

Most metals are toxic, including aluminum, arsenic, beryllium, bismuth, cadmium, chromium, cobalt, copper, iron, lead, manganese, mercury, nickel, selenium, strontium, thallium, tin, titanium, and zinc. Metals differ from other toxic substances in that they are totally nondegradable, which means they are virtually indestructible in the environment. Some of these metals, such as chromium and iron, are essential nutrients in our diets, but in higher doses they can cause a range of adverse impacts on the body, including nervous system and kidney damage, creation of mutations, and induction of tumors.

Metals may be inhaled, as is often the case with lead, for example, and they may be ingested. How well they are absorbed in the body depends somewhat on the particular metal in question and the particular form that it exists in. For example, liquid mercury is not very toxic, and most of what is ingested is excreted from the body. Mercury vapor, on the other hand, is highly toxic. As a vapor it enters the lungs, where it diffuses into the bloodstream. When blood containing mercury reaches the brain, the mercury can pass into the brain, where it causes serious damage to the central nervous system. By contrast, lead does not pose much of a threat as a vapor since it has such a low vapor pressure, and it is most dangerous when it is dissolved into its ionic form, Pb^{2+}. Lead dissolved in blood is transferred to vital organs, including the kidneys and brain, and it readily passes from a pregnant woman to her fetus. Children and fetuses are the most at risk since their brains are growing rapidly and exposure to lead can cause severe and permanent brain damage. Most human exposure to lead results from inhalation of aerosols created when tetraethyllead is added to gasoline. More will be said about lead in Chapter 7.

The most important route for the elimination of metals once they are inside a person is via the kidneys. In fact, kidneys can be considered to be complex filters whose primary purpose is to eliminate toxic substances from the body. The kidneys contain millions of excretory units called nephrons, and chemicals that are toxic to the kidneys are called *nephrotoxins*. Cadmium, lead, and mercury are examples of nephrotoxic metals.

Pesticides

The term *pesticide* is used to cover a range of chemicals that kill organisms that humans consider undesirable. Pesticides can be delineated as insecticides, herbicides, rodenticides, and fungicides.

There are three main groups of synthetic organic insecticides: *organochlorines* (also known as *chlorinated hydrocarbons*), *organophosphates,* and *carbamates.* The most widely known organochlorine pesticide is DDT (dichlorodiphenyl-trichloroethane), which has been widely used to control insects that carry such diseases such as malaria (mosquitoes), typhus (body lice), and plague (fleas). By contributing to the control of these diseases, DDT is credited with saving literally millions of lives worldwide. In spite of its more recent reputation as a dangerous pesticide, in terms of human toxicity DDT is considered to be relatively safe. In fact, organochlorine insecticides in general are highly toxic to insects, but their acute human toxicity is relatively low. It was DDT's impact on food chains, rather than its toxicity to humans, that led to its ban in the developed countries of the world (but it is still used in developing countries).

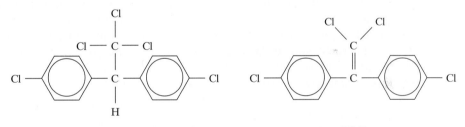

DDT: para-dichlorodiphenyltrichloroethane   DDE

Organochlorine pesticides, such as DDT, have two properties that cause them to be particularly disruptive to food chains. They are very *persistent,* which means they last a long time in the environment before being broken down into other substances, and they are quite *soluble* in hydrocarbon solvents, which means they easily accumulate in fatty tissue. The accumulation of organochlorine pesticides in fatty tissue means that organisms at successively higher trophic levels in a food chain are consuming food that has successively higher concentrations of pesticide. At the top of the food chain, body concentrations of these pesticides are the highest, and it is there that organochlorine toxicity has been most recognizable. Birds, for example, are high on the food chain, and it was the adverse effect of DDT on their reproductive success that focused attention on this particular pesticide. DDT, and its metabolite, DDE (dichlorodiphenyldichloroethene), interferes with the enzyme that regulates the distribution of calcium in birds, resulting in eggs with shells that are too thin to support the weight of the parent. The resulting difficulty to reproduce has been shown to affect a number of species, including peregrine falcons, bald eagles, ospreys, and brown pelicans. As environmental levels of DDT and DDE subside, some of these species are enjoying a resurgence in numbers. The bald eagle has been taken off the endangered species list in some areas, such as the Pacific Northwest and around the Great Lakes.

Other widely used organochlorines included methoxychlor, chlordane, heptachlor, aldrin, dieldrin, endrin, endosulfan, and Kepone®. Animal studies have shown that dieldrin, heptachlor, and chlordane produce liver cancers, and aldrin, dieldrin, and endrin have been shown to cause birth defects in mice and hamsters. Workers' exposure to Kepone in a manufacturing plant in Virginia showed severe neurological damage, and the plant was ultimately closed. Given the ecosystem disruption, their potential long-term health effects in humans (e.g., cancer), and the biological resistance to these pesticides that many insect species have developed, organochlorines have largely been replaced with organophosphates and carbamates.

The organophosphates, such as parathion, malathion, diazinon, TEPP (tetraethyl prophosphate), and dimethoate, are effective against a wide range of insects and they are not persistent. However, they are much more acutely toxic to humans than the organochlorines that they have replaced. They are rapidly absorbed through the skin, lungs, and gastrointestinal tract and hence, unless proper precautions are taken, they are very hazardous to those who use them. Humans exposed to excessive amounts have shown a range of symptoms, including tremor, confusion, slurred speech, muscle twitching, and convulsions.

The third category of insecticides, carbamates, are derived from carbamic acid, H_2NCOOH. They are similar to organophosphates in that they are short-lived in the environment. Therefore, they share the advantage of not being around long enough to bioaccumulate in food chains, but they also share the disadvantage of high human toxicity. Popular carbamate pesticides include propoxur, carbaryl, and aldicarb. Acute human exposure to carbamates has lead to a range of symptoms such as nausea, vomiting, blurred vision, and in extreme cases, convulsions and death.

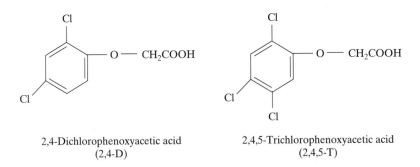

2,4-Dichlorophenoxyacetic acid
(2,4-D)

2,4,5-Trichlorophenoxyacetic acid
(2,4,5-T)

Chlorinated hydrocarbons are also used as herbicides. The chlorophenoxy compounds, 2,4,5-T and 2,4-D, are among the most well known because they were used as defoliants in the Vietnam war. Mixed together, they were called Agent Orange. They can kill broad-leafed plants without harming grasses, and they have been used in controlling excessive growth of aquatic plants in lakes and reservoirs. The herbicide 2,4,5-T has been banned in part because the manufacturing process that produces 2,4,5-T also produces a highly toxic side-product, dioxin. Dioxins also enter the environment as products of combustion from incinerators, and more will be said about them in Chapter 9. Other herbicides include paraquat, which acquired some fame as the pesticide of choice for destroying marijuana, and metolachlor, which is commonly used on soybeans and corn.

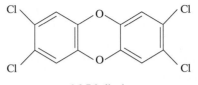

2,3,7,8-dioxin

Volatile Organic Compounds

Volatile organic compounds (VOCs) are among the most commonly found contaminants in groundwater. They are often used as solvents in industrial processes, and a number of them are either known or suspected carcinogens or mutagens. Their volatility means they are not often found in concentrations above a few μg/L in surface waters, but in groundwater their concentrations can be hundreds or thousands of times higher. Their volatility also suggests the most common method of treatment, which is to aerate the water to encourage them to vaporize.

Five VOCs are especially toxic, and their presence in drinking water is cause for special concern: vinyl chloride, tetrachloroethylene, trichloroethylene, 1,2-dichloroethane, and carbon tetrachloride.

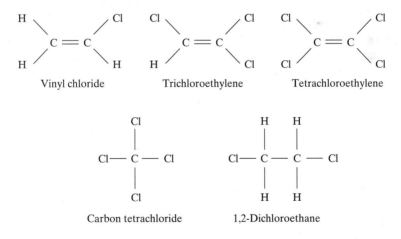

The most toxic of the five is *vinyl chloride* (chloroethylene). It is a known human carcinogen used primarily in the production of polyvinyl chloride resins. *Tetrachloroethylene* is used as a solvent, as a heat transfer medium, and in the manufacture of chlorofluorocarbons. It causes tumors in animals, but there is inadequate evidence to call it a human carcinogen. Of the five, it is the one most commonly found in groundwater. *Trichloroethylene* (TCE) is a solvent that was quite commonly used to clean everything from electronics parts to jet engines and septic tanks. It is a suspected carcinogen and it is among the most frequently found contaminants in groundwater. *1,2-Dichloroethane* is a metal degreaser that is also used in the manufacture of a number of products, including vinyl chloride, tetraethyllead, fumigants, varnish removers, and soap compounds. Though it is not a known carcinogen, high levels of exposure are known to cause injury to the central nervous system, liver, and kidneys. It is also a common groundwater contaminant that is quite soluble, making it one of the more difficult to remove by air stripping. *Carbon tetrachloride* was a common household cleaning agent that is now more often used in grain fumigants, fire extinguishers, and solvents. It is very toxic if ingested; only a few milliliters can produce death. It is relatively insoluble in water, however, and so it is only occasionally found in contaminated groundwater.

5.4 STATUS OF SURFACE WATER QUALITY

As a way to help monitor progress toward the goals of the Clean Water Act, states are required to submit water quality assessments of their rivers, lakes, and estuaries. The starting point for these assessments is a designation of the *beneficial uses* that individual bodies of water shall be required to support. The EPA provides guidance on these beneficial uses, as shown in Table 5.8. The Clean Water Act authorizes the states to set their own water quality standards that accompany these beneficial uses, as long as those standards comply with the "fishable and swimmable" goals of the Act.

The water quality assessments that the states must submit identify sources of pollution and the fraction of the resource that is "impaired" by pollutants. A body of water is said to be impaired when at least one of the designated beneficial uses, such as recreational swimming or fish consumption, is not being supported by the quality of water. The water quality assessments submitted in 1992 showed that 38 percent of the

TABLE 5.8   Beneficial Uses of Surface Water

Beneficial use	Descriptor
Aquatic life support	The waterbody provides suitable habitat for survival and reproduction of desirable fish, shellfish, and other aquatic organisms
Fish consumption	The waterbody supports a population of fish free from contamination that could pose a human health risk to consumers
Shellfish harvesting	The waterbody supports a population of shellfish free from toxicants and pathogens that could pose a human health risk to consumers
Drinking water supply	The waterbody can supply safe drinking water with conventional treatment
Primary contact recreation	People can swim in the waterbody without risk of adverse human health effects (such as catching waterborne diseases from raw sewage contamination)
Secondary contact recreation	People can perform activities on the water (such as kayaking) without risk of adverse human health effects from occasional contact with the water
Agriculture	The water quality is suitable for irrigating fields or watering livestock

Source: U.S. EPA (1994).

assessed miles of river, and 44 percent of the assessed area of lakes, were impaired for one reason or another.

The most important pollutants that cause impairment of water quality in rivers and streams are shown in Figure 5.8*a*, and the leading sources of those pollutants are given in Figure 5.8*b*. The highest-ranked problem is siltation caused by solids washed off of plowed fields, logging sites, urban areas, and stream banks when it rains. The resulting sediment load can smother habitats needed by fish and other aquatic organisms. In most states the leading source of siltation is agricultural runoff, but in some states where logging is a major activity, silviculture (forest activity) is the leading cause.

The pollutants and sources of impairment for lakes and reservoirs are shown in Figure 5.7. Lakes and reservoirs are most impaired by metals and nutrients coming from nonpoint sources. Agricultural runoff causes the most problems, but urban runoff is also very important.

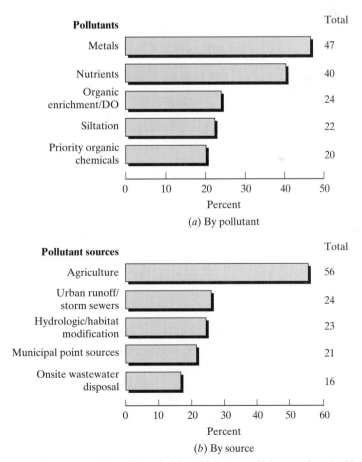

FIGURE 5.7   Extent of impairment by pollution in lakes: (a) Percent of lake acres impaired by pollutants, (b) percent of lake acres impaired by source of pollution. (U.S. EPA, 1994)

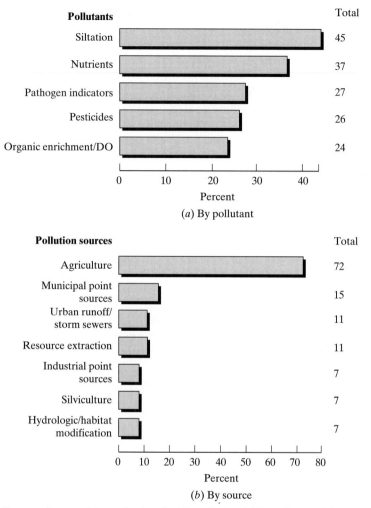

FIGURE 5.8 Percent of assessed river miles impaired by pollution: (a) By pollutants, (b) by sources of pollution. (U.S. EPA, 1994)

5.5 BIOCHEMICAL OXYGEN DEMAND

Surface water is obviously highly susceptible to contamination. It has historically been the most convenient sewer for industry and municipalities alike, while at the same time it is the source of the majority of our water for all purposes. One particular category of pollutants, oxygen-demanding wastes, has been such a pervasive surface-water problem, affecting both moving water and still water, that it will be given special attention.

When biodegradable organic matter is released into a body of water, microorganisms, especially bacteria, feed on the wastes, breaking them down into simpler organic and inorganic substances. When this decomposition takes place in an *aerobic* environment—that is, in the presence of oxygen—the process produces nonobjectionable, stable end products such as carbon dioxide (CO_2), sulfate (SO_4), orthophosphate (PO_4), and nitrate (NO_3). A simplified representation of aerobic decomposition is given by the following:

Aerobic Decomposition

$$\text{Organic matter} + O_2 \xrightarrow{\text{Microorganisms}} CO_2 + H_2O + \text{New cells} + \text{Stable products } (NO_3, PO_4, SO_4, \dots)$$

When insufficient oxygen is available, the resulting *anaerobic* decomposition is performed by completely different microorganisms. They produce end products that can be highly objectionable, including hydrogen sulfide (H_2S), ammonia (NH_3), and methane (CH_4). Anaerobic decomposition can be represented by the following:

Anaerobic Decomposition

$$\text{Organic matter} \xrightarrow{\text{Microorganisms}} CO_2 + H_2O + \text{New cells} + \text{Unstable products } (H_2S, NH_3, CH_4, \dots)$$

The methane produced is physically stable, biologically degradable, and is a potent greenhouse gas. When emitted from bodies of water, it is often called swamp gas. It is also generated in the anaerobic environment of landfills, where it is sometimes collected and used as an energy source.

The amount of oxygen required by microorganisms to oxidize organic wastes aerobically is called the *biochemical oxygen demand* (BOD). BOD may have various units, but most often it is expressed in milligrams of oxygen required per liter of wastewater (mg/L). The biochemical oxygen demand, in turn, is made up of two parts: the *carbonaceous oxygen demand* (CBOD), and the *nitrogenous oxygen demand* (NBOD). Those distinctions will be clarified later.

Five-day BOD Test

The total amount of oxygen that will be required for biodegradation is an important measure of the impact that a given waste stream will have on the receiving body of water. While we could imagine a test in which the oxygen required to degrade *completely* a sample of waste would be measured, for routine purposes such a test would take too long to be practical (at least several weeks would be required). As a result, it has become standard practice simply to measure and report the oxygen demand over a shorter, restricted period of five days, realizing that the ultimate demand is considerably higher.

The five-day BOD, or BOD_5, is the total amount of oxygen consumed by microorganisms during the first five days of biodegradation. In its simplest form, a BOD_5 test would involve putting a sample of waste into a stoppered bottle and measuring the concentration of dissolved oxygen (DO) in the sample at the beginning of the test and again five days later. The difference in DO divided by the volume of waste would be the five-day BOD. Light must be kept out of the bottle to keep algae from adding oxygen by photosynthesis, and the stopper is to keep air from replenishing DO that has been removed by biodegradation. To standardize the procedure, the test is run at a fixed temperature of 20 °C. Since the oxygen demand of typical waste is several hundred milligrams per liter, and since the saturated value of DO for water at 20 °C is only 9.1 mg/L, it is usually necessary to dilute the sample to keep final DO above zero. If during the five days the DO drops to zero, the test is invalid since more oxygen would have been removed had more been available.

The five-day BOD of a diluted sample is given by

$$BOD_5 = \frac{DO_i - DO_f}{P} \tag{5.3}$$

where

DO_i = the initial dissolved oxygen (DO) of the diluted wastewater

DO_f = the final DO of the diluted wastewater, 5 days later

P = the dilution fraction = $\dfrac{\text{volume of wastewater}}{\text{volume of wastewater plus dilution water}}$

A standard BOD bottle holds 300 mL, so P is just the volume of wastewater divided by 300 mL.

EXAMPLE 5.1   Unseeded Five-day BOD Test

A 10.0-mL sample of sewage mixed with enough water to fill a 300-mL bottle has an initial DO of 9.0 mg/L. To help assure an accurate test, it is desirable to have at least a 2.0-mg/L drop in DO during the five-day run, and the final DO should be at least 2.0 mg/L. For what range of BOD_5 would this dilution produce the desired results?

Solution   The dilution fraction is $P = 10/300$. To get at least a 2.0-mg/L drop in DO, the minimum BOD needs to be

$$BOD_5 \geq \frac{DO_i - DO_f}{P} = \frac{2.0 \text{ mg/L}}{(10/300)} = 60 \text{ mg/L}$$

To assure at least 2.0 mg/L of DO remaining after five days requires that

$$BOD_5 \leq \frac{(9.0 - 2.0) \text{ mg/L}}{(10/300)} = 210 \text{ mg/L}$$

So this dilution will be satisfactory for BOD_5 values between 60 and 210 mg/L. ∎

So far we have assumed that the dilution water added to the waste sample has no BOD of its own, which would be the case if pure water were added. In some cases it is necessary to seed the dilution water with microorganisms to assure that there is an adequate bacterial population to carry out the biodegradation. In such cases, to find the BOD of the waste itself, it is necessary to subtract the oxygen demand caused by the seed from the demand in the mixed sample of waste and dilution water.

To be able to sort out the effect of seeded dilution water from the waste itself, two BOD bottles must be prepared, one containing just the seeded dilution water and the other containing the mixture of both the wastewater and seeded dilution water (Figure 5.9). The change in DO in the bottle containing just seeded dilution water

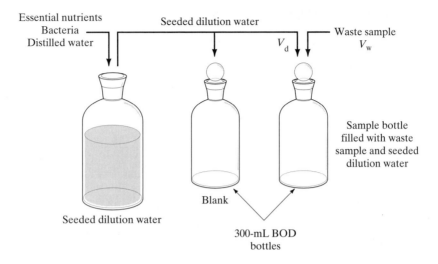

FIGURE 5.9   Laboratory test for BOD using seeded dilution water.

(called the "blank") as well as the change in DO in the mixture are then noted. The oxygen demand of the waste itself (BOD_w) can then be determined as follows:

$$BOD_m V_m = BOD_w V_w + BOD_d V_d \qquad (5.4)$$

where

BOD_m = BOD of the mixture of wastewater and seeded dilution water

BOD_w = BOD of the wastewater alone

BOD_d = BOD of the seeded dilution water alone (the blank)

V_w = the volume of wastewater in the mixture

V_d = the volume of seeded dilution water in the mixture

V_m = the volume of the mixture = $V_d + V_w$

Let P = the fraction of the mixture that is wastewater = V_w/V_m so that $(1 - P)$ = the fraction of the mixture that is seeded dilution water = V_d/V_m. Rearranging (5.4) gives

$$BOD_w = BOD_m\left(\frac{V_m}{V_w}\right) - BOD_d\left(\frac{V_d}{V_w} \times \frac{V_m}{V_m}\right) \qquad (5.5)$$

where the last term has been multiplied by unity (V_m/V_m). A slight rearrangement of (5.5) yields

$$BOD_w = \frac{BOD_m}{(V_w/V_m)} - BOD_d\frac{(V_d/V_m)}{(V_w/V_m)} \qquad (5.6)$$

Substituting the definitions of P and $(1 - P)$ into (5.6) gives

$$BOD_w = \frac{BOD_m - BOD_d(1 - P)}{P} \tag{5.7}$$

Since

$$BOD_m = DO_i - DO_f \quad \text{and} \quad BOD_d = B_i - B_f$$

where

B_i = initial DO in the seeded dilution water (blank)

B_f = final DO in the seeded dilution water

our final expression for the BOD of the waste itself is thus

$$BOD_w = \frac{(DO_i - DO_f) - (B_i - B_f)(1 - P)}{P} \tag{5.8}$$

EXAMPLE 5.2   A seeded BOD test

A test bottle containing just seeded dilution water has its DO level drop by 1.0 mg/L in a five-day test. A 300-mL BOD bottle filled with 15 mL of wastewater and the rest seeded dilution water (sometimes expressed as a dilution of 1:20) experiences a drop of 7.2 mg/L in the same time period. What would be the five-day BOD of the waste?

Solution   The dilution factor P is

$$P = 15/300 = 0.05$$

Using (5.8), the five-day BOD of the waste would be

$$BOD_5 = \frac{7.2 - 1.0(1 - 0.05)}{0.05} = 125 \text{ mg/L} \qquad \blacksquare$$

Modeling BOD as a First-order Reaction

Suppose we imagine a flask with some biodegradable organic waste in it. As bacteria oxidize the waste, the amount of organic matter remaining in the flask will decrease with time until eventually it all disappears. Another way to describe the organic matter in the flask is to say as time goes on, the amount of organic matter already oxidized goes up until finally all of the original organic matter has been oxidized. Figure 5.10 shows these two equivalent ways to describe the organic matter. We can also describe oxygen demand from those same two perspectives. We could say that the remaining demand for oxygen to decompose the wastes decreases with time until there is no more demand, or we could say the amount of oxygen demand already exerted, or utilized, starts at zero and rises until all of the original oxygen demand has been satisfied.

Translating Figure 5.10 into a mathematical description is straightforward. To do so, it is often assumed that the rate of decomposition of organic wastes is proportional

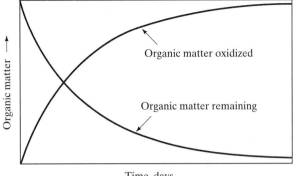

FIGURE 5.10   Two equivalent ways to describe the time dependence of organic matter in a flask.

to the amount of waste that is left in the flask. If we let L_t represent the amount of oxygen demand left after time t, then, assuming a first-order reaction, we can write

$$\frac{dL_t}{dt} = -kL_t \tag{5.9}$$

where k = the BOD reaction constant (time^{-1}).

The solution to (5.9) is

$$L_t = L_0 e^{-kt} \tag{5.10}$$

where L_0 is the *ultimate carbonaceous oxygen demand.* It is the total amount of oxygen required by microorganisms to oxidize the carbonaceous portion of the waste to simple carbon dioxide and water. (Later we will see that there is an additional demand for oxygen associated with the oxidation of nitrogen compounds.) The ultimate carbonaceous oxygen demand is the sum of the amount of oxygen already consumed by the waste in the first t days (BOD$_t$), plus the amount of oxygen remaining to be consumed after time t. That is,

$$L_0 = \text{BOD}_t + L_t \tag{5.11}$$

Combining (5.10) and (5.11) gives us

$$\text{BOD}_t = L_0(1 - e^{-kt}) \tag{5.12}$$

A graph of Eqs. (5.10) and (5.12) is presented in Figure 5.11. If these two figures are combined, the result would look exactly like Figure 5.10. Notice that oxygen demand can be described by the BOD remaining (you might want to think of L_t as how much oxygen demand is *L*eft at time t), as in Figure 5.11a, or equivalently as oxygen demand already satisfied (or utilized, or exerted), BOD$_t$, as in Figure 5.11b. Also notice how the five-day BOD is more easily described using the BOD utilized curve.

Sometimes the analysis leading to (5.12) is made using logarithms to the base 10 rather than the base e, as they were here. The relationship equivalent to (5.12), but in base 10, is

$$\text{BOD}_t = L_0(1 - 10^{-Kt}) \tag{5.13}$$

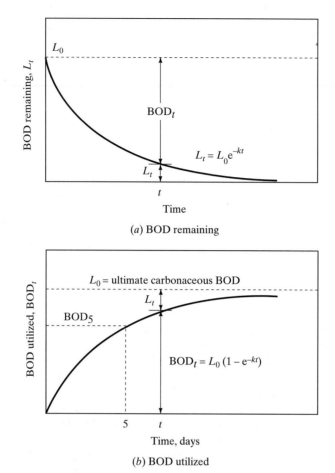

FIGURE 5.11 Idealized carbonaceous oxygen demand: (*a*) The BOD remaining as a function of time, and (*b*) the oxygen already consumed as a function of time.

where upper case K is the reaction rate coefficient to the base 10. It is easy to show that

$$k = K \ln 10 = 2.303K \tag{5.14}$$

EXAMPLE 5.3   Estimating L_0 from BOD_5

The dilution factor P for an unseeded mixture of waste and water is 0.030. The DO of the mixture is initially 9.0 mg/L, and after five days it has dropped to 3.0 mg/L. The reaction rate constant k has been found to be 0.22 day^{-1}.

a. What is the five-day BOD of the waste?
b. What would be the ultimate carbonaceous BOD?
c. What would be the remaining oxygen demand after five days?

Solution

a. From (5.3), the oxygen consumed in the first five days is

$$BOD_5 = \frac{DO_i - DO_f}{P} = \frac{9.0 - 3.0}{0.030} = 200 \text{ mg/L}$$

b. The total amount of oxygen needed to decompose the carbonaceous portion of the waste can be found by rearranging (5.12):

$$L_0 = \frac{BOD_5}{(1 - e^{-kt})} = \frac{200}{(1 - e^{-0.22 \times 5})} = 300 \text{ mg/L}$$

c. After five days, 200 mg/L of oxygen demand out of the total 300 mg/L would have already been used. The remaining oxygen demand would therefore be (300 − 200) mg/L = 100 mg/L. ■

The BOD Reaction Rate Constant k

The BOD reaction rate constant k is a factor that indicates the rate of biodegradation of wastes. As k increases, the rate at which dissolved oxygen is used increases, although the ultimate amount required, L_0, does not change. The reaction rate will depend on a number of factors, including the nature of the waste itself (for example, simple sugars and starches degrade easily while cellulose does not), the ability of the available microorganisms to degrade the wastes in question (it may take some time for a healthy population of organisms to be able to thrive on the particular waste in question), and the temperature (as temperatures increase, so does the rate of biodegradation).

Some typical values of the BOD reaction rate constant, at 20 °C, are given in Table 5.9. Notice that raw sewage has a higher rate constant than either well-treated sewage or polluted river water. This is because raw sewage contains a larger proportion of easily degradable organics that exert their oxygen demand quite quickly, leaving a remainder that decays more slowly.

The rate of biodegradation of wastes increases with increasing temperature. To account for these changes, the reaction rate constant k is often modified using the following equation:

$$k_T = k_{20}\theta^{(T-20)} \tag{5.15}$$

TABLE 5.9   Typical values for the BOD rate constant k at 20 °C

Sample	k (day^{-1})[a]	K (day^{-1})[b]
Raw sewage	0.35–0.70	0.15–0.30
Well-treated sewage	0.12–0.23	0.05–0.10
Polluted river water	0.12–0.23	0.05–0.10

[a]Lowercase k reaction rates to the base e.
[b]Uppercase K reaction rates to the base 10.
Source: Davis and Cornwell (1991).

where k_{20} is the reaction rate constant at the standard 20 °C laboratory reference temperature, and k_T is the reaction rate at a different temperature T (expressed in °C). The most commonly used value for θ is 1.047, and although θ is somewhat temperature dependent, that single value will suffice for our purposes.

EXAMPLE 5.4   Temperature Dependence of BOD$_5$

In Example 5.3 the wastes had an ultimate BOD equal to 300 mg/L. At 20 °C, the five-day BOD was 200 mg/L and the reaction rate constant was 0.22/day. What would the five-day BOD of this waste be at 25 °C?

Solution   First we will adjust the reaction rate constant with (5.15) using a value of θ equal to 1.047:

$$k_{25} = k_{20}\theta^{(T-20)} = 0.22 \times (1.047)^{(25-20)} = 0.277/\text{day}$$

So, from (5.12),

$$\text{BOD}_5 = L_0(1 - e^{-k5}) = 300(1 - e^{-0.277\times5}) = 225 \text{ mg/L}$$

Notice that the five-day BOD at 25 °C is somewhat higher than the 20 °C value of 200 mg/L. The same total amount of oxygen is required at either temperature, but as temperature increases, it gets used sooner. ∎

Nitrification

So far, it has been assumed that the only oxygen demand is associated with the biodegradation of the carbonaceous portion of the wastes. There is a significant additional demand, however, caused by the oxidation of nitrogen compounds that we will now briefly describe.

Nitrogen is the critical element required for protein synthesis and, hence, is essential to life. When living things die or excrete waste products, nitrogen that was tied to complex organic molecules is converted to ammonia by bacteria and fungi. Then, in aerobic environments, nitrite bacteria (*Nitrosomonas*) convert ammonia to nitrite (NO_2^-), and nitrate bacteria (*Nitrobacter*) convert nitrite to nitrate (NO_3^-). This process, called nitrification, can be represented with the following two reactions:

$$2NH_3 + 3O_2 \xrightarrow{\text{Nitrosomonas}} 2NO_2^- + 2H^+ + 2H_2O \tag{5.16}$$

$$2NO_2^- + O_2 \xrightarrow{\text{Nitrobacter}} 2NO_3^- \tag{5.17}$$

This conversion of ammonia to nitrate requires oxygen, so nitrification exerts its own oxygen demand. Thus, we have the combination of oxygen requirements. The oxygen needed to oxidize organic carbon to carbon dioxide is called the *carbonaceous oxygen demand* (CBOD), while the oxygen needed to convert ammonia to nitrate is called the *nitrogenous oxygen demand* (NBOD).

Nitrification is just one part of the biogeochemical cycle for nitrogen. In the atmosphere nitrogen is principally in the form of molecular nitrogen (N_2) with a small but important fraction being nitrous oxide (N_2O). (Nitrous oxide is a greenhouse gas that will be considered again in Chapter 8.) Nitrogen in the form of N_2 is unusable by plants and must first be transformed into either ammonia (NH_3) or nitrate (NO_3^-) in the process called *nitrogen fixation.* Nitrogen fixation occurs during electrical storms when N_2 oxidizes, combines with water, and is rained out as HNO_3. Certain bacteria and blue-green algae are also capable of fixing nitrogen. Under anaerobic conditions, certain denitrifying bacteria are capable of reducing NO_3 back into NO_2 and N_2, completing the nitrogen cycle.

While the entire nitrogen cycle obviously is important, our concern in this section is with the nitrification process itself, in which organic-nitrogen in waste is converted to ammonia, ammonia to nitrite, and nitrite to nitrate. Figure 5.12 shows this sequential process, starting with all of the nitrogen bound up in organic form and weeks later ending with all of the nitrogen in the form of nitrate. Notice that the conversion of ammonia to nitrite does not begin right away, which means the nitrogenous biochemical oxygen demand does not begin to be exerted until a number of days have passed.

Figure 5.13 illustrates the carbonaceous and nitrogenous oxygen demands as they might be exerted for typical municipal wastes. Notice that the NBOD does not normally begin to exert itself for at least five to eight days, so most five-day tests are not affected by nitrification. In fact, the potential for nitrification to interfere with the standard measurement for CBOD was an important consideration in choosing the standard five-day period for BOD tests. To avoid further the nitrification complication, it is now an accepted practice to modify wastes in a way that will inhibit nitrification during that five-day period.

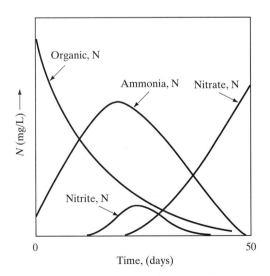

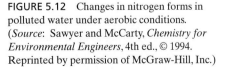

FIGURE 5.12   Changes in nitrogen forms in polluted water under aerobic conditions. (*Source*: Sawyer and McCarty, *Chemistry for Environmental Engineers*, 4th ed., © 1994. Reprinted by permission of McGraw-Hill, Inc.)

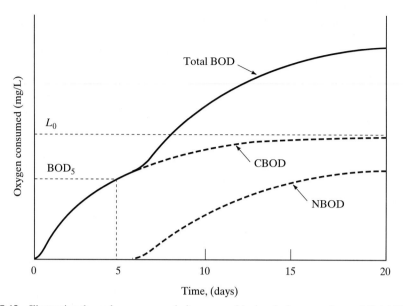

FIGURE 5.13 Illustrating the carbonaceous and nitrogenous biochemical oxygen demand. Total BOD is the sum of the two.

A stoichiometric analysis of (5.16) and (5.17) allows us to quantify the oxygen demand associated with nitrification, as the following example illustrates.

EXAMPLE 5.5   Nitrogenous Oxygen Demand

Some domestic wastewater has 30 mg/L of nitrogen either in the form of organic nitrogen or ammonia. Assuming that very few new cells of bacteria are formed during the nitrification of the waste (that is, the oxygen demand can be found from a simple stoichiometric analysis of the nitrification reactions given above), find

a. The ultimate nitrogenous oxygen demand

b. The ratio of the ultimate NBOD to the concentration of nitrogen in the waste.

Solution

a. Combining the two nitrification reactions (5.16) and (5.17) yields

$$NH_3 \ + \ 2O_2 \ \longrightarrow \ NO_3^- \ + \ H^+ \ + \ H_2O$$

The molecular weight of NH_3 is 17, and the molecular weight of O_2 is 32. The foregoing reaction indicates that one g-mol of NH_3 (17 g) requires two g-mole of O_2 ($2 \times 32 = 64$ g). Since 17 g of NH_3 contains 14 g of N, and the concentration of N is 30 mg/L, we can find the final, or ultimate, NBOD:

$$NBOD = 30 \text{ mg N/L} \times \frac{17 \text{g NH}_3}{14 \text{ g N}} \times \frac{64 \text{ g O}_2}{17 \text{ g NH}_3} = 137 \text{ mg O}_2/\text{L}$$

b. The oxygen demand due to nitrification divided by the concentration of nitrogen in the waste is

$$\frac{137 \text{ mg O}_2/\text{L}}{30 \text{ mg N/L}} = 4.57 \text{ mg O}_2/\text{mg N} \qquad \blacksquare$$

The total concentration of organic and ammonia nitrogen in wastewater is known as the *total Kjeldahl nitrogen,* or TKN. As was demonstrated in the preceding example, the nitrogenous oxygen demand can be estimated by multiplying the TKN by 4.57. This is a result work noting:

$$\text{Ultimate NBOD} \approx 4.57 \times \text{TKN} \qquad (5.18)$$

Since untreated domestic wastewaters typically contain approximately 15–50 mg/L of TKN, the oxygen demand caused by nitrification is considerable, ranging from roughly 70 to 230 mg/L. For comparison, typical raw sewage has an ultimate carbonaceous oxygen demand of 250–350 mg/L.

Other Measures of Oxygen Demand

In addition to the CBOD and NBOD measures already presented, there are two other indicators that are sometimes used to describe the oxygen demand of wastes. These are the *theoretical oxygen demand* (ThOD) and the *chemical oxygen demand* (COD).

As was described in Chapter 2, the theoretical oxygen demand is the amount of oxygen required to oxidize completely a particular organic substance, as calculated from simple stoichiometric considerations. Stoichiometric analysis, however, for both the carbonaceous and nitrogenous components, tends to overestimate the amount of oxygen actually consumed during decomposition. The explanation for this discrepancy is based on a more careful understanding of how microorganisms actually decompose waste. While there is plenty of food for bacteria, they rapidly consume waste and in the process convert some of it to cell tissue. As the amount of remaining wastes diminishes, bacteria begin to draw on their own tissue for the energy they need to survive, a process called endogenous respiration. Eventually, as bacteria die they become the food supply for other bacteria; all the while protozoa act as predators consuming both living and dead bacteria. Throughout this sequence, more and more of the original waste is consumed until finally all that remains is some organic matter, called humus, that stubbornly resists degradation. The discrepancy between theoretical and actual oxygen demands is explained by carbon still bound up in humus. The calculation of theoretical oxygen demand is of limited usefulness in practice since it presupposes a particular, single pollutant with known chemical formula, and even if that is the case, the demand is overestimated.

Some organic matter, such as cellulose, phenols, benzene, and tannic acid, resists biodegradation. Other types of organic matter, such as pesticides and various industrial chemicals, are nonbiodegradable because they are toxic to microorganisms. The chemical oxygen demand, COD, is a measured quantity that does not depend either on the ability of microorganisms to degrade the waste or on knowledge of the particular substances in question. In a COD test, a strong chemical oxidizing agent is used to oxidize the organics rather than relying on microorganisms to do the job. The COD test is much quicker than a BOD test, taking only a matter of hours. However, it does not distinguish between the oxygen demand that will actually be felt in a natural environment due to biodegradation, and the chemical oxidation of inert organic matter. It also does not provide any information on the rate at which actual biodegradation will take place. The measured value of COD is higher than BOD, though for easily biodegradable matter the two will be quite similar. In fact, the COD test is sometimes used as a way to estimate the ultimate BOD.

5.6 THE EFFECT OF OXYGEN-DEMANDING WASTES ON RIVERS

The amount of dissolved oxygen in water is one of the most commonly used indicators of a river's health. As DO drops below 4 or 5 mg/L, the forms of life that can survive begin to be reduced. In the extreme case, when anaerobic conditions exist, most higher forms of life are killed or driven off. Noxious conditions, including floating sludges, bubbling, odorous gases, and slimy fungal growths, then prevail.

A number of factors affect the amount of DO available in a river. Oxygen-demanding wastes remove DO; photosynthesis adds DO during the day, but those plants remove oxygen at night; and the respiration of organisms living in the water as well as in sediments removes oxygen. In addition, tributaries bring their own oxygen supplies, which mix with those of the main river. In the summer, rising temperatures reduce the solubility of oxygen while lower flows reduce the rate at which oxygen enters the water from the atmosphere. In the winter, ice may form, blocking access to new atmospheric oxygen. To model properly all of these effects and their interactions is a difficult task. A simple analysis, however, can provide insight into the most important parameters that affect DO. We should remember, however, that our results are only a first approximation to reality.

The simplest model of the oxygen resources in a river focuses on two key processes: the removal of oxygen by microorganisms during biodegradation, and the replenishment of oxygen through reaeration at the interface between the river and the atmosphere. In this simple model, it is assumed that there is a continuous discharge of waste at a given location on the river. As the water and wastes flow downriver, it is assumed that they are uniformly mixed at any given cross section of river, and it is assumed that there is no dispersion of wastes in the direction of flow. These assumptions are part of what is referred to as the *point-source, plug flow* model, illustrated in Figure 5.14.

Deoxygenation

The rate of deoxygenation at any point in the river is assumed to be proportional to the BOD remaining at that point. That is,

$$\text{Rate of deoxygenation} = k_d L_t \tag{5.19}$$

where

k_d = the deoxygenation rate constant (day^{-1})

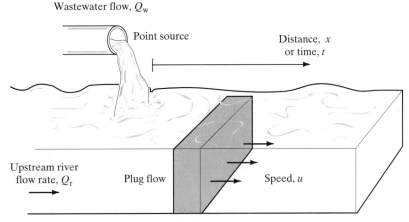

Wastewater flow, Q_w

Point source

Distance, x
or time, t

Upstream river
flow rate, Q_r

Plug flow

Speed, u

FIGURE 5.14   The point-source, plug flow model for dissolved-oxygen calculations.

L_t = the BOD remaining t (days) after the wastes enter the river, (mg/L)

The deoxygenation rate constant k_d is often assumed to be the same as the (temperature adjusted) BOD rate constant k obtained in a standard laboratory BOD test. For deep, slowly moving rivers, this seems to be a reasonable approximation, but for turbulent, shallow, rapidly moving streams, the approximation is less valid. Such streams have deoxygenation constants that can be significantly higher than the values determined in the laboratory.

Substituting (5.10), which gives BOD remaining after time t, into (5.19) gives

$$\text{Rate of deoxygenation} = k_d L_0 e^{-k_d t} \qquad (5.20)$$

where L_0 is the BOD of the mixture of streamwater and wastewater at the point of discharge. Assuming complete and instantaneous mixing,

$$L_0 = \frac{Q_w L_w + Q_r L_r}{Q_w + Q_r} \qquad (5.21)$$

where

L_0 = ultimate BOD of the mixture of streamwater and wastewater (mg/L)

L_r = ultimate BOD of the river just upstream of the point of discharge (mg/L)

L_w = Ultimate BOD of the wastewater (mg/L)

Q_r = volumetric flow rate of the river just upstream of the discharge point (m³/s)

Q_w = volumetric flow rate of wastewater (m³/s)

EXAMPLE 5.6   Downstream BOD

A wastewater treatment plant serving a city of 200,000 discharges 1.10 m³/s of treated effluent having an ultimate BOD of 50.0 mg/L into a stream that has a flow of 8.70 m³/s and a BOD of its own, equal to 6.0 mg/L. The deoxygenation constant, k_d is 0.20/day.

   a. Assuming complete and instantaneous mixing, estimate the ultimate BOD of the river just downstream from the outfall.

   b. If the stream has constant cross section so that is flows at a fixed speed equal to 0.30 m/s, estimate the BOD remaining in the stream at a distance 30,000 m downstream.

Solution

a. The BOD of the mixture of effluent and stream water can be found using (5.21):

$$L_0 = \frac{1.10 \text{ m}^3/\text{s} \times 50.0 \text{ mg/L} + 8.70 \text{ m}^3/\text{s} \times 6.0 \text{ mg/L}}{(1.10 + 8.70) \text{ m}^3/\text{s}} = 10.9 \text{ mg/L}$$

b. At a speed of 0.30 m/s, the time required for the waste to reach a distance 30,000 m downstream would be

$$t = \frac{30.000 \text{ m}}{0.30 \text{ m/s}} \times \frac{\text{hr}}{3600 \text{ s}} \times \frac{\text{day}}{24 \text{ hr}} = 1.16 \text{ days}$$

So the BOD remaining at that point, 30 km downstream, would be

$$L_t = L_0 e^{-k_d t} = 10.9 \, e^{-(0.2/\text{d} \times 1.16\text{d})} = 8.7 \text{ mg/L} \qquad \blacksquare$$

Reaeration

The rate at which oxygen is replenished is assumed to be proportional to the difference between the actual DO in the river at any given location and the saturated value of dissolved oxygen. This difference is called the oxygen deficit, D:

$$\text{Rate of reaeration} = k_r D \tag{5.22}$$

where

$$k_r = \text{reaeration constant (time}^{-1)}$$
$$D = \text{dissolved oxygen deficit} = (DO_s - DO) \tag{5.23}$$
$$DO_s = \text{saturated value of dissolved oxygen}$$
$$DO = \text{actual dissolved oxygen at a given location downstream}$$

   The reaeration constant, k_r, is very much dependent on the particular conditions in the river. A fast-moving, whitewater river will have a much higher reaeration constant than a sluggish stream or a pond. Many attempts have been made empirically to relate key stream parameters to the reaeration constant, with one of the most commonly used formulations being the following (O'Connor and Dobbins, 1958):

$$k_r = \frac{3.9 \, u^{1/2}}{H^{3/2}} \tag{5.24}$$

where

$$k_r = \text{reaeration coefficient at 20\,°C (day}^{-1})$$
$$u = \text{average stream velocity (m/s)}$$

$$H = \text{average stream depth (m)}$$

Typical values of the reaeration constant k_r for various bodies of water are given in Table 5.10. Adjustments to the reaeration rate constant for temperatures other than 20 °C can be made using (5.15) but with a temperature coefficient θ equal to 1.024.

The solubility of oxygen in water DO_s was first introduced in Chapter 2, where it was noted that the saturated value of dissolved oxygen varies with temperature, atmospheric pressure, and salinity. Table 5.11 gives representative values of the solubility of oxygen in water at various temperatures and chloride concentrations.

Both the wastewater that is being discharged into a stream and the stream itself are likely to have some oxygen deficit. If we assume complete mixing of the two, we can calculate the initial deficit of the polluted river using a weighted average based on their individual concentrations of dissolved oxygen:

$$D_0 = DO_s - \frac{Q_w DO_w + Q_r DO_r}{Q_w + Q_r} \tag{5.25}$$

where

D_0 = initial oxygen deficit of the mixture of river and wastewater

DO_s = saturated value of DO in water at the temperature of the river

TABLE 5.10   Typical Reaeration Constants for Various Bodies of Water

Water body	Range of k_r at 20 °C (day^{-1})[a]
Small ponds and backwaters	0.10–0.23
Sluggish streams and large lakes	0.23–0.35
Large streams of low velocity	0.35–0.46
Large streams of normal velocity	0.46–0.69
Swift streams	0.69–1.15
Rapids and waterfalls	> 1.15

[a]Base e.
Source: Tchobanoglous and Schroeder (1985).

TABLE 5.11   Solubility of Oxygen in Water (mg/L) at 1 atm pressure

Temperature (°C)	Chloride concentration in water (mg/L)			
	0	5000	10,000	15,000
0	14.62	13.73	12.89	12.10
5	12.77	12.02	11.32	10.66
10	11.29	10.66	10.06	9.49
15	10.08	9.54	9.03	8.54
20	9.09	8.62	8.17	7.75
25	8.26	7.85	7.46	7.08
30	7.56	7.19	6.85	6.51

Source: Thomann and Mueller (1987).

$DO_w = $ DO in the wastewater

$DO_r = $ DO in the river just upstream of the wastewater discharge point

---

EXAMPLE 5.7   Initial Oxygen Deficit

The waste water in Example 5.6 has a dissolved oxygen concentration of 2.0 mg/L and a discharge rate of 1.10 m³/s. The river that is receiving this waste has DO equal to 8.3 mg/L, a flow rate 8.70 m³/s, and a temperature of 20 °C. Assuming complete and instantaneous mixing, estimate the initial dissolved oxygen deficit of the mixture of wastewater and river water just downstream from the discharge point.

Solution   The initial amount of dissolved oxygen in the mixture of waste and river would be

$$DO = \frac{1.10 \text{ m}^3/\text{s} \times 2.0 \text{ mg/L} + 8.70 \text{ m}^3/\text{s} \times 8.3 \text{ mg/L}}{(1.10 + 8.70) \text{ m}^3/\text{s}} = 7.6 \text{ mg/L}$$

The saturated value of dissolved oxygen DO_s at 20 °C is given in Table 5.11 as 9.09 mg/L, so the initial deficit would be

$$D_0 = 9.09 \text{ mg/L} - 7.6 \text{ mg/L} = 1.5 \text{ mg/L} \qquad \blacksquare$$

The Oxygen Sag Curve

The deoxygenation caused by microbial decomposition of wastes and oxygenation by reaeration are competing processes that are simultaneously removing and adding oxygen to a stream. Combining the two equations (5.20) and (5.22) yields the following expression for the rate of increase of the oxygen deficit:

Rate of increase of the deficit = Rate of deoxygenation − Rate of oxygenation

$$\frac{dD}{dt} = k_d L_0 e^{-k_d t} - k_r D \qquad (5.26)$$

which has the solution

$$D = \frac{k_d L_0}{k_r - k_d} \left(e^{-k_d t} - e^{-k_r t} \right) + D_0 e^{-k_r t} \qquad (5.27)$$

Since the deficit D is the difference between the saturation value of dissolved oxygen DO_s and the actual value DO, we can write the equation for the DO as

$$DO = DO_s - \left[\frac{k_d L_0}{k_r - k_d} \left(e^{-k_d t} - e^{-k_r t} \right) + D_0 e^{-k_r t} \right] \qquad (5.28)$$

Equation (5.28) is the classic *Streeter-Phelps oxygen sag equation* first described in 1925. A plot of this DO is given in Figure 5.15. As can be seen, there is a stretch of river immediately downstream of the discharge point where the DO drops rapidly. At the *critical point* downstream, dissolved oxygen reaches its minimum value and river conditions are at their worst. Beyond the critical point, the remaining organic matter in the

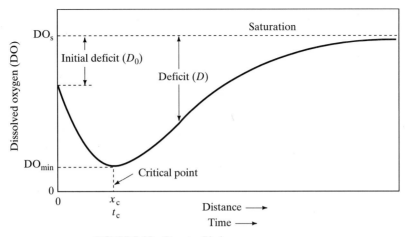

FIGURE 5.15   Streeter-Phelps oxygen sag curve.

river has diminished to the point where oxygen is being added to the river by reaeration faster than it is being withdrawn by decomposition, and the river begins to recover.

There are a few more things we can add to this analysis of oxygen deficit. For the special case where $k_r = k_d$, the denominator in (5.27) goes to zero, which is mathematically unacceptable, so the equation needs to be rederived under those conditions. Under these circumstances the solution to (5.26) becomes

$$D = (k_d L_0 t + D_0)e^{-k_d t} \tag{5.29}$$

If the stream has a constant cross-sectional area and is traveling at a speed u, then time and distance downstream are related by

$$x = ut \tag{5.30}$$

where

x = distance downstream

u = stream speed

t = elapsed time between discharge point and distance x downstream

Equation (5.27) can be rewritten as

$$D = \frac{k_d L_0}{k_r - k_d}\left(e^{-k_d x/u} - e^{-k_r x/u}\right) + D_0 e^{-k_r x/u} \tag{5.31}$$

The location of the critical point, and the corresponding minimum value of DO is of obvious importance. It is at this point where stream conditions are at their worst. Setting the derivative of the oxygen deficit equal to zero, and solving for the critical time, yields

$$t_c = \frac{1}{k_r - k_d}\ln\left\{\frac{k_r}{k_d}\left[1 - \frac{D_0(k_r - k_d)}{k_d L_0}\right]\right\} \tag{5.32}$$

The maximum deficit can then be found by substituting the value obtained for the critical time, t_c, into (5.28).

The oxygen sag curve should make some intuitive sense, even without the mathematical analysis. Near the outfall, there is so much organic matter being degraded that the rate of removal of oxygen from the water is higher than the rate that it can be returned by reaeration, so the dissolved oxygen drops. As we move further downstream there is less and less organic matter remaining, so the rate of removal of oxygen keeps dropping as well. At the critical point the rate of removal of oxygen equals the rate of addition of oxygen by reaeration. Beyond the critical point, reaeration begins to dominate, returning oxygen to the river at a faster rate than the bacteria remove it so the dissolved oxygen begins its climb back to the saturation value. Figure 5.16 shows the rate of deoxygenation, the rate of reaeration, and the oxygen sag curve.

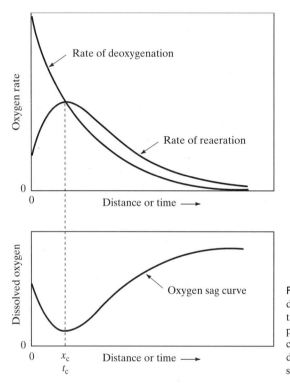

FIGURE 5.16   While the rate of deoxygenation exceeds the rate of reaeration the DO in the river drops. At the critical point those rates are equal. Beyond the critical point, reaeration exceeds decomposition, the DO curve climbs toward saturation, and the river recovers.

EXAMPLE 5.8   Streeter-Phelps Oxygen Sag Curve

Just below the point where a continuous discharge of pollution mixes with a river, the BOD is 10.9 mg/L and DO is 7.6 mg/L. The river and waste mixture has a temperature of 20 °C, a deoxygenation constant k_d of 0.20/day, an average flow speed of 0.30 m/s, and an average depth of 3.0 m. (In other words, this is just a continuation of the problem started in Examples 5.6 and 5.7).

   a. Find the time and distance downstream at which the oxygen deficit is a maximum.

   b. Find the minimum value of DO.

Solution   From Table 5.11, the saturation value of DO at 20 °C is 9.1 mg/L, so the initial deficit is

$$D_0 = 9.1 - 7.6 = 1.5 \text{ mg/L}$$

To estimate the reaeration constant, we can use the O'Connor and Dobbins relationship given in (5.24):

$$k_r = \frac{3.9 \, u^{1/2}}{H^{3/2}} = \frac{3.9 \, (0.30)^{1/2}}{(3.0)^{3/2}} = 0.41/\text{day}$$

a. Using (5.32), we can find the time at which the deficit is a maximum:

$$t_c = \frac{1}{k_r - k_d} \ln \left\{ \frac{k_r}{k_d} \left[1 - \frac{D_0(k_r - k_d)}{k_d L_0} \right] \right\}$$

$$= \frac{1}{(0.41 - 0.20)} \ln \left\{ \frac{0.41}{0.20} \left[1 - \frac{1.5 \, (0.41 - 0.20)}{0.20 \times 10.9} \right] \right\} = 2.67 \text{ days}$$

so the critical distance downstream would be

$$x_c = ut_c = 0.30 \text{ m/s} \times 3600 \text{ s/hr} \times 24 \text{ hr/d} \times 2.67 \text{ d} = 69,300 \text{ m} = 69.3 \text{ km}$$

which is about 43 miles

b. The maximum deficit can be found from (5.27):

$$D = \frac{k_d L_0}{k_r - k_d} \left(e^{-k_d t} - e^{-k_r t} \right) + D_0 e^{-k_r t}$$

$$= \frac{0.20 \times 10.9}{(0.41 - 0.20)} \left(e^{-0.20 \times 2.67} - e^{-0.41 \times 2.67} \right) + 1.5 e^{-0.41 \times 2.67} = 3.1 \text{ mg/L}$$

so the minimum value of DO will be the saturation value minus this maximum deficit:

$$\text{DO}_{\min} = (9.1 - 3.1) \text{ mg/L} = 6.0 \text{ mg/L}$$ ∎

In the preceding example, the lowest value of DO was found to be 6.0 mg/L, an amount sufficient for most aquatic life. If the amount of BOD added to the river is excessive, the oxygen sag curve may drop below a minimum acceptable level, leading to a stretch of river with unhealthful conditions, as shown in Figure 5.17. Fish that cannot tolerate water with such oxygen will be driven elsewhere or they will die. The variety of animals that can inhabit the unhealthy region is diminished and less desirable forms take over, forming thick mats of fungi, filamentous bacteria, sludge, and blood worms that blanket the bottom. The extreme case of excessive pollution is one in which the dissolved oxygen is driven to zero, creating a nearly lifeless, anaerobic stretch of river. Decomposition continues at a much slower rate by anaerobic microbes, releasing noxious and toxic gases such as hydrogen sulfide and ammonia.

Figure 5.17 may seem to suggest that the cause of an unhealthful drop in DO is an increase in the BOD from a waste source. That may be the case, but the oxygen sag curve is also sensitive to other sorts of changes. The DO curve changes with the

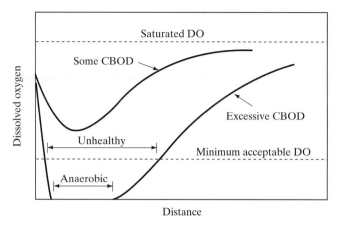

FIGURE 5.17   As a river gets more polluted, the oxygen sag curve drops below an acceptable level, and in the extreme case anaerobic conditions can occur.

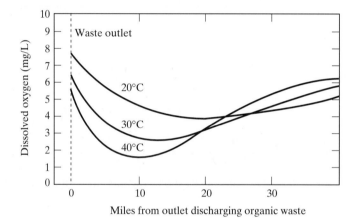

FIGURE 5.18   Changes in the oxygen sag curve as temperature increases. At higher temperatures the minimum DO is lower.

seasons, with the temperature, and with the time of day, even if the pollutant load is constant. In summer months, for example, river flows are usually diminished so the amount of waste dilution decreases, the BOD of the river/waste mix goes up, and the minimum DO drops.

The effect of temperature on the oxygen sag curve is also important (Figure 5.18). As temperatures rise, wastes decompose faster and the rate of deoxygenation increases. At the same time, the saturated value of DO drops, so reaeration slows down. The combination of these effects causes the critical point downstream to be reached sooner and the minimum value of DO to be lower as well. Thus a stream that may have sufficient DO in colder months may have an unacceptable deficit in the warmer months of summer. It also illustrates the potential adverse impact caused by thermal pollution from power plants. A river that might have been able to accept a certain sewage load without adverse effects could have unacceptably low oxygen levels when a power plant is added.

Photosynthesis also affects DO. Algae and other aquatic plants add DO during the daytime hours, while photosynthesis is occurring, but at night their continued respiration draws it down again. The net effect is a diurnal variation that can lead to elevated levels of DO in the late afternoon and depressed concentrations at night. For a lake or a slow-moving stream that is already overloaded with BOD and choked with algae, it is not unusual for respiration to cause offensive, anaerobic conditions late at night, even though the river seems fine during the day.

There are other factors not included in our simple oxygen sag model. Accumulated sludge along the bottom contributes to the oxygen demand; tributaries contribute their own BOD and DO; multiple sources cause multiple dips in the sag curve; nonpoint sources contribute pulses of BOD and other contaminants when it rains; and this model has not included the effects of nitrification. Nitrification causes a second dip in the oxygen sag curve a little further downstream as ammonia and organic nitrogen convert to nitrite and nitrate. If there is not enough DO downstream for nitrification to proceed, nitrogen remains as ammonia, which is toxic.

Modeling the impacts of BOD on the oxygen resources in a stream or river is an important part of the permitting process for new sources. Stream models can help determine the maximum amount of additional BOD that will be allowed, which, in turn, affects facility siting decisions and the extent of on-site waste treatment that will be required.

5.7 WATER QUALITY IN LAKES AND RESERVOIRS

All lakes gradually accumulate silt and organic matter as they undergo a natural aging process known as *eutrophication*. A young lake is characterized by a low nutrient content and low plant productivity. Such *oligotrophic* ("few foods") lakes gradually acquire nutrients from their drainage basins, which enables increased aquatic growth. Over time, the increased biological productivity causes the water to become murky with phytoplankton, while decaying organic matter contributes to the depletion of available dissolved oxygen. The lake becomes *eutrophic* ("well fed"). As the accumulating silt and organic debris causes the lake to get shallower and warmer, more plants take root along the shallow edges, and the lake slowly transforms into a marsh or bog.

While such eutrophication is a natural process that may take thousands of years, it is possible to accelerate greatly the rate of change through human activities. This is called *cultural eutrophication*. Municipal wastewater, industrial wastes, and runoff from fertilized agricultural lands add nutrients that stimulate algal growth and degrade water quality. Algal blooms die and decay, causing unsightly, odorous clumps of rotting debris along the shoreline and thick mats of dead organic matter in the lake. The decomposition of dead algae uses up available oxygen, resulting in the same sort of oxygen depletion problems already mentioned for streams. Among the first casualties are coldwater fish, whose temperature sensitivity forces them to stay in the colder bottom waters of a lake where the least amount of oxygen is available. In some lakes there are periods of time when anaerobic conditions prevail near the bottom. Not only are organisms at risk for lack of oxygen, but the toxicity of the water increases as hydrogen sulfide and metals, such as iron and manganese, which are normally tied up as precipitates in sediments, are dissolved and released into the lake.

The EPA lake and reservoir assessments mentioned in Section 5.4 indicate that more than half of U.S. lakes are degraded by eutrophication caused by nutrient enrichment from human activities. As Figure 5.7 points out, the most common source of lake and reservoir pollution is runoff from agricultural and urban nonpoint sources. Unfortunately, these are the most difficult sources to control.

Lakes and reservoirs are susceptible to a host of other pollution problems besides eutrophication. Unlike streams, which can transport pollutants "away," the slow dispersal rates in lakes let contamination accumulate quickly, which would make them seem to be less likely to be used as chemical dumping grounds. Unfortunately, that does not seem to be the case. For example, 99 percent of the shoreline along the Great Lakes is not meeting the EPA's criteria for unimpaired beneficial uses because of toxic organic chemical pollution (U.S. EPA, 1994). Lakes and reservoirs are also vulnerable to effects of sulfuric and nitric acids that are rained out of the sky.

Controlling Factors in Eutrophication

There are many factors that control the rate of production of algae, including the availability of sunlight to power the photosynthetic reactions, and the concentration of nutrients required for growth.

The amount of light available is related to the transparency of the water, which is in turn a function of the level of eutrophication. An oligotrophic lake, such as Lake Tahoe, may have enough sunlight to allow significant rates of photosynthesis to take place at a depth of 100 m or more, while eutrophic lakes may be so murky that photosynthesis is restricted to a thin layer of water very near the surface. The top layer of water in a lake, where plants produce more oxygen by photosynthesis than they remove by respiration, is called the *euphotic zone.* Below that lies the *profundal zone.* The transition between the two zones is designated the *light compensation level.* The light compensation level corresponds roughly to a depth at which light intensity is about 1 percent of full sunlight.

While the amount of sunlight available can be a limiting factor in algal growth, it is not something that we could imagine controlling as a way to slow eutrophication. The more obvious approach is to try to reduce the supply of nutrients. The list of nutrients that we might consider controlling, however, is long since it could include every nutrient known to be essential to plant growth. While that list would include carbon, nitrogen, phosphorus, sulfur, calcium, magnesium, potassium, sodium, iron, manganese, zinc, copper, boron, plus some other essential nutrients, the problem is greatly simplified by focusing on the two that most often limit algal growth: phosphorus and nitrogen.

Justus Liebig, in 1840, first formulated the idea that "growth of a plant is dependent on the amount of foodstuff that is presented to it in minimum quantity." This has come to be known as *Liebig's law of the minimum.* In essence, this law states that algal growth will be limited by the nutrient which is least available relative to its needs; therefore, the quickest way to control eutrophication would be to identify the *limiting nutrient* and reduce its concentration.

Liebig's law also implies that reductions in a nonlimiting nutrient will not provide effective control unless its concentration can be reduced to the point where it becomes the limiting nutrient. Thus, for example, reducing phosphorus loading by eliminating phosphates in detergents will have little effect in a region with nitrogen-limited surface water, but the same reductions could be very effective where phosphorus is the limiting nutrient. As lakes eutrophy, the dominant species of algae are often blue-green *Cyanophyta,* which have the unusual characteristic of being able to obtain their nitrogen directly from the atmosphere rather than the water. In addition, nitrogen enters water as "fallout" from combustion sources, particularly fossil-fuel-fired power plants. With the atmosphere as their rather unlimited nitrogen supply, most freshwater systems are phosphorus limited.

To help illustrate the relative amounts of nitrogen and phosphorus that are required for algal growth, consider the following frequently used representation of algal photosynthesis (Stumm and Morgan, 1981):

$$106\,CO_2 + 16NO_3^- + HPO_4^{2-} + 122H_2O + 18H^+ \longrightarrow C_{106}H_{263}O_{110}N_{16}P + 138O_2 \quad (5.33)$$

Using a simple stoichiometric analysis and remembering the atomic weights of nitrogen (14) and phosphorus (31), the ratio of the mass of nitrogen to phosphorus in this algae is

$$\frac{N}{P} = \frac{16 \times 14}{1 \times 31} = 7.2$$

As a first approximation, then, it takes about 7 times more nitrogen than phosphorus to produce a given mass of algae. As a rough guideline, when the concentration (mg/L) of nitrogen in water is more than 10 times the concentration of phosphorus, the body of water will probably be phosphorus limited. When it is less than 10:1 it will probably be nitrogen limited (Thomann and Mueller, 1987). Most marine waters have an N/P ratio less than 5 and are nitrogen limited.

Sawyer (1947) suggests that phosphorus concentrations in excess of 0.015 mg/L and nitrogen concentrations above 0.3 mg/L are sufficient to cause blooms of algae. These are in line with more recent estimates that suggest that 0.010 mg/L of phosphorus are "acceptable" while 0.020 mg/L are "excessive" (Vollenweider, 1975).

A Simple Phosphorus Model

Suppose we want to estimate the phosphorus concentration that would be expected in a completely mixed lake under steady-state conditions given some combination of phosphorus sources and sinks. By comparing the calculated phosphorus level with phosphorus concentrations that are generally considered "acceptable," we would be able to estimate the amount of phosphorus control needed to prevent a eutrophication problem. In phosphorus-limited lakes, 0.010 mg/L of phosphorus seems to be low enough to prevent excessive algal production, while 0.020 mg/L is excessive. Following the analysis presented in Thomann and Mueller (1987), we will make a first cut at this modeling exercise by making a number of simplifying assumptions and then performing a mass balance (introduced in Chapter 1) for phosphorus on the resulting idealized lake.

Consider the simple model shown in Figure 5.19. Phosphorus is shown entering the lake from a point source as well as from the incoming stream flow. Phosphorus is assumed to leave the lake either as part of the exiting streamflow or by settling into sediments. It would be easy to add other sources, including nonpoint sources, if their phosphorus loading rates could be estimated, but that will not be done here.

If we assume equal incoming and outgoing stream-flow rates, negligible volumetric flow input from the point source, a well-mixed lake, and steady-state conditions representing a seasonal or annual average, then we can write the following phosphorus balance:

$$\text{Rate of addition of } P = \text{Rate of removal of } P$$

$$QC_{\text{in}} + S = QC + v_s AC \tag{5.34}$$

where

S = the rate of addition of phosphorus from the point source (g/s)

C = the concentration of phosphorus in the lake (g/m^3)

C_{in} = the concentration of phosphorus in the incoming streamflow (g/m^3)

Q = the stream inflow and outflow rate (m^3/s)

v_s = the phosphorus settling rate (m/s)

A = the surface area of the lake (m^2)

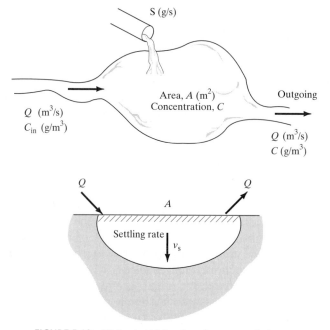

FIGURE 5.19 Well-mixed lake phosphorus mass balance.

which results in a steady-state concentration of

$$C = \frac{QC_{in} + S}{Q + v_s A} \tag{5.35}$$

The settling rate, v_s, is an empirically determined quantity that is difficult to predict with any confidence. Thomann and Mueller cite studies that suggest settling rates are around 10 to 16 m/yr.

EXAMPLE 5.9   Phosphorus Loading in a Lake

A phosphorus-limited lake with surface area equal to $80 \times 10^6 \, m^2$ is fed by a 15.0 m³/s stream that has a phosphorus concentration of 0.010 mg/L. In addition, effluent from a point source adds 1 g/s of phosphorus. The phosphorus settling rate is estimated at 10 m/yr.

   a. Estimate the average total phosphorus concentration.

   b. What rate of phosphorus removal at the wastewater treatment plant would be required to keep the concentration of phosphorus in the lake at an acceptable level of 0.010 mg/L?

Solution

   a. The phosphorus loading from the incoming stream is

$$QC_{in} = 15.0 \, m^3/s \times 0.010 \, mg/L \times \frac{1 \, g/m^3}{mg/L} = 0.15 \, g/s$$

Adding the 1.0 g/s from the point source gives a total phosphorus input rate of 1.15 g/s. The estimated settling rate is

$$v_s = \frac{10 \, m/yr}{365 \, d/yr \times 24 \, hr/d \times 3600 \, s/hr} = 3.17 \times 10^{-7} \, m/s$$

Using (5.35), the steady-state concentration of total phosphorus would be

$$C = \frac{QC_{in} + S}{Q + v_s A} = \frac{1.15 \, g/s}{15 \, m^3/s + 3.17 \times 10^{-7} \, m/s \times 80 \times 10^6 \, m^2}$$

$$= 0.028 \, g/m^3 = 0.028 \, mg/L$$

which is above the 0.010 mg/L needed to control eutrophication.

   b. To reach 0.010 mg/L, the phosphorus loading from the point source must be

$$S = C(Q + v_s A) - QC_{in}$$

$$= 0.010 \, g/m^3 (15 \, m^3/s + 3.17 \times 10^{-7} m/s \times 80 \times 10^6 \, m^2) - 15 \, m^3/s \times 0.010 \, g/m^3$$

$$= 0.25 \, g/s$$

The point-source effluent currently supplies 1.0 g/s, so 75 percent removal of phosphorus is needed. ∎

This model is based on a number of unrealistic assumptions. For example, due to a phenomenon called *thermal stratification,* which will be described in the next section, the assumption that the lake is well mixed is usually reasonable only during certain times of the year and in certain parts of the lake. The assumption of steady state ignores the dynamic behavior of lakes as weather and seasons change. We have also assumed that phosphorus is the controlling nutrient and that a simple measure of its concentration will provide an adequate indication of the potential for eutrophication. Finally, the model assumes a constant phosphorus settling rate that does not depend on such factors as the fraction of incoming phosphorus that is in particulate form, the movement of phosphorus both to sediments and from sediments, and physical parameters of the lake such as the ratio of lake volume to streamflow (the hydraulic detention time). Despite these unrealistic assumptions, the model has been shown in practice to produce very useful results (Thomann and Mueller, 1987).

Thermal Stratification

As we have seen, nutrients stimulate algal growth and the subsequent death and decay of that algae can lead to oxygen depletion. This oxygen depletion problem is worsened by certain physical characteristics of lakes, which we will now consider.

As mentioned earlier, one of the most unusual properties of water is the fact that its density does not monotonically increase as the temperature drops. Instead, it has a maximum point at 4 °C, as shown in Figure 5.20. One result of this density maximum is that ice floats because the water surrounding it is slightly warmer and denser. If water were like other liquids, ice would sink and it would be possible for lakes to freeze solid from the bottom up. Fortunately, this is not the case.

Above 4 °C the density of water decreases with temperature. As a result, a lake warmed by the sun during the summer will tend to have a layer of warm water floating on the top of the denser, colder water below. Conversely, in the winter, if the lake's surface drops below 4 °C, it will create a layer of cold water that floats on top of the more dense, 4 °C water below. These density differences between surface water and the water nearer to the bottom inhibit vertical mixing in the lake, causing a very stable layering effect known as *thermal stratification.*

Figure 5.21 shows the stratification that typically occurs in a deep lake, in the temperate zone, during the summer. In the upper layer, known as the *epilimnion,* the

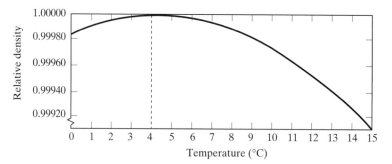

FIGURE 5.20   The density of water reaches a maximum at 4 °C.

warm water is completely mixed by the action of wind and waves, causing an almost uniform temperature profile. The thickness of the epilimnion varies from lake to lake and from month to month. In a small lake it may be only a meter or so deep, whereas in large lakes it may extend down 20 m or more. Below the epilimnion is a transition layer called the *thermocline,* or *metalimnion,* in which the temperature drops rather quickly. Most swimmers have experienced a thermocline's sudden drop in temperature when diving down into a lake. Below the thermocline is a region of cold water called the *hypolimnion.*

In terms of mixing, summer stratification creates essentially two separate lakes—a warm lake (epilimnion) floating on top of a cold lake (hypolimnion). The separation is quite stable and becomes increasingly so as the summer progresses. Once summer stratification begins, the lack of mixing between the layers causes the epilimnion, which is absorbing solar energy, to warm even faster, creating an even greater density difference. This difference is enhanced in a eutrophic lake since in such lakes most, if not all, of the absorption of solar energy occurs in the epilimnion.

As the seasons progress and winter approaches, the temperature of the epilimnion begins to drop and the marked stratification of summer begins to disappear. Sometime in the fall, perhaps along with a passing storm that stirs things up, the stratification will disappear, the temperature will become uniform with depth, and complete mixing of the lake becomes possible. This is called the *fall overturn.* Similarly, in climates that are cold enough for the surface to drop below 4 °C, there will be a winter stratification, followed by a *spring overturn* when the surface warms up enough to allow complete mixing once again.

Stratification and Dissolved Oxygen

Dissolved oxygen, one of the most important water quality parameters, is greatly affected by both eutrophication and thermal stratification. Consider, for example, two different stratified lakes, one oligotrophic and one eutrophic. In both lakes, the waters of the epilimnion can be expected to be rich in DO since oxygen is readily available from reaeration and photosynthesis. The hypolimnion, on the other hand, is cut off from the oxygen-rich epilimnion by stratification. The only source of new oxygen in the hypolimnion will be photosynthesis, but that will only happen if the water is clear enough to allow the euphotic zone to extend below the thermocline. That is, the

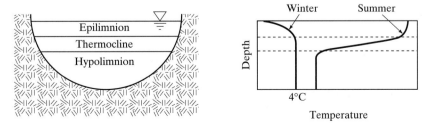

FIGURE 5.21   Thermal stratification of a lake showing winter and summer stratification temperature profiles.

hypolimnion of the clear, oligotrophic lake at least has the possibility of having a source of oxygen while that of the eutrophic lake does not.

In addition, the eutrophic lake is rich in nutrients and organic matter. Algal blooms suddenly appear and die off, leaving rotting algae that washes onto the beaches or sinks to the bottom. The rain of organic debris into the hypolimnion leads to increased oxygen demands there. Thus, not only is there inherently less oxygen available in the hypolimnion, but there is also more demand for oxygen due to decomposition, especially if the lake is eutrophic. Once summer stratification sets in, DO in the hypolimnion will begin dropping, driving fish out of the colder bottom regions of the lake and into the warmer, more oxygen-rich surface waters. As lakes eutrophy, fish that require cold water for survival are the first victims. In the extreme case, the hypolimnion of a eutrophic lake can become anaerobic during the summer, as is suggested in Figure 5.22.

During the fall and spring overturns, which may last for several weeks, the lake's waters become completely mixed. Nutrients from the bottom are distributed throughout the lake and oxygen from the epilimnion becomes mixed with the oxygen-poor hypolimnion. The lake, in essence, takes a deep breath of air.

In winter, demands for oxygen decrease as metabolic rates decrease, while at the same time the ability of water to hold oxygen increases. Thus, even though winter stratification may occur, its effects tend not to be as severe as those in the summer. If ice forms, however, both reaeration and photosynthesis may cease to provide oxygen, and fish my die.

Tropical lakes may remain stratified indefinitely since they do not experience the degree of seasonal variation that allows lakes to overturn. When a tropical lake eutrophies the hypolimnion goes anaerobic and tends to stay that way. Decomposition in the hypolimnion produces gases such as hydrogen sulfide (H_2S), methane (CH_4), and carbon dioxide (CO_2), which tend to remain dissolved in the bottom waters, especially if the lake is deep and bottom pressure is high. In volcanically formed crater lakes, underground springs of carbonated groundwater can also contribute CO_2 to the hypolimnion. When lake bottoms become saturated with these dissolved gases, there is danger that some event, such as an earthquake or landslide, can trigger a sudden and dangerous gas bubble to erupt from the surface. Such an event occurred in Lake Monoun in Cameroon, Africa, in 1984, and another gaseous eruption occurred in Lake Nyos, Cameroon, two years later. Each eruption belched a cloud of heavier-than-air carbon dioxide that hugged the ground, rolled into valleys, and, by displacing oxygen,

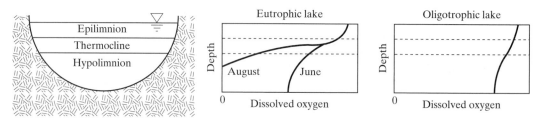

FIGURE 5.22 Dissolved oxygen curves for eutrophic and oligotrophic lakes during summer thermal stratification.

asphyxiated farm animals and people. The bubble of CO_2 emerging from Lake Nyos is said to have created a toxic fountain that shot 150 m into the air. The Lake Monoun event killed 37 people, and Lake Nyos killed more than 1700 people, some as far away as 25 km (Monastersky, 1994). Both events occurred in August, the coolest and rainiest time of the year, when thermal stratification is the least stable.

Acidification of Lakes

As was indicated in Chapter 2, all rainfall is naturally somewhat acidic. As was demonstrated there, pure water in equilibrium with atmospheric carbon dioxide forms a weak solution of carbonic acid (H_2CO_3) with a pH of about 5.6. As a result, only rainfall with pH less than 5.6 is considered to be "acid rain." It is not unusual for rainfall in the northeastern United States to have pH between 4.0 and 5.0, and for acid fogs in southern California to have pH less than 3.0. There is little question that such pH values are caused by anthropogenic emissions of sulfur and nitrogen oxides formed during the combustion of fossil fuels, as will be discussed in Chapter 7. Since some sulfur oxides are actually particles that can settle out of the atmosphere without precipitation, the popular expression "acid rain" is more correctly described as *acid deposition.*

The effects of acid deposition on materials, terrestrial ecosystems, and aquatic ecosystems are still only partially understood, but some features are emerging quite clearly. Acids degrade building materials, especially limestone, marble (a form of limestone), various commonly used metals such as galvanized steel, and certain paints. In fact, the increased rate of weathering and erosion of building surfaces and monuments was one of the first indications of adverse impacts from acid rain. Terrestrial ecosystems, especially forests, seem to be experiencing considerable stress due to acid deposition, with reductions in growth and increased mortality being especially severe in portions of the eastern United States, Canada, and northern Europe. It is the impact of acidification on aquatic ecosystems, however, that will be described here.

Bicarbonate Buffering

Aquatic organisms are very sensitive to pH. Most are severely stressed if pH drops below 5.5, and very few are able to survive when pH falls below 5.0. Moreover, as pH drops, certain toxic minerals, such as aluminum, lead, and mercury, which are normally insoluble and hence relatively harmless, enter solution and can be lethal to fish and other organisms.

It is important to note however, that adding acid to a solution may have little or no effect on pH, depending on whether or not the solution has *buffers.* Buffers are substances capable of neutralizing added hydrogen ions. The available buffering of an aquatic ecosystem is not only a function of the chemical characteristics of the lake itself, but also of nearby soils through which water percolates as it travels from land to the lake. Thus, information on the pH of precipitation alone, without taking into account the chemical characteristics of the receiving body of water and surrounding soils, is a poor indicator of the potential effect of acid rain on an aquatic ecosystem.

Most lakes are buffered by bicarbonate (HCO_3^-), which is related to carbonic acid (H_2CO_3) by the following reaction:

$$H_2CO_3 \;\rightleftharpoons\; H^+ \;+\; HCO_3^- \qquad\qquad (5.36)$$

Carbonic acid   Bicarbonate

We have already encountered this reaction in Chapter 2, where the carbonate system was first introduced. Some bicarbonate results from the dissociation of carbonic acid, as is suggested in (5.36), and some comes from soils. Consider what happens to a lake containing bicarbonate when hydrogen ions (acid) are added. As reaction (5.36) suggests, some of the added hydrogen ions will react with bicarbonate to form neutral carbonic acid. To the extent that this occurs, the addition of hydrogen ions does not show up as an increase in hydrogen ion concentration so the pH may only change slightly. That is, bicarbonate is a buffer.

Notice that the reaction of hydrogen ions with bicarbonate removes bicarbonate from solution so, unless there is a source of new bicarbonate, its concentration will decrease as more acid is added. At some point, there may be so little bicarbonate left that relatively small additional inputs of acid will cause pH to decrease rapidly. This phenomenon leads to one way to classify lakes in terms of their acidification. As shown in Figure 5.23, a *bicarbonate* lake shows little decrease in pH as hydrogen ions are added, until the pH drops to about 6.3. As pH drops below this point, the bicarbonate buffering is rapidly depleted and the lake enters a transitional phase. As shown in the figure, transitional lakes with pH between about 5.0 and 6.0 are very sensitive to small changes in acid. Below pH 5.0, the lakes are unbuffered and chronically acidic (Wright, 1984).

The implications of Figure 5.23 are worth repeating. Acid precipitation may have little or no effect on a lake's pH up until the point where the natural buffering is exhausted. During that stage, there may be no evidence of harm to aquatic life and the lake may appear to be perfectly healthy. Continued exposure to acidification beyond that point, however, can result in a rapid drop in pH that can be disastrous to the ecosystem.

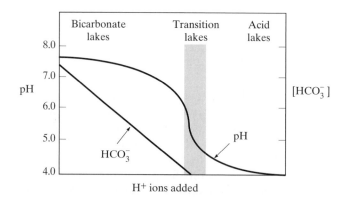

FIGURE 5.23 Bicarbonate buffering strongly resists acidification until pH drops below 6.3. As more H^+ ions are added, pH decreases rapidly. (After Henriksen, 1980)

In a Norwegian study, 684 lakes were categorized by their water chemistry into bicarbonate lakes, transition lakes, and acid lakes, and for each category observations of fish populations were made (Henriksen, 1980; Wright, 1984). Four fish-status categories were chosen: barren (indicative of long-term, chronic failures in fish reproduction), sparse populations (which reflect occasional failures), good populations, and overpopulated. The results, shown in Figure 5.24, clearly show the dependency of fish populations on lake acidification.

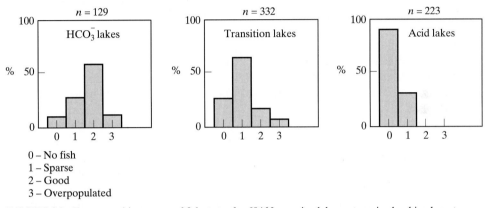

0 – No fish
1 – Sparse
2 – Good
3 – Overpopulated

FIGURE 5.24 Frequency histograms of fish status for 684 Norwegian lakes categorized as bicarbonate, transition, or acid lakes. (Wright, 1984)

Importance of the Local Watershed

If there is a source of bicarbonate to replace that which is removed during acidification, then the buffering ability of bicarbonate can be extensive. Consider the ability of limestone ($CaCO_3$) to provide buffering:

$$H^+ + \underset{\text{Limestone}}{CaCO_3} \longrightarrow Ca^{2+} + \underset{\text{Bicarbonate}}{HCO_3^-} \tag{5.37}$$

$$\underset{\text{Bicarbonate}}{HCO_3^-} + H^+ \longrightarrow \underset{\text{Carbonic acid}}{H_2CO_3} \longrightarrow CO_2 + H_2O \tag{5.38}$$

As this pair of reactions suggests, limestone reacts with acids (hydrogen ions) to form bicarbonate, and the bicarbonate neutralizes even more hydrogen ions as it converts eventually to carbon dioxide. These reactions show how limestone is such an effective buffer to acidification. They also show why limestone and marble monuments and building facades are deteriorating these days under the attack of acid deposition.

The bicarbonate that is formed in (5.37) can replenish the lake's natural bicarbonate buffers. Calcareous lakes, which have an abundance of calcium carbonate, are thus invulnerable to acidification. Copying this natural phenomenon, one way to temporarily mitigate the effects of acidification is artificially to treat vulnerable lakes with limestone ($CaCO_3$). This approach is being vigorously pursued in a number of countries that are most impacted by acid rain, including Sweden, Norway, Canada, and the United States (Shepard, 1986).

The ability of nearby soils to buffer acid deposition is an extremely important determinant of whether or not a lake will be subject to acidification. Soil systems derived from calcareous (limestone) rock, for example, are better able to assimilate acid deposition than soils derived from granite bedrock. The depth and permeability of soils are also important characteristics. Thin, relatively impermeable soils provide little contact between soil and runoff, which reduces the ability of natural soil buffers to affect acidity. The size and shape of the watershed itself also affect a lake's vulnerability. Steep slopes and a small watershed area create conditions in which the runoff has little time to interact with soil buffers, increasing the likelihood of lake acidification. Even the type of vegetation growing in the watershed can affect acidification. Rainwater that falls through a forest canopy interacts with natural materials in the trees, such as sap, and its pH is affected. Deciduous foliage tends to decrease acidity, while water dripping from evergreen foliage tends to be more acidic than the rain itself.

In other words, to a large extent, the characteristics of the watershed itself will determine the vulnerability of a lake to acid. The most vulnerable lakes will be in areas with shallow soil of low permeability, granite bedrock, a steep watershed, and a predominance of conifers. Using factors such as these enables scientists to predict areas where lakes are potentially most sensitive to acid rain. One such prediction, using bedrock geology as the criterion, produced the map shown in Figure 5.25*a*. Figure 5.25*b* shows the average pH of rainfall across North America. Large areas in the northeastern portion of the continent, especially those near the Canadian border, have the unfortunate combination of vulnerable soils and very acidic precipitation.

Mobilization of Aluminum

While the carbonate buffering system is the most important one at pH values above 6, for lower pH other compounds become increasingly important. At pH below 5, aluminum may become the dominant buffering effect in the soil. Aluminum is the third most abundant element in the earth's outer crust, but it rarely occurs in solution in any significant concentration unless the pH is very low, in which cases reactions such as the following occur. For the clay mineral kaolinite, $Al_2Si_2O_5(OH)_4$;

$$Al_2Si_2O_5(OH)_4 \ + \ 6\,H^+ \ \longrightarrow \ 2Al^{3+} \ + \ 2Si\,(OH)_4 \ + \ H_2O \qquad (5.39)$$
Kaolinite

and for the commonly occurring gibbsite, $Al(OH)_3$,

$$Al(OH)_3 \ + \ 3\,H^+ \ \longrightarrow \ Al^{3+} \ + \ 3\,H_2O \qquad (5.40)$$
Gibbsite

In each of these reactions, the removal of hydrogen ions provides the buffering capacity.

Of even greater importance than the buffering ability of aluminum is the resulting liberation of highly toxic Al^{3+} ions that enter solution. Aluminum that is normally bound up in soil minerals, and hence is harmless to living organisms, is leached from

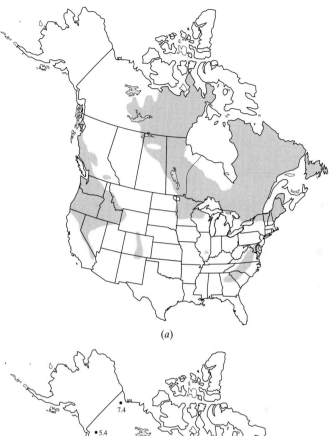

(a)

(b)

FIGURE 5.25 (a) Regions in North America containing lakes that would be sensitive to potential acidification by acid precipitation (shaded areas), based on bedrock geology. (EPA, 1984, based on Galloway and Cowling, 1978); (b) pH of wet deposition in 1982. (Interagency Task Force on Acid Precipitation, 1983)

the soil by acid deposition and moves with runoff into streams and lakes. The toxic action of aluminum on fish seems to be a combination of effects, but the most important is its ability to cause mucous clogging of the gills, leading to respiratory distress and death. It has been found to be toxic to fish at concentrations as low as 0.1 mg/L, and it has been implicated in fish kills in field observations, field experiments, and laboratory studies (U.S. EPA, 1984). Especially troubling is the fact that aluminum can cause fish kills at a moderate value of pH that would, by itself, not be considered harmful.

5.8 GROUNDWATER

Groundwater is the source of about one-third of this country's drinking water and, excluding water for power plant cooling, it supplies almost 40 percent of our total water withdrawals. In rural areas, almost all of the water supply comes from groundwater, and more than one-third of our 100 largest cities depend on it for at least part of their supply. Historically, the natural filtering resulting from water working its way through the subsurface was believed to provide sufficient protection from contamination to allow untreated groundwater to be delivered to customers. Then Love Canal and other dramatic incidents in the 1970s made us realize that our groundwater was contaminated with hazardous substances from hundreds of thousands of leaking underground storage tanks, industrial waste pits, home septic systems, municipal and industrial landfills, accidental chemical spills, careless use of solvents, illegal "midnight" dumping, and widespread use of agricultural chemicals. Groundwater contamination became the environmental issue of the 1980s.

In response to growing pressure to do something about hazardous wastes, Congress passed the *Comprehensive Environmental Response, Compensation, and Liability Act* (CERCLA, but more commonly known as Superfund) in 1980 to deal with already contaminated sites, and in 1980 and 1984, it strengthened the *Resource Conservation and Recovery Act* (RCRA), which controls the manufacturing, transportation, and disposal of newly produced hazardous substances. These two acts, CERCLA cleaning up problems of the past and RCRA helping avoid problems in the future, have had major impact. The billions of dollars that have been spent attempting to clean up contaminated soil and groundwater, along with the very high costs of handling and disposing of new hazardous wastes, have stimulated interest in pollution prevention as the environmental theme of the future.

Once contaminated, groundwater is difficult, if not impossible, to restore. A recent study by the National Research Council (1994) estimates that there are between 300,000 and 400,000 sites in the United States that may have contaminated soil or groundwater requiring some form of remediation. The bulk of these sites are leaking underground storage tanks that have been costing around $100,000 each to clean up; more complex sites are costing an average of $27 million each (U.S. EPA, 1993). If we decide to clean up all of these sites, the estimated cost would be between $480 billion and $1 trillion (NRC, 1994). These enormous costs have called into question the goal of cleaning up such sites to drinking water quality, especially if the groundwater is unlikely to ever be needed for such purposes.

5.9 AQUIFERS

Rainfall and snow melt can flow into rivers and streams, return to the atmosphere by evaporation or transpiration, or seep into the ground to become part of the subsurface, or underground, water. As water percolates down through cracks and pores of soil and rock, it passes through a region called the *unsaturated zone,* which is characterized by the presence of both air and water in the spaces between soil particles. Water in the unsaturated zone, called *vadose water,* is essentially unavailable for use. That is, it cannot be pumped, though plants certainly use soil water that lies near the surface. In the *saturated* zone, all spaces between soil particles are filled with water. Water in the saturated zone is called *groundwater,* and the upper boundary of the saturated zone is called the *water table.* There is a transition region between these two zones called the *capillary fringe,* where water rises into small cracks as a result of the attraction between water and rock surfaces. Figure 5.26 illustrates these regions of the subsurface.

An *aquifer* is a saturated geologic layer that is permeable enough to allow water to flow fairly easily through it. An aquifer sits on top of a *confining bed* or, as it is sometimes called, an *aquitard* or an *aquiclude,* which is a relatively impermeable layer that greatly restricts the movement of groundwater. The two terms, aquifer and confining bed, are not precisely defined and are often used in a relative sense.

The illustration in Figure 5.26 shows an *unconfined* aquifer. A *water table well* drilled into the saturated zone of such an aquifer will have water at atmospheric pressure at the level of the water table. Groundwater also occurs in *confined aquifers,* which are aquifers sandwiched between two confining layers, as shown in Figure 5.27. Water in a confined aquifer can be under pressure so that a well drilled into it will have water naturally rising above the upper surface of the confined aquifer, in which case the well is called an *artesian well.* A line drawn at the level to which water would rise in an artesian well defines a surface called the *piezometric surface* or the *potentiometric surface.* In some circumstances, enough pressure may exist in a confined aquifer to cause water in a well to rise above the ground level and flow without pumping. Such a well is called a

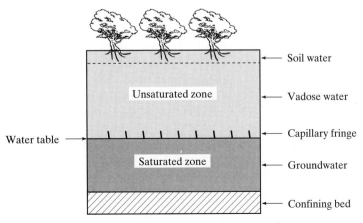

FIGURE 5.26   Identification of subsurface regions.

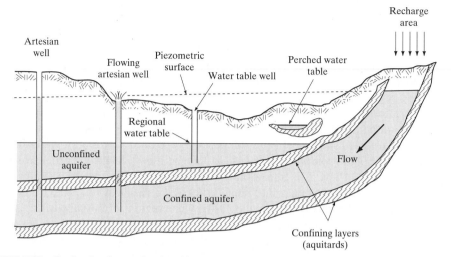

FIGURE 5.27 Confined and unconfined aquifers, a perched water table, water table well, and artesian wells.

flowing artesian well. Also shown in Figure 5.27 is a local impermeable layer in the midst of an unsaturated zone, above the main body of groundwater. Downward percolating water is trapped above this layer, creating a *perched water table.*

The amount of water that can be stored in a saturated aquifer depends on the *porosity* of the soil or rock that makes up the aquifer. Porosity (η) is defined to be the ratio of the volume of voids (openings) to the total volume of material:

$$\text{Porosity } (\eta) = \frac{\text{Volume of voids}}{\text{Total volume of solids and voids}} \tag{5.41}$$

While porosity describes the water-bearing capacity of a geologic formation, it is not a good indicator of the total amount of water that can be removed from that formation. Some water will always be retained as a film on rock surfaces and in very small cracks and openings. The volume of water that can actually be drained from an *unconfined* aquifer per unit of area per unit decline in the water table is called the *specific yield,* or the *effective porosity.* Representative values of porosity and specific yield for selected materials are given in Table 5.12. For confined aquifers, the estimated yield is somewhat affected by the pressure released as water is removed, so a different term, the *storage coefficient,* is used to describe the yield.

EXAMPLE 5.10   Specific yield

For an aquifer of sand, having characteristics given in Table 5.12, what volume of water would be stored in a saturated column with cross-sectional area equal to 1.0 m² and depth 2.0 m? How much water could be extracted from that volume?

Solution   The volume of material is 1.0 m² × 2.0 m = 2.0 m³ so the volume of water stored would be

$$\text{Volume of water} = \text{Porosity} \times \text{Volume of material}$$

TABLE 5.12   Representative Values of Porosity and Specific Yield.

Material	Porosity (%)	Specific Yield (%)
Clay	45	3
Sand	34	25
Gravel	25	22
Gravel and sand	20	16
Sandstone	15	8
Limestone, shale	5	2
Quartzite, granite	1	0.5

Source: Linsley et al. (1992).

$$= 0.34 \times 2.0 \text{ m}^3 = 0.68 \text{ m}^3$$

The amount that could actually be removed would be

$$\text{Yield} = \text{Specific yield} \times \text{Volume of material} = 0.25 \times 2.0 \text{ m}^3 = 0.5 \text{ m}^3 \qquad \blacksquare$$

5.10 HYDRAULIC GRADIENT

In an unconfined aquifer, the slope of the water table, measured in the direction of the steepest rate of change, is called the *hydraulic gradient.* It is important because groundwater flow is in the direction of the gradient and at a rate proportional to the gradient. To define it more carefully we need to introduce the notion of the hydraulic head. As shown in Figure 5.28, the hydraulic head is the vertical distance from some reference datum plane (usually taken to be sea level) to the water table. It has dimensions of

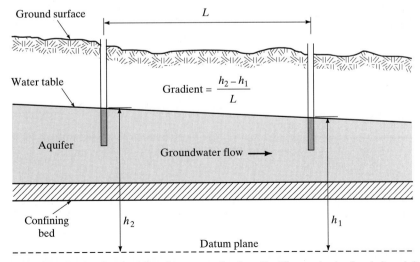

FIGURE 5.28   Showing head and gradient in an unconfined aquifer. If groundwater flow is from left to right in the plane of the page, then the gradient is $\Delta h/L$.

length, such as "meters of water" or "feet of water." If we imagine two wells directly in line with the groundwater flow, the gradient would be simply the difference in head divided by the horizontal distance between them. In an analogous way, the gradient for a confined aquifer is the slope of the piezometric surface along the direction of maximum rate of change.

If the flow in Figure 5.28 is from left to right in the plane of the page, the gradient would be

$$\text{Hydraulic gradient} = \frac{\text{Change in head}}{\text{Horizontal distance}} = \frac{h_2 - h_1}{L} \qquad (5.42)$$

Notice that the gradient is dimensionless as long as numerator and denominator both use the same units. In a microscopic sense, gradient can be expressed as

$$\text{Hydraulic gradient} = \frac{dh}{dL} \qquad (5.43)$$

If we imagine looking down onto the groundwater flow of Figure 5.28, we could imagine *streamlines* showing the direction of flow, and *equipotential lines,* perpendicular to the streamlines, which represent locations with equal head. The combination of streamlines and equipotential lines creates a two-dimensional *flow net,* as shown in Figure 5.29. If we happened to have two wells aligned along a streamline as shown in Figure 5.29, the gradient would be simply the difference in head divided by the distance between equipotentials.

Figure 5.29 represents the special case of two wells along the same streamline. In the more general case, it is possible to estimate the gradient using measurements made at three wells using the following simple graphical procedure (Heath, 1983): Begin by finding the locations and hydraulic heads for three nearby wells, as shown in Figure 5.30. Then

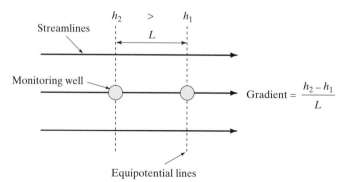

FIGURE 5.29 A two-dimensional flow net consists of streamlines and equipotential lines. If two wells happen to fall on a streamline, the gradient is just $\Delta h/L$.

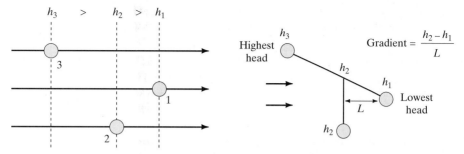

FIGURE 5.30   Using three wells to determine the gradient: Drawing a line between the wells with highest and lowest head and finding the spot along that line corresponding to the third well establishes the direction of equipotentials.

1. Draw a line between the two wells with the highest and lowest head and divide that line into equal intervals. Identify the location on the line where the head is equal to the head of the third (intermediate head) well.

2. Draw a line between the intermediate-head well and the spot on the above line that corresponds to the head at the intermediate well. This is an *equipotential line,* meaning that the head anywhere along the line should be approximately constant. Groundwater flow will be in a direction perpendicular to this line.

3. Draw a line perpendicular to the equipotential line through the well with the lowest (or the highest) head. This is a *flow line,* which means groundwater flow is in a direction parallel to this line.

4. Determine the gradient as the difference in head between the head on the equipotential and the head at the lowest (or highest) well, divided by the distance from the equipotential line to that well.

EXAMPLE 5.11   Estimating the Hydraulic Gradient from Three Wells

Two wells are drilled 200 m apart along an east-west axis. The west well has a total head of 30.2 meters and the east well has a 30.0 m head. A third well located 100 m due south of the east well has a total head of 30.1 m. Find the magnitude and direction of the hydraulic gradient.

Solution   The locations of the wells are shown in Figure 5.31*a*. In Figure 5.31*b*, a line has been drawn between the wells with the highest (west, 30.2 m) and lowest (east, 30.0 m) head, and the location along that line with head equal to the intermediate well (south, 30.1 m) has been indicated. The line corresponding to a 30.1-m equipotential has been drawn.

In Figure 5.31*c*, streamlines perpendicular to the equipotential line have been drawn through the east well. The direction of the gradient is thus at a 45 ° angle from the southwest. The distance between the equipotential and the east well is easily determined by geometry to be $L = 100/\sqrt{2}$. So the gradient is

$$\text{Hydraulic gradient} = \frac{(30.1 - 30.0)\text{m}}{100\sqrt{2}\text{ m}} = 0.00141$$

Notice that the gradient is dimensionless.   ∎

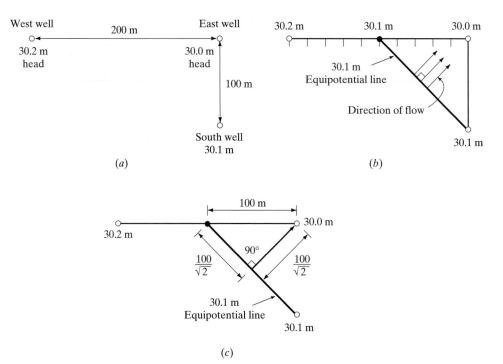

FIGURE 5.31   Finding the gradient for the well field in Example 5.11.

5.11 DARCY'S LAW

The basic equation governing groundwater flow was first formulated by the French hydraulic engineer, Henri Darcy, in 1856. Based on laboratory experiments in which he studied the flow of water through sand filters, Darcy concluded that flow rate Q is proportional to the cross sectional area A times the hydraulic gradient (dh/dL):

$$Q = KA\left(\frac{dh}{dL}\right) \tag{5.44}$$

where

Q = flow rate (m^3/day)

K = hydraulic conductivity, or coefficient of permeability (m/day)

A = cross-sectional area (m^2)

$\left(\dfrac{dh}{dL}\right)$ = the hydraulic gradient

Equation (5.44) is known as *Darcy's law* for flow through porous media. Darcy's law assumes linearity between flow rate and hydraulic gradient, which is a valid

TABLE 5.13   Approximate Values of Hydraulic Conductivity

Material	Conductivity	
	(gpd/ft^2)	(m/day)
Clay	0.01	0.0004
Sand	1000	41
Gravel	100,000	4100
Gravel and sand	10,000	410
Sandstone	100	4.1
Limestone, shale	1	0.041
Quartzite, granite	0.01	0.0004

Source: Linsley et al. (1992).

assumption in most, but not all, circumstances. It breaks down when flow is turbulent, which may occur in the immediate vicinity of a pumped well. It is also invalid when water flows through extremely fine-grained materials, such as colloidal clays, and it should only be used when the medium is fully saturated with water. It also depends on temperature. Some approximate values of the constant of proportionality K are given Table 5.13. It should be appreciated, however, that these values for the hydraulic conductivity are very rough. Conductivities can easily vary over several orders of magnitude for any given category of material, depending on differences in particle orientation and shape as well as relative amounts of silt and clay that might be present.

Aquifers that have the same hydraulic conductivity throughout are said to be *homogeneous,* while those in which hydraulic conductivity differs from place to place are *heterogeneous.* Not only may hydraulic conductivity vary from place to place within the aquifer, but it may also depend on the direction of flow. It is quite common, for example, to have higher hydraulic conductivities in the horizontal direction than in the vertical. Aquifers that have the same hydraulic conductivity in any flow direction are said to be *isotropic,* while those in which conductivity depends on direction are *anisotropic.* While it is mathematically convenient to assume that aquifers that are both homogeneous and isotropic, they rarely, if ever, are.

EXAMPLE 5.12   Flow Through an Aquifer

A confined aquifer 20.0 m thick has two monitoring wells spaced 500 m apart along the direction of groundwater flow. The difference in water level in the wells is 2.0 m (the difference in piezometric head). The hydraulic conductivity is 50 m/day. Estimate the rate of flow per meter of distance perpendicular to the flow.

Solution   Figure 5.32 summarizes the data. The gradient is

$$\left(\frac{dh}{dL} \right) = \frac{2.0 \text{ m}}{500 \text{ m}} = 0.004$$

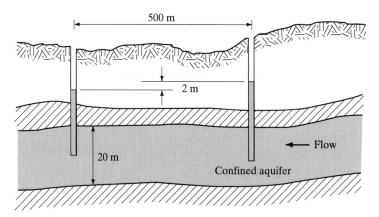

FIGURE 5.32 Example 5.12, flow through a confined aquifer.

Using Darcy's law, with an arbitrary aquifer width of 1 m, yields

$$Q = KA\left(\frac{dh}{dL}\right)$$

$$= 50 \text{ m/d} \times 1.0 \text{ m} \times 20.0 \text{ m} \times 0.004 = 4.0 \text{ m}^3/\text{day per meter of width}$$ ∎

Flow Velocity

It is often important to be able to estimate the rate at which groundwater is moving through an aquifer, especially when a toxic plume exists upgradient from a water supply well. If we combine the usual relationship between flow rate, velocity, and cross-sectional area,

$$Q = Av \tag{5.45}$$

with Darcy's law, we can solve for velocity:

$$\text{Darcy velocity } v = \frac{Q}{A} = \frac{KA(dh/dL)}{A} = K\frac{dh}{dL} \tag{5.46}$$

The velocity given in (5.46) is known as the *Darcy velocity*. It is not a "real" velocity in that, in essence, it assumes that the full cross-sectional area A is available for water to flow through. Since much of the cross-sectional area is made up of solids, the actual area through which all of the flow takes place is much smaller, and as a result, the *real groundwater velocity is considerably faster than the Darcy velocity.*

As suggested in Figure 5.33, consider the cross section of an aquifer to be made up of voids and solids, with A representing the total cross-sectional area and A' being the area of voids filled with water. Letting v' be the actual *average linear velocity* (sometimes called the *seepage velocity*), we can rewrite (5.46) as

$$Q = Av = A'v' \tag{5.47}$$

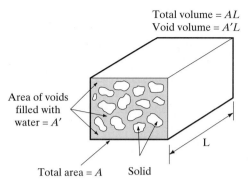

Total volume = AL
Void volume = $A'L$

Area of voids
filled with
water = A'

L

Total area = A Solid

FIGURE 5.33   The cross sectional area available for flow, A', is less than the overall cross-sectional area of the aquifer A.

Solving for v' and introducing an arbitrary length of aquifer L gives

$$v' = \frac{Av}{A'} = \frac{ALv}{A'L} = \frac{\text{Total volume} \times v}{\text{Void volume}} \tag{5.48}$$

But recall that the ratio of void volume to total volume is just the porosity η introduced in (5.41). Therefore, the actual average linear velocity through the aquifer is the Darcy velocity divided by porosity:

$$v' = \frac{\text{Darcy velocity}}{\text{Porosity}} = \frac{v}{\eta} \tag{5.49}$$

or, using (5.46),

$$\text{Average linear velocity} \quad v' = \frac{K}{\eta}\left(\frac{dh}{dL}\right) \tag{5.50}$$

EXAMPLE 5.13   A Groundwater Plume

Suppose the aquifer in Example 5.12 has become contaminated upgradient of the two wells. Consider the upgradient well as a *monitoring* well whose purpose is to provide early detection of the approaching plume to help protect the second, drinking-water well. How long after the monitoring well is contaminated would you expect the drinking-water well to be contaminated? Make the following three assumptions (each of which will be challenged later):

1. Ignore dispersion or diffusion of the plume (that is, it does not spread out).
2. Assume the plume moves at the same speed as the groundwater.
3. Ignore the "pulling" effect of the drinking-water well.

The aquifer has a porosity of 35 percent.

Solution   The Darcy velocity is given by (5.46):

$$\text{Darcy velocity } v = K\frac{dh}{dL} = 50 \text{ m/d} \times 0.004 = 0.20 \text{ m/d}$$

The average linear velocity is the Darcy velocity divided by porosity:

$$\text{Average linear velocity} \quad v' = \frac{0.20 \text{ m/d}}{0.35} = 0.57 \text{ m/d}$$

so the time to travel the 500 m distance would be

$$t = \frac{500 \text{ m}}{0.57 \text{ m/d}} = 877 \text{ days } = 2.4 \text{ yr}$$

As this example illustrates, groundwater moves very slowly.   ∎

5.12 CONTAMINANT TRANSPORT

A few comments on the assumptions made in Example 5.13 are in order. The first assumption was that there would be no dispersion or diffusion, so the contamination would move forward with a sharp front, the so-called *plug flow* case. The second was that the plume moves at the same speed as the groundwater flow. The third was that the drinking-water well did not pull the plume toward it, which would make the plume speed up as it approached the well. In reality, all three of these assumptions need to be examined.

Dispersion and Diffusion

When there is a difference in concentration of a solute in groundwater, molecular *diffusion* will tend to cause movement from regions of high concentration to regions where the concentration is lower. That is, even in the absence of groundwater movement, a blob of contaminant will tend to diffuse in all directions, blurring the boundary between it and the surrounding groundwater. A second process that causes a contaminant plume to spread out is *dispersion*. Since a contaminant plume follows irregular pathways as it moves, some finding large pore spaces in which it can move quickly while other portions of the plume have to force their way through more confining voids, there will be a difference in speed of an advancing plume that tends to cause the plume to spread out. Since diffusion and dispersion both tend to smear the edges of the plume, they are sometimes linked together and simply referred to as hydrodynamic dispersion.

Since contamination spreads out as it moves, it does not arrive all at once at a given location downgradient. This effect is easily demonstrated in the laboratory by establishing a steady-state flow regime in a column packed with a homogeneous granular material and then introducing a continuous stream of a nonreactive tracer, as shown in Figure 5.34a. If the tracer had no dispersion, a plot of concentration versus the time that it takes to leave the column would show a sharp jump. Instead, the front arrives smeared out, as shown in Figure 5.34b.

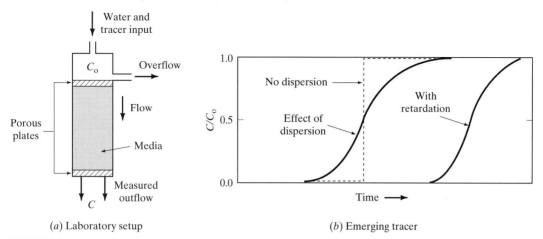

(*a*) Laboratory setup   (*b*) Emerging tracer

FIGURE 5.34   Dispersion and retardation as a continuous feed of tracer passes through a column. With no dispersion, the tracer emerges all at once. With retardation and dispersion, the tracer smears out and emerges with some delay.

While the column experiment described in Figure 5.34 illustrates longitudinal dispersion, there is also some dispersion normal to the main flow. Figure 5.35 shows what might be expected if an instantaneous (pulse) source of contaminant is injected into a flowfield, such as might occur with an accidental spill that contaminates groundwater. As the plume moves downgradient, dispersion causes the plume to spread in the longitudinal as well as orthogonal directions.

Retardation

A second assumption used in Example 5.13 was that the contaminants would move at the same speed as the groundwater, which may or may not be the case in reality. As contaminants move through an aquifer, some are *absorbed* by solids along the way, and some are *adsorbed* (that is, adhere to the surface of particles). The general term *sorption* applies to both processes. The ratio of total contaminant in a unit volume of aquifer to the contaminant dissolved in groundwater is called the *retardation factor.* For example, the retardation factor for chloride ions is 1, which means all of it is dissolved in groundwater. A retardation factor of 5 for some other contaminant means 20 percent is dissolved in groundwater and 80 percent is sorbed to the aquifer solids.

When the amount sorbed reaches equilibrium in the aquifer, the retardation factor takes on a more intuitive meaning: It is the ratio of the average velocity of groundwater v' to the velocity of the sorbed material, v_s:

$$\text{Retardation factor} = R = \frac{\text{Average groundwater velocity, } v'}{\text{Velocity of the sorbed material, } v_s} \geq 1 \quad (5.51)$$

So, for example, a retardation factor of 5 means the groundwater travels five times as far in a given period of time as does the contaminant.

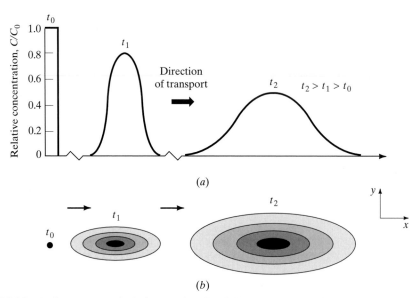

FIGURE 5.35 An instantaneous (pulse) source in a flow field creates a plume that spreads as it moves downgradient. (*a*) In one dimension. (*b*) In two dimensions (darker colors mean higher concentrations). (*Source*: *Ground Water Contamination* by Bedient/Rifai/Newell, © 1994. Reprinted by permission of Prentice-Hall, Inc., Upper Saddle River, NJ.)

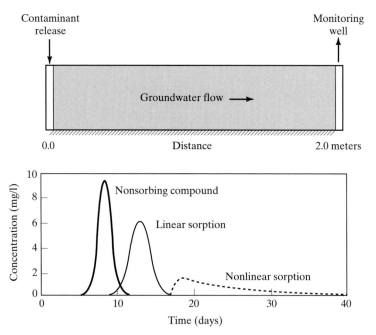

FIGURE 5.36 Influence of sorption on contaminant transport. A pulse of contaminant at $t = 0$ shows up in the monitoring well at different times depending on the extent of sorption. (*Source:* National Research Council, 1994)

Figure 5.36 illustrates the impact of sorption on the time required for a pulse of contaminant to make its way to a monitoring well. A nonsorbing compound shows modest dispersion and emerges rather quickly. When a substance sorbs at a rate proportional to its concentration, it is said to exhibit linear sorption, and such compounds show retardation, but as shown in the figure the pulse maintains its symmetrical shape. For nonlinear sorption, a long tail is shown, suggesting that much larger volumes of water will be required to flush the system.

It is relatively straightforward to perform retardation experiments in the laboratory, producing results such as are shown in Figure 5.36. In an actual aquifer, however, retardation experiments are much more difficult. In one such in situ experiment (Roberts et al., 1986), a number of organic solutes, including carbon tetrachloride (CTET) and tetrachloroethylene (PCE), were injected into the groundwater, along with a chloride tracer, which is assumed to move at the same rate as the groundwater itself. The position of the three plumes roughly 21 months later, as indicated by two-dimensional contours of depth-averaged concentrations, is shown in Figure 5.37. As can be seen, the center of the chloride plume has moved roughly 60 m away from the point of injection ($x = 0$, $y = 0$), the CTET has moved about 25 m, and the PCE has moved only a bit over 10 m.

Using actual plume measurements such as are shown in Figure 5.37, Roberts et al. (1986) determined retardation factors for CTET and PCE, as well as for bromoform (BROM), dichlorobenzene (DCB), and hexachlorethane (HCE). As shown in Figure 5.38, retardation factors are not constants, but instead appear to increase over time and eventually reach a steady-state value.

In a conventional pump-and-treat system for groundwater cleanup, retardation can greatly increase the cost of the system since pumping must continue for a longer period of time. Figure 5.39 shows an estimate of the operating cost of such a system as a function of retardation.

5.13 CONE OF DEPRESSION

Another assumption made in Example 5.13 was that pumping water from the aquifer would not affect the hydraulic gradient. That is, in fact, not the case, since there must be a gradient toward the well in order to provide flow to the well. Moreover, the faster the well is pumped, the steeper the gradient will be in the vicinity of the well. When a well is pumped, the water table in an unconfined aquifer, or the piezometric surface for a confined aquifer, forms a *cone of depression* in the vicinity of the well such as is shown in Figure 5.40.

If we make enough simplifying assumptions, we can use Darcy's law to derive an expression for the shape of the cone of depression, as follows. We will assume that pumping has been steady for a long enough time that the shape of the cone is no longer changing; that is, we assume equilibrium conditions. In addition, we will assume that the original water table is horizontal. If we also assume that the drawdown is small relative to the depth of the aquifer and that the well draws from the entire depth of the aquifer, then the flow to the well is horizontal and radial. Under these conditions, and

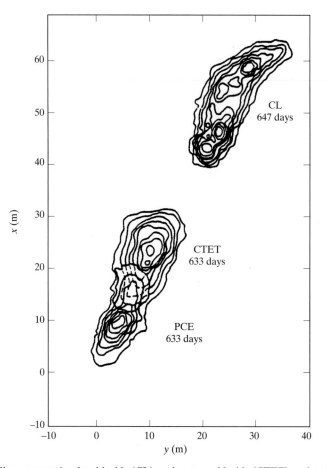

FIGURE 5.37   Plume separation for chloride (CL), carbon tetrachloride (CTET), and tetrachloroethylene (PCE) 21 months after injection. (*Source:* Roberts, Goltz, and Mackay, 1986, "A natural gradient experiment on solute transport in a sand aquifer, 3, Retardation estimates and mass balances for organic solutes," *Water Resources Research* 22 (13):2047-2058, © by the American Geophysical Union.)

using Figures 5.40 and 5.41, we can write Darcy's law for flow passing through a cylinder of radius r, depth h, and hydraulic gradient (dh/dr) as

$$Q = KA\frac{dh}{dr} = K2\pi rh\frac{dh}{dr} \tag{5.52}$$

where the cross-sectional area of an imaginary cylinder around the well is $2\pi rh$, K is the hydraulic conductivity, and (dh/dr) is the slope of the water table at radius r. The flow through the cylinder toward the well equals the rate at which water is being pumped from the well, Q.

Rearranging Equation (5.52) into an integral form,

$$\int_r^{r_1} Q\frac{dr}{r} = 2\pi K\int_h^{h_1} h\,dh \tag{5.53}$$

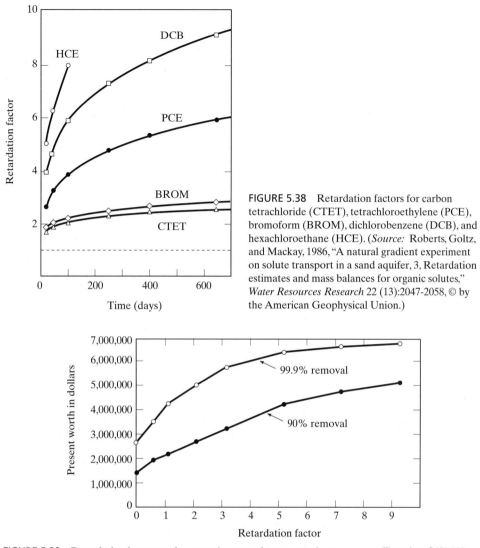

FIGURE 5.38   Retardation factors for carbon tetrachloride (CTET), tetrachloroethylene (PCE), bromoform (BROM), dichlorobenzene (DCB), and hexachloroethane (HCE). (*Source:* Roberts, Goltz, and Mackay, 1986, "A natural gradient experiment on solute transport in a sand aquifer, 3, Retardation estimates and mass balances for organic solutes," *Water Resources Research* 22 (13):2047-2058, © by the American Geophysical Union.)

FIGURE 5.39   Retardation increases the operating cost of a pump-and-treat system. (Based on $650,000 initial capital cost, $180,000 annual O&M, 3.5 percent discount factor, 25-year equipment life, and one pore-volume pumping per year.) (Source: National Research Council, 1994)

Notice how the limits have been set up. The radial term is integrated between two arbitrary values, r and r_1, corresponding to heads h and h_1. Integrating (5.53) gives

$$Q \ln \left(\frac{r_1}{r} \right) = \pi K \left(h_1^2 - h^2 \right) \tag{5.54}$$

so

Unconfined aquifer: $$Q = \frac{\pi K (h_1^2 - h^2)}{\ln (r_1/r)} \tag{5.55}$$

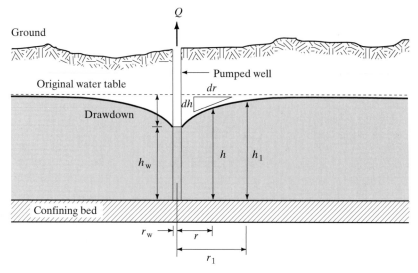

FIGURE 5.40 Cone of depression in an unconfined aquifer.

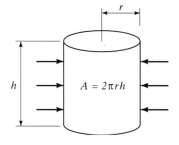

FIGURE 5.41 Flow to the well passes through a cylinder of area
A = 2πrh.

Equation (5.55), for an unconfined aquifer, can be used in several ways. It can be used to estimate the shape of the cone of depression for a given pumping rate, or it can be rearranged and used to determine the aquifer hydraulic conductivity K. To estimate K, two observation wells capable of measuring heads h and h_1 are set up at distances r and r_1 from a well pumping at a known rate Q. The data obtained can then be used in (5.55), as the following example illustrates. Actually, since equilibrium conditions take fairly long to be established, and since more information about the characteristics of the aquifer can be obtained with a transient study, this method of obtaining K is not commonly used. A similar, but more complex, model based on a transient analysis of the cone of depression is used (see, for example, Freeze and Cherry, 1979).

EXAMPLE 5.14   Determining K from the cone of depression

Suppose a well 0.30 m in diameter has been pumped at a rate of 6000 m³/day for a long enough time that steady-state conditions apply. An observation well located 30 m from the pumped well has been drawn down by 1.0 m and another well at 100 m is drawn down by 0.50 m. The well extends completely through an unconfined aquifer 30.0 m thick.

   a. Determine the hydraulic conductivity K.
   b. Estimate the drawdown at the well.

Solution   It helps to put the data onto a drawing, as has been done in Figure 5.42.

   a. Rearranging (5.55) for K and then inserting the quantities from the figure gives

$$K = \frac{Q \ln (r_1/r)}{\pi(h_1^2 - h^2)}$$

$$= \frac{6000 \text{ m}^3/\text{d} \ln (100/30.0)}{\pi\left[(29.5)^2 - (29.0)^2\right] \text{m}^2} = 78.6 \text{ m/d}$$

   b. To estimate the drawdown, let $r = r_w = 0.30/2 = 0.15$ m, and let us use the first observation well for $r_1 = 30$ m and $h_1 = 29$ m. Using (5.55) to find the head at the outer edge of the well h_w gives

$$Q = \frac{\pi K(h_1^2 - h_w^2)}{\ln (r_1/r_w)} = \frac{\pi(78.6 \text{ m/d})(29.0^2 - h_w^2)\text{m}^2}{\ln (30.0/0.15)} = 6000 \text{ m}^3/\text{d}$$

Solving for h_w yields

$$h_w = 26.7 \text{ m}$$

so the drawdown would be $30.0 - 26.7 = 3.3$ m.   ■

Equation (5.55) was derived for an unconfined aquifer. The derivation for a confined aquifer is similar, but now the height of the cylinder at radius r, through which

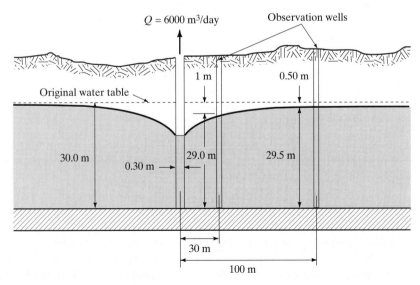

FIGURE 5.42   Data from Example 5.14 used to determine hydraulic conductivity.

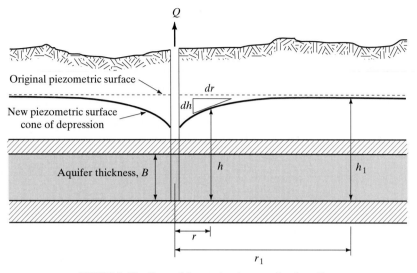

FIGURE 5.43   Cone of depression for a confined aquifer.

water flows to reach the well, is a constant equal to the thickness of the aquifer, B. Also, as shown in Figure 5.43, the cone of depression now appears in the piezometric surface.

With the same assumptions as were used for the unconfined aquifer, we can write Darcy's law as

$$Q = K2\pi rB \left(\frac{dh}{dr} \right) \tag{5.56}$$

which integrates to

$$\text{Confined aquifer:} \qquad Q = \frac{2\pi KB(h_1 - h)}{\ln (r_1/r)} \tag{5.57}$$

5.14 CAPTURE-ZONE CURVES

The most common way to begin the cleanup of contaminated groundwater is to install extraction wells. As an extraction well pumps water from the contaminated aquifer, the water table (or piezometric surface, if it is a confined aquifer) in the vicinity of the well is lowered, creating a hydraulic gradient that draws the plume toward the well. With properly located extraction wells, the polluted groundwater can be pumped out of the aquifer, cleaned in an above-ground treatment facility, and used or returned to the aquifer. The process is referred to as *pump-and-treat* technology.

Figure 5.44 shows the effect of an extraction well located in a region with a uniform and steady regional groundwater flow that is parallel to and in the direction of the negative x-axis. As water is extracted, the natural streamlines are bent toward the

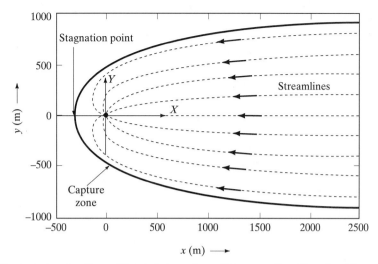

FIGURE 5.44 A single extraction well located at $x = 0, y = 0$, in an aquifer with regional flow along the x-axis. The capture zone is the region in which all flow lines converge on the extraction well. Drawn for $(Q/Bv) = 2000$. (Javendal and Tsang, 1986)

well as shown in the figure. The outer envelope of the streamlines that converge on the well is called the *capture-zone* curve. Groundwater inside the capture zone is extracted; water outside the capture zone is not. Flow lines outside of the capture zone may curve toward the well, but the regional flow associated with the natural hydraulic gradient is strong enough to carry that groundwater past the well.

Javandel and Tsang (1986) have developed the use of capture-zone type curves as an aid to the design of extraction well fields for aquifer cleanup. Their analysis is based on an assumed ideal aquifer (that is, one that is homogeneous, isotropic, uniform in cross section, and infinite in width). Also, it is either confined or unconfined with an insignificant drawdown relative to the total thickness of the aquifer. They assume extraction wells that extend downward through the entire thickness of the aquifer and are screened to extract uniformly from every level. These are very restrictive assumptions that are unlikely ever to be satisfied in any real situation; nonetheless, the resulting analysis does give considerable insight into the main factors that affect more realistic, but complex, models.

For a single extraction well located at the origin of the coordinate system shown in Figure 5.44, Javandel and Tsang derive the following relationship between the x and y coordinates of the envelope surrounding the capture zone:

$$y = \pm\frac{Q}{2Bv} - \frac{Q}{2\pi Bv}\tan^{-1}\frac{y}{x} \tag{5.58}$$

where

$\quad\quad\quad B$ = aquifer thickness (m)

$\quad\quad\quad v$ = Darcy velocity, conductivity $\times$ gradient (m/day)

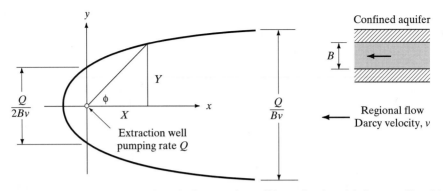

FIGURE 5.45   Capture-zone curve for a single extraction well located at the origin in an aquifer with regional flow velocity v, thickness B, and pumping rate Q.

$$Q = \text{pumping rate from well } (\text{m}^3/\text{day})$$

Equation (5.58) can be rewritten in terms of an angle ϕ (radians) drawn from the origin to the x,y coordinate of interest on the capture-zone curve, as shown in Figure 5.45. That is,

$$\tan \phi = \frac{y}{x} \tag{5.59}$$

so that, for $0 \le \phi \le 2\pi$,

$$y = \frac{Q}{2Bv}\left(1 - \frac{\phi}{\pi}\right) \tag{5.60}$$

Equation (5.60) makes it easy to predict some important measures of the capture zone. For example, as x approaches infinity, $\phi = 0$ and $y = Q/(2Bv)$, which sets the maximum total width of the capture zone at $2\,[Q/(2Bv)] = Q/Bv$. For $\phi = \pi/2$, $x = 0$ and y becomes equal to $Q/(4Bv)$. Thus the width of the capture zone along the y-axis is $Q/(2Bv)$, which is only half as broad as it is far from the well. These relationships are illustrated in Figure 5.45.

The width of the capture zone is directly proportional to the pumping rate Q and inversely proportional to the product of the regional (without the effect of the well) Darcy flow velocity v and the aquifer thickness B. Higher regional flow velocities therefore require higher pumping rates to capture the same area of plume. Usually there will be some maximum pumping rate determined by the acceptable amount of drawdown at the well that restricts the size of the capture zone. Assuming that the aquifer characteristics have been determined and the plume boundaries defined, one way to use capture-zone type curves is first to draw the curve corresponding to the maximum acceptable pumping rate. Then by superimposing the plume onto the capture zone curve (drawn to the same scale), it can be determined whether or not a single well will be sufficient to extract the entire plume and, if it is, where the well can be located. Figure 5.46 suggests the approach, and the following example illustrates its use for a very idealized plume.

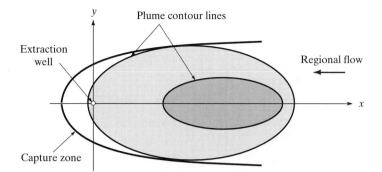

FIGURE 5.46   Superimposing the plume onto a capture-zone type curve for a single extraction well.

EXAMPLE 5.15   A single extraction well

Consider a confined aquifer having thickness 20 m, hydraulic conductivity 1.0×10^{-3} m/s, and a regional hydraulic gradient equal to 0.002. The maximum pumping rate has been determined to be 0.004 m^3/s. The aquifer has been contaminated and for simplicity, consider the plume to be rectangular, with width 80 m. Locate a single extraction well so that it can totally remove the plume.

Solution   Let us first determine the regional Darcy velocity:

$$v = K\frac{dh}{dx} = 1.0 \times 10^{-3} \, \text{m/s} \times 0.002 = 2.0 \times 10^{-6} \, \text{m/s}$$

Now find the critical dimensions of the capture zone. Along the y-axis, its width is

$$\frac{Q}{2Bv} = \frac{0.004 \, \text{m}^3/\text{s}}{2 \times 20 \, \text{m} \times 2.0 \times 10^{-6} \, \text{m/s}} = 50 \, \text{m}$$

and, at an infinite distance upgradient, the width of the capture zone is

$$\frac{Q}{Bv} = 100 \, \text{m}$$

So the 80-m-wide plume will fit within the capture zone if the well is located some distance downgradient from the front edge. Using Figure 5.47 as a guide, we can determine the distance x that must separate the plume from the well. From (5.60), with $y = 40$ m,

$$y = \frac{Q}{2Bv}\left(1 - \frac{\phi}{\pi}\right) = 40 = 50\left(1 - \frac{\phi}{\pi}\right)$$

so the angle, in radians to the point where the plume just touches the capture zone, is

$$\phi = 0.2\pi \quad \text{rad}$$

and, from figure 5.47,

$$x = \frac{y}{\tan \phi} = \frac{40}{\tan (0.2\pi)} = 55 \, \text{m}$$

The extraction well should be placed 55 m ahead of the oncoming plume and directly in line with it.   ■

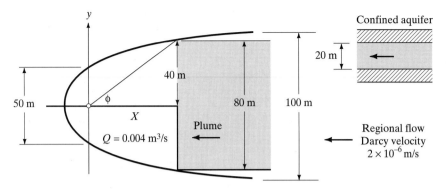

FIGURE 5.47 Example problem with single extraction well.

The single-well solution found in Example 5.15 is not necessarily a good one. The extraction well is far downgradient from the plume, which means a large volume of clean groundwater must be pumped before any of the contaminated plume even reaches the well. That can add years of pumping time and raise total costs considerably before the aquifer is rehabilitated.

A better solution would involve more extraction wells placed closer to the head of the plume. Javandel and Tsang have derived capture-zone type curves for a series of n optimally placed wells, each pumping at the same rate Q, lined up along the y-axis. Optimality is defined to be the maximum spacing between wells that will still prevent any flow from passing between them. The separation distance for two wells has been determined to be $Q/(\pi Bv)$. If the wells are any farther apart than this, some of the flow can pass between them and not be captured. With this optimal spacing, the two wells will capture a plume as wide as $Q/(Bv)$ along the y-axis and as wide as $2Q/Bv$ a long distance upgradient from the wells, as shown in Figure 5.48a. Analogous parameters for the case of three optimally spaced wells are given in Figure 5.48b.

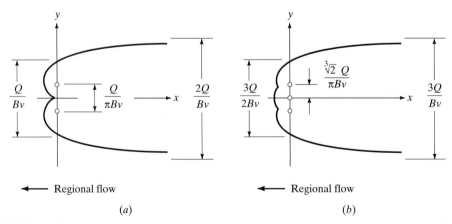

FIGURE 5.48 Capture-zone type curves for optimally spaced wells along the y-axis, each pumping at the rate Q: (a) two wells; (b) three wells.

A general equation for the positive half of the capture-zone type curve for n optimally spaced wells arranged symmetrically along the y-axis is:

$$y = \frac{Q}{2Bv}\left(n - \frac{1}{\pi}\sum_{i=1}^{n}\phi_i\right) \tag{5.61}$$

where ϕ_i is the angle between a horizontal line through the ith well and a spot on the capture-zone curve.

To demonstrate use of these curves, let us redo Example 5.15, but this time we will use two wells.

EXAMPLE 5.16   Capture zone for two wells

Consider the same plume that was described in Example 5.15—that is, it is rectangular with width 80 m in a confined aquifer with thickness $B = 20$ m and Darcy velocity $v = 2.0 \times 10^{-6}$ m/s.

a. If two optimally located wells are aligned along the leading edge of the plume, what minimum pumping rate Q would assure complete plume capture? How far apart should the wells be?

b. If the plume is 1000 m long and the aquifer porosity is 0.30, how long would it take to pump an amount of water equal to the volume of water contained in the plume? Notice that it would take much, much longer to pump out the whole plume than this estimate would suggest since there will be some uncontaminated groundwater removed with the plume and *we are ignoring retardation*. The effect of retardation on the amount of pumping needed to clean an aquifer will be considered later.

Solution

a. The plume width along the y-axis (also the leading edge of the plume) is 80 m so, from Figure 5.48,

$$\frac{Q}{Bv} = \frac{Q}{20 \text{ m} \times 2.0 \times 10^{-6} \text{ m/s}} = 80 \text{ m}$$

$$Q = 0.0032 \text{ m}^3/\text{s} \quad \text{(each)}$$

From Figure 5.48, the optimal spacing between two wells is given as

$$\text{Optimal separation} = \frac{Q}{\pi Bv}$$

$$= \frac{0.0032 \text{ m}^3/\text{s}}{\pi \times 20 \text{ m} \times 2.0 \times 10^{-6} \text{ m/s}} = 25.5 \text{ m}$$

These dimensions are shown in Figure 5.49.

b. The volume of contaminated water in the plume is the porosity times the plume volume:

$$V = 0.30 \times 80 \text{ m} \times 20 \text{ m} \times 1000 \text{ m} = 480{,}000 \text{ m}^3$$

At a total pumping rate of $2 \times 0.0032 \text{ m}^3/\text{s} = 0.0064 \text{ m}^3/\text{s}$, it would take

$$t = \frac{480{,}000 \text{ m}^3}{0.0064 \text{ m}^3/\text{s} \times 3600 \text{ s/hr} \times 24 \text{ hr/day} \times 365 \text{ day/yr}} = 2.4 \text{ yr}$$

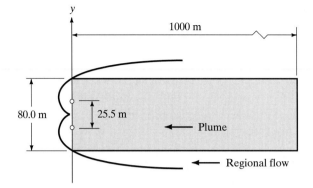

FIGURE 5.49 Example problem with two extraction wells.

to pump a volume of water equal to that contained in the plume. Again note, however, that the time required to pump the actual plume would be much greater than 2.4 years, depending on retardation and the fraction of the water pumped that is actually from the plume. ∎

Since real contaminant plumes and aquifers are so much more complex than the ideal ones considered in this chapter, designing a well field is a much more difficult task than has been presented here. Interested readers are referred to more advanced texts such as those by Gupta (1989) or Bedient et al. (1994).

5.15 CONTROL OF GROUNDWATER PLUMES

We have seen how a column of extraction wells lined up in front of an approaching groundwater plume can be used to intercept the pollution, preventing it from contaminating the aquifer any further downgradient. Another way to protect downgradient uses, such as a drinking-water well, is to use some combination of *extraction wells* and *injection wells*. Extraction wells are used to lower the water table (or piezometric surface), creating a hydraulic gradient that draws the plume to the wells. Injection wells raise the water table and push the plume away. Through careful design of the location and pumping rates of such wells, the hydraulic gradient can be manipulated in such a way that plumes can be kept away from drinking-water wells and drawn toward extraction wells. Extracted, contaminated groundwater can then be treated and either reinjected back into the aquifer, reused, or released into the local surface water system.

Figure 5.50 shows two strategies to control a contaminant plume that endangers a nearby well, here called a production well. In Figure 5.50*a* an injection well pushes the plume away from the production well, and an extraction well pulls it away from the production well. Additional wells would be placed in the plume itself to extract the contaminants. In Figure 5.50*b*, a different strategy is used. The injection well is used to push the plume into the vicinity of the extraction well so that the plume can be removed more quickly from the aquifer.

Manipulating the hydraulic gradient to control and remove a groundwater plume is called *hydrodynamic control*. The well field used to create hydrodynamic control of a

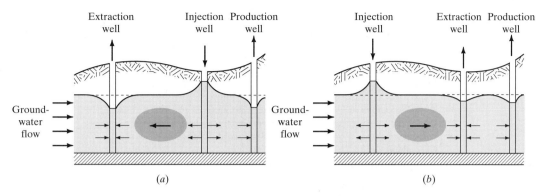

FIGURE 5.50 Manipulating the hydraulic gradient with multiple wells: (*a*) the injection and extraction wells push and pull the plume away from the production well; (*b*) the injection well pushes the plume into the extraction well.

contaminant plume can be anything from a single extraction well, properly placed and pumping at the right rate, to a complex array of extraction wells and injection wells.

Hydrodynamic control of groundwater plumes is an effective way to protect production wells. Construction costs are relatively low, and if the original well field is not sufficient to control the plume adequately, it is always possible to add additional wells when necessary. In addition, the environmental disturbance on the surface is minimal. On the negative side, the operation and maintenance costs can be high since the wells must be pumped for many years.

An alternate approach to protecting production wells is literally to surround the plume with a wall of impermeable material that extends from the surface down to the aquitard. A number of such *physical containment* schemes involving different materials and construction techniques are possible. Perhaps the most common is a *slurry cut-off wall* in which a narrow trench, about 1 to 2 m wide, is dug around the plume, down to bedrock, and then backfilled with a relatively impermeable mixture of soil and bentonite. The slurry wall keeps the plume from migrating off-site while other remediation measures are applied to clean up the aquifer.

5.16 CONTAMINANTS IN GROUNDWATER

In the discussion of surface water contamination, the focus was on controlling the spread of infectious diseases, reducing the oxygen demand of the wastes, and removing nutrients to help reduce eutrophication. Surface waters are, of course, also subject to contamination by a range of toxic chemicals, including pesticides, metals, PCBs, and so forth, but those pollutants were mentioned only briefly in that context. In the study of groundwater contamination, however, toxic chemicals are the principal pollutants of concern.

The 25 most frequently detected contaminants found in groundwater at hazardous waste sites are listed in Table 5.14. Nine of these contaminants are inorganic: lead (Pb), chromium (Cr), zinc (Zn), arsenic (As), cadmium (Cd), manganese (Mn), copper (Cu), barium (Ba), and nickel (Ni). Lead, arsenic, and cadmium are not only

TABLE 5.14   The 25 Most Frequently Detected Groundwater Contaminants At Hazardous Waste Sites

Rank	Compound	Common Sources
1	Trichloroethylene	Dry cleaning; metal degreasing
2	Lead	Gasoline (prior to 1975); mining; construction material (pipes); manufacturing
3	Tetrachloroethylene	Dry cleaning; metal degreasing
4	Benzene	Gasoline; manufacturing
5	Toluene	Gasoline; manufacturing
6	Chromium	Metal plating
7	Methylene chloride	Degreasing; solvents; paint removal
8	Zinc	Manufacturing; mining
9	1,1,1-Trichloroethane	Metal and plastic cleaning
10	Arsenic	Mining; manufacturing
11	Chloroform	Solvents
12	1,1-Dichloroethane	Degreasing; solvents
13	1,2-Dichloroethene, trans-	Transformation product of 1,1,1-trichloroethane
14	Cadmium	Mining; plating
15	Manganese	Manufacturing; mining; occurs in nature as oxide
16	Copper	Manufacturing; mining
17	1,1-Dichloroethene	Manufacturing
18	Vinyl chloride	Plastic and record manufacturing
19	Barium	Manufacturing; energy production
20	1,2-Dichloroethane	Metal degreasing; paint removal
21	Ethylbenzene	Styrene and asphalt manufacturing; gasoline
22	Nickel	Manufacturing; mining
23	Di(2-ethylhexy)phthalate	Plastics manufacturing
24	Xylenes	Solvents; gasoline
25	Phenol	Wood treating; medicines

Source: National Research Council (1994).

high on the list, but they are also among the most toxic chemicals found in groundwater. The rest of the top 25 are organic chemicals.

Nonaqueous-phase Liquids

Many of the organic chemicals listed in Table 5.14 do not dissolve very well in water. They are called nonaqueous-phase liquids (NAPLs), and their presence makes the task of restoring an aquifer to drinking-water quality very difficult or even impossible. NAPLs are generally divided into two categories: Those that are more dense than water are called dense NAPLs, or DNAPLs, while those that are less dense than water are called light NAPLs, or LNAPLs.

Examples of DNAPLs include chlorinated solvents, such as trichloroethylene and tetrachloroethylene, polychlorinated biphenyls (PCBs), pesticides such as chlordane, and polycyclic aromatic hydrocarbons (PAHs). Many of the LNAPLs are fuel hydrocarbons, including crude oil, gasoline, and benzene. Table 5.15 lists some important NAPLs along with their solubility in water and their specific gravity.

As LNAPLs enter the unsaturated zone, some may dissolve into water in the pores, some may volatilize and become mixed with air in other pore spaces, and some

TABLE 5.15   Examples of Nonaqueous-Phase Liquids

Pollutant	Specific gravity	Aqueous solubility (mg/L)
DNAPLs		
Carbon tetrachloride	1.58	7.57×10^2
Trichloroethylene	1.47	1.10×10^3
Tetrachloroethylene	1.63	1.50×10^2
Phenol	1.24	9.3×10^4
LNAPLs		
Benzene	0.873	1.75×10^3
Toluene	0.862	5.35×10^2
p-Xylene	0.861	1.98×10^2

may adsorb onto soil particles. When LNAPLs reach the water table, they do not dissolve well and instead spread out to form a layer of contaminant floating on top of the saturated zone, as suggested in Figure 5.51a. DNAPLs also sink, dissolve, adsorb, and volatilize in the unsaturated zone, but when they reach the water table they keep on sinking until they reach a layer of relatively impermeable material. There they form pools that can overflow and sink to the next impermeable layer, as suggested in Figure 5.51b. Portions of DNAPLs tend to form small globules that become trapped in pore spaces, making them virtually impossible to remove by pumping. As they remain lodged in tiny crevices, and as they accumulate in pools on the bottom of aquifers, they slowly dissolve into passing groundwater. Their solubility is so low, however, that removal by dissolution can take centuries. Attempts to remove DNAPLs by extracting groundwater can be nearly impossible.

Table 5.15 lists aqueous solubilities for a selection of NAPLs, and those can be used to make a first-cut estimate of the length of time required to remove NAPLs by groundwater pumping. The solubilities found in Table 5.15 tend to be much higher than those actually found in contaminated aquifers, which tend to be more like 10 percent of the theoretical value (National Research Council, 1994). Consider the following example.

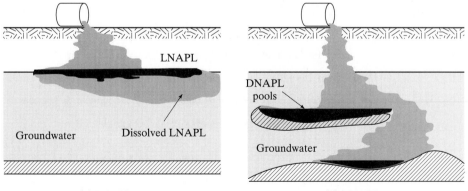

(a) LNAPLs   (b) DNAPLs

FIGURE 5.51   Nonaqueous-phase liquids do not dissolve very well in groundwater: (a) LNAPLs float on top of groundwater; (b) DNAPLs form pools.

EXAMPLE 5.17   Estimated time to remove a TCE spill[1]

Suppose 1 m³ of aquifer is contaminated with 30 L of trichloroethylene (TCE). The aquifer has porosity of 0.3, groundwater moves through it with an actual speed of 0.03 m/day, and the TCE has a dissolved concentration equal to 10 percent of its aqueous solubility.

    a. Find the mass of dissolved TCE and the mass of undissolved DNAPL.

    b. Estimate the time for TCE to be removed.

Solution

    a. From Table 5.15 the aqueous solubility of TCE is given as 1100 mg/L, but the actual dissolved TCE is only 10 percent of that, or 110 mg/L. The porosity of the aquifer is 0.3, so the volume of fluid in 1 m³ of aquifer is 0.3 m³. The amount of dissolved TCE is therefore

$$\text{Dissolved TCE} = 110 \text{ mg/L} \times 0.3 \text{ m}^3 \times 10^3 \text{ L/m}^3 = 33{,}000 \text{ mg} = 33 \text{ g}$$

Table 5.15 indicates that the specific gravity of TCE is 1.47. That is, it is 1.47 times the 1 = kg/L density of water. The total mass of TCE in the aquifer is therefore

$$\text{Total TCE} = 30 \text{ L} \times 1.47 \times 1 \text{ kg/L} \times 10^3 \text{ g/kg} = 44.100 \text{ g}$$

Since 33 g are dissolved in groundwater, the remaining 44,097 g is NAPL mass. That is, 99.92 percent of the TCE has not dissolved.

    b. If we picture the 1 m³ of aquifer as a cube 1 m on a side, then the rate at which fluid leaves the cube is

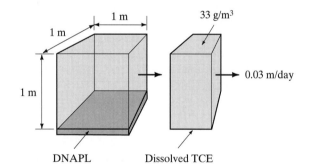

$$\text{Fluid leaving} = 1 \text{ m}^2 \times 0.03 \text{ m/day} = 0.03 \text{ m}^3/\text{day}$$

taking away an amount of TCE equal to

$$\text{TCE flux through 1 m}^2 = 1 \text{ m}^2 \times 0.03 \text{ m/day} \times 33 \text{ g/m}^3 = 0.99 \text{ g/day}$$

So the time needed to remove all 44,100 g of TCE would be

$$\text{Time to remove TCE} = \frac{44{,}100 \text{ g}}{0.99 \text{ g/day} \times 365 \text{ day/yr}} = 122 \text{ years}$$ ∎

---

[1]Based on an example given in National Research Council, (1994).

While Example 5.17 yields an estimate of over a century to clean this aquifer completely, in actuality it would probably take even longer. As the globules of TCE dissolve, their surface area decreases and the rate at which they dissolve drops, so the actual time required would be significantly longer than 122 years.

5.17 CONVENTIONAL PUMP-AND-TREAT SYSTEMS

Pump-and-treat technology for aquifer cleanup is based on extracting contaminated groundwater and then treating it above ground. Treated effluent can then be used for beneficial purposes or returned to the aquifer. There are an estimated 3000 pump-and-treat sites in operation in the United States, which represents approximately three-quarters of all groundwater remediation projects.

The cost of cleaning up an aquifer and the length of time required to do so escalate dramatically as the level of cleanup desired increases. A hypothetical illustration of the cost-and-time-escalations for a conventional pump-and-treat system is presented in Figure 5.52. The example corresponds to a 190-million-liter plume with an average concentration of 1 mg/L of TCE, using EPA cost estimates, including a 4 percent discount factor. Achieving 80 percent contaminant removal is estimated to take 15 years and cost $2.8 million (present value, including capital, operation, and maintenance costs). With a 99.99 percent removal goal, the system would cost $6 million and take 84 years to complete (National Research Council, 1994). Obviously, there are many assumptions and uncertainties in a calculation such as this one, but the implications are clear.

One assumption in Figure 5.52 is that the site can be cleaned up using pump and treat technology. A recent study by the Committee on Ground Water Cleanup

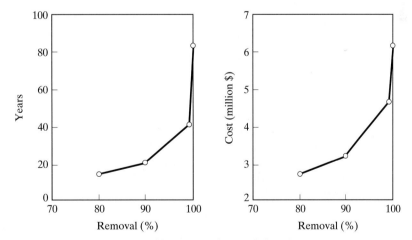

FIGURE 5.52 Impact of cleanup goal on (present value) cost and duration of a conventional pump-and-treat system. Upper data points correspond to 99 percent and 99.99 percent removal of TCE. Drawn for a plume volume of 190×10^6 L, a pumping rate of 380 L/min, a retardation factor of 4.8 (TCE), and a 4 percent discount factor. (*Source:* Based on data in National Research Council, 1994).

TABLE 5.16 Relative Ease of Cleaning Up Contaminated Aquifers as a Function of Contaminant Chemistry and Hydrogeology

Hydrogeology	Contaminant Chemistry					
	Moblile Dissolved (degrades/volatilizes)	Mobile, Dissolved	Strongly Sorbed, Dissolved (degrades/volatilizes)	Strongly Sorbed, Dissolved	Separate Phase LNAPL	Separate Phase DNAPL
Homogeneous, single layer	1[a]	1–2	2	2–3	2–3	3
Homogeneous, multiple layers	1	1–2	2	2–3	2–3	3
Heterogeneous, single layer	2	2	3	3	3	4
Heterogeneous, multiple layers	2	2	3	3	3	4
Fractured	3	3	3	3	4	4

[a]Relative ease of cleanup, where 1 is easiest and 4 is most difficult.
Source: National Research Council, (1994).

Alternatives for the National Research Council evaluated 77 contaminated sites where pump and treat systems are operating (National Research Council, 1994) and found only limited circumstances in which such systems would be likely to achieve cleanup goals in a reasonable period of time at a reasonable cost. They established a site rating system based on contaminant chemistry and hydrogeology that attempts to rate sites by their relative ease of cleanup. Four ratings were developed, with Category 1 sites being the easiest to clean up and Category 4 sites the most difficult. Category 1 sites are characterized by a single-layer, homogeneous aquifer contaminated with pollutants that are mobile, dissolve easily, and are to some extent removed naturally by degradation or volatilization. Pump-and-treat technology should be able to restore groundwater to drinking-water quality in these rare sites. At the other extreme, Category 4 sites have complex hydrogeology involving multiple layers of heterogeneous materials with complex contaminant pathways that are difficult to predict. The contaminants in Category 4 are NAPLs, with DNAPLs being the most difficult. Category 4 sites are unlikely to achieve drinking-water-quality goals. Of the 77 sites studied by the National Research Council Committee, only 2 were placed in Category 1, while 42 were ranked Category 4. Table 5.16 shows the characteristics of each of the four categories.

Attempting to achieve groundwater cleanup to drinking-water standards using conventional pump and treat technology appears to be a futile goal for most contaminated aquifers.

5.18 ADDITIONAL REMEDIATION TECHNOLOGIES

The limited effectiveness of conventional pump and treat systems is motivating a search for other technologies that can be used to augment or replace these conventional systems. Two of the most promising are soil vapor extraction systems and in situ bioremediation. A number of other approaches are also being developed.

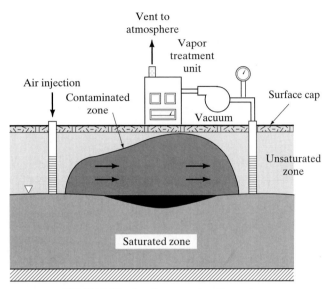

FIGURE 5.53   A soil vapor extraction system.

Soil Vapor Extraction

Soil vapor extraction (SVE) systems are designed to remove organic vapors from the unsaturated zone. As shown in Figure 5.53, the suction side of a blower pulls soil vapors up extraction wells and sends them to a vapor treatment unit. The extraction wells consist of slotted, plastic pipe set into a permeable packing material. They may be placed vertically or horizontally depending on circumstances. Horizontal systems are often used when the contamination is near the surface. The system may include an air injection system that helps push vapors toward collection wells. The air injection system can be connected to the treatment system, creating a closed-loop system. A soil cap can help keep the vapors from venting directly to the atmosphere.

Soil vapor extraction systems are quite effective at removing volatile organic compounds that have leaked from underground storage tanks, such as leaky gasoline storage tanks at service stations. They may also help remove some NAPLs in the unsaturated zone. A combination of SVE and pump and treat can help remove LNAPLs that tend to sit on the surface of the saturated zone. When the pumping system lowers the water table, some of the LNAPLs are left behind. LNAPLs that remain sorbed onto particles that are now in an unsaturated zone can then be affected by the soil vapor extraction system.

The performance of SVE systems is highly dependent on the characteristics of the subsurface and the contaminants. Low-permeability zones in the subsurface are not easily flushed, so contaminants located there will be difficult to remove, and contaminants that are strongly sorbed to subsurface particles will also resist removal. With highly volatile substances located in permeable soils, SVE systems perform well.

Soil vapor extraction systems can be augmented with *air sparging systems,* as shown in Figure 5.54. An air sparging system consists of a compressor that drives air through an injection well into the saturated zone. The injected air rises through the contaminant plume and captures volatile compounds as it moves. Adsorbed contaminants

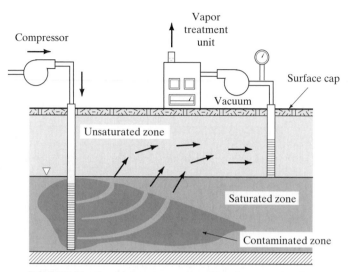

FIGURE 5.54   Combining air sparging with soil vapor extraction.

can desorb directly into the rising air stream, and dissolved volatile compounds can move from the liquid phase into the vapor phase to be vacuumed out by the SVE system. Some air sparging systems have been shown to be very effective at removing substantial quantities of volatile organic compounds, such as benzene and gasoline.

Air sparging can also be done using a horizontal injection distribution system that lies under the contamination with a horizontal suction system that is above the contamination, as shown in Figure 5.55.

In Situ Bioremediation

In situ systems are designed to degrade subsurface pollution in place without the need to capture and deliver contaminants to an above-ground treatment system. By treating contaminants in situ, the need for excavation, above-ground treatment, and transportation to a disposal site is eliminated, and the risk of human exposure to hazardous chemicals is greatly reduced. There are other advantages as well, including the potential to degrade compounds that are sorbed to subsurface materials.

In situ bioremediation is based on stimulating the growth of microorganisms, primarily bacteria, that are indigeneous to the subsurface and that can biodegrade contaminants. If provided with oxygen (or other electron acceptors) and nutrients, microorganisms can degrade a number of common soil and groundwater contaminants, especially petroleum-based hydrocarbons, to carbon dioxide and water. If contamination is close to the land surface, the treatment system can be as simple as an infiltration gallery that allows nutrient- and oxygen-rich water to percolate down through the contamination zone. For deeper contamination, an injection system, such as is shown in Figure 5.56, can deliver the amended water to the desired locations. The oxygen needed for aerobic decomposition can be supplied by bubbling air or oxygen into the injected water or by adding hydrogen peroxide. The added nutrients are often

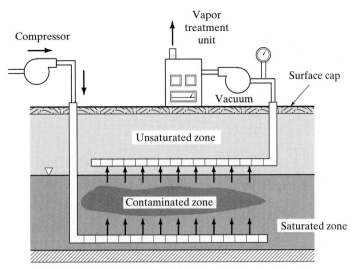

FIGURE 5.55   Air sparging with horizontal wells.

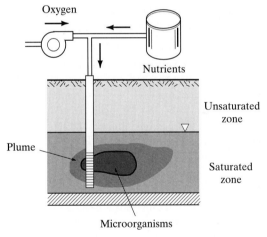

FIGURE 5.56   In situ bioremediation.

nitrogen and phosphorus when the contaminants are hydrocarbons. Extraction wells may be added to the system to help control the flow of the injected amended water.

In situ bioremediation has also been applied to more difficult contaminants than hydrocarbons, including chlorinated solvents and metals. Biodegradation of chlorinated compounds can be accomplished either aerobically or anaerobically. The aerobic pathway involves methanotrophic bacteria, which derive their energy from methane. To stimulate their growth, methane is injected into the contaminated aquifer. In the process of consuming methane, methanotrophs produce an enzyme that has the ability to transform hazardous contaminants that are ordinarily resistant to biodegradation. Field studies have demonstrated high rates of transformation for some chlorinated compounds. Semprini et al. (1990) stimulated indigenous methanotrophic bacteria by injecting dissolved methane and oxygen into an aquifer under the Moffett Naval Air Station in California. Their experiment showed the following amounts of biodegradation in just a matter of days: trichloroethylene, 20 to 30 percent; *cis*-1,2-dichloroethylene, 45 to 55 percent; *trans*-1,2-dichloroethylene, 80 to 90 percent; and vinyl chloride, 90 to 95 percent.

Chlorinated solvents can also be degraded anaerobically. A different consortium of bacteria are involved, and the nutrients that need to be supplied include an electron donor such as methanol, glucose, or acetate. Some concern has been expressed for intermediate products of this biodegradation, such as vinyl chloride, that are themselves hazardous. Complete degradation to simple ethene and ethane is possible, but anaerobic processes proceed much more slowly than their aerobic counterparts, and such degradation may take years in the field.

PROBLEMS

5.1. In a standard five-day BOD test,
  (a) Why is the BOD bottle stoppered?
  (b) Why is the test run in the dark (or in a black bottle)?
  (c) Why is it usually necessary to dilute the sample?
  (d) Why is it sometimes necessary to seed the sample?
  (e) Why isn't ultimate BOD measured?

5.2. Incoming wastewater, with BOD_5 equal to about 200 mg/L, is treated in a well-run secondary treatment plant that removes 90 percent of the BOD. You are to run a five-day BOD test with a standard 300-mL bottle, using a mixture of treated sewage and dilution water (no seed). Assuming the initial DO is 9.2 mg/L,
  a. Roughly what maximum volume of treated wastewater should you put in the bottle if you want to have at least 2.0 mg/L of DO at the end of the test (filling the rest of the bottle with water)?
  b. If you make the mixture half water and half treated wastewater, what DO would you expect after five days?

5.3. A standard five-day BOD test is run using a mix consisting of four parts distilled water and one part wastewater (no seed). The initial DO of the mix is 9.0 mg/L and the DO after five days is determined to be 1.0 mg/L. What is BOD_5?

5.4. A BOD test is to be run on a sample of wastewater that has a five-day BOD of 230 mg/L. If the initial DO of a mix of distilled water and wastewater is 8.0 mg/L and the test requires a decrease in DO of at least 2.0 mg/L with at least 2.0 mg/L of DO remaining at the end of the five days, what range of dilution factors (P) would produce acceptable results? In 300-mL bottles, what range of wastewater volumes could be used?

5.5. The following data have been obtained in a BOD test that is made to determine how well a wastewater treatment plant is operating:

	Initial DO (mg/L)	Final DO (mg/L)	Volume of wastewater (mL)	Volume of dilution water (mL)
Untreated sewage	6.0	2.0	5	295
Treated sewage	9.0	4.0	15	285

What percentage of the BOD is being removed by this treatment plant? If this is a secondary treatment plant that is supposed to remove 85% of the BOD, would you say it is operating properly?

5.6. Figure P5.6 shows a plot of BOD remaining versus time for a sample of effluent taken from a wastewater treatment plant.

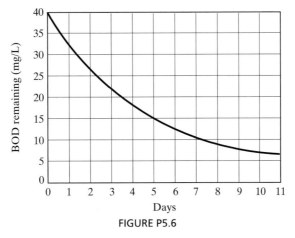

FIGURE P5.6

(a) What is the ultimate BOD (L_0)?

(b) What is the five-day BOD?

(c) What is L_5?

5.7. If the BOD_5 for some wastewater is 200 mg/L and the ultimate BOD is 300 mg/L, find the reaction rate constants k (base e) and K (base 10).

5.8. A BOD test is run using 100 mL of treated wastewater mixed with 200 mL of pure water. The initial DO of the mix is 9.0 mg/L. After 5 days, the DO is 4.0 mg/L. After a long period of time, the DO is 2.0 mg/L and it no longer seems to be dropping. Assuming nitrification has been inhibited so the only BOD being measured is carbonaceous,

(a) What is the five-day BOD of the wastewater?

(b) Assuming no nitrification effects, estimate the ultimate carbonaceous BOD.

(c) What would be the remaining BOD after five days have elapsed?

(d) Estimate the reaction rate constant k (day^{-1}).

5.9. Suppose you are to measure the BOD removal rate for a primary wastewater treatment plant. You take two samples of raw sewage on its way into the plant and two samples of the effluent leaving the plant. Standard five-day BOD tests are run on the four samples, with no seeding, producing the following data:

Sample	Source	Dilution	DO_i(mg/L)	DO_f(mg/L)
1	Raw	1:30	9.2	2.2
2	Raw	1:15	9.2	?
3	Treated	1:20	9.0	2.0
4	Treated	?	9.0	>0

(a) Find BOD_5 for the raw and treated sewage, and the percent removal of BOD in the treatment plant.

(b) Find the DO that would be expected in Sample 2 at the end of the test.

(c) What would be the maximum volume of treated sewage for sample 4 that could be put into the 300-mL BOD bottle and still have the DO after five days remain above 2 mg/L?

5.10. A standard BOD test is run using seeded dilution water. In one bottle, the waste sample is mixed with seeded dilution water giving a dilution of 1:30. Another bottle, the blank, contains just seeded dilution water. Both bottles begin the test with DO at the saturation value of 9.2 mg/L. After five days, the bottle containing waste has DO equal to 2.0 mg/L, while that containing just seeded dilution water has DO equal to 8.0 mg/L. Find the five-day BOD of the waste.

5.11. A mixture consisting of 30 mL of waste and 270 mL of seeded dilution water has an initial DO of 8.55 mg/L; after five days, it has a final DO of 2.40 mg/L. Another bottle containing just the seeded dilution water has an initial DO of 8.75 mg/L and a final DO of 8.53 mg/L. Find the five-day BOD of the waste.

5.12. Some wastewater has a BOD_5 of 150 mg/L at 20 °C. The reaction rate k at that temperature has been determined to be 0.23/day.

(a) Find the ultimate carbonaceous BOD.

(b) Find the reaction rate coefficient at 15 °C.

(c) Find BOD_5 at 15 °C.

5.13. Some waste has a five-day BOD at 20 °C equal to 210 mg/L and an ultimate BOD of 350 mg/L. Find the five-day BOD at 25 °C.

5.14. A clever approach for finding L_0 involves daily measurements of BOD (that is, BOD_1, BOD_2, BOD_3). A straight line is fitted to a plot of BOD_{t+1} vs. BOD_t and the intersection point of that line with a line drawn through the origin with slope = 1 is then found. That intersection point occurs where $BOD_{t+1} = BOD_t$; that is, it is the point where BOD is no longer changing; hence it is L_0, as shown in Figure P5.14.

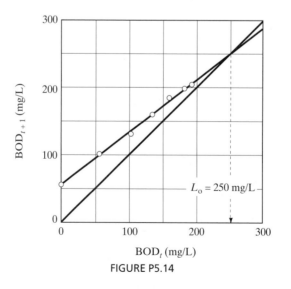

FIGURE P5.14

The following are BOD data for the sample waste graphed in P.5.14, along with three additional wastes. Determine L_0 for each of the additional wastes using this graphical procedure:

t (day)	Example BOD$_t$	BOD$_{t+1}$	t (day)	Waste 1 BOD$_t$	Waste 2 BOD$_t$	Waste 3 BOD$_t$
0	0	57	0	0	0	0
1	57	102	1	62	38	41
2	102	134	2	104	72	79
3	134	160	3	142	104	101
4	160	184	4	179	123	121
5	184	199	5	200	142	140
6	199	207	6	222	151	152
7	207		7	230	167	159

5.15. Show that the procedure in Problem 5.14 works by verifying that a plot of BOD$_{t+1}$ vs. BOD$_t$ is linear; that is

$$BOD_{t+1} = a\ BOD_t + b$$

At the intersection point where BOD$_{t+1}$ = BOD$_t$, BOD utilized is no longer increasing so it is equal to L_0.

5.16. Suppose some wastewater has a BOD$_5$ equal to 180 mg/L and a reaction rate k equal to 0.22/day. It also has a total Kjeldahl nitrogen content (TKN) of 30 mg/L.

(a) Find the ultimate carbonaceous oxygen demand (CBOD).

(b) Find the ultimate nitrogenous oxygen demand (NBOD).

(c) Find the remaining BOD (nitrogenous plus carbonaceous) after five days have elapsed.

5.17. Suppose some pond water contains 10.0 mg/L of some algae, which can be represented by the chemical formula $C_6H_{15}O_6N$. Using the following reactions:

$$C_6H_{15}O_6N\ +\ 6\,O_2\ \longrightarrow\ 6\,CO_2\ +\ 6\,H_2O\ +\ NH_3$$
$$NH_3\ +\ 2\,O_2\ \longrightarrow\ NO_3^-\ +\ H^+\ +\ H_2O$$

(a) Find the theoretical carbonaceous oxygen demand (see Example 2.2).

(b) Find the total theoretical (carbonaceous plus nitrogenous) oxygen demand.

5.18. For a solution containing 200 mg/L of glycine [$CH_2(NH_2)COOH$] whose oxidation can be represented as

$$2\,CH_2(NH_2)COOH\ +\ 3\,O_2\ \longrightarrow\ 4\,CO_2\ +\ 2\,H_2O\ +\ 2\,NH_3$$
$$NH_3\ +\ 2\,O_2\ \longrightarrow\ NO_3^-\ +\ H^+\ +\ H_2O$$

(a) Find the theoretical CBOD.

(b) Find the ultimate NBOD.

(c) Find the total theoretical BOD.

5.19. A sample contains 200 mg/L of casein ($C_8H_{12}O_3N_2$). Calculate the theoretical CBOD, NBOD, and total BOD. If none of the NBOD is exerted in the first five days and $k = 0.25$/day, estimate the five-day BOD.

5.20. An approximate empirical formula for bacterial cells is $C_5H_7O_2N$. What would be the total carbonaceous and nitrogenous oxygen demand for 1 g of such cells?

5.21. A wastewater treatment plant discharges 1.0 m³/s of effluent having an ultimate BOD of 40.0 mg/L into a stream flowing at 10.0 m³/s. Just upstream from the discharge point, the stream has an ultimate BOD of 3.0 mg/L. The deoxygenation constant k_d is estimated at 0.22/day.

    (a) Assuming complete and instaneous mixing, find the ultimate BOD of the mixture of waste and river just downstream from the outfall.

    (b) Assuming a constant cross-sectional area for the stream equal to 55 m², what ultimate BOD would you expect to find at a point 10,000 m downstream?

5.22. The wastewater in Problem 5.21 has DO equal to 4.0 mg/L when it is discharged. The river has its own DO, just upstream from the outfall, equal to 8.0 mg/L. Find the initial oxygen deficit of the mixture just downstream from the discharge point. The temperatures of sewage and river are both 15 °C.

5.23. A single source of BOD causes an oxygen sag curve with a minimum downstream DO equal to 6.0 mg/L. If the BOD of the waste is doubled (without increasing the waste flow rate), what would be the new minimum downstream DO? In both cases assume that the initial oxygen deficit just below the source is zero and the saturated value of DO is 10.0 mg/L. (Note that when the initial deficit is zero, the deficit at any point is proportional to the initial BOD.)

5.24. The oxygen sag caused by a cannery reaches a minimum DO equal to 3.0 mg/L. Upstream from the cannery, the river DO is saturated at 10.0 mg/L and it has no BOD of its own. Just downstream from the discharge point, the DO is still essentially saturated (i.e., consider the initial oxygen deficit to be zero so the downstream deficit is proportional to initial BOD). By what percentage should the BOD of the cannery waste be reduced to assure a healthy stream with at least 5.0 mg/L DO everywhere?

5.25. Two point sources of BOD along a river (A and B) cause the oxygen sag curve shown in Figure P5.25.

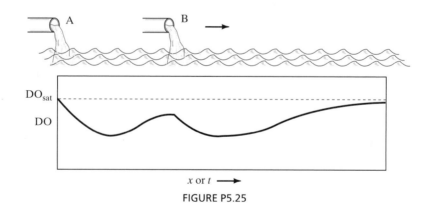

FIGURE P5.25

    (a) Sketch the rate of reaeration vs. distance downriver.

    (b) Sketch L_t (that is, the BOD remaining) as a function of distance downriver.

5.26. Untreated sewage with a BOD of 240 mg/L is sent to a wastewater treatment plant where 50 percent of the BOD is removed. The river receiving the effluent has the oxygen sag curve shown in Figure P5.26 (the river has no other sources of BOD). Notice that downstream is expressed both in miles and days required to reach a given spot.

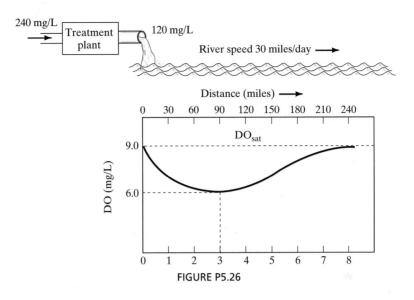

FIGURE P5.26

(a) Suppose the treatment plant breaks down and it no longer removes any BOD. Sketch the new oxygen sag curve a long time after the breakdown. Label the coordinates of the critical distance downriver.

(b) Sketch the oxygen sag curve as it would have been only four days after the breakdown of the treatment plant.

5.27. Suppose the only source of BOD in a river is untreated wastes that are being discharged from a food processing plant. The resulting oxygen sag curve has a minimum value of DO, somewhere downstream, equal to 3.0 mg/L (see Figure P5.27). Just below the discharge point, the DO of the stream is equal to the saturation value of 10.0 mg/L.

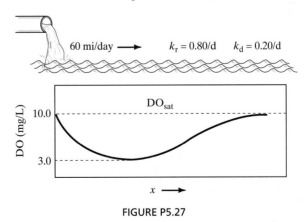

FIGURE P5.27

(a) By what percent should the BOD of the wastes be reduced to assure a healthy stream with at least 5.0 mg/L of DO everywhere? Would a primary treatment plant be sufficient to achieve this reduction?

(b) If the stream flows 60 miles per day and it has a reaeration coefficient k_r equal to 0.80/day and a deoxygenation coefficient k_d of 0.20/day, how far downstream (miles) would the lowest DO occur?

(c) What ultimate BOD (L_0 mg/L) of the mixture of river and wastes just downstream from the discharge point would cause the minimum DO to be 5.0 mg/L?

(d) Sketch the oxygen sag curve before and after treatment recommended in (a), labeling critical points (DO_{min} location and value).

5.28. The ultimate BOD of a river just below a sewage outfall is 50.0 mg/L and the DO is at the saturation value of 10.0 mg/L. The deoxygenation rate coefficient k_d is 0.30/day and the reaeration rate coefficient k_r is 0.90/day. The river is flowing at the speed of 48.0 miles per day. The only source of BOD on this river is this single outfall.

(a) Find the critical distance downstream at which DO is a minimum.

(b) Find the minimum DO.

(c) If a wastewater treatment plant is to be built, what fraction of the BOD would have to be removed from the sewage to assure a minimum of 5.0 mg/L everywhere downstream?

5.29. If the river in Problem 5.28 has an initial oxygen deficit, just below the outfall, of 2.0 mg/L, find the critical distance downstream at which DO is a minimum and find that minimum DO.

5.30. A city of 200,000 people deposits 37 cubic feet per second (cfs) of sewage having a BOD of 28.0 mg/L and 1.8 mg/L of DO into a river that has a flow rate of 250 cfs and a flow speed of 1.2 ft/s. Just upstream of the release point, the river has a BOD of 3.6 mg/L and a DO of 7.6 mg/L. The saturation value of DO is 8.5 mg/L. The deoxygenation coefficient k_d is 0.61/day and the reaeration coefficient k_r is 0.76/day. Assuming complete and instantaneous mixing of the sewage and river find

(a) The initial oxygen deficit and ultimate BOD just downstream of the outfall

(b) The time and distance to reach the minimum DO

(c) The minimum DO

(d) The DO that could be expected 10 miles downstream

5.31. For the following waste and river characteristics just upstream from the outfall, find the minimum downstream DO that could be expected:

Parameter	Wastewater	River
Flow (m^3/s)	0.3	0.9
Ultimate BOD (mg/L)	6.4	7.0
DO (mg/L)	1.0	6.0
k_d (day^{-1})	—	0.2
k_r (day^{-1})	—	0.37
Speed (m/s)	—	0.65
DO_{sat} (mg/L)	8.0	8.0

5.32. Redo Example 5.8 at a temperature of 30 °C.

5.33. Just downstream of the outfall from a point source of pollution the DO of a river is 6 mg/L and the mix of river and wastes has a BOD of 20 mg/L. The saturation value of DO is 9 mg/L. The deoxygenation constant is $k_d = 0.20$/day.

(a) Estimate the reaeration coefficient using the O'Connor and Dobbins relationship (5.24), assuming that the river speed is 0.25 m/s and the average stream depth is 3 m.

(b) Find the critical time downstream at which minimum DO occurs.

(c) Find the minimum DO downstream.

(d) If the outfall is the only source of BOD, what percent removal of BOD would be needed to assure a minimum DO of 5 mg/L?

5.34. An often used chemical representation of algae is $C_{106}H_{263}O_{110}N_{16}P$.

(a) Determine the mass (mg) of each element in 1 g of algae.

(b) Suppose there are 0.10 mg of N and 0.04 mg of P available for algal production per liter of water. Assuming adequate amounts of the other nutrients, which is the limiting nutrient?

(c) What mass of algae could be produced (milligrams algae per liter of water)?

(d) If the nitrogen source could be cut by 50 percent, how much algae (mg/L) could be produced?

(e) If the phosphorus source could be cut by 50 percent, how much algae could be produced?

5.35. Suppose the N and P content of some algae is as shown in the following table. The third column shows milligrams of nutrient available per liter of water.

Nutrient	Milligrams per gram of Algae	Milligrams available per Liter
Nitrogen	60	0.12
Phosphorus	10	0.03

(a) What percent reduction in nitrogen is needed to control algal production to 1.0 mg/L?

(b) What percent reduction in phosphorus is needed to control algal production to 1.0 mg/L?

5.36. Consider a lake with $100 \times 10^6 \, m^2$ of surface area for which the only source of phosphorus is the effluent from a wastewater treatment plant. The effluent flow rate is 0.4 m³/s and its phosphorus concentration is 10.0 mg/L (10.0 g/m³). The lake is also fed by a stream having 20 m³/s of flow with no phosphorus. If the phosphorus settling rate is estimated to be 10 m/yr, estimate the average phosphorus concentration in the lake. What level of phosphorus removal at the treatment plant would be required to keep the average lake concentration below 0.010 mg/L?

5.37. A 50 cm³ sample of dry soil from an aquifer weighs 100 g. When it is poured into a graduated cylinder, it displaces 35 cm³ of water.

(a) What is the porosity of the soil?

(b) What is the average density of the actual solids contained in the soil?

5.38. Using data from Table 5.12, what volume of water would be removed from an unconfined gravel aquifer 10,000 m² in area if the water table drops by 1 m? What fraction of the water contained in that portion of the aquifer would have been removed?

5.39. Consider three monitoring wells, each located at the vertex of an equilateral triangle. The distance between any pair of wells is 300 m. The head at each well, referenced to some common datum, is as follows: well 1, 100 m; well 2, 100.3 m; well 3, 100.3 m. Sketch the well field and find the magnitude and direction of the hydraulic gradient.

5.40. Three wells are located in an x,y plane at the following coordinates: well 1, (0,0); well 2, (100 m, 0); and well 3, (100 m, 100 m). The ground surface is level and the distance from the surface to the water table for each well is as follows: well 1, 10 m; well 2, 10.2 m; well 3, 10.1 m. Sketch the well field and find the hydraulic gradient.

5.41. The aquifer described in Problem 5.39 has a gradient of 0.00115, a hydraulic conductivity of 1000 m/d, and a porosity of 0.23.

  (a) What is the Darcy velocity?

  (b) What is the average linear velocity of the groundwater?

  (c) If the front edge of a plume is perfectly straight, how long would it take to travel to well 1 after first arriving (simultaneously) at wells 2 and 3, assuming a retardation factor of 2?

5.42. A 750-m section of river runs parallel to a channel 1000 m away (see Figure P5.42). An aquifer connecting the two has hydraulic conductivity equal to 7.0 m/day and an average thickness of 10 m. The surface of the river is 5.0 m higher than the surface of the channel. Estimate the rate of seepage from the river to the channel.

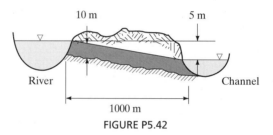

10 m   5 m

River   Channel

1000 m

FIGURE P5.42

5.43. Based on observations of the rate of travel between two observation wells of a tracer element having retardation factor equal to 1.0, it is determined that the average linear flow velocity in an aquifer is 1.0 m/day when the hydraulic gradient is 0.0005. A sample of the aquifer is tested and found to have a porosity of 0.20. Estimate the hydraulic conductivity.

5.44. Derive the expression for the cone of depression in a confined aquifer as given in the chapter in (5.57).

5.45. Drawdown at a 0.20-m-diameter, fully penetrating well, which has been pumping at the rate of 1000 m³/day for a long enough time that steady-state conditions have been reached, is determined to be 0.70 m. The aquifer is unconfined and 10.0 m thick. An observation well 10 m away has been drawn down by 0.20 m. Determine the hydraulic conductivity of the aquifer.

5.46. A confined aquifer 30.0 m thick has been pumped from a fully penetrating well at a steady rate of 5000 m³/day for a long time. Drawdown at an observation well 15 m from the pumped well is 3.0 m and drawdown at a second observation well 150 m away is 0.30 m. Find the hydraulic conductivity of the aquifer.

5.47. Derive the following expression for the length of time required for groundwater to flow from an observation well to a pumped well in a confined aquifer:

$$t = \frac{\pi B \eta}{Q}(R^2 - r_w^2)$$

where

B = thickness of the confined aquifer

η = aquifer porosity

R = radial distance from the observation well to the pumped well

r_w = radius of the pumped well

Q = pumping rate

Hint: Combine (5.56), $Q = 2\pi rBK\,(dh/dr)$, with the average linear velocity from Eq. (5.50):

$$v' = -\frac{K}{\eta}\left(\frac{dh}{dr}\right) = \frac{dr}{dt}$$

5.48. For the aquifer described in Problem 5.46, and using the equation given in Problem 5.47, determine the time required for groundwater to travel from the observation well 15 m away to the pumped well with diameter 0.40 m. The porosity is 0.30.

5.49. A *stagnation point* in a capture-zone type curve is a spot where groundwater would have no movement. For the case of a single extraction well, the stagnation point is located where the capture-zone curve crosses the x-axis. Use the fact that for small angles $\tan\theta \approx \theta$ to show that the x-axis intercept of the capture-zone curve for a single well is $x = -Q/(2Bv\pi)$.

5.50. Suppose a spill of $0.10\ \text{m}^3$ of trichloroethylene (TCE) distributes itself evenly throughout an aquifer 10.0 m thick, forming a rectangular plume 2000 m long and 250 m wide (see Figure P5.50). The aquifer has porosity 0.40, hydraulic gradient 0.001, and hydraulic conductivity 0.001 m/s.

FIGURE P5.50

(a) Given the solubility of TCE, could this much TCE be totally dissolved in the aquifer? What would be the concentration of TCE (mg/L) in this idealized groundwater plume?

(b) Using capture-zone type curves, design an extraction field to pump out the plume under the assumption that the wells are all lined up along the leading edge of the plume, with each well to be pumped at the same rate, not to exceed $0.003\ \text{m}^3/\text{s}$ per well. What is the smallest number of wells that could be used to capture the whole plume? What minimum pumping rate would be required for each well?

(c) What would the optimal spacing be between the wells (at that minimum pumping rate)?

5.51. Starting with (5.61), show that the width of the capture zone along the y-axis for n optimally spaced wells is equal to $nQ/(2Bv)$.

5.52. A single well is to be used to remove a symmetrical oblong plume of contaminated groundwater in an aquifer 20.0 m thick with porosity 0.30, hydraulic conductivity 1.0×10^{-4} m/s, and hydraulic gradient 0.0015. With the plume and capture-zone curve superimposed as shown in Figure P5.52, the angle from the well to the point where the two just touch is 45°, and the width of the plume is 100.0 m. What pumping rate would create these conditions?

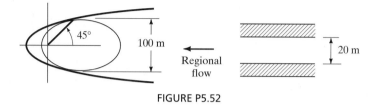

FIGURE P5.52

5.53. A cubic meter of a gravel-and-sand aquifer has been contaminated with 20 L of tetrachloroethylene. If the amount of tetrachloroethylene dissolved in aquifer water is 20 percent of its aqueous solubility,

(a) How much tetrachloroethylene is dissolved?

(b) How much remains as undissolved DNAPL mass?

(c) If the aquifer has a gradient of 0.001, use porosity and hydraulic conductivity data for gravel-and-sand aquifers to estimate the average linear velocity of the groundwater.

(d) How long would it take to remove the tetrachloroethylene?

REFERENCES

Bedient, P. B., H. S. Rifai, and C. J. Newell, 1994, *Ground Water Contamination, Transport and Remediation,* Prentice Hall, Englewood Cliffs, NJ.

Davis, M. L. and D. A. Cornwell, 1991, *Introduction to Environmental Engineering,* 2nd ed., McGraw-Hill, New York.

Frederick, K. D., and J. C. Hanson, 1982, *Water for Western Agriculture,* Resources for the Future, Washington, DC.

Freeze, R. A., and J. A. Cherry, 1979, *Groundwater,* Prentice Hall, Englewood Cliffs, NJ.

Galloway, J. N., and E. B. Cowling, 1978, The effects of precipitation on aquatic and terrestrial ecosystems, A proposed precipitation chemistry network, *Journal of the Air Pollution Control Association,* 28(3).

Gleick, P. H., 1993, *Water in Crisis; A Guide to the World's Fresh Water Resources,* Oxford University Press, New York.

Gupta, R. S., *Hydrology and Hydraulic Systems,* Prentice Hall, Englewood Cliffs, NJ.

Hammer, M. J., and M. J. Hammer, Jr., 1996, *Water and Wastewater Technology,* 3rd ed., Prentice Hall, Englewood Cliffs, NJ.

Heath, R. C., 1983, *Basic Ground-Water Hydrology,* U.S. Geological Survey Water-Supply Paper 2220, Washington, DC.

Henriksen, A., 1980, *Proceedings of the International Conference on the Ecological Impact of Acid Precipitation,* D. Drablos and A. Tollan (eds.).

Interagency Task Force on Acid Precipitation, 1983, *Annual Report 1983 to the President and Congress,* Washington, DC.

Javendel, I., and C. Tsang, 1986, Capture-zone type curves: A tool for aquifer cleanup, *Ground Water,* 24(5):616–625.

Linsley, R. K., J. B. Franzini, D. L. Freyberg, G. Tchobanoglous, 1992, *Water-Resources Engineering,* 4th ed., McGraw-Hill, NY.

Monastersky, R., 1994, Cameroon's killer lakes: A rising threat, *Science News,* 145:215.

National Research Council, 1994, *Alternatives for Ground Water Cleanup,* National Academy Press, Washington, DC.

O'Connor, D. J., and W. E. Dobbins, 1958, Mechanism of reaeration in natural streams, *Transactions of the American Society of Civil Engineers,* 153:641.

Postel. S., 1996, Forging a sustainable water strategy, *State of the World 1996,* L. Brown (ed.), W. W. Norton, New York.

Reisner, M., 1986, *Cadillac Desert, The American West and Its Disappearing Water,* Viking, New York.

Roberts, P. V., M. N. Goltz, and D. M. Mackay, 1986, A natural gradient experiment on solute transport in a sand aquifer, 3, retardation estimates and mass balances for organic solutes, *Water Resources Research,* 22(13):2047–2058.

Sawyer, C. N., 1947, Fertilization of lakes by agricultural and urban drainage, *Journal of the New England Water Works Association,* 41(2).

Sawyer, C. N., P. L. McCarty, and G. F. Parkin, 1994, *Chemistry for Environmental Engineering,* 4th ed., McGraw-Hill, New York.

Semprini, L., P. V. Roberts, G. D. Hopkins, and P. L. McCarty, 1990, a field evaluation of in situ biodegradation of chlorinated ethanes: Part 2—results of biostimulation and biotransformation experiments, *Ground Water,* 28(5):715–727.

Shepard, M., 1986, Restoring life to acidified lakes, *EPRI Journal,* April/May.

Shiklomanov, I. A., 1993, World fresh water resources in *Water in Crisis: A Guide to the World's Fresh Water Resources,* P. H. Gleick (ed.), Oxford University Press, New York.

Skogerboe, G. V, and J. P. Law, 1971, *Research Needs for Irrigation Return Flow Quality Control,* Project No. 13030, U.S. Environmental Protection Agency, Washington, DC.

Streeter, N. W., and E. B. Phelps, 1925, U.S. Public Health Service Bulletin No. 146.

Stumm, W., and J. J. Morgan, 1981, *Aquatic Chemistry,* 2nd ed., Wiley, New York.

Tchobanoglous, G., and E. D. Schroeder, 1985, *Water Quality,* Addison-Wesley, Reading, MA.

Thomann, R. V., and J. A. Mueller, 1987, *Principles of Surface Water Quality Modeling and Control,* Harper & Row, New York.

UNEP, 1993, *Environmental Data Report, 1993–94,* United Nations Environment Programme, Blackwell, Oxford.

U.S. EPA 1984, *Acid Deposition Phenomenon and its Effects, Critical Assessment Review Papers,* A. P. Atshuller and R. A. Linthurst, eds., EPA-600/8-83-016bF.

U.S. EPA, 1993, *Cleaning up the Nation's Waste Sites: Markets and Technology Trends,* EPA 542-R-92-012, Office of Solid Waste and Emergency Response, Environmental Protection Agency, Washington, DC.

U.S. EPA, 1994, *The Quality of Our Nation's Water: 1992,* Environmental Protection Agency, EPA841-S-94-002, Washington, DC.

U.S. EPA, 1996, *Strategic Plan for the Office of Research and Development,* Environmental Protection Agency, EPA/600/R-96/059, Washington, DC.

U.S. GS, 1984, *Estimated Use of Water in the United States, 1980,* U.S. Geological Survey, Department of the Interior, Washington DC.

U.S. GS, 1985, *Study and Interpretation of the Chemical Characteristics of Natural Water,* Water-Supply Paper 2254, U.S. Geological Survey, Department of the Interior, Washington, DC.

U.S. Water Resources Council, 1978, *The Nation's Water Resources, 1975–2000,* Second National Water Assessment, Washington, DC.

Vollenweider, R. A., 1975, Input-output models with special reference to the phosphorus loading concept in limnology, *Schweiz Z. Hydrol.,* 37:53–83.

Wright, R. F., 1984, Norwegian models for surface water chemistry: An overview, in *Modeling of Total Acid Precipitation Impacts,* J. L. Schnoor, ed., Butterworth, Boston.

C H A P T E R   6

Water Quality Control

"Water, water, everywhere, nor any drop to drink." —*The Rime of the Ancient Mariner,* Samuel Taylor Coleridge

6.1 INTRODUCTION

One of the first things that world travelers worry about is whether it is safe to drink the water and whether uncooked foods washed in local water are safe to eat. The unfortunate answer in most places is no. For over 2 billion people in the developing countries of the world, access to safe drinking water is simply not possible today. The rest of us usually assume (correctly, in most circumstances) that water coming out the tap is clean and safe. That important luxury is the result of the coordinated efforts of scientists, engineers, water plant operators, and regulatory officials.

Our traditional confidence in the quality of drinking water in the United States, however, has been shaken of late. In 1993, 400,000 people in Milwaukee became ill, and more than 100 died, from an intestinal parasite, *Cryptosporidium,* in their drinking water. In the same year, residents of Manhattan and others in the Washington, D.C. area were told to boil their water when surprising numbers of *E. coli* bacteria began to show up in their drinking water, despite heavy doses of chlorination. Compounding the chlorination problem has been the realization that byproducts of the disinfection process, called *trihalomethanes,* may be causing on the order of 10,000 cancer cases per year in the United States. The seriousness of these problems has led the Environmental Protection Agency's Office of Research and Development to identify drinking water disinfection as one of its six highest-priority research topics in its 1996 Strategic Plan (U.S. EPA, 1996).

Two complementary approaches are necessary to protect the quality of our water. Legislative bodies provide the laws that regulatory agencies use to define acceptable emissions and establish standards that govern the minimum quality of water for its many beneficial uses. The scientific and engineering community provides the technical guidance needed by legislators and regulators, as well as the technology that is used to achieve those standards. This chapter will explore both.

6.2 MUNICIPAL WATER AND WASTEWATER SYSTEMS

There are two critical systems that combine to break the carrier-feces-water-victim sequence responsible for the spread of waterborne diseases. The first is the water collection, treatment, and distribution system that provides safe drinking water. The principal legislation in the United States that regulates drinking water quality is the Safe Drinking Water Act (SDWA). The second is the wastewater collection and treatment system that removes contaminants before the effluents are released back into the local stream, lake, estuary, or coastal waters. The primary responsibility of these two systems is to kill pathogens before and after water is used. Wastewater treatment systems also reduce BOD and nutrient loading on the receiving water, and some remove toxic chemicals. The Clean Water Act (CWA) governs the regulation of wastewater effluents and water quality in the receiving body of water.

As shown in Figure 6.1, municipal systems may get their water from a local stream, reservoir, or groundwater system. Larger cities tend to rely heavily on surface water supplies, while small community water systems more often take advantage of groundwater. In the United States, about half of the drinking water comes from surface water supplies and the other half from groundwater. Water treatment plants filter and disinfect the water before distributing it to customers.

After water is used in households and businesses, it is collected in a sanitary sewer system and sent to the local wastewater treatment facility. Industrial wastewater may be treated and released directly into the receiving body of water, or it may use the municipal sanitary sewer system. In the latter case, the industrial effluent often must receive some pretreatment before it can be disposed of in the sanitary sewer system. Discharges of any wastewater in the United States are regulated under provisions of the Clean Water Act. Sources must obtain permits issued under the *National Pollutant Discharge Elimination System* (NPDES). An NPDES permit requires the discharger to meet certain technology-based effluent limits and perform effluent monitoring.

Also shown in Figure 6.1 is a storm sewer system that collects runoff from urban streets. In older cities, the stormwater sewer lines join the sanitary system and the combination of wastewaters flows to the municipal wastewater treatment plant. There are an estimated 1100 of these combined sewer systems in use today in the United States, serving some 43 million people. These *combined systems* are unsatisfactory when it rains since they often end up carrying more wastewater than the local treatment system can handle. When that happens a portion of the flow, which includes raw sewage, must be diverted around the treatment plant and released directly into the receiving water. The result is contaminated shorelines that must be posted with warnings after almost every storm. Separating these combined systems is immensely expensive. The preferred approach has been to create massive reservoirs, usually underground, that

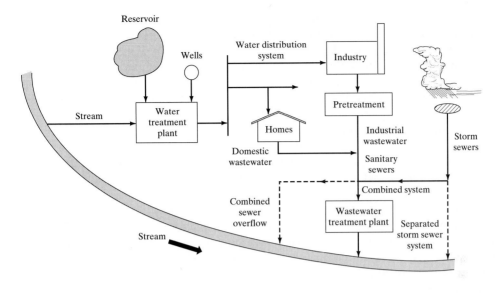

FIGURE 6.1   Water and wastewater systems. In older parts of cities, storm and sanitary sewers are combined, which can lead to untreated wastewater discharges during rainy conditions.

store the combined flow until the storm passes, after which time the reservoir is slowly drained back into the sanitary sewer system.

In newer cities, and in newer sections of old cities, the stormwater collection system is completely separated from the sanitary sewer system to avoid the problem of overloading. Separated systems are not without their own problems, however. Runoff from streets during both wet and dry periods is passed untreated into the local receiving water.

6.3 THE SAFE DRINKING WATER ACT

Legislation to protect drinking water quality in the United States began with the Public Health Service Act of 1912. In this Act, the nation's first water quality standards were created. These standards slowly evolved over the years, but it was not until the passage of the *Safe Drinking Water Act* (SDWA) of 1974 that federal responsibility was extended beyond interstate carriers to include all community water systems serving 15 or more outlets, or 25 or more customers. This original Safe Drinking Water Act had two basic thrusts: (1) It required the EPA to establish national standards for drinking-water quality, and (2) it required the operators of some 200,000 public water systems in the country to monitor the quality of water being delivered to customers and to treat that water, if necessary, to assure compliance with the standards.

Twelve years later, in 1986, strengthening amendments were added to the SDWA. These required the EPA to quicken the pace of standard setting, required the equivalent of filtration for all surface water supplies, and required disinfection for all water systems. The regulations also recognized the potential for lead contamination in water distribution systems, including the plumbing systems in homes and nonresidential facilities. To help prevent lead poisoning, the 1986 amendments require the use of "lead-free" pipe, solder, and flux for all such systems (in the past, solder normally contained about 50 percent lead; now it may contain no more than 0.2 percent lead). In 1996, Congress added a requirement for all large municipal water systems to report annually to their customers data on the quality of their drinking water. Many states already do this. Other provisions require additional controls on harmful pollutants, such as *Cryptosporidium,* and revisions to drinking-water standards.

Drinking-water standards fall into two categories: *primary standards,* which specify *maximum contaminant levels* (MCLs) based on health-related criteria, and *secondary standards,* which are unenforceable guidelines based on both aesthetics such as taste, odor, and color of drinking water, as well as nonaesthetic characteristics such as corrosivity and hardness. In setting MCLs, the EPA is required to balance the public health benefits of the standard against what is technologically and economically feasible. In this way, MCLs are quite different from National Ambient Air Quality Standards, which must be set at levels that protect public health regardless of cost or feasibility. In some circumstances, such as when a pollutant is difficult to detect and measure, a treatment technique (TT) is required rather than an actual MCL. For example, since detection of pathogens such as viruses, *Giardia,* and *Cryptosporidium* is difficult, the standard simply specifies water filtration as the requirement.

The Environmental Protection Agency also sets unenforceable *maximum contaminant level goals* (MCLGs). These goals are set at levels that present no known or anticipated health effects, including a margin of safety, regardless of technological feasibility or cost. Since exposure to any amount of a cancer-causing chemical is deemed to pose some risk to the individual, the MCLGs for carcinogens are generally set to zero. The Safe Drinking Water Act requires the EPA to review the actual MCLs periodically to determine whether they can be brought closer to the desired MCLGs.

Chemical Standards

Contaminants for which MCLs and MCLGs are established are classified as inorganic chemicals, organic chemicals, radionuclides, or microbiological-turbidity contaminants. Under the impetus of the SDWA, the list of substances in each category is continuously being updated both in terms acceptable concentrations and the number of substances regulated. A list of primary drinking-water standards is given in Table 6.1.

Inorganic chemicals include highly toxic metals, such as arsenic, cadmium, lead, and mercury; nitrites (NO_2) and nitrates (NO_3), which can cause methemoglobinemia ("blue-baby syndrome"); fluoride, which is purposely added to water to help prevent dental caries but which can cause mottling of teeth if the exposure is excessive (the standard is temperature dependent); and asbestos fibers, which are especially dangerous when inhaled, although their danger is less certain for ingestion (the MCL is written in terms millions of fibers per liter with lengths greater than 10 μm).

Organic chemical contaminants for which MCLs have been promulgated are conveniently classified using the following three groupings:

TABLE 6.1 Primary Drinking-Water Maximum Contaminant Levels (mg/L) Organized by Category

Inorganic Chemicals

Arsenic	0.05	Mercury	0.002
Barium	2.	Nickel	0.1
Cadium	0.005	Nitrate (as N)	10.
Chromium (total)	0.1	Nitrite (as N)	1.
Copper	TT[a]	Nitrate + nitrite	10.
Fluoride[b]	4.0	Selemium	0.05
Lead	TT[a]	Thallium	0.002
Asbestos	7 million fibers/liter (longer than 10 μm)		

Volatile Organic Chemicals

Benzene	0.005	Ethylbenzene	0.7
Carbon tetrachloride	0.005	Monochlorobenzene	0.1
p-Dichlorobenzene	0.075	Tetrachloroethylene	0.005
o-Dichlorobenzene	0.6	1,2,4-Trichlorobenzene	0.07
1,2-Dichloroethane	0.005	1,1,1-Trichloroethane	0.2
1,1-Dichloroethylene	0.007	1,1,2-Trichloroethane	0.005
cis-1,2-Dichloroethylene	0.07	Trichloroethylene	0.005
trans-1,2-Dichloroethylene	0.1	Vinyl chloride	0.002
1,2-Dichloropropane	0.005		

Synthetic Organic Chemicals

Acrylamide	TT[a]	Glyphosate	0.7
Adipate (diethylhexyl)	0.4	Heptachlor	0.0004
Alachlor	0.002	Heptachlor epoxide	0.0002
Atrazine	0.003	Hexachlorobenzene	0.001
Benzo-a-pyrene	0.0002	Hexachlorocyclopentadiene	0.05
Carbofuran	0.04	Lindane	0.0002
Chlordane	0.002	Methoxychlor	0.04
Dalapon	0.2	Oxylamyl (Vydate)	0.2
Dibromochloropropane	0.0002	Pentachlorophenol	0.001
Di(ethylhexyl)adipate	0.4	Picloram	0.5
Di(ethylhexyl)phthlate	0.006	Polychlorinated byphenyls	0.0005
Dichloro-methane	0.005	Simazine	0.004
Dinoseb	0.007	Styrene	0.1
Diquat	0.02	Toluene	1.
Endothall	0.1	Toxaphene	0.003
Endrin	0.002	Xylenes (total)	10.
Epichlorohydrin	TT[a]	2,4-D	0.07
Ethylene dibromide	0.00005	2,4,5-TP (Silvex)	0.05
2,3,7,8-TCDD (Dioxin)	0.00000003		

Disinfection By-Products (interim)

Total trihalomethanes	0.10

Radionuclides

Radium 226	20 pCi/L	Beta particle and photon radioactivity	4 mrem/yr
Radium 228	20 pCi/L	Radon	300 pCi/L
Gross alpha particle activity	15 pCi/L	Uranium	20 μg/L

[a]Treatment technique (TT) requirement rather than an MCL.
[b]Many states require public notification if flouride is in excess of 2.0 mg/L.
Source: Hammer and Hammer, 1996.

1. *Synthetic organic chemicals* (SOCs) are compounds used in the manufacture of a wide variety of agricultural and industrial products. They include primarily insecticides and herbicides.

2. *Volatile organic chemicals* (VOCs) are synthetic chemicals that readily vaporize at room temperature. These include degreasing agents, paint thinners, glues, dyes, and some pesticides. Representative chemicals include benzene, carbon tetrachloride, 1,1,1-trichloroethane (TCA), trichlorethylene(TCE), and vinyl chloride.

3. *Trihalomethanes* (THMs) are the byproducts of water chlorination. They include chloroform ($CHCl_3$), bromodichloromethane ($CHBrCl_2$), dibromochloromethane ($CHBr_2Cl$), and bromoform ($CHBr_3$).

A list of organic chemicals for which MCLs have been established is included in Table 6.1.

Radionuclides

Radioactivity in public drinking-water supplies is another category of contaminants regulated by the Safe Drinking Water Act. Some radioactive compounds, or *radionuclides,* are naturally occurring substances such as radon and radium-226, which are often found in groundwater, while others, such as strontium-90 and tritium, are surface water contaminants resulting from atmospheric nuclear weapons testing fallout. The MCLs for most radionuclides are expressed as picocuries per liter (pCi/L), where 1 pCi corresponds to 2.2 radioactive decays per minute (1 Ci is the decay rate of 1 gram of radium). For example, the gross alpha particle activity MCL (including radium-226 but excluding radon and uranium) is 15 pCi/L. The MCL for beta particle and photon radioactivity is an annual dose either to the whole body or to any particular organ, of 4 mrem/yr.

The most important radionuclide associated with drinking water is dissolved radon gas. It is a colorless, odorless, and tasteless gas that occurs naturally in some groundwater. It is an unusual contaminant because the danger arises not from drinking radon-contaminated water, but from breathing the gas after it has been released into the air. When radon-laden water is heated or agitated, such as occurs in showers or washing machines, the dissolved radon gas is released. As will be discussed in Chapter 7, inhaled radon gas is thought to be an important cause of lung cancer.

Microbiological Standards

Another category of primary MCLs is *microbiological contaminants.* While it would be desirable to evaluate the safety of a given water supply by individually testing for specific pathogenic microorganisms, such tests are too difficult to perform on a routine basis or to be used as a standard. Instead, a much simpler technique is used, based on testing water for evidence of any fecal contamination. In this test, coliform bacteria (typically *Escherichia coli*) are used as indicator organisms whose presence suggests that the water is contaminated. Since the number of coliform bacteria excreted in feces is on the order of 50 million per gram and the concentration of coliforms in untreated domestic wastewater is usually several million per 100 mL, it would be highly unlikely that water contaminated with human wastes would have no coliforms. That conclusion is the basis for the drinking-water standard for microbiological contaminants, which specifies that for large water systems (serving more than 1000 people), no more than 5 percent of the test samples can show any coliforms; for smaller systems testing fewer than 40 samples per month, no more than one sample can be test positive.

The assumption that the absence of coliforms implies an absence of pathogens is based primarily on the following two observations. First, in our society it is the excreta from relatively few individuals that adds pathogens to a wastestream, while the entire population contributes coliforms. Thus the number of coliforms should far exceed the number of pathogens. Second, for many of the waterborne diseases that have plagued humankind, the survival rate of pathogens outside the host is much lower than the survival rate of coliforms. The combination of these factors suggests that, statistically speaking, the ratio of pathogens to coliforms should be sufficiently small that we can conclude that it is extremely unlikely that a sample of water would contain a pathogen without also containing numerous coliforms.

This approach to testing for microbiological purity has been quite effective, but it is not an absolutely certain measure. For example, some nonbacterial pathogens, notably viruses and *Giardia* cysts, survive considerably longer outside of their hosts than coliform bacteria, which increases the probability of encountering pathogens without accompanying coliforms. However, as suggested by the approximations given in Table 6.2 for viruses, the risk of such an encounter is still considered to be acceptably small under normal circumstances.

Microbiological drinking water standards provide an example of the specification of a treatment technique (TT) rather than an actual MCL. Since direct testing for pathogens is impractical, the primary drinking-water standard for viruses, *Giardia lambia,* and *Cryptosporidium* is stated as a requirement for filtration (or, in special circumstances, watershed controls).

The coliform test is also used to assess the safety of water-contact recreational activities, with many states recommending a limit of 1000 coliforms per 100 mL. However, proper interpretation of a coliform test made on surface water is complicated by the fact that fecal coliforms are discharged by animals as well as humans. Thus a high fecal coliform count is not necessarily an indication of human contamination. When it is important to distinguish between human and animal contamination, more sophisticated testing can be performed. Such testing is based on the fact that the ratio of fecal coliform to fecal streptococci is different in human and animal discharges.

TABLE 6.2 Virus-Coliform Ratios for Sewage and Polluted Surface Water

	Virus	Coliform	Virus/Coliform Ratio
Sewage	500/100 mL	46×10^6/100 mL	1:92,000
Polluted surface water	1/100 mL	5×10^4/100 mL	1:50,000

Source: Robeck et al. (1962).

TABLE 6.3 Secondary Standards For Drinking Water

Contaminant	Level	Contaminant effects
Aluminum	0.05–0.2 mg/L	Water discoloration
Chloride	250 mg/L	Taste, pipe corrosion
Color	15 color units	Aesthetic
Copper	1 mg/L	Taste, porcelain staining
Corrosivity	Noncorrosive	Pipe leaching of lead
Fluoride	2.0 mg/L	Dental fluorosis
Foaming agents	0.5 mg/L	Aesthetic
Iron	0.3 mg/L	Taste, laundry staining
Manganese	0.05 mg/L	Taste, laundry staining
Odor	3 threshold odor number	Aesthetic
pH	6.5–8.5	Corrosive
Silver	0.1 mg/L	Skin discoloration
Sulfate	250 mg/L	Taste, laxative effects
Total dissolved solids	500 mg/L	Taste, corrosivity, detergents
Zinc	5 mg/L	Taste

Secondary Standards

Secondary standards are nonenforceable, maximum contaminant levels intended to protect "public welfare." Public welfare criteria include factors such as taste, color, corrosivity, and odor, rather than health effects. The limits suggested in Table 6.3 for about half of the contaminants are in large part based on taste. Excessive sulfate is undesirable because of its laxative effect; iron and manganese are objectionable because of taste and their ability to stain laundry and fixtures; foaming and color are visually upsetting; excessive fluoride causes a brownish discoloration of teeth; and odor from various dissolved gases may make water unacceptable to the drinker.

6.4 WATER TREATMENT SYSTEMS

The purpose of water treatment systems is to bring raw water up to drinking-water quality. The particular type of treatment equipment required to meet these standards will depend to some extent on the source of water. About half of the drinking water in the United States comes from groundwater, and half from surface water. Most large cities rely more heavily on surface water, while most small towns or communities depend more on groundwater. Surface water tends to have more turbidity and a much greater chance of microbial contamination, so filtration is almost always a necessity. Groundwater, on the other hand, is uncontaminated and has relatively little suspended solids, so filtration is less important. Groundwater, however, may have objectionable dissolved gases that need to be removed, and hardness (ions of calcium and magnesium) removal is usually needed.

As suggested in Figure 6.2, a typical treatment plant for surface water might include the following sequence of steps:

Screening to remove relatively large floating and suspended debris.

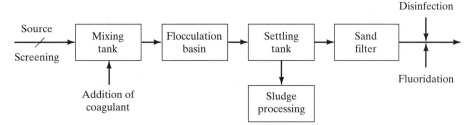

FIGURE 6.2   Schematic of a typical water treatment plant for surface water. Softening may be required as an additional step for groundwater.

Mixing the water with chemicals that encourage suspended solids to coagulate into larger particles, which will settle more easily.

Flocculation, which is the process of gently mixing the water and coagulant, allowing the formation of large particles of floc.

Sedimentation, in which the flow is slowed enough so that gravity will cause the floc to settle, and *filtration* in which the effluent is cleaned.

Sludge processing, in which the mixture of solids and liquids collected from the settling tank is dewatered and disposed of.

Disinfection of the liquid effluent to ensure that the water is free of harmful pathogens. Hardness removal can be added to this generalized flow diagram if needed.

Coagulation and Flocculation

Raw water may contain suspended particles of color, turbidity, and bacteria that are too small to settle in a reasonable time period and cannot be removed by simple filtration. The object of coagulation is to alter these particles in such a way as to allow them to adhere to each other. Thus they can grow to a size that will allow removal by sedimentation and filtration. Coagulation is considered to be a *chemical* treatment process that destabilizes colloidal particles (particles in the size range of about 0.001 to 1 μm), as opposed to the *physical* treatment operations of flocculation, sedimentation, and filtration that follow.

Most colloids of interest in water treatment remain suspended in solution because they have a net negative surface charge that cause the particles to repel each other. The intended action of the coagulant is to neutralize that charge, allowing the particles to come together to form larger particles that can be more easily removed from the raw water. The usual coagulant is alum $Al_2(SO_4)_3 \cdot 18\ H_2O$, although $FeCl_3$, $FeSO_4$, and other coagulants, such as polyelectrolytes, can be used. Since the intention here is simply to introduce the concepts of water treatment and leave the complexities for more specialized books, let us just look at the reactions involving alum. Alum ionizes in water producing Al^{3+} ions, some of which neutralize the negative charges on the colloids. Most of the aluminum ions, however, react with alkalinity in the water (bicarbonate) to form insoluble aluminum hydroxide, $Al(OH)_3$. The aluminum hydroxide adsorbs positive ions from solution and forms a precipitate of Al $(OH)_3$ and adsorbed sulfates. The overall reaction is

$$Al_2(SO_4)_3 \cdot 18 H_2O + 6 HCO_3^- \rightleftharpoons 2 Al(OH)_3 \downarrow + 6 CO_2 + 18 H_2O + 3 SO_4^{2-}$$

Alum   Aluminum   Sulfates

                                                        hydroxide   (6.1)

If insufficient bicarbonate is available for this reaction to occur, the pH must be raised, usually by adding lime, $Ca(OH)_2$, or sodium carbonate, Na_2CO_3.

Coagulants are added to the raw water in a chamber that has rapidly rotating paddles to mix the chemicals. Detention times in the rapid mix tank are typically less than one-half minute. Flocculation follows in a tank that provides gentle agitation for approximately one-half hour. During this time, the precipitating aluminum hydroxide attracts colloidal particles, forming a plainly visible floc. The mixing in the flocculation tank must be done very carefully. It must be sufficient to encourage particles to make contact with each other, enabling the floc to grow in size, but it cannot be so vigorous that the fragile floc particles will break apart. Mixing also helps keep the floc from settling in this tank, rather than in the sedimentation tank that follows. Figure 6.3 shows a cross section of a mixing tank followed by a sedimentation tank.

Sedimentation and Filtration

After flocculation, the water flows through a sedimentation basin, or clarifier. A sedimentation basin is a large circular, or rectangular, concrete tank designed to hold the water for a long enough time to allow most of the suspended solids to settle out. Typical detention times range from 1 to 10 hours. The longer the detention time, the bigger and more expensive the tank must be, but, correspondingly, the better will be the tank's performance. Solids that collect on the bottom of the tank may be removed manually by periodically shutting down the tank and washing out the collected sludge, or the tank may be continuously and mechanically cleaned using a bottom scraper. The effluent from the tank is then filtered.

One of the most widely used filtration units is called a *rapid-sand filter,* which consists of a layer of carefully sieved sand on top of a bed of graded gravels. The pore openings between grains of sand are often greater than the size of the floc particles that are to be removed, so much of the filtration is accomplished by means other than simple straining. Adsorption, continued flocculation, and sedimentation in the pore spaces are also important removal mechanisms. When the filter becomes clogged with

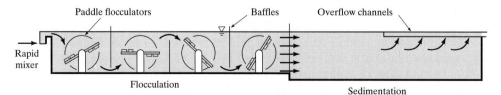

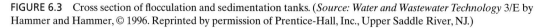

FIGURE 6.3 Cross section of flocculation and sedimentation tanks. (*Source: Water and Wastewater Technology* 3/E by Hammer and Hammer, © 1996. Reprinted by permission of Prentice-Hall, Inc., Upper Saddle River, NJ.)

particles, which occurs roughly once a day, the filter is shut down for a short period of time and cleaned by forcing water backward through the sand for 10 to 15 minutes. After cleaning, the sand settles back in place and operation resumes.

Disinfection

During coagulation, settling, and filtration, practically all of the suspended solids, most of the color, and all but a few percent of the bacteria are removed. The final step is disinfection, to kill any remaining pathogenic organisms, and fluoridation, to help control dental caries.

Chlorination using chlorine gas (Cl_2), sodium hypochlorite (NaOCl), or calcium hypochlorite [$Ca(OCl)_2$], is the most commonly used method of disinfection in this country. Chlorine is a powerful oxidizing agent that is easy to use, inexpensive, and reliable. The precise mechanism by which chlorine kills microorganisms is not well understood. It is thought that its strong oxidizing power destroys the enzymatic processes necessary for cell life. Though chlorination is completely effective against bacteria, its effectiveness is less certain with protozoan cysts, most notably those of *Giardia lamblia* and *Cryptosporidium,* or with viruses. In almost all circumstances, complete treatment, including coagulation, sedimentation, filtration, as well as disinfection, is the treatment technique (TT) required to meet the microbiological standards of the Clean Water Act.

Using chlorine gas to illustrate the chemical reactions occurring during chlorination, the key reaction is

$$Cl_2 \; + \; H_2O \; \Longleftrightarrow \; HOCl \; + \; H^+ \; + \; Cl^- \qquad (6.2)$$
$$\text{Chlorine gas} \qquad\qquad \text{Hypochlorous acid}$$

The hypochlorous acid formed (HOCl) is the prime disinfecting agent. Its dissociation is pH dependent, yielding less effective hypochlorite (OCl^-) ions at higher pH values:

$$HOCl \; \Longleftrightarrow \; H^+ \; + \; OCl^- \qquad (6.3)$$
$$\text{Hypochlorous acid} \qquad\qquad \text{Hypochlorite}$$

Together, HOCl and OCl^- are called the *free available chlorine.*

A principal advantage of chlorination over other forms of disinfection is that a chlorine residual is created that can protect the treated water after leaving the treatment plant. This guards against possible contamination that might occur in the water distribution system. To increase the lifetime of this residual, some systems add ammonia to the treated water, forming *chloramines* (NH_2Cl, $NHCl_2$, NCl_3). Chloramines, although they are less effective as oxidants than HOCl, are more persistent. Residual chlorine that exists as chloramine is referred to as *combined* available residual chlorine.

A disadvantage of chlorination is the formation of *trihalomethanes* (THMs), such as the carcinogen chloroform ($CHCl_3$). THMs are created when chlorine combines with natural organic substances, such as decaying vegetation, that may be present in the water itself. One approach to reducing THMs is to remove more of the organics before any chlorination takes place. It was common practice in the past, for example, to chlorinate the incoming raw water before coagulation and filtration, but it has since been realized that that contributes to formation of THMs. By eliminating this step and by increasing the organic removal during treatment before chlorination, some degree of THM control is achieved. In the future, the actual removal of THMs from the treated water, perhaps by aeration or adsorption on activated carbon, may become necessary.

The problem of THMs is helping spur interest in alternatives to chlorination as the preferred method of disinfection. Alternative disinfectants include chloramines, chlorine dioxide, and ozone. Each has the advantage of not creating THMs, but there are uncertainties and known disadvantages to each that have restricted their more widespread use. Chloramines, as mentioned previously, are less certain disinfectants than free chlorine, so chloramination is usually used in combination with some other more effective method of disinfection. Chlorine dioxide (ClO_2), on the other hand, is a potent bactericide and viricide, and it does form a residual capable of protecting water in the distribution system. However, there is concern for certain toxic chlorate and chlorite substances that it may create, and it is a very costly method of disinfection. Ozonation involves the passage of ozone (O_3) through the water. Ozone is a very power disinfectant that is even more effective against cysts and viruses than chlorine, and it has the added advantage of leaving no taste or odor problems. Although ozonation is widely used in European water treatment facilities, it has the disadvantages of not forming a protective residual in the treated water, and it is more expensive than chlorination.

Hardness and Alkalinity

The presence of multivalent cations, most notably calcium and magnesium ions, is referred to as water *hardness*. Groundwater is especially prone to excessive concentrations of these ions. Hardness causes two distinct problems. First, the reaction between hardness and soap produces a sticky, gummy deposit called "soap curd" (the ring around the bathtub). Essentially all home cleaning activities, from bathing and grooming to dishwashing and laundering, are made more difficult with hard water.

While the introduction of synthetic detergents has decreased, but not eliminated, the impact of hardness on cleaning, the second problem, that of scaling, remains significant. When hard water is heated, calcium carbonate ($CaCO_3$) and magnesium hydroxide [$Mg(OH)_2$] readily precipitate out of solution, forming a rocklike scale that clogs hot water pipes and reduces the efficiency of water heaters, boilers, and heat exchangers. Pipes filled with scale must ultimately be replaced, usually at great expense. Heating equipment that has scaled up not only transmits heat less readily, thus increasing fuel costs, but also is prone to failure at a much earlier time. For both of these reasons, if hardness is not controlled at the water treatment plant itself, many individuals and industrial facilities find it worth the expense to provide their own water softening.

Hardness is defined as the concentration of all multivalent metallic cations in solution. The principal ions causing hardness in natural water are calcium (Ca^{2+}) and magnesium (Mg^{2+}). Others, including iron (Fe^{2+}), manganese (Mn^{2+}), strontium (Sr^{2+}), and aluminum (Al^{3+}), may be present, though in much smaller quantities.

It is conventional practice to calculate hardness (as well as alkalinity) using the concepts of *equivalents* and *equivalent weights*. Chemists in the eighteenth century noted, for example, that 8 g of oxygen combine with 1 g of hydrogen (to form H_2O), while 3 g of carbon combine with 1 g of hydrogen (to form CH_4). They inferred from this that 8 g of oxygen ought to combine with 3 g of carbon (which it does in CO_2). In this example, the *equivalent weight* (EW) of oxygen would be 8 g, while that of carbon would be 3 g, and one equivalent weight of oxygen would combine with one equivalent weight of carbon (to form CO_2). This idea of equivalents, when first proposed, was not taken very seriously because it is complicated by the fact that individual elements can have more than one equivalent weight, depending on the reaction in question (e.g., carbon forms CO as well as CO_2). With the understanding of the structure of the atom that came later, however, it regained favor as a convenient method of handling certain chemical computations.

The *equivalent weight* of a substance is its atomic or molecular weight divided by a number n that relates to its valence or ionic charge. For the reactions of interest in hardness and alkalinity calculations, n is simply the ionic charge. For compounds, it is the number of hydrogen ions that would be required to replace the cation:

$$\text{Equivalent weight (EW)} = \frac{\text{Atomic or molecular weight}}{n} \tag{6.4}$$

Thus, for example, to find the equivalent weight of $CaCO_3$ we note that it would take two hydrogen ions to replace the cation(Ca^{2+}). Its equivalent weight is therefore

$$\text{Equivalent weight of } CaCO_3 = \frac{(40 + 12 + 3 \times 16)}{2} = \frac{100}{2}$$

$$\text{Equivalent weight of } CaCO_3 = 50.0\,\text{g/eq} = 50.0\,\text{mg/meq} \tag{6.5}$$

where (mg/meq) are the units of milligrams per milliequivalent. For the calcium ion itself (Ca^{2+}), which has an atomic weight of 40.1 and a charge of 2, the equivalent weight is

$$\text{Equivalent weight of } Ca^{2+} = \frac{40.1}{2} \approx 20.0\,\text{mg/meq}$$

Table 6.4 lists the equivalent weights for a number of common chemicals important in water hardness calculations.

In measuring hardness, the concentrations of the multivalent cations are converted to *mg/L as $CaCO_3$* using the following expression:

$$\text{mg/L of } X \text{ as } CaCO_3 = \frac{\text{Concentration of } X(\text{mg/L}) \times 50.0\,\text{mg } CaCO_3/\text{meq}}{\text{Equivalent weight of } X(\text{mg/meq})} \tag{6.6}$$

The *total hardness* (TH) as $CaCO_3$ is the sum of each individual hardness:

$$\text{Total hardness} = Ca^{2+} + Mg^{2+} \tag{6.7}$$

where it has been assumed in (6.7) that calcium and magnesium are the only two multivalent cations with appreciable concentrations. The following example shows how to work with these units.

TABLE 6.4 Equivalent Weights For Selected Elements, Radicals, and Compounds

Name	Symbol or formula	Atomic or molecular weight	Charge n	Equivalent weight (mg/meq)
Aluminum	Al^{3+}	27.0	3	9.0
Calcium	Ca^{2+}	40.1	2	20.0
Hydrogen	H^+	1.0	1	1.0
Magnesium	Mg^{2+}	24.3	2	12.2
Potassium	K^+	39.1	1	39.1
Sodium	Na^+	23.0	1	23.0
Bicarbonate	HCO_3^-	61.0	1	61.0
Carbonate	CO_3^{2-}	60.0	2	30.0
Chloride	Cl^-	35.5	1	35.5
Hydroxyl	OH^-	17.0	1	17.0
Nitrite	NO_2^-	46.0	1	46.0
Nitrate	NO_3^-	62.0	1	62.0
Orthophosphate	PO_4^{3-}	95.0	3	31.7
Sulfate	SO_4^{2-}	96.0	2	48.0
Calcium carbonate	$CaCO_3$	100.0	2	50.0

EXAMPLE 6.1 Total Hardness as $CaCO_3$

A sample of groundwater has 100 mg/L of Ca^{2+} and 10 mg/L of Mg^{2+}. Express its hardness in units of meq/L and mg/L as $CaCO_3$.

Solution The contribution of calcium in meq/L is

$$\frac{100\,mg/L}{20.0\,mg/meq} = 5.0\,meq/L$$

In mg/L as $CaCO_3$, the calcium concentration is

$$5.0\ meq/L \times 50\ mg\ CaCO_3/meq = 250.0\ mg/L\ as\ CaCO_3$$

Table 6.4 gives 12.2 mg/meq for magnesium; so at 10 mg/L its hardness is

$$\frac{10\,mg/L}{12.2\,mg/meq} = 0.82\ meq/L$$

which is

$$0.82\ meq/L \times 50\ mg\ CaCO_3/meq = 41.0\ mg/L\ as\ CaCO_3$$

The total hardness is

$$5.0\ meq/L + 0.82\ meq/L = 5.8\ meq/L$$

or

$$(250.0 + 41.0)\ mg/L = 291\ mg/L\ as\ CaCO_3 \qquad \blacksquare$$

TABLE 6.5 Hardness Classifications of Water

Description	Hardness	
	meq/L	mg/L as $CaCO_3$
Soft	< 1	< 50
Moderately hard	1–3	50–150
Hard	3–6	150–300
Very hard	> 6	> 300

Source: Tchobanoglous and Schroeder (1985).

While public acceptance of hardness is dependent on the past experiences of individual consumers, hardness above about 150 mg/L as $CaCO_3$ is noticed by most consumers. Although there are no absolute distinctions, the qualitative classification of hardness given in Table 6.5 is often used. Groundwater frequently has hardness over 300 mg/L as $CaCO_3$, which would be classified as "very hard." Surface water has less opportunity to dissolve calcium and magnesium minerals so it is usually "soft."

It is useful at times to separate total hardness, which is almost entirely the combination of calcium and magnesium cations, into two components:*carbonate hardness* (CH), associated with the anions HCO_3^- and CO_3^{2-}, and *noncarbonate hardness* (NCH), associated with other anions. If carbonate hardness exceeds the total hardness, then CH is given the same value as TH. Carbonate hardness is especially important since it leads to scaling, as the following reaction suggests:

$$Ca^{2+} + 2HCO_3^- \longrightarrow CaCO_3 + CO_2 + H_2O \qquad (6.8)$$

$$\text{Bicarbonate} \qquad \text{Calcium carbonate}$$

Carbonate hardness is sometimes referred to as "temporary hardness" because it can be removed by simply heating the water.

Another important characteristic of water is its *alkalinity*, which is a measure of the water's ability to absorb hydrogen ions without significant pH change. That is, alkalinity is a measure of the buffering capacity of water. In most natural water, the total amount of H^+ that can be neutralized is dominated by the carbonate buffering system described in Chapter 2. Thus,

$$\text{Alkalinity (mol/L)} = [HCO_3^-] + 2[CO_3^{2-}] + [OH^-] - [H^+] \qquad (6.9)$$

where concentrations in the brackets are [mol/L]. Notice that the concentration of carbonate $[CO_3^{2-}]$ is multiplied by 2 since each ion can neutralize two H^+ ions. This assumes that the concentrations are being measured in molarity units (mol/L). More often, the concentrations are measured in terms of equivalents, or in mg/L as $CaCO_3$, in which case the 2 is already accounted for in the conversions, so concentrations are added directly:

$$\text{Alkalinity (meq/L)} = (HCO_3^-) + (CO_3^{2-}) + (OH^-) - (H^+) \qquad (6.10)$$

where the quantities in parentheses are concentrations in meq/L or mg/L as $CaCO_3$. The following example demonstrates these alkalinity calculations.

EXAMPLE 6.2   Calculating Alkalinity

A sample of water at pH 10.0 has 32.0 mg/L of CO_3^{2-} and 56.0 mg/L of HCO_3^-. Find the alkalinity as $CaCO_3$.

Solution   The equivalent weight of carbonate is given in Table 6.4 as 30 mg/meq. Converting 32.0 mg/L of carbonate into mg/L as $CaCO_3$ gives

$$(CO_3^{2-}) = 32.0 \text{ mg/L} \times \frac{1}{30.0 \text{ mg/meq}} \times 50.0 \text{ mg } CaCO_3/\text{meq} = 53.3 \text{ mg/L as } CaCO_3$$

The equivalent weight of bicarbonate (HCO_3^-) is given in Table 6.4 as 61.0 mg/meq, so its concentration is

$$(HCO_3^-) = 56.0 \text{ mg/L} \times \frac{1}{61.0 \text{ mg/meq}} \times 50.0 \text{ mg } CaCO_3/\text{meq}$$

$$= 45.9 \text{ mg/L as } CaCO_3$$

The pH is 10, so $[H^+] = 1 \times 10^{-10}$ mol/L, and its EW is 1 mg/meq so it is

$$(H^+) = 1 \times 10^{-10} \text{ mol/L} \times 1 \text{ g/mol} \times \frac{10^3 \text{ mg/g}}{1 \text{ mg/meq}} \times 50.0 \text{ mg } CaCO_3/\text{meq}$$

$$= 5.0 \times 10^{-6} \text{ mg/L as } CaCO_3$$

Since $[H^+][OH^-] = 1 \times 10^{-14}$, then $[OH^-] = 1 \times 10^{-4}$ mol/L. The hydroxyl concentration is therefore

$$(OH^-) = 1 \times 10^{-4} \text{ mol/L} \times 17.0 \text{ g/mol} \times \frac{10^3 \text{ mg/g}}{17.0 \text{ mg/meq}} \times 50.0 \text{ mg } CaCO_3/\text{meq}$$

$$= 5.0 \text{ mg/L as } CaCO_3$$

Total alkalinity, then, is just (CO_3^{2-}) + (HCO_3^-) + (OH^-) − (H^+)

$$\text{Alkalinity} = 53.3 + 45.9 + 5.0 - 5.0 \times 10^{-6} = 104.2 \text{ mg/L as } CaCO_3 \qquad \blacksquare$$

For nearly neutral water (pH around 6 to 8) the concentrations of (H^+) and (OH^-) are insignificant and alkalinity is determined entirely by the carbonates:

$$\text{Alkalinity (meq/L)} = (HCO_3^-) + (CO_3^{2-}) \qquad (6.11)$$

EXAMPLE 6.3   Chemical Analysis of a Sample of Water

An analysis of a sample of water with pH 7.5 has produced the following concentrations (mg/L):

Cations (mg/L)		Anions (mg/L)	
Ca^{2+}	80	Cl^-	100
Mg^{2+}	30	SO_4^{2-}	201
Na^+	72	HCO_3^-	165
K^+	6		

Find the total hardness (TH), the carbonate hardness (CH), the noncarbonate hardness (NCH), and the alkalinity, all expressed as $CaCO_3$. Find the total dissolved solids (TDS) in mg/L.

Solution It is helpful to set up a table in which each of the concentrations can be expressed in terms of $CaCO_3$. A calculation for (Ca^{2+}) shows concentration

$$(Ca^{2+}) = 80 \text{ mg/L} \times \frac{1}{20.0 \text{ mg/meq}} \times 50.0 \text{ mg } CaCO_3/\text{meq} = 200 \text{ mg/L as } CaCO_3$$

Other rows of the table are found in a similar way. Equivalent weights are from Table 6.4.

Ion	mg/L	mg/meq	mg/L as $CaCO_3$
Ca^{2+}	80	20.0	200.0
Mg^{2+}	30	12.2	123.0
Na^+	72	23.0	156.5
K^+	6	39.1	7.7
Cl^-	100	35.5	140.8
SO_4^{2-}	201	48.0	209.4
HCO_3^-	165	61.0	135.2

As a first check on the chemical analysis, we can compare the sum of the concentrations of cations and anions as $CaCO_3$ (or as meq/L) to see if they are nearly equal.

$$\Sigma \text{ cations} = 200.0 + 123.0 + 156.5 + 7.7 = 487.2 \text{ mg/L as } CaCO_3$$
$$\Sigma \text{ anions} = 140.8 + 209.4 + 135.2 = 485.4 \text{ mg/L as } CaCO_3$$

which is quite close. The difference would probably be associated with small concentrations of other ions as well as measurement error.

a. The total hardness (TH) is the sum of the multivalent cations, (Ca^{2+}) and (Mg^{2+}):

$$TH = 200.0 + 123.0 = 323.0 \text{ mg/L as } CaCO_3$$

so this is very hard water.

b. The carbonate hardness (CH) is that portion of total hardness associated with carbonates, which in this case is just bicarbonate HCO_3^-:

$$CH = 135.2 \text{ mg/L as } CaCO_3$$

c. The noncarbonate hardness (NCH) is the difference between the total hardness and the carbonate hardness:

$$NCH = TH - CH = 323.0 - 135.2 = 187.8 \text{ mg/L as } CaCO_3$$

d. Since the pH is nearly neutral, the concentrations of (H^+) and (OH^-) are negligible, so the alkalinity is given by just the bicarbonate:

$$\text{Alkalinity} = (HCO_3^-) = 135.2 \text{ mg/L as } CaCO_3$$

e. The total dissolved solids (TDS) is simply the sum of the cation and anion concentrations expressed in mg/L:

$$\text{TDS} = 80 + 30 + 72 + 6 + 100 + 201 + 165 = 654 \text{ mg/L} \qquad\blacksquare$$

It is sometimes helpful to display the ionic constituents of water using a bar graph such as that shown in Figure 6.4.

Softening

Hard water causes scaling of pipes and makes laundering more difficult, so many water treatment plants provide water softening. Surface waters seldom have hardness levels above 200 mg/L as $CaCO_3$, so softening is not usually part of the treatment process. For groundwater, however, where hardness levels are sometimes over 1000 mg/L, it is quite common. There are two popular approaches to softening water: the *lime-soda process* and the *ion-exchange process*. Either may be used in a central treatment plant prior to distribution, but individual home units use the ion-exchange process.

In the lime-soda process, either quick lime (CaO) or hydrated lime [$Ca(OH)_2$] is added to the water, raising the pH to about 10.3 and converting soluble bicarbonate ions (HCO_3^-) into insoluble carbonate (CO_3^{2-}). The carbonate then precipitates out as $CaCO_3$, as is suggested in the following reaction:

$$Ca(HCO_3)_2 + Ca(OH)_2 \longrightarrow 2\,CaCO_3 \downarrow + 2\,H_2O \qquad (6.12)$$

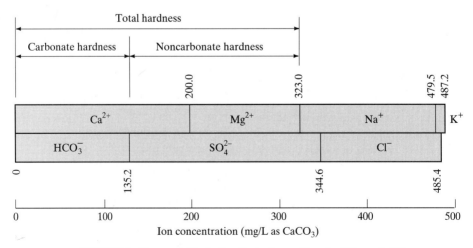

FIGURE 6.4 Bar graph illustrating the ionic constituents of Example 6.3.

Similarly, magnesium ions can be removed by forming the precipitate $Mg(OH)_2$:

$$Mg(HCO_3)_2 + 2\,Ca(OH)_2 \longrightarrow 2\,CaCO_3 \downarrow + Mg(OH)_2 \downarrow + 2\,H_2O \qquad (6.13)$$

If insufficient natural bicarbonate alkalinity (HCO_3^-) is available to cause reaction (6.12) or (6.13), it may be necessary to add carbonate in the form of soda ash (Na_2CO_3). The reaction for magnesium would then be

$$MgSO_4 + Ca(OH)_2 + Na_2CO_3 \longrightarrow Mg(OH)_2 \downarrow + CaCO_3 \downarrow + Na_2SO_4 \qquad (6.14)$$

The additional carbonate required makes the process of noncarbonate hardness removal more expensive than the removal of carbonate hardness.

Most of the precipitates formed are removed in a sedimentation basin. Particles too finely divided to settle can later contribute to the clogging of filters and distribution piping, so the water is often recarbonated with CO_2 to convert those carbonate particles into soluble bicarbonates. Hardness reduction by precipitation can bring the total hardness down to as low as 40 mg/L as $CaCO_3$.

The alternative to softening by precipitation uses an ion-exchange process. The exchange process involves replacement of the unwanted ions in solution with ions of a different species that are attached to an insoluble resin. The ion-exchange process can be used to remove nitrate ions, metal ions, and other unwanted ions as well as the ions responsible for hardness. Laboratory-quality deionized water is obtained in this way, but the processing required to obtain such high-quality water is too expensive to use in municipal treatment plants.

In the ion-exchange process, hard water is forced through a column containing solid resin beads made of naturally occurring clays called *zeolites*, or the resins can be synthetically produced. In an ion-exchange unit, the resin removes Ca^{2+} and Mg^{2+} ions from the water and replaces them with sodium ions, which form soluble salts. Using $Ca(HCO_3)_2$ as an example, the reaction can be represented as follows:

$$Ca(HCO_3)_2 + Na_2R \longrightarrow CaR + 2\,NaHCO_3 \qquad (6.15)$$

where R represents the solid ion-exchange resin. The calcium reacts with the resin and is removed from the water as CaR. The alkalinity (HCO_3^-) remains unchanged. The sodium salts that are formed do not cause hardness, but the dissolved sodium ions remain in the treated water and may be harmful to individuals with heart problems.

The hardness removal is essentially 100 percent effective as long as the ion-exchange medium has sodium remaining. When the sodium is depleted, the ion-exchange bed must be regenerated by removing it from service and backwashing it with a solution of NaCl, forming new Na_2R. The wastewater produced during regeneration must be properly disposed of since it contains a high concentration of chlorides. The regeneration reaction involving Ca can be represented as

$$CaR + 2\,NaCl \longrightarrow Na_2R + CaCl_2 \qquad (6.16)$$

The ion-exchange process can be used in waste treatment as well as water treatment. In such cases, the process can enable recovery of valuable chemicals for reuse, or harmful ones for disposal. For example, it is often used to recover chromic acid from metal finishing waste, for reuse in chrome-plating baths. It is even used for the removal of radioactivity.

Desalination

There are a number of approaches to removing salts from water, including distillation, reverse osmosis, electrodialysis, freezing, and ion exchange. Of these, multistage distillation and reverse osmosis are the two technologies most commonly used. Between them they account for 87 percent of the desalination capacity worldwide (excluding on-board desalination used on ships), as shown in Figure 6.5.

Desalination by distillation is based on the fact that salts do not evaporate with water, and hence if water can be boiled and the vapor condensed, the vapor will be pure water. Over half of the world's desalination capacity is based on this approach. Figure 6.6*a* shows a simple, one-stage distillation device. Notice that the incoming cold salt water passes through a heat exchanger that cools and condenses the steam. The heat exchanger also raises the temperature of the incoming salty water, reducing the external energy needed to bring it to a boil. By stringing many of these stages together in a *multistage flash* (MSF) distillation plant, the overall energy efficiency is increased. Distillation is especially appropriate for desalinizing sea water (35,000 mg/L salt) since the distillation process is nearly independent of the dissolved solids concentration in the water. Disposal of the concentrated salt residue, however, does affect costs, making desalination of seawater somewhat more expensive than brackish water. Gleick (1993) estimates that the average MSF cost in the Middle East (where most of the plants are) to desalinate brackish water is $1.33 per cubic meter; the mean cost for seawater is $1.87 per m^3. For comparison, urban users in the western United States pay approximately $0.30 per m^3 while farmers pay closer to $0.03 per m^3.

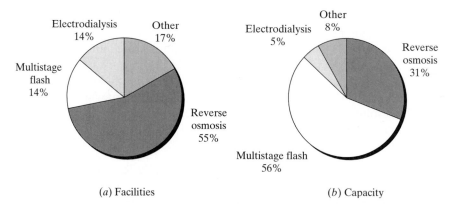

(*a*) Facilities   (*b*) Capacity

FIGURE 6.5   Worldwide desalination by technology, 1990. (*a*) Percentage, by number of facilities; (*b*) percentage, by treatment capacity. On-board ship desalinators are not included in these counts. (Based on Gleick, 1993)

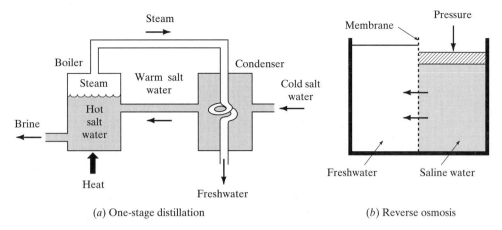

(*a*) One-stage distillation   (*b*) Reverse osmosis

FIGURE 6.6   The two most widely used desalination technologies: (*a*) multi- stage distillation (one stage shown), and (*b*) reverse osmosis.

Desalination by *reverse osmosis* (RO) is the other most commonly used process. In fact, there are more RO plants worldwide than all of the other technologies combined, but they tend to be smaller in capacity than distillation plants so their combined desalination capacity is much less than multistage flash. The key to reverse osmosis is a very thin membrane that is able to pass water molecules but not salt ions. In Figure 6.6*b* a two-chamber RO unit is illustrated. When pressure is applied to the saline water on one side of the membrane, water molecules are forced through the membrane but the passage of salts is blocked. The performance of an RO unit does depend on salinity, so they are often used on brackish waters. Gleick (1993) estimates that the average RO cost in the Middle East to desalinate brackish water is $0.93/m^3 (cheaper than MSF), while the mean cost for seawater is $2.12/m^3 (more than MSF).

6.5 WASTEWATER TREATMENT

Municipal wastewater is typically over 99.9 percent water. The characteristics of the remaining portion vary somewhat from city to city, with the variation depending primarily on inputs from industrial facilities that mix with the somewhat predictable residential flows. Given the almost limitless combinations of chemicals found in wastewater, it is too difficult to list them individually. Instead, they are often described by a few general categories, as has been done in Table 6.6.

In Table 6.6, a distinction is made between *total dissolved solids* (TDS) and *suspended solids* (SS). The sum of the two is *total solids* (TS). The suspended solids portion is, by definition, the portion of total solids that can be removed by a membrane filter (having a pore size of about 1.2 μm). The remainder (TDS) that cannot be filtered includes dissolved solids, colloidal solids, and very small suspended particles.

Wastewater treatment plants are usually designated as providing *primary, secondary,* or *advanced* treatment, depending on the degree of purification. Primary treatment plants utilize physical processes, such as screening and sedimentation, to remove

TABLE 6.6 Composition of untreated domestic wastewater

Constituent	Abbreviation	Concentration (mg/L)
5-day biochemical oxygen demand	BOD_5	100–300
Chemical oxygen demand	COD	250–1000
Total dissolved solids	TDS	200–1000
Suspended solids	SS	100–350
Total Kjeldahl nitrogen	TKN	20–80
Total phosphorus (as P)	TP	5–20

pollutants that will settle, float, or that are too large to pass through simple screening devices. This is followed by disinfection. Primary treatment typically removes about 35 percent of the BOD and 60 percent of the suspended solids. In the early 1970s, the sewage of about 50 million people in the United States was receiving no better treatment than this. While the most visibly objectionable substances are removed in primary treatment and some degree of safety is provided by the disinfection, the effluent still has enough BOD to cause oxygen depletion problems and enough nutrients, such as nitrogen and phosphorus, to accelerate eutrophication.

The Clean Water Act (CWA), in essence, requires at least secondary treatment for all publicly owned treatment works (POTWs) by stipulating that such facilities provide at least 85 percent BOD removal (with possible case-by-case variances that allow lower percentages for marine discharges). This translates into an effluent requirement of 30 mg/L for both five-day BOD and suspended solids (monthly average). In secondary treatment plants, the physical processes that make up primary treatment are augmented with processes that involve the microbial oxidation of wastes. Such biological treatment mimics nature by utilizing microorganisms to oxidize the organics, with the advantage being that the oxidation can be done under controlled conditions in the treatment plant itself, rather than in the receiving body of water. When properly designed and operated, secondary treatment plants remove about 90 percent of the BOD and 90 percent of the suspended solids.

While the main purpose of primary treatment (in addition to disinfecting the wastes) is to remove objectionable solids, and the principal goal of secondary treatment is to remove most of the BOD, neither is effective at removing nutrients, dissolved material, or biologically resistant (refractory) substances. For example, typically no more than half of the nitrogen and one-third of the phosphorus are removed during secondary treatment. This means the effluent can still be a major contributor to eutrophication problems. In circumstances where either the raw sewage has particular pollutants of concern or the receiving body of water is especially sensitive, so-called advanced treatment (previously called *tertiary* treatment) may be required. Advanced treatment processes are varied and specialized, depending on the nature of the pollutants that must be removed. In most circumstances, advanced treatment follows primary and secondary treatment, although in some cases, especially in the treatment of industrial waste, it may completely replace those conventional processes.

An example flow diagram for a wastewater treatment plant that provides primary and secondary treatment is illustrated in Figure 6.7.

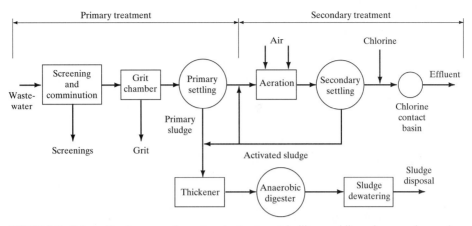

FIGURE 6.7   Schematic of an example wastewater treatment facility providing primary and secondary treatment using the activated sludge process.

Primary Treatment

As suggested in Figure 6.7, primary treatment begins with simple screening. Screening removes large floating objects such as rags, sticks, old shoes, and whatever else that might otherwise damage the pumps or clog small pipes. Screens vary, but typically consist of parallel steel bars spaced anywhere from 2 cm to 7 cm apart, perhaps followed by a wire mesh screen with smaller openings. One way to avoid the problem of disposal of materials collected on screens is to use a device called a comminuter, which grinds those coarse materials into small enough pieces that they can be left right in the wastewater flow.

After screening, the wastewater passes into a grit chamber, where it is held for a few minutes. The *detention* time (the tank volume divided by the flow rate) is chosen to be long enough to allow sand, grit, and other heavy material to settle out but is too short to allow lighter, organic materials to settle. By collecting only these heavier materials, the disposal problem is simplified since those materials are usually nonoffensive and, after washing, can be easily disposed of in a landfill.

From the grit chamber, the sewage passes to a primary settling tank (also known as a "sedimentation basin" or a "primary clarifier"), where the flow speed is reduced sufficiently to allow most of the suspended solids to settle out by gravity. Detention times of approximately two to three hours are typical, resulting in the removal of from 50 to 65 percent of the suspended solids and 25 to 40 percent of the BOD. Primary settling tanks are either round or rectangular and their behavior is similar to that of the clarifiers already described for water treatment facilities. The solids that settle, called *primary sludge* or raw sludge, are removed for further processing, as is the grease and scum that floats to the top of the tank. If this is just a primary treatment plant, the effluent at this point is chlorinated to destroy bacteria and help control odors. Then it is released.

Sizing of primary settling tanks (clarifiers) is based on several parameters that have been found to be the key determinants of successful performance. One such design parameter is the *overflow rate*, which is the ratio of the average daily flow through the tank divided by the surface area. Most settling tanks are circular, with raw sewage entering at the center of the circle and treated effluent leaving the perimeter of the tank as it overflows across a weir. If the tank is too small and wastewater flows too quickly toward the overflow weir, the solids will be moving too fast to consolidate and settle. Overflow rates in the range of 15 to 30 m³/day per m² of surface area are typical. Another parameter is the detention time of wastewater in the tank. Detention time is the tank volume divided by the influent flow rate, which is numerically equal to the time that would be required to fill an empty tank with wastewater entering the tank at the daily average flow rate. To see how these minimal specifications contribute to the sizing of a primary clarifier, consider the following example.

EXAMPLE 6.4   Sizing a Primary Clarifier

A town of 30,000 sends 0.5 m³ per person per day to the wastewater treatment plant. A circular primary clarifier is to be designed to have an average detention time of 2.5 hours and an average overflow rate of 20 m³/day per square meter. What should the dimensions of the clarifier be?

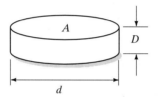

Solution   At 0.5 m³/day the daily flow for 30,000 people would be 15,000 m³/day. The overflow rate is defined to be the ratio of flow to area, so

$$\text{Overflow rate} = \frac{\text{Flow rate}}{\text{Area}} = \frac{15,000 \text{ m}^3/\text{day}}{A \text{ (m}^2)} = 20 \frac{\text{m}^3/\text{day}}{\text{m}^2}$$

The area is therefore

$$A = \frac{15,000}{20} = 750 \text{ m}^2 = \frac{\pi}{4} d^2$$

Solving for the diameter of this circular tank gives

$$d = \sqrt{\frac{4 \times 750 \text{ m}^2}{\pi}} = 30.9 \text{ m}$$

The detention time is the ratio of volume to flow rate, and volume is area times depth:

$$\text{Detention time} = \frac{\text{Volume}}{\text{Flow rate}} = \frac{750 \text{ m}^2 \times D \text{ (m)}}{15,000 \text{ m}^3/\text{day}} \times 24 \text{ hr/day} = 1.2 \, D \text{ (hr)} = 2.5 \text{ hr}$$

Solving for the depth,

$$D = 2.5/1.2 = 2.1 \text{ m}$$

As it turns out, this depth is about the minimum that would be considered good design practice.

■

Maximum flow rates over weirs per lineal meter of weir, minimum depths of tanks, constraints related to hour-to-hour flow variations as well as seasonal variations, and so forth complicate the design of an actual clarifier, so Example 6.4 is merely a first cut at specifying the tank.

Secondary (Biological) Treatment

The main purpose of secondary treatment is to provide additional BOD and suspended solids removal, beyond what is achievable by simple sedimentation. There are three commonly used approaches, all of which take advantage of the ability of microorganisms to convert organic wastes into stabilized, low-energy compounds. Two of these approaches, the *trickling filter* (and its variations) and the *activated sludge* process, sequentially follow normal primary treatment. The third, *oxidation ponds* (or lagoons), however, can provide equivalent results without preliminary treatment.

Trickling Filters. A trickling filter consists of a rotating distribution arm that sprays liquid wastewater over a circular bed of "fist size" rocks or other coarse materials (Figure 6.8). The spaces between the rocks allow air to circulate easily so that aerobic conditions can be maintained. Of course, the size of the openings is such that there is no actual filtering taking place, so the name *trickling filter* is somewhat of a misnomer. Instead, the individual rocks in the bed are covered by a layer of biological slime that adsorbs and consumes the wastes trickling through the bed. This slime consists mainly of bacteria, but it may also include fungi, algae, protozoa, worms, insect larvae, and snails. The accumulating slime periodically slides off individual rocks and is collected at the bottom of the filter, along with the treated wastewater, and passed on to the secondary settling tank, where it is removed. Not shown is a provision for returning some of the effluent from the filter back into the incoming flow. Such recycling not only enables more effluent organic removal, but it also provides a way to keep the biological slimes from drying out and then dying during low-flow conditions.

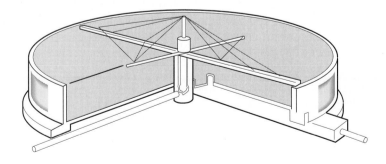

FIGURE 6.8   Cross section of a trickling filter. (*Source: Water and Wastewater Technology* 3/E by Hammer and Hammer, © 1996. Reprinted by permission of Prentice-Hall, Inc., Upper Saddle River, NJ.)

If ordinary rocks are used in the bed of a trickling filter, structural problems caused by their weight tend to restrict the bed depth to about 3 m. To offset the shallow depth, the diameter must be increased. Diameters as great as 60 m are not unusual. However, plastic media are becoming increasingly popular as a replacement for rocks, since in the same volume they can be designed to achieve greater surface areas for slime growth, and their lightness allows much deeper beds. The combination allows equivalent treatment to rock beds, but with much smaller land area requirements. They can also be designed to be less prone to plugging by the accumulating slime, and modestly higher rates of BOD removal are possible. These filters, made with plastic media, are sometimes referred to as *biological towers*.

Rotating Biological Contactor. Trickling filters (and biological towers) are examples of devices that rely on microorganisms that grow on the surface of rocks, plastic, or other media. A variation on this *attached growth* idea is provided by the *rotating biological contactor* (RBC). An RBC consists of a series of closely spaced, circular, plastic disks, typically 3.6 m in diameter, that are attached to a rotating horizontal shaft. The bottom 40 percent of each disk is submersed in a tank containing the wastewater to be treated. The biomass film that grows on the surface of the disks moves into and out of the wastewater as the RBC rotates. While the microorganisms are submerged in the wastewater, they adsorb organics; while they are rotated out of the wastewater, they are supplied with needed oxygen. By placing modular RBC units in series, treatment levels that exceed conventional secondary treatment can be achieved (Figure 6.9). These devices have been used in the United States only since 1969, and although early units suffered from assorted mechanical problems, they are now generally accepted. They are easier to operate under varying load conditions than trickling filters, since it is easier to keep the solid medium wet at all times.

Activated Sludge. Trickling filters were first used in 1893 and they have been used successfully ever since, but they do cost more to build, are more temperature sensitive, and remove less BOD than the activated sludge plants that were later developed.

The example wastewater treatment plant flow diagram given in Figure 6.7 was drawn to illustrate the activated sludge process. As indicated there, the key biological unit in the process is the aeration tank, which receives effluent from the primary clarifier. It also receives a mass of recycled biological organisms from the secondary settling

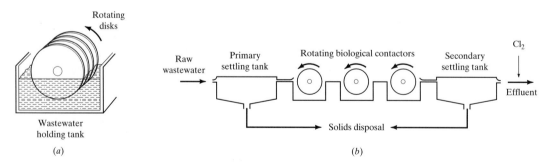

FIGURE 6.9 Rotating biological contactor cross section and treatment system: (*a*) RBC cross section; (*b*) RBC included in a secondary wastewater treatment system.

tank, known as *activated sludge*. To maintain aerobic conditions, air or oxygen is pumped into the tank and the mixture is kept thoroughly agitated. After about six to eight hours of agitation, the wastewater (now referred to as the "mixed liquor") flows into the secondary settling tank, where the solids, mostly bacterial masses, are separated from the liquid by subsidence. A portion of those solids is returned to the aeration tank to maintain the proper bacterial population there, while the remainder must be processed and disposed of.

By allowing greater contact between microorganisms and wastewater in a given volume of space, activated sludge tanks can take up considerably less land area than trickling filters with equivalent performance. They are also less expensive to construct than trickling filters, have fewer problems with flies and odors, and can achieve higher rates of BOD removal. They do, however, require more energy for pumps and blowers, and hence have higher operating costs. Figure 6.10 shows representative values of BOD, suspended solids, total nitrogen, and total phosphorus as wastewater passes through primary sedimentation, biological aeration, and secondary settling.

Sludge Treatment

The processes described thus far have the purpose of removing solids and BOD from the wastewater before the liquid effluent is released to a convenient, nearby body of water. What remains to be disposed of is a mixture of solids and water, called sludge. The collection, processing, and disposal of sludge can be the most costly and complex aspect of wastewater treatment.

The quantity of sludge produced may be as high as 2 percent of the original volume of wastewater, depending somewhat on the treatment process being used. Since sludge can be as much as 97 percent water, and since the cost of disposal will be related to the volume of sludge being processed, one of the primary goals of sludge treatment is to separate as much of the water from the solids as possible. The other goal is to stabilize the solids so that they are no longer objectionable or environmentally damaging.

The traditional method of sludge processing utilizes anaerobic digestion. That is, it depends on bacteria that thrive in the absence of oxygen. Anaerobic digestion is slower than aerobic digestion, but has the advantage that only a small percentage of

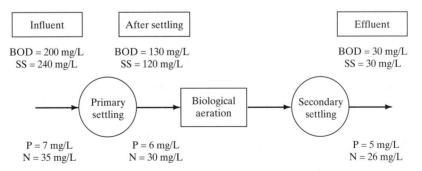

FIGURE 6.10   Approximate concentrations of BOD_5, suspended solids, total nitrogen, and total phosphorus as wastewater passes through a secondary wastewater treatment plant. (*Source: Water and Wastewater Technology* 3/E by Hammer and Hammer, © 1996. Reprinted by permission of Prentice-Hall, Inc., Upper Saddle River, NJ.)

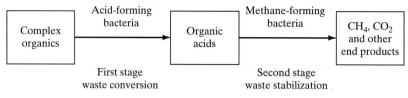

First stage
waste conversion

Second stage
waste stabilization

FIGURE 6.11 The two phases of anaerobic digestion.

the wastes are converted into new bacterial cells. Instead, most of the organics are converted into carbon dioxide and methane gas. The digestion process is complex, but can be summarized by the two steps shown in Figure 6.11. In the first phase, complex organics, such as fats, proteins, and carbohydrates, are biologically converted into simpler organic materials, mostly organic fatty acids. The bacteria that perform this conversion are commonly referred to as *acid formers*. They are relatively tolerant to changes in temperature and pH, and they grow much faster than the *methane formers* that carry out the second stage of digestion.

Methane-forming bacteria slowly convert organic acids into CO_2, CH_4, and other stable end products. These bacteria are very sensitive to temperature, pH, toxins, and oxygen. If their environmental conditions are not just right, the rate at which they convert organic acids to methane slows, and organic acids begin to accumulate, dropping the pH. A positive feedback loop can be established in which the acid formers continue to produce acid while the methane formers, experiencing lower and lower pH, become more and more inhibited. When this occurs, the digester is said to have gone sour, and massive doses of lime may be required to bring it back to operational status.

Most treatment plants that incorporate anaerobic digestion for sludge stabilization use a two-stage digester such as shown in Figure 6.12. Sludge in the first stage is thoroughly mixed and heated to increase the rate of digestion. Typical retention times are between 10 and 15 days. The second stage tank is neither heated nor mixed and is

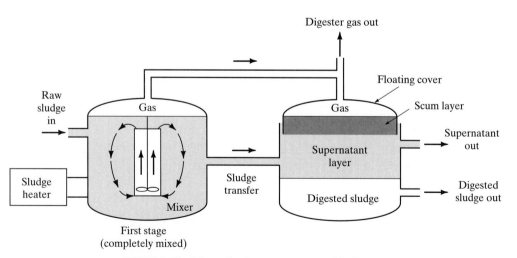

FIGURE 6.12 Schematic of a two-stage anaerobic digester.

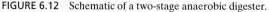

likely to have a floating cover to accommodate the varying amount of gas being stored. Stratification occurs in the second stage, which allows a certain amount of separation of liquids (called supernatant) and solids, as well as the accumulation of gas. The supernatant is returned to the main treatment plant for further BOD removal, and the settled sludge is removed, dewatered, and disposed of. The gas produced in the digester is about 60 percent methane, which is a valuable fuel with many potential uses within the treatment plant. The methane may be used to heat the first stage of the digester, and it can run in an engine/generator set to power pumps, compressors, and other electrical equipment.

Digested sludge removed from the second stage of the anaerobic digester is still mostly liquid. The solids have been well digested, so there is little odor. The most popular way of dewatering has been to pump the sludge onto large sludge drying beds, where evaporation and seepage remove the water. Other methods include the use of vacuum filters, filter presses, centrifuges, or incinerators. The digested and dewatered sludge is potentially useful as a soil conditioner, but most often it is simply trucked away and disposed of in a landfill.

Oxidation Ponds

Oxidation ponds are large, shallow ponds, typically 1 to 2 m deep, where raw or partially treated sewage is decomposed by microorganisms. The conditions are similar to those that prevail in a eutrophic lake. Oxidation ponds can be designed to maintain aerobic conditions throughout, but more often the decomposition taking place near the surface is aerobic, while that near the bottom is anaerobic. Such ponds, having a mix of aerobic and anaerobic conditions, are called *facultative ponds.* In ponds, the oxygen required for aerobic decomposition is derived from surface aeration and algal photosynthesis; deeper ponds, called *lagoons,* are mechanically aerated. A schematic diagram of the reactions taking place in a facultative pond is given in Figure 6.13.

Oxidation ponds can be designed to provide complete treatment of raw sewage, but they require a good deal of space. These ponds have been used extensively in small communities where land constraints are not so critical. The amount of pond surface area required is considerable, with 1 hectare per 240 people (1 acre per 100 people) often being recommended, although in areas with warm climates and mild winters, such as in the southwestern United States, about half that area is often used (Viessman and Hammer, 1985).

Ponds are easy to build and manage, they accommodate large fluctuations in flow, and they can provide treatment that approaches that of conventional biological systems but at a much lower cost. The effluent, however, may contain undesirable concentrations of algae and, especially in the winter when less oxygen is liberated by photosynthesis, they may produce unpleasant odors. Used alone, they also have the disadvantage that the effluent may not meet the EPA secondary treatment requirement of 30 mg/L BOD_5 and suspended solids. Their simplicity and effectiveness in destroying pathogenic organisms, however, make these ponds especially useful in developing countries.

Oxidation ponds are also used to augment secondary treatment, in which case they are often called *polishing ponds.*

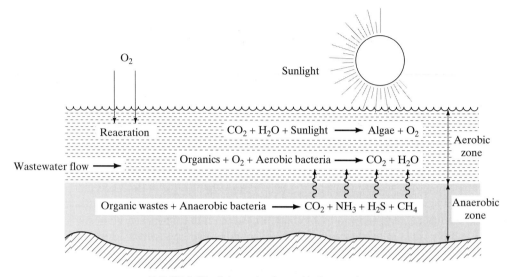

FIGURE 6.13 Schematic of an oxidation pond.

Nutrient Removal

It used to be that anything that followed conventional primary and biological treatment was considered to be advanced treatment. As such, extra stages designed to remove nutrients, especially nitrogen and phosphorus, were considered part of advanced treatment. Nutrient removal is now so common, however, that the term *advanced* is now used to describe additional steps taken to remove various toxic substances, such as metals and other hazardous wastes. Those treatment technologies will be described later in this chapter.

Nitrogen Removal. As bacteria decompose waste, nitrogen that was bound up in complex organic molecules is released as ammonia nitrogen. Subsequent oxidation of ammonia requires oxygen, which, if it occurs in the receiving body of water, contributes to oxygen depletion problems. In addition, nitrogen is an important nutrient for algal growth and, as Figure 6.10 indicates, only about 30 percent is normally removed in a conventional secondary treatment facility. To avoid these oxygen demand and eutrophication problems, treatment plants need to be augmented to achieve higher rates of nitrogen removal.

Increased nitrogen control utilizes aerobic bacteria to convert ammonia (NH_4^+) to nitrate (NO_2^-), which is nitrification, followed by an anaerobic stage in which different bacteria convert nitrates to nitrogen gas (N_2), which is denitrification. The overall process then is referred to as *nitrification/denitrification.*

The nitrification step actually occurs in two stages. Ammonia is converted to nitrites (NO_2^-) by *Nitrosomonas,* while *Nitrobacter* oxidize nitrites to nitrates, as was described in Section 5.5. The combination of steps can be summarized by

$$NH_4^+ \;+\; 2\,O_2 + \xrightarrow{\text{Bacteria}} NO_3^- \;+\; 2H^+ \;+\; H_2O \qquad (6.17)$$

Nitrification does not begin to be important until domestic wastewater is at least five to eight days old. Thus, if this method of nitrogen control is to be used, the wastewater must be kept in the treatment plant for a much longer time than would normally be the case. Detention times of 15 days or more are typically required. If reaction (6.17) takes place in the treatment plant rather than in the receiving body of water, at least the oxygen demand for nitrification is satisfied. The nitrogen, however, remains in the effluent, and if the process were to stop here, that nitrogen could go on to contribute to unwanted algal growth. To avoid this, a denitrification step is required.

The second phase of the nitrification/denitrification process is anaerobic denitrification:

$$2\,NO_3^- \; + \; \text{Organic matter} \; \xrightarrow{\text{Bacteria}} \; N_2 \uparrow \; + \; CO_2 \; + \; H_2O \qquad (6.18)$$

which releases harmless, elemental nitrogen gas. The energy to drive this reaction comes from the organic matter indicated in (6.18). Since this denitrification process occurs after waste treatment, there may not be enough organic material left in the waste stream to supply the necessary energy so an additional source, usually methanol (CH_3OH), must be provided.

Phosphorus Removal. Only about 30 percent of the phosphorus in municipal wastewater is removed during conventional primary and biological treatment. Since phosphorus is very often the limiting nutrient, its removal from the waste stream is especially important when eutrophication is a problem.

Phosphorus in wastewater exists in many forms, but all of it ends up as orthophosphate ($H_2PO_4^-$, HPO_4^{2-}, and PO_4^{3-}). Removing phosphates is most often accomplished by adding a coagulant, usually alum [$Al_2(SO_4)_3$] or lime [$Ca(OH)_2$]. The pertinent reaction involving alum is

$$Al_2(SO_4)_3 \; + \; 2\,PO_4^{3-} \; \rightarrow \; 2\,AlPO_4 \downarrow \; + \; 3SO_4^{2-} \qquad (6.19)$$

Alum is sometimes added to the aeration tank when the activated sludge process is being used, thus minimizing the need for additional equipment.

6.6 HAZARDOUS WASTES

As infectious diseases such as cholera and typhoid have been brought under control in the industrialized countries, there has been a shift in attention toward noninfectious human health problems, such as cancer and birth defects, that are induced by toxic substances. Dramatic incidents in the late 1970s such as occurred at Love Canal, where hazardous substances from an abandoned dump site oozed into backyards and basements, and Times Beach, Missouri, where tens of thousands of gallons of dioxin-laced oil were carelessly sprayed onto the dusty streets, helped motivate the coming decade of hazardous waste controls.

In response to growing public pressure to do something about hazardous wastes, Congress passed the *Comprehensive Environmental Response, Compensation, and Liability Act* (CERCLA) in 1980 to deal with *already contaminated* sites, and in 1980 and 1984 it strengthened the *Resource Conservation and Recovery Act* (RCRA), which controls *new sources* of hazardous waste. In essence, CERCLA deals with problems of the past, while RCRA attempts to prevent future problems. These two laws have provided the driving force behind almost all efforts at hazardous waste control.

What is a hazardous waste? Unfortunately, the answer is largely a matter of definition, with various pieces of environmental legislation defining it somewhat differently. Hazardous waste is defined in RCRA §1004(5) as "anything which, because of its quantity, concentration, or physical, chemical, or infectious characteristics may cause, or significantly contribute to, an increase in mortality; or cause an increase in serious irreversible, or incapacitating reversible, illness; or pose a substantial present or potential hazard to human health and the environment when improperly treated, stored, transported, or disposed of, or otherwise managed." At times a distinction needs to be made between a hazardous substance and a hazardous waste. A hazardous substance, or material, has some commercial value, while a hazardous waste is a material that has been used, spilled, or is no longer needed.

More specifically, the EPA uses two procedures to define wastes as hazardous. One is based on a listing of substances and industrial process wastes that have been designated as hazardous. This list is published in the Code of Federal Regulations (CFR). The second defines a substance as hazardous if it possesses any of the following four characteristic attributes: *reactivity, ignitability, corrosivity,* or *toxicity.* Briefly,

> *Ignitable* substances are easily ignited and burn vigorously and persistently. Examples include volatile liquids, such as solvents, whose vapors ignite at relatively low temperatures (defined as 60 °C or less).
>
> *Corrosive* substances include liquids with pH less than 2 or greater than 12.5, and those that are capable of corroding metal containers.
>
> *Reactive* substances are unstable under normal conditions. They can cause explosions and/or liberate toxic fumes, gases, and vapors when mixed with water.
>
> *Toxic* substances are harmful or fatal when ingested or absorbed. Toxicity of a waste is determined using an extraction test that has been designed to simulate the expected leaching action that might occur, for example, in a sanitary landfill. A substance is designated as being toxic if the extract from the test contains any of a long list of chemicals in concentrations greater than allowed. The EPA changed the extraction test procedure in 1990 from one called the EP Toxicity Test to the Toxicity Characteristic Leaching Procedure (TCLP) test.

Note that the term *hazardous substance* is more all-encompassing than *toxic substance,* though the two are often used interchangeably.

While the four characteristic attributes just given provide guidelines for considering a substance to be hazardous, the EPA maintains an extensive list of specific hazardous wastes. EPA *listed wastes* are organized into three categories: *source-specific wastes, generic wastes,* and *commercial chemical products.* Source-specific wastes include sludges and wastewaters from treatment and production processes in *specific industries,* such as petroleum refining and wood preserving. The list of generic wastes includes wastes from common manufacturing and industrial processes, such as solvents used in degreasing operations. The third list contains specific chemical products, such as benzene, creosote, mercury, and various pesticides. All listed wastes are presumed to

TABLE 6.7 Examples of Hazardous Wastes Generated by Business and Industries

Waste generators	Waste types
Chemical manufacturers	Strong acids and bases
	Spent solvents
	Reactive wastes
Vehicle Maintenance shops	Heavy metal paint wastes
	Ignitable wastes
	Used lead acid batteries
	Spent solvents
Printing industry	Heavy metal solutions
	Waste inks
	Spent solvents
	Spent electroplating wastes
	Ink sludes containing heavy metals
Leather products manufacturing	Waste toluene and benzene
Paper industry	Paint wastes containing heavy metals
	Ignitable solvents
	Strong acids and bases
Construction industry	Ignitable paint wastes
	Spent solvents
	Strong acids and bases
Cleaning agents and cosmetics manufacturing	Heavy metal dusts
	Ignitable wastes
	Flammable solvents
	Strong acids and bases
Furniture and wood manufacturing and refinishing	Ignitable wastes
	Spent solvents
Metal Manufacturing	Paint wastes containing heavy metals
	Strong acids and bases
	Cyanide wastes
	Sludges containing heavy metals

be hazardous regardless of their concentrations. There are well over 1000 chemicals listed by the EPA as hazardous.

Examples of the types of hazardous wastes generated by various businesses and industries are given in Table 6.7.

6.7 HAZARDOUS MATERIALS LEGISLATION

There are many federal environmental laws that regulate hazardous materials in the United States. Table 6.8 gives a very brief description of the most important of these, along with the agency in charge of enforcement. In this section, we will focus on the three acts that are most crucial to the current management programs for hazardous materials. The first is the Toxic Substances Control Act (TSCA), which authorizes the EPA to regulate new chemicals as they are being developed as well as existing chemi-

---

TABLE 6.8 Environmental Laws Controlling Hazardous Substances

---

Atomic Energy Act (Nuclear Regulatory Commission)—Regulates nuclear energy production and nuclear waste disposal.

Clean Air Act (EPA)—Regulates the emission of hazardous air pollutants.

Clean Water Act (EPA)—Regulates the discharge of hazardous pollutants into the nation's surface water.

Comprehensive Environmental Response, Compensation, and Liability Act (Superfund) (EPA)—Provides for the cleanup of inactive and abandoned hazardous waste sites.

Emergency Planning and Community Right-to-Know Act (EPA)—Requires written emergency response plans for chemical releases and establishes the Toxic Release Inventory.

Federal Insecticide, Fungicide, and Rodenticide Act (EPA)—Regulates the manufacture, distribution, and use of pesticides and the conduct of research into their health and environmental effects.

Hazardous Materials Transportation Act (Department of Transportation)—Regulates the transportation of hazardous materials.

Marine Protection, Research, and Sanctuaries Act (EPA)—Regulates waste disposal at sea.

Occupational Safety and Health Act (Occupational Safety and Health Administration)—Regulates hazards in the workplace, including worker exposure to hazardous substances.

Pollution Prevention Act (EPA)—Establishes priority of prevention, recycling, treatment, and waste minimization, before disposal in landfills.

Resource Conservation and Recovery Act (EPA)—Regulates hazardous waste generation, storage, transportation, treatment, and disposal.

Safe Drinking Water Act (EPA)—Regulates contaminant levels in drinking water and the disposal of wastes into injection wells.

Surface Mining Control and Reclamation Act (Department of the Interior)—Regulates the environmental aspects of mining (particularly coal) and reclamation.

Toxic Substances Control Act (EPA)—Regulates the manufacture, use, and disposal of specific chemicals.

---

cals when there are new concerns for their toxicity. The second is the Resource Conservation and Recovery Act (RCRA), which provides guidelines for prudent management of new and future hazardous substances. The third is the Comprehensive Environmental Response, Compensation, and Liability Act (CERCLA), which deals primarily with mistakes of the past—inactive and abandoned hazardous waste sites.

Toxic Substances Control Act

The Toxic Substances Control Act (TSCA) gives the EPA the authority to gather information on the toxicity of particular chemicals, to assess whether those chemicals cause unreasonable risks to humans and the environment, and to institute appropriate controls when deemed necessary. TSCA covers the full life cycle of specific chemicals, from premanufacturing to manufacturing, importation, processing, distribution, use, and disposal.

TSCA gives the EPA the authority to compile an inventory of existing chemical substances that are being manufactured or imported for commercial uses. The inventory contains information on well over 60,000 chemical substances. Entities that are about to manufacture or import new chemicals that are not in the inventory must give the EPA a 90-day advance notification of their intent to do so. Included in the notification must be the identity and molecular structure of the chemical, an estimate of the amounts to be manufactured and the proposed uses, the byproducts resulting from manufacture, processing, use, and disposal of the chemical, and test data relating to the health and environmental effects of the chemical. Similarly, significant new uses of chemicals that have already been approved for other purposes must also be approved by the EPA. Based on the data submitted, the EPA may prohibit the manufacture or importation until adequate data are developed to show that the chemical will not present an unreasonable risk of injury to health or the environment. A manufacturer may also be required to recall a substance that is later found to present an unreasonable risk.

TSCA also contains a provision that requires an exporter to notify the EPA when certain chemicals are going to be exported. The EPA, in turn, notifies the importing country's government of the export and provides available information on the chemicals, but in general it does not restrict such exports. However, chemicals that are manufactured for export only can be covered by TSCA if the EPA finds that international use poses an unreasonable risk within the United States.

Both TSCA and the RCRA regulate the disposal of hazardous substances. The key distinction between the two is that TSCA deals with individual, specific chemicals (e.g., PCBs), while RCRA generally deals with combinations of chemicals contained in waste streams.

Resource Conservation and Recovery Act

The Resource Conservation and Recovery Act (RCRA) regulates the generation, storage, transportation, treatment, and disposal of hazardous substances. It is our single most important law dealing with the management of hazardous waste and it is perhaps the most comprehensive piece of legislation that the EPA has ever promulgated. Its origins are in the 1965 Solid Waste Disposal Act, which was the first federal law to address the enormous problem of how to dispose safely of household, municipal, commercial, and industrial refuse. Congress amended that law in 1970 with the passage of the Resource Recovery Act, and finally RCRA itself was passed in 1976. Revisions to RCRA were made in 1980 and again in 1984. The 1984 amendments, referred to as the *Hazardous and Solid Waste Amendments* (HSWA), significantly expand the scope of RCRA, particularly in the area of land disposal.

The portions of RCRA that regulate hazardous wastes are contained in a section designated as Subtitle C. A major section of RCRA, Subtitle D, controls solid-waste disposal, and those provisions will be presented in Chapter 9.

Transportation, Storage, and Disposal. The key concept in RCRA is that hazardous substances must be properly managed from the moment they are generated until their ultimate disposal. This step-by-step management is often referred to as the *cradle-to-grave* approach, and it has three key elements:

1. A tracking system, in which a *manifest* document accompanies any waste that is transported from one location to another;

2. A *permitting* system, which helps assure safe operation of facilities that treat, store, or dispose of hazardous wastes;

3. A system of controls and restrictions governing the *disposal* of hazardous wastes onto, or into, the land.

About 96 percent of U.S. hazardous wastes are treated or disposed of at the site where they were originally generated (U.S. EPA 1987b). The remaining 4 percent still represents a substantial volume of material that is transported from the source to *treatment, storage,* or *disposal* (TSD) facilities. To help eliminate improper handling of such transported wastes, the EPA requires generators to prepare a *hazardous waste manifest* that must accompany the waste. The manifest identifies the type and quantity of waste, the generator, the transporter, and the TSD facility to which the waste is being shipped. One copy of the manifest is sent to the EPA when the waste leaves the generator, and another when the waste arrives at a TSD facility. The generator, who is ultimately responsible for the waste tracking system, also receives a copy of the manifest after the waste arrives at the TSD facility. Figure 6.14 illustrates the manifest system.

Under RCRA, the treatment, storage, and disposal facilities that accept hazardous waste must first obtain permits from the EPA. The permitting process has been established to give the EPA power to enforce a number of standards and requirements, including the power to inspect facilities and to bring civil actions against violators of any of RCRA's provisions. To help assure proper management of hazardous wastes in the longer term, RCRA regulations require that TSD facilities acquire sufficient financial assurance mechanisms to ensure that the facility can be properly maintained after it ultimately closes, including a provision for 30 years of facility maintenance, security measures, and groundwater monitoring.

The Hazardous and Solid Waste Amendments of 1984. The 1984 Hazardous and Solid Waste Amendments (HSWA) to RCRA represented a turning point in the relationship between Congress and the Environmental Protection Agency. Before HSWA, Congress allowed considerable discretion to the EPA with regard to the implementation of environmental laws. During the early 1980s, however, it was perceived that the EPA had become politicized and was not meeting Congressional expectations, especially with regard to implementation of CERCLA. When the EPA decided to allow the disposal of liquid hazardous waste in land disposal facilities, Congress decided that the EPA had gone too far and it was time to return policy control back to Congress, which it did with HSWA.

HSWA has four main provisions. First, it imposed severe restrictions on land disposal of hazardous waste. Second, it significantly reduced the level below which the rate of generation of hazardous waste was exempt from RCRA provisions. Third, it began the regulation of hundreds of thousands of leaking underground storage tanks (LUSTs). Fourth, it closed a loophole that had allowed some hazardous waste sites to be exempt from either RCRA or CERCLA regulations.

In the past, about 80 percent of U.S. hazardous wastes were disposed of on land, either by pumping them underground using deep injection wells, burying them in landfills, or containing them in surface impoundments. The 1984 Hazardous and Solid Waste Amendments to RCRA, however, significantly restrict disposal of wastes on land. In

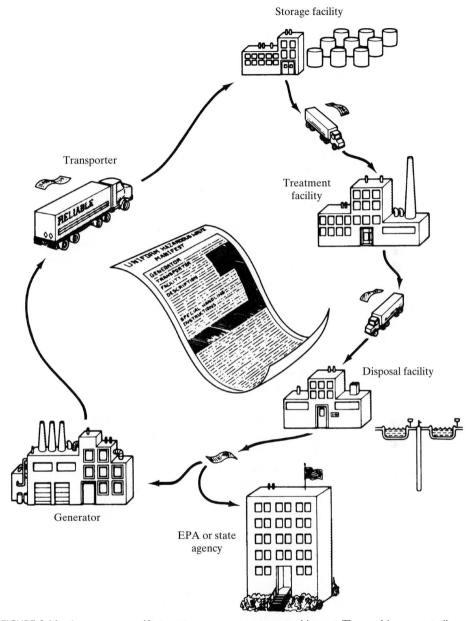

FIGURE 6.14   A one-page manifest must accompany every waste shipment. The resulting paper trail documents the waste's progress through treatment, storage, and disposal, providing a mechanism for alerting the generator and the EPA if irregularities occur. (U.S. EPA, 1986)

fact, land disposal is banned unless the EPA determines that for a particular site and type of waste, there will be no migration of hazardous constituents for as long as the wastes remain hazardous. HSWA has been written in such a way that if the EPA does not follow through with such a determination within a specified period of time, then the land ban is automatically imposed. This "hammer" provision was specifically intended to keep the EPA from dragging its feet on implementation and is a direct result of the conflict between Congress and the EPA during the Reagan administration.

To the extent that land disposal can continue to be used, new landfills and surface impoundments are required to have double liners, leachate collection systems, and groundwater monitoring facilities to help assure long-term containment. Given the ongoing problem of groundwater contamination from landfills, all liquid hazardous wastes are banned. Further details on land disposal technologies are described later in this chapter and again in Chapter 9.

Waste Reduction. It is clear that rising economic costs of hazardous waste management are dictating a fresh look at the very origins of the problem itself. Is it really necessary for us to generate over one ton of hazardous waste per person per year in order to provide the goods and services that we have come to expect? It is almost always cheaper and simpler to control the amount of pollution generated in the first place than to attempt to find engineering fixes that deal with it after it has been created. The 1984 amendments to RCRA recognize this important role of waste reduction by stating that "the Congress hereby declares it to be the national policy of the United States that, wherever feasible, the generation of hazardous waste is to be reduced or eliminated as expeditiously as possible."

Figure 6.15 suggests a priority system for ranking various approaches to hazardous waste reduction. The first priority is to find ways to *eliminate* uses of hazardous substances. Elimination might be achieved by changing manufacturing processes or by substituting products that can satisfy the same need without creating hazardous wastes (an example would be the substitution of concrete posts for toxic, creosote-treated wooden posts). The next priority is to *reduce* the amounts generated. Again, manufacturing process changes can be important. The third strategy is to *recycle* hazardous substances such as solvents and acids to maximize their use before treatment and disposal becomes necessary. Finally, hazardous substances can be *treated* to reduce their volume

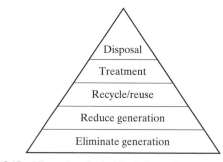

FIGURE 6.15   Hierarchy of priorities in hazardous waste management.

and toxicity. Only after all of those options have been exhausted should disposal be considered.

Eliminating, reducing, and recycling hazardous substances reduces treatment and disposal costs, and often generators can find it in their own economic best interest to seek out these source reduction opportunities as a way to reduce the overall cost of doing business. RCRA, in fact, requires generators to certify on their hazardous waste manifests that they have taken steps to reduce the volume and toxicity of the hazardous waste that they are creating.

The Comprehensive Environmental Response, Compensation, and Liabilities Act

The Comprehensive Environmental Response, Compensation, and Liabilities Act (CERCLA) was enacted in 1980 to deal with abandoned hazardous waste sites. It is more often referred to as the Superfund law as a result of one of its key provisions, in which a $1.6 billion trust fund was created from special crude oil and chemical company taxes. In 1986, CERCLA was reauthorized and the trust fund was increased to $8.5 billion, with half of the money coming from petroleum and chemical companies and most of the rest coming from a special corporate environmental tax augmented by general revenues. The revisions to CERCLA enacted in 1986 are designated as the *Superfund Amendments and Reauthorization Act* (SARA).

CERCLA is an unusual environmental law in that it does not involve the usual regulations and permitting requirements. Instead, it is focused on identifying hazardous waste sites, preparing cleanup plans, and then forcing those deemed to be responsible parties (RPs) to pay for the remediation. Under CERCLA, the EPA can deal with both short-term, emergency situations triggered by the actual or potential release of hazardous substances into the environment, as well as with long-term problems involving abandoned or uncontrolled hazardous waste sites for which more permanent solutions are required.

Short-term *removal actions* can be taken at any site where there is an imminent threat to human health or the environment, such as might occur during a spill or fire, or when wastes that have been illegally disposed of are discovered ("midnight dumping"). In addition to removing and disposing of hazardous substances and securing the endangered area, the EPA may take more extensive actions as necessary, such as providing alternate drinking-water supplies to local residents if their drinking water has been contaminated, or even temporarily relocating residents.

More complex and extensive problems that are not immediately life threatening are handled under the EPA's *remedial* programs, which are outlined in Figure 6.16. Central to the Superfund process is the creation of a National Priorities List (NPL) of sites that are eligible for federally financed remedial activities. The NPL identifies the worst sites in the nation based on such factors as the quantities and toxicity of wastes involved, the exposure pathways, the number of people potentially exposed, and the importance and vulnerability of the underlying groundwater. In other words, a *risk assessment* of the sort that was described in Chapter 4 is used to rank potential NPL sites.

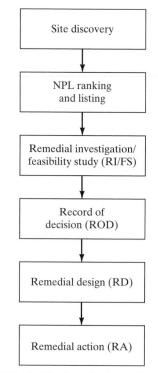

FIGURE 6.16 The Superfund remedial process.

Once a site is listed, the EPA begins a *remedial investigation/feasibility study* (RI/FS) to determine appropriate remedial actions for the site. The remedial investigation phase involves gathering information needed to characterize the contamination problem at the site, including environmental and public health risks. The feasibility study that follows uses the remedial investigation information to identify, evaluate, and select cleanup alternatives. These alternative cleanup approaches are analyzed based on their relative effectiveness and cost. The RI/FS process culminates with the signing of a *record of decision* (ROD), in which the remedial action that has been selected is set forth. After the ROD has been signed, the detailed *remedial design* of the selected alternative takes place. For complex sites, the entire process from listing to the beginning of the actual cleanup itself has often taken five years or more.

The final *remedial action* may consist of a number of short-term steps, such as installing surface water runoff controls, excavating soil, building below-grade containment walls, and capping the site. As we saw in Chapter 5, most remedial actions have involved pump-and-treat technologies, which have serious limitations and which often require decades of pumping. The cleanup of a typical NPL site is a slow, complex, expensive process, which has led some to question whether it is really worth all that effort.

One of the most important and guiding policies of CERCLA is that those parties who are responsible for hazardous waste problems will be forced to pay the entire cost of cleanup. *Responsible parties* (RPs) may be the historic owners or operators of the site, any generators who disposed of wastes at the site (whether legally or not), or even the transporters who brought wastes to the site. The courts have recognized the concept of *retroactive, strict,* and *joint-and-several* liability for costs of cleanup. *Retroactive* covers problems created before Superfund was enacted; *strict liability* refers to the fact that liability does not depend on whether the RPs were negligent or diligent in their disposal practices; and *joint-and-several* means, in essence, that if damages cannot be individually apportioned by the responsible parties themselves, then each and every party is subject to liability for the entire cleanup cost. In other words, no matter when their contribution occurred, no matter how careful they were, and no matter how little an individual RP may have contributed to the overall problem, they can, theoretically, be liable for the entire cost of the cleanup. If responsible parties do not voluntarily perform the appropriate response actions, the EPA can use Superfund monies to clean up the site. If they do so, then the EPA is empowered to collect three times the cleanup cost from the RPs.

The retroactive, strict, and joint-and-several provisions of Superfund have been heavily criticized. Banks and lending institutions worry that they could be liable for cleanup costs just because they loaned money to firms that were named as RPs. Insurance companies are fighting lawsuits when RPs claim they were covered for cleanup costs under their general liability insurance. Municipalities that have been named as RPs when their landfills have leaked do not feel they should be liable for enormous cleanup costs. To many, it seems unfair to be held responsible for disposal practices that were perfectly legal in the past, especially when a single party's contributions to the problem were minimal. Finally, it seems particularly unfair to some that owners of property are at risk for cleanup costs even though they were not responsible in any way for the contamination.

CERCLA has also been criticized because it requires that the cleanup solution selected provides a permanent solution to the contamination problem. For some sites, it might be argued that temporary measures can cost very little while virtually eliminating health risks. For example, consider a site located some distance from residences, having contaminated soil but not contaminated groundwater. It can be argued that capping the site, surrounding it with fences and signs, and monitoring the soil and groundwater would be a much better use of resources than a permanent site remediation project. But it might not be allowed under CERCLA.

Brownfields. In an effort to address some of the liability concerns and fairness issues of CERCLA, as well as to help revitalize communities affected by contamination, the EPA launched its *Brownfields Economic Redevelopment Initiative* in 1995. Brownfields are defined to be "abandoned, idled, or under-used industrial and commercial facilities where expansion or redevelopment is complicated by real or perceived environmental contamination." The Initiative is based on the belief that public fear of contamination and lender concerns for liability are causing companies to leave "brownfields" in the urban core and head for "greenfields" outside the cities. It is hoped that implementation of EPA's Brownfields Action Agenda will help reverse the spiral of unaddressed contamination, declining property values, and increased unemployment often found in inner-city industrial areas (U.S. EPA, 1996).

Key components of the Brownfields initiative include removing eligible sites from the Superfund site tracking system, known as the Comprehensive Environmental Response, Compensation, and Liability Information System (CERCLIS). Initial actions removed 27,000 such sites out of the 40,000 that were being tracked. Removing sites from CERCLIS is intended to convey assurance that the EPA will no longer pursue Superfund action at these sites. A second, very important component of the Initiative is a clarification of liability issues that frequently cause potential buyers and lenders to avoid contaminated properties. For example, the EPA will not take action against owners when hazardous substances enter the property from aquifers contaminated elsewhere, as long as the landowner did not in any way cause the problem and as long as the property owner did not have a contractual arrangement with the polluter.

The Brownfields issue is helping to focus attention on risk-based corrective actions in which the most feasible remediation technologies are coupled with best management practices to provide cost-effective cleanup without compromising the protection of public health, water quality, and the environment.

6.8 HAZARDOUS WASTE TREATMENT TECHNOLOGIES

Even with a much more vigorous hazardous waste reduction program, as RCRA requires, there will still be large quantities of hazardous wastes that will require treatment and disposal. In the past, there was little treatment, and disposal was most often on land. In both SARA (Superfund) and the 1984 Hazardous and Solid Waste Amendments of RCRA, emphasis is on the development and use of alternative and innovative treatment technologies that result in permanent destruction of wastes or a reduction in toxicity, mobility, and volume. Land disposal is greatly restricted under the 1984 RCRA amendments.

Treatment technologies are often categorized as being physical, chemical, biological, thermal, or stabilization/fixation. These categories are reasonably well defined, though there is room for confusion when technologies have overlapping characteristics.

> *Chemical, biological, and physical wastewater treatment processes* are currently the most commonly used methods of treating aqueous hazardous waste. Chemical treatment transforms waste into less hazardous substances using such techniques as pH neutralization, oxidation or reduction, and precipitation. Biological treatment uses microorganisms to degrade organic compounds in the waste stream. Physical treatment processes include gravity separation, phase change systems such as air and steam stripping of volatiles from liquid wastes, and various filtering operations, including carbon adsorption.
>
> *Thermal destruction processes* include incineration, which is increasingly becoming a preferred option for the treatment of hazardous wastes, and pyrolysis, which is the chemical decomposition of waste brought about by heating the material in the absence of oxygen.
>
> *Fixation/stabilization techniques* involve removal of excess water from a waste and solidifying the remainder either by mixing it with a stabilizing agent, such as

Portland cement, or vitrifying it to create a glassy substance. Solidification is most often used on inorganic sludges.

Choosing an appropriate technology to use in any given situation is obviously beyond the scope of this text. Not only are there many different kinds of hazardous wastes, in terms of their chemical makeup, but the treatability of the wastes depends on their form. A technology suitable for treating PCBs in sludges, for example, may not be appropriate for treating the same contaminant in dry soil. Table 6.9 gives a partial listing of available treatment technologies appropriate for a variety of types of hazardous

TABLE 6.9 A partial list of treatment technologies for various hazardous waste streams.

Treatment process	Hazardous waste streams												Form of waste		
	Corrosives	Cyanides	Halogenated solvents	Nonhalogenated organics	Chlorinated organics	Other organics	Oily wastes	PCBs	Aqueous with metals	Aqueous with organics	Reactives	Contaminated soils	Liquids	Solids/Sludges	Gases
Separation/filtration		×	×	×	×	×			×	×			×		
Carbon adsorption									×	×	×		×		
Air and steam stripping			×	×	×	×				×			×		×
Electrolytic recovery									×				×		
Ion exchange	×								×	×			×		
Membranes									×	×			×		
Chemical precipitation	×								×				×		
Chemical oxidation/reduction		×							×				×		
Ozonation		×		×		×						×	×		×
Evaporation			×	×	×	×							×	×	
Solidification	×	×										×	×	×	
Liquid injection incineration			×	×	×	×	×						×		×
Rotary kilns			×	×	×	×	×	×				×	×	×	×
Fluidized bed incineration			×	×	×	×	×	×				×	×	×	×
Pyrolysis			×	×	×	×						×	×	×	
Molten glass			×	×	×	×	×			×			×	×	×

Source: Based on Freeman (1989)

waste streams along with the applicable form of waste (liquid, gaseous, solids/sludges.) For a more complete list, as well as detailed descriptions of each technology, see Freeman (1989).

Physical Treatment

Sedimentation. The simplest physical treatment systems that separate solids from liquids take advantage of gravity settling and natural flotation. Special sedimentation tanks and clarification tanks are designed to encourage solids to settle so they can be collected as a sludge from the bottom of the tank. Some solids will float naturally to the surface and can be removed with a skimming device. It is also possible to encourage flotation by introducing finely divided bubbles into the waste stream. The bubbles collect particles as they rise and the combination can be skimmed from the surface. Separated sludges can then be further concentrated by evaporation, filtration, or centrifugation. An example of a vacuum filter is shown in Figure 6.17.

Adsorption. Physical treatment can also be used to remove small concentrations of hazardous substances dissolved in water that would never settle out. One of the most commonly used techniques for removing organics involves the process of *adsorption*, which is the physical adhesion of chemicals onto the surface of a solid. The effectiveness of the adsorbent is directly related to the amount of surface area available to attract the molecules or particles of contaminant. The most commonly used adsorbent is a very porous matrix of *granular activated carbon* (GAC), which has an enormous

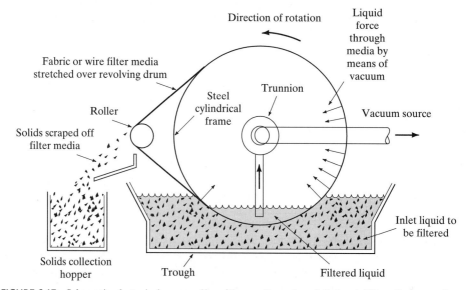

FIGURE 6.17   Schematic of a typical vacuum filter. (*Source: Hazardous & Industrial Waste Treatment & Disposal* by Haas and Vamos, © 1995. Reprinted by permission of Prentice-Hall, Inc., Upper Saddle River, NJ.)

surface area (on the order of 1000 m^2/g). A single handful of GAC has an internal surface area of about one acre.

Granular activated carbon treatment systems usually consist of a series of large vessels partially filled with adsorbent. Contaminated water enters the top of each vessel, trickles down through the GAC, and is released at the bottom. After a period of time, the carbon filter becomes clogged with adsorbed contaminants and must be either replaced or regenerated. Regeneration can be an expensive, energy-intensive process, usually done off site. During regeneration, the contaminants are usually burned from the surface of the carbon granules, although in some cases a solvent is used to remove them. Carbon filters that cannot be regenerated due to their contaminant composition must be properly managed for disposal.

Aeration. For chemicals that are relatively volatile, another physical process, *aeration,* can be used to drive the contaminants out of solution. These stripping systems typically use air, although in some circumstances steam is used. In the most commonly used air stripper, contaminated water is sprayed downward through packing material in a tower, while air is blown upward, carrying away the volatiles with it. Such a *packed tower* can easily remove over 95 percent of the volatile organic compounds (VOCs), including such frequently encountered ones as trichloroethylene, tetrachloroethylene, trichloroethane, benzene, toluene, and other common organics derived from solvents. There is another type of stripper, called an *induced-draft stripper,* which does not use a blower or packing material. In the induced-draft tower, a carefully engineered series of nozzles spray contaminated water horizontally through the sides of a chamber. Air passing through the chamber draws off the volatiles. Induced-draft strippers are cheaper to build and operate, but their performance is much lower than a packed tower.

By combining air stripping with GAC, many volatile and nonvolatile organic compounds can be removed from water to nondetectable levels. By passing contaminated water first through the air stripper, most of the volatile organics are removed before reaching the GAC system, which extends the life of the carbon before regeneration or replacement is required.

The volatiles removed in an air stripper are, in some circumstances, released directly to the atmosphere. When discharge into the atmosphere is unacceptable, a GAC treatment system can be added to the exhaust air, as shown in Figure 6.18.

Other Physical Processes. Other physical processes that are sometimes used to treat hazardous wastes include reverse osmosis, ion exchange, and electrodialysis. *Reverse osmosis* devices use pressure to force contaminated water against a semipermeable membrane. The membrane acts as a filter, allowing the water to be pushed through its pores but restricting the passage of larger molecules that are to be removed. *Ion exchange* is a process wherein ions to be removed from the waste stream are exchanged with ions associated with a special exchange resin. Ion exchange has already been mentioned in the context of water softening, where calcium and magnesium ions are replaced with sodium ions from the exchange resin. In the context of hazardous wastes, ion exchange is often used to remove toxic metal ions from solution. *Electrodialysis* uses ion-selective membranes and an electric field to separate anions

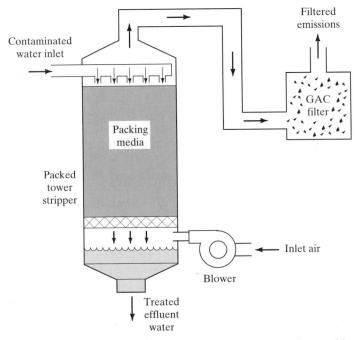

FIGURE 6.18 An air-stripping tower followed by a granular activated carbon filter provides effective removal of VOCs. To remove nonvolatiles, the treated water coming out of the tower may be pumped through other carbon adsorbers.

and cations in solution. In the past, electrodialysis was most often used for purifying brackish water, but it is now finding a role in hazardous waste treatment. Metal salts from plating rinses are sometimes removed in this way.

Chemical Treatment

Chemically treating hazardous waste has the potential advantage of not only converting it to less hazardous forms, but can also produce useful byproducts in some circumstances. By encouraging resource recovery, the treatment cost can sometimes be partially offset by the value of the end products produced.

Neutralization. There are many chemical processes that can be used to treat hazardous wastes, and the process decision depends primarily on the characteristic of the waste. For example, recall that one of RCRA's categories of hazardous waste is anything corrosive—that is, having a pH of less than 2 or more than 12.5. Such wastes can be chemically *neutralized*. Acidic wastewaters are usually neutralized with slaked lime [$Ca(OH)_2$] in a continuously stirred chemical reactor. The rate of addition of lime is controlled with a feedback control system that monitors pH and adjusts the feed rate accordingly.

Alkaline wastewaters may be neutralized by adding acid directly or by bubbling in gaseous CO_2, forming carbonic acid (H_2CO_3). The advantage of CO_2 is that it is

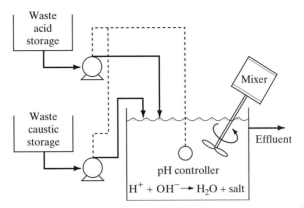

FIGURE 6.19   Simultaneous neutralization of acid and caustic waste. (U.S. EPA, 1987a)

often readily available in the exhaust gas from any combustion process at the treatment site. Simultaneous neutralization of acid and caustic waste can be accomplished in the same vessel, as is suggested by Figure 6.19.

Chemical Precipitation. The ability to adjust pH is important not only for waste neutralization, but also because it facilitates other chemical processes that actually remove undesirable substances from the waste stream. For example, a common method for removing heavy metals from a liquid waste is via *chemical precipitation,* which is pH dependent. By properly adjusting pH, the solubility of toxic metals can be decreased, leading to formation of a precipitate that can be removed by settling and filtration.

Frequently, the precipitation involves the use of lime, $Ca(OH)_2$, or caustic (NaOH) to form metal hydroxides. For example, the following reaction suggests the use of lime to form the hydroxide of a divalent metal (M^{2+}):

$$M^{2+} \;+\; Ca(OH)_2 \;\rightarrow\; M(OH)_2 \;+\; Ca^{2+} \qquad (6.20)$$

Metal hydroxides are relatively insoluble in basic solutions, and, as shown in Figure 6.20, they are *amphoteric*—that is, they have some pH at which their solubility is a minimum. Since each metal has its own optimum pH, it is tricky to control precipitation of a mix of different metals in the same waste. For a waste containing several metals, it may be necessary to use more than one stage of precipitation to allow different values of pH to control the removal of different metals.

While hydroxide precipitation using lime is the most common metal removal process, even lower concentrations of metals in the effluent can be obtained by precipitating the metals as sulfides. As can be seen in Figure 6.20, metal sulfides are considerably less soluble than metal hydroxides. A disadvantage of sulfide precipitation is the potential formation of odorous and toxic hydrogen sulfide gas.

Chemical Reduction-Oxidation. *Reduction-oxidation* (redox) reactions provide another important chemical treatment alternative for hazardous wastes. When electrons are removed from an ion, atom, or molecule, the substance is *oxidized;* when electrons are added, it is *reduced.* Both oxidation and reduction occur in the same

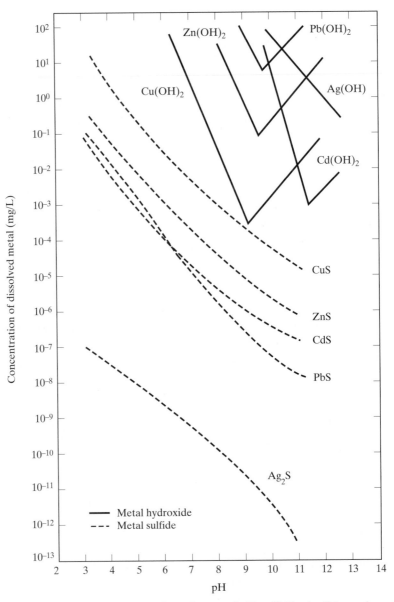

FIGURE 6.20 Chemical precipitation of metals can be controlled by pH. Metal sulfides are less soluble than metal hydroxides. (U.S. EPA, 1980)

reaction, hence the abbreviation redox. One of the most important redox treatment processes is the reduction of hexavalent chromium (Cr VI) to trivalent chromium (Cr III) in large electroplating operations. Sulfur dioxide is often used as the reducing agent, as shown in the following reactions :

$$SO_2 + H_2O \rightarrow H_2SO_3 \tag{6.21}$$

$$2\,CrO_3 + 3\,H_2SO_3 \rightarrow Cr_2(SO_4)_3 + 3\,H_2O \tag{6.22}$$

The trivalent chromium formed in reaction (6.22) is much less toxic and more easily precipitated than the original hexavalent chromium. Notice that the chromium in reaction (6.22) is reduced from an oxidation state of +6 to +3, while the sulfur is oxidized from +4 to +6.

Another important redox treatment involves the oxidation of cyanide wastes, which are also common in the metal finishing industry. In the following reactions, cyanide is first converted to a less toxic cyanate using alkaline chlorination (pH above 10); further chlorination oxidizes the cyanate to simple carbon dioxide and nitrogen gas. Nearly complete destruction of cyanide results.

$$NaCN + Cl_2 + 2\,NaOH \rightarrow NaCNO + 2\,NaCl + H_2O \tag{6.23}$$

$$2\,NaCNO + 3\,Cl_2 + 4\,NaOH \rightarrow 2\,CO_2 + N_2 + 6\,NaCl + 2\,H_2O \tag{6.24}$$

Wastes that can be treated via redox oxidation include benzene, phenols, most organics, cyanide, arsenic, iron and manganese; those that can be successfully treated using reduction treatment include chromium (VI), mercury, lead, silver, chlorinated organics like PCBs, and unsaturated hydrocarbons (U.S. EPA, 1988).

Ultraviolet Radiation/Oxidation. A promising approach to destruction of dissolved organic compounds is oxidation, which converts organics into simple carbon dioxide and water. A number of oxidants are possible, including hydrogen peroxide, oxygen, ozone, and potassium permanganate. One of the most potent oxidizers is the hydroxyl free radical ($OH^\bullet$), which easily initiates the oxidation of hydrogen-containing organic molecules. The formation of hydroxyl can be enhanced by exposing an oxidant, such as hydrogen peroxide (H_2O_2) or ozone (O_3), to ultraviolet (UV) light, as the following reactions suggest:

$$O_3 + h\nu + H_2O \rightarrow H_2O_2 + O_2 \tag{6.25}$$

$$H_2O_2 + h\nu \rightarrow 2\,OH^\bullet \tag{6.26}$$

where $h\nu$ represents a photon of light with frequency ν. Low-pressure mercury vapor lamps produce photons of about the right wavelength (254 nm) needed to convert the hydrogen peroxide in (6.26). The initial oxidization of an organic molecule using the hydroxyl can be represented as

$$OH^\bullet + HR \rightarrow H_2O + R^\bullet \rightarrow \cdots \rightarrow CO_2 \tag{6.27}$$

where $R^\bullet$ represents a carbon-based radical. The rate of destruction of various chlorinated organics using oxidation by ozone alone is compared with destruction using ozone and UV light together in Figure 6.21.

Biological Treatment

Virtually all municipal wastewater treatment plants in the United States and a large number of industrial systems rely on biological treatment processes to decompose

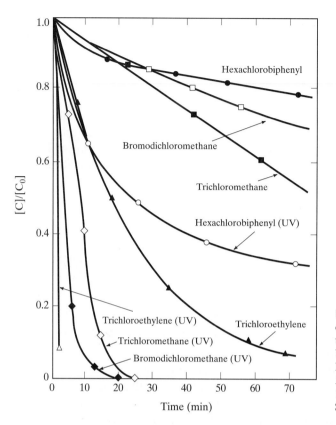

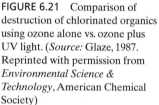

FIGURE 6.21 Comparison of destruction of chlorinated organics using ozone alone vs. ozone plus UV light. (*Source:* Glaze, 1987. Reprinted with permission from *Environmental Science & Technology*, American Chemical Society)

organic wastes. Biological treatment systems use microorganisms, mainly bacteria, to metabolize organic material, converting it to carbon dioxide, water, and new bacterial cells. Since biological systems rely on living organisms to transform wastes, considerable care must be exercised to assure that conditions are conducive to life. Microbes need a source of carbon and energy, which they can get from the organics they consume, as well as nutrients such as nitrogen and phosphorus. They are sensitive to pH and temperature, and some need oxygen.

As living organisms, microbes are susceptible to toxic substances, which at first glance makes biological treatment of hazardous wastes seem an unlikely choice. Surprisingly, though, most hazardous organics are amenable to biological treatment provided that the proper distribution of organisms can be established and maintained. For any given organic substance, there may be some organisms that will find that substance to be an acceptable food supply, while others may find it toxic. Moreover, organisms that flourish with the substance at one concentration may die when the concentration is increased beyond some critical level. Finally, even though a microbial population may have been established that can handle a particular kind of organic waste, it may be destroyed if the characteristics of the waste are changed too rapidly. If changes are made slowly enough, however, selection pressures may allow the microbial consortium to adjust to the new conditions and thereby remain effective.

Aqueous Waste Treatment. It is convenient to consider biological treatment of various sorts of wastewaters, including leachates from hazardous waste landfills, separately from in situ biological treatment of soils and groundwater. When liquid hazardous wastes can be conveyed to the treatment facility, it is possible to control the characteristics of the waste that reach the biological portion of the facility, increasing the likelihood of a successful degradation process.

Biological treatment is often just one step in an overall treatment system. As suggested in Figure 6.22, an example system would include a chemical treatment stage to oxidize and precipitate some of the toxics, followed by physical treatment to separate the resulting solids from the waste stream. The effluent from the physical treatment step may then be conditioned to give it the right pH and nutrient supply needed by the microorganisms in the biological treatment step.

The biological treatment stage itself utilizes processes already described for municipal wastewater treatment plants. After biological treatment, further sedimentation and clarification followed by carbon adsorption can be used to polish the effluent. Inorganic sludges produced during chemical processing and organic sludges from the biotreatment stage are separated from the liquid waste stream and treated. These sludges must be dewatered and disposed of in accordance with RCRA regulations since they are likely to be hazardous themselves.

Waste Incineration

Waste incineration is being strongly advocated by the EPA as its technology of choice for many types of hazardous wastes. Incineration is particularly effective with organic wastes, not only in soils but in other solids, gases, liquids, and slurries (thin mixtures of liquids and solids) and sludges (thick mixtures) as well. Carcinogens, mutagens, teratogens,

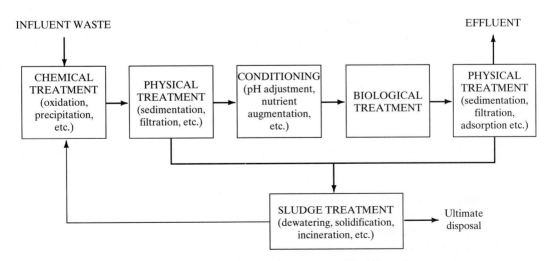

FIGURE 6.22   General flow diagram for treatment of liquid hazardous waste.

and pathological wastes can all be completely detoxified in a properly operated incinerator. Incinerators are not, however, capable of destroying inorganic compounds, although they can concentrate them in ash, making transportation and disposal more efficient. In addition, metals that volatilize at temperatures below 2000 °F pose a particular problem since, once vaporized, they are difficult to remove using conventional air pollution control equipment.

The principal measure of an incinerator's performance is known as the *destruction and removal efficiency* (DRE). A DRE of 99.99 percent, for example (commonly called "four nines DRE"), means that one molecule of an organic compound is released to the air for every 10,000 molecules entering the incinerator. RCRA requires a minimum DRE of 99.99 percent for most organic compounds, and a DRE of 99.9999 percent (six nines) for dioxins and dibenzofurans. The Toxic Substances Control Act regulations cover thermal destruction of PCBs, and although they are written somewhat differently from RCRA, in essence they require a 99.9999 percent DRE. More will be said about incineration and these particularly toxic compounds in the solid waste management chapter of this book (Chapter 9).

As is the case for all combustion processes, the most critical factors that determine combustion completeness are (1) the temperature in the combustion chamber, (2) the length of time that combustion takes place, (3) the amount of turbulence, or degree of mixing, and (4) the amount of oxygen available for combustion. Controlling these factors, which is crucial to obtaining the high levels of performance required by law, is made especially difficult in hazardous waste incinerators because of the variability of the wastes being burned. In addition to combustion controls, stack gas cleaning systems similar to those that are described in Chapter 7 are a necessary part of the system. Proper operation and maintenance of these complex incineration systems requires highly trained personnel, diligent and qualified supervisory staff, and an alert governmental agency to assure compliance with all regulations.

While there are a number of types of hazardous waste incinerators, there are only two principal designs that account for most of the existing units in operation: the *liquid injection incinerator,* and the *rotary kiln incinerator.* Liquid injection incinerators are the most common even though they are usable only for gases, liquids, and slurries thin enough to be pumped through an atomizing nozzle. The nozzle emits tiny droplets of waste that are mixed with air and an auxiliary fuel such as natural gas or fuel oil. The resulting gaseous mixture is burned at a very high temperature. The atomizing nozzle used in a liquid injection incinerator must be designed to accommodate the particular characteristics of the expected waste stream, which limits the types of waste that any given incinerator can treat.

The rotary kiln incinerator is more versatile than the liquid injection type, being capable of handling gases, liquids, sludges, and solids of all sorts, including drummed wastes. Figure 6.23 shows a diagram of such an incinerator. The main unit consists of a slightly inclined, rotating cylinder perhaps 2 to 5 m in diameter and 3 to 10 m long. Wastes and auxiliary fuel are introduced into the high end of the kiln, and combustion takes place while the cylinder slowly rotates. The rotation helps increase turbulence, which improves combustion efficiency. Partially combusted waste gases are passed to a secondary combustion chamber for further oxidation. Rotary kiln incinerators are commercially available as mobile units and fixed installations.

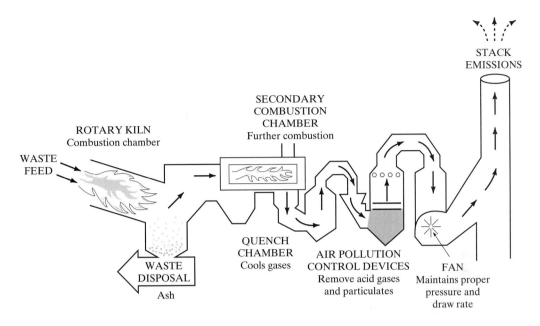

FIGURE 6.23 A rotary kiln hazardous waste incinerator (USEPA, 1988).

In spite of numerous controls, hazardous waste incinerators have the potential to emit amounts of noxious gases that may be unacceptable to neighbors. Emissions may include unburned organic compounds from the original waste, various *products of incomplete combustion* (PIC) formed in the incinerator itself, odors, carbon monoxide, nitrogen and sulfur oxides, hydrogen chloride, and particulates. The unburned ash and sludge from the air pollution control devices are considered hazardous wastes themselves and must be treated as such. If they are transported off site, then not only are there hazardous materials transported into the facility, but there are some leaving as well. The perception of potentially adverse impacts associated with incineration has made the siting of these facilities an extremely difficult task.

Liquid injection incinerators have been built into two ships, the *Vulcanus I* and the *Vulcanus II*, for hazardous waste incineration at sea, but public opposition has prevented their use. Proponents of incineration at sea point out the advantages of incineration, the difficulties associated with land-based siting, and the reductions in human risk that would come with incineration far from any centers of population. Opposition has focused on the chances of hazardous waste spills and the near impossibility of controlling such a spill should it occur. Also, the small amounts of unburned wastes routinely released, as well as those that might be released should the incinerator ever be improperly operated, could have unknown implications for marine life.

An ongoing debate is being conducted between advocates of incineration as a way to avoid land disposal (with its potential for toxic leakage into groundwater) and a concerned public that has learned to be suspicious of any complex and potentially dangerous technology.

6.9 LAND DISPOSAL

Land disposal techniques include landfills, surface impoundments, underground injection wells, and waste piles. About 5 percent of the hazardous waste that we dispose of on land in the United States is placed in specially designed landfills. About 35 percent is disposed of in diked surface impoundments such as pits, ponds, lagoons, and basins. About 60 percent is disposed of using deep underground injection wells. Waste piles, which are noncontainerized accumulations of solid hazardous waste typically used for temporary storage, account for less than 1 percent of our disposal volume (U.S. EPA, 1986).

Historically, land disposal has been the traditional method of getting rid of hazardous wastes in this country. Unfortunately, many of these disposal sites have been poorly engineered and monitored, and the results have sometimes been tragic as well as very expensive to rectify. They have been used extensively in the past because they were the most convenient and inexpensive method of disposal. However, remediation at older sites that have leaked toxics into the soil and groundwater has proven to be tremendously costly, and the originally perceived economic advantage of land disposal is now seen to have been shortsighted.

As was mentioned earlier, the 1984 Hazardous and Solid Waste Amendments (HSWA) to RCRA ban unsafe, untreated wastes from land disposal (U.S. EPA, 1991). The HSWA provisions are referred to as land disposal restrictions (LDRs), where land disposal includes placement of hazardous waste in

- Landfills
- Surface impoundments
- Waste piles
- Injection wells
- Land treatment facilities
- Salt domes or salt bed formations
- Underground mines or caves
- Concrete vaults or bunkers, intended for disposal purposes.

HWSA requires that the EPA assess all hazardous wastes to determine whether land disposal can be used and, if so, under what conditions. Without an EPA determination on a particular waste, that waste is automatically banned from land disposal. RCRA goes on to provide new restrictions and standards for those land disposal facilities that will be allowed to accept hazardous substances, including the following:

- Banning liquids from landfills
- Banning underground injection of hazardous waste within 1/4 mile of a drinking-water well
- Requiring more stringent structural and design conditions for landfills and surface impoundments, including two or more liners, leachate collection systems above and between the liners, and groundwater monitoring
- Requiring cleanup or corrective action if hazardous waste leaks from a facility

- Requiring information from disposal facilities on pathways of potential human exposure to hazardous substances
- Requiring location standards that are protective of human health and the environment (for example, allowing disposal facilities to be constructed only in suitable hydrogeologic settings).

Landfills. In accordance with these new, more stringent RCRA requirements, the design and operation of hazardous waste landfills has become much more sophisticated. A hazardous waste landfill is now designed as a modular series of three-dimensional control cells. By incorporating separate cells it becomes possible to segregate wastes so that only compatible wastes are disposed of together. Arriving wastes are placed in an appropriate cell and covered at the end of each working day with a layer of cover soil.

Beneath the hazardous wastes there must be a double-liner system to stop the flow of liquids, called *leachate,* from entering the soil and groundwater beneath the site. The upper liner must be a *flexible-membrane lining* (FML) usually made of sheets of plastic or rubber. Commonly used plastics include polyvinyl chloride (PVC) , high-density polyethylene (HDPE), and chlorinated polyethylene (CPE). Rubber FMLs include chlorosulfonated polyethylene (CSPE) and ethylene propylene diene monomer (EPDM). Depending on the material chosen for the FML, the thickness is typically anywhere from 0.25 mm (10 mil) to over 2.5 mm (100 mil). The lower liner is usually an FML, but recompacted clay at least 3/ft thick is also considered acceptable.

Leachate that accumulates above each liner is collected in a series of perforated drainage pipes and pumped to the surface for treatment. To help reduce the amount of leachate formed by precipitation seeping into the landfill, a low-permeability cap is placed over completed cells. When the landfill is finally closed, a cap that may consist of an FML along with a layer of compacted clay is placed over the entire top with enough slope to assure drainage from the wastes.

The landfill must also include monitoring facilities. The groundwater flowing beneath the site should be tested with monitoring wells placed upgradient and downgradient from the site. There may only need to be one upgradient well to test the "natural" quality of the groundwater before it flows under the site, but there should be at least three or more monitoring wells placed downgradient to assure detection of any leakage from the site. In addition, the soil under the site, above the water table, should be tested using devices called suction lysimeters.

A cross section of a completed hazardous waste landfill is shown in Figure 6.24.

Surface Impoundments. Surface impoundments are excavated or diked areas used to store liquid hazardous wastes. Usually, storage is temporary unless the impoundment has been designed eventually to be closed as a landfill. Impoundments have been popular because they have been cheap and because wastes remain accessible, allowing some treatment to take place during storage. Typical treatment technologies used in surface impoundments include neutralization, precipitation, settling, and biodegradation.

Historically, surface impoundments have typically been poorly constructed and monitored. In a survey of 180,000 surface impoundments, the EPA estimated that prior

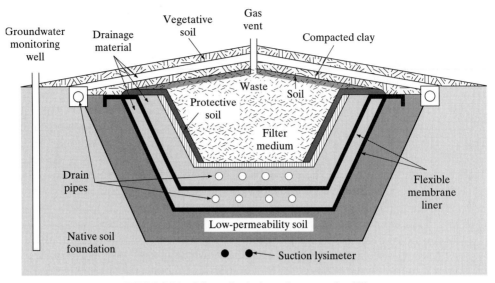

FIGURE 6.24   Schematic of a hazardous waste landfill.

to 1980 only about one-fourth were lined and fewer than 10 percent had monitoring programs (U.S. EPA, 1984). The same survey also found that surface impoundments were usually poorly sited. More than half were located over very thin or very permeable soils that would allow easy transport of leachate to groundwater. Over three-fourths of the impoundments were located over very thick and permeable aquifers that would allow relatively rapid dispersion of contaminants should they reach the water table. Moreover, about 98 percent of the surface impoundments were located less than one mile from sources of high-quality drinking water.

As a result of these poor siting, construction, and management problems, surface impoundments are the principal source of contamination in a large number of Superfund sites. Recent EPA regulations require new surface impoundments, or expansions to existing impoundments, to have two or more liners, a leachate-collection system, and monitoring programs similar to those required for landfills. However, the legacy of past practices will undoubtedly take billions of dollars and decades of time to remediate.

Underground Injection. The most popular way to dispose of liquid hazardous wastes has been to force them underground through deep injection wells (Figure 6.25). To help assure that underground drinking-water supplies will not become contaminated, injection wells used to dispose of hazardous industrial wastes are required to extend below the lowest formation containing underground sources of drinking water. Typical injection depths are more than 700 m below the surface. Since the main concern with underground injection is the potential for contaminating underground drinking-water supplies, the regulation of such systems has come under the Safe Drinking Water Act of 1974.

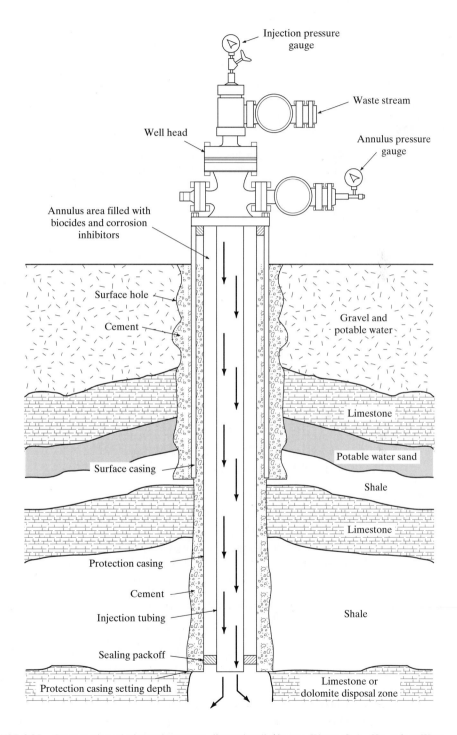

FIGURE 6.25 Cross section of a hazardous waste disposal well. (*Source:* Wentz, C. A., *Hazardous Waste Management.* Copyright © 1989 by McGraw-Hill, Inc. Used with permission of the publisher.)

Unfortunately, a number of hazardous waste injection wells have had leakage problems, so such wells cannot be considered entirely safe. Regulations covering construction, operation, and monitoring of injection wells are becoming more stringent, and as is the case for all land disposal options, continued reliance on this technology is being discouraged.

PROBLEMS

6.1. A sample of groundwater has 150/mg/L of Ca^{2+} and 60 mg/L of Mg^{2+}. Find the total hardness expressed in milliequivalents per liter (meq/L) and mg/L as $CaCO_3$. Using Table 6.5, how would this water be classified (e.g., soft, hard, etc.)?

6.2 For a solution with pH equal to 9.0, express the concentrations of H^+ and OH^- in meq/L and mg/L as $CaCO_3$.

6.3 A sample of water at pH 10.5 has 39.0 mg/L of CO_3^{2-} and 24.5 mg/L of HCO_3^-.

    a. Ignoring the contribution of $[H^+]$ and $[OH^-]$ to alkalinity, what is the alkalinity as $CaCO_3$?

    b. Including the contribution of $[H^+]$ and $[OH^-]$, find the alkalinity as $CaCO_3$.

6.4 A chemical analysis of a surface water yields the following data:

Ion	Concentration mg/L
Ca^{2+}	90
Mg^{2+}	30
Na^+	72
K^+	6
Cl^-	120
SO_4^{2-}	225
HCO_3^-	165
pH	7.5

    a. Determine the alkalinity expressed as $CaCO_3$.

    b. Determine the hardness as $CaCO_3$.

    c. Estimate the total dissolved solids.

6.5 The Cl^- concentration in the water analysis given in Problem 6.4 has been questioned. What concentration of Cl^- (mg/L) would make the cations and anions balance?

6.6 A sample of water has the following concentration of ions (and the pH is near neutral):

Cations	mg/L	Anions	mg/L
Ca^{2+}	95.0	HCO_3^-	160.0
Mg^{2+}	26.0	SO_4^{2-}	135.0
Na^+	15.0	Cl^-	73.0

    a. What is the total hardness (TH)?

    b. What is the carbonate hardness (CH)?

    c. What is the noncarbonate hardness (NCH)?

   d. What is the alkalinity?

   e. What is the total dissolved solids concentration?

   f. Draw an ion concentration bar graph.

6.7 A sample of water has the following concentrations of ions:

Cations	mg/L	Anions	mg/L
Ca^{2+}	40.0	HCO_3^-	110.0
Mg^{2+}	10.0	SO_4^{2-}	67.2
Na^+	?	Cl^-	11.0
K^+	7.0		

   a. Assuming no other constituents are missing, use an anion-cation balance to estimate the concentration of Na^+.

   b. What is the total hardness (TH)?

   c. Draw an ion concentration bar graph.

6.8 A reverse osmosis plant desalinates 5×10^6 L/day of feedwater containing 1500 mg/L of salts yielding 3×10^6 L/day of product water with 75 mg/L of salt (see Figure P6.8). What would the salt concentration be in the brine?

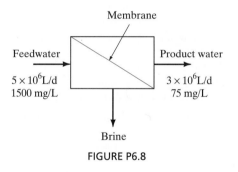

Feedwater

5×10^6 L/d
1500 mg/L

Membrane

Product water

3×10^6 L/d
75 mg/L

Brine

FIGURE P6.8

6.9 A rectangular primary clarifier for a domestic wastewater plant is to be designed to settle 2000 m³/day with an overflow rate of 32 m³/m²-day. The tank is to be 2.4 m deep and 4.0 m wide. How long should it be and what detention time would it have?

6.10 A final settling tank for a 2-million-gallon-per-day (2 mgd) activated-sludge treatment plant it to be designed to have an average overflow rate of 800 g/day-ft². The tank needs to have a minimum detention time of 2.0 hr and to allow proper settling it must be at least 11 ft deep. If the tank is circular, what should its diameter and depth be?

REFERENCES

Freeman, H. M. (ed.), 1989, *Standard Handbook of Hazardous Waste Treatment and Disposal,* McGraw-Hill, New York.

Glaze, W. H., 1987, Drinking water treatment with ozone, *Environmental Science and Technology,* 21(3).

Gleick, P. H., 1993, *Water in Crisis: A Guide to the World's Fresh Water Resources,* Oxford University Press, New York.

Haas, C. N., and R. J. Vamos, 1995, *Hazardous and Industrial Waste Treatment,* Prentice Hall, Englewood Cliffs, NJ.

Hammer, M. J. and M. J. Hammer, Jr., 1996, *Water and Wastewater Technology,* 3rd ed., Prentice Hall, Englewood Cliffs, NJ.

Robeck, G. G., N. A. Clarke, and K. A. Dostal, 1962, Effectiveness of water treatment processes in virus removal, *Journal of the American Water Works Association,* 54:1275–1290.

Tchobanoglous, G., and E. D. Schroeder, 1985, *Water Quality,* Addison-Wesley, Reading, MA.

U.S. EPA, 1980, *Summary Report, Control and Treatment Technology for Metal Finishing Industry,* Environmental Protection Agency, EPA 625/8–80–003, Washington DC.

U.S. EPA, 1984, *Surface and Impoundment Assessment National Report,* Environmental Protection Agency, EPA–570/9–84–002, Washington, DC.

U.S. EPA 1986, *Solving the Hazardous Waste Problem: EPA's RCRA Program.* Office of Solid Waste, Environmental Protection Agency, Washington, DC.

U.S. EPA, 1987a, *A Compendium of Technologies Used in the Treatment of Hazardous Wastes,* Center for Environmental Research Information, EPA/625/8–87/014, Environmental Protection Agency, Washington DC.

U.S. EPA, 1987b, *The Hazardous Waste System,* Environmental Protection Agency, Office of Solid Waste and Emergency Reponse, Washington, DC.

U.S. EPA, 1988, *Hazardous Waste Incineration: Questions and Answers,* Environmental Proteciton Agency, Office of Solid Waste, EPA/530–SW–88–018, Washington, DC.

U.S. EPA, 1988b. *Technology Screening Guide for Treatment of CERCLA Soils and Sludges,* Environmental Protection Agency, Office of Solid Waste and Emergency Response, EPA/2–88/004, Washington, DC.

U.S. EPA, 1991, *Land Disposal Restrictions, Summary of Requirements,* U.S. Environmental Protection Agency, OSWER 9934.0–1A, Washington, DC.

U.S. EPA, 1996, *Brownfields Action Agenda,* EPA/500/F–95/001, U.S. Environmental Protection Agency, Washington, DC.

Viessman, W., Jr., and M. J. Hammer, 1985, *Water Supply and Pollution Control,* 4th ed., Harper & Row, New York.

Wentz, C. A., 1989, *Hazardous Waste Management,* McGraw-Hill, New York.

Air Pollution

"From a particulate exposure standpoint, a 2-percent decrease in environmental tobacco smoke (passive smoking) would be equivalent to eliminating all the coal-fired power plants in the country."

—Kirk R. Smith, East-West Center, Program on Environment, Honolulu, 1993

7.1 INTRODUCTION

Air pollution is certainly not a new phenomenon. Indeed, early references to it date to the Middle Ages, when smoke from burning coal was already considered such a serious problem that in 1307, King Edward I banned its use in lime kilns in London. In more recent times, though still decades ago, several serious episodes focused attention on the need to control the quality of the air we breathe. The worst of these occurred in London, in 1952. A week of intense fog and smoke resulted in over 4000 excess deaths that were directly attributed to the pollution. In the United States the most alarming episode occurred during a four-day period in 1948 in Donora, Pennsylvania, when 20 deaths and almost 6000 illnesses were linked to air pollution. At the time, Donora had a population of only 14,000, making this the highest per capita death rate ever recorded for an air pollution episode.

Those air pollution episodes were the results of exceptionally high concentrations of sulfur oxides and particulate matter, the primary constituents of *industrial smog* or *sulfurous smog*. Sulfurous smog is caused almost entirely by combustion of fossil fuels, especially coal, in stationary sources such power plants and smelters. In contrast, the air pollution problem in many cities is caused by emissions of carbon monoxide, oxides of nitrogen, and various volatile organic compounds, which swirl around in the atmosphere reacting with each other and with sunlight to form *photochemical smog*. Although stationary sources also contribute to photochemical smog, the problem is most closely associated with motor vehicles. A major effect of efforts in the United States to control both sulfurous smog and photochemical smog has been the elimination of those dramatic, peak concentrations of pollution that were responsible for the air pollution episodes just mentioned. In their place, however, is the more insidious problem of morbidity and mortality increases associated with long-term exposure to lower concentrations of pollution. The human toll is much more difficult to document, but estimates place the current excess deaths caused by air pollution (mostly small particles) at several tens of thousands per year in the United States alone.

Much of the work on air pollution in the last few decades has centered on a small set of six substances, called *criteria pollutants*, that have been identified as contributors to both sulfurous and photochemical smog problems. The sources, transport, effects, and methods of controlling these criteria pollutants will be a principal focus of this chapter.

More recently, attention has been shifting toward the characterization and control of a growing list of especially hazardous air pollutants, many of which we are exposed to in our homes and workplaces, where we spend roughly 90 percent of our time. As the quote at the begining of this chapter suggests, modest improvements in indoor air quality can improve public health as much as major reductions in the traditional outdoor sources, which have been the focus of most of the scientific and political efforts of the past 50 years.

In the next chapter we will discuss the emissions and impacts of carbon dioxide, chlorofluorocarbons, and other trace gases that are affecting global climate and causing stratospheric ozone depletion. In a number of ways these gases are so different from the usual air pollutants that they deserve special treatment.

7.2 OVERVIEW OF EMISSIONS

There are many sources of the gases and particulate matter that pollute our atmosphere. Substances that are emitted directly into the atmosphere are called *primary* pollutants, while others that are created by various physical processes and chemical reactions that take place in the atmosphere are called *secondary* pollutants. For example, nitrogen oxides and hydrocarbons emitted when fuels are burned are primary pollutants, but the ozone that is created when those chemicals react with each other in the atmosphere is a secondary pollutant.

The sources of primary pollutant emissions can be conveniently categorized by the processes that create them. Most primary pollutants enter the atmosphere as a result of either combustion, evaporation, or grinding and abrasion. Volatile substances such as gasoline, paints, and cleaning fluids enter the atmosphere by evaporation; dust kicked up when land is plowed and asbestos fibers that flake off of pipe insulation are examples of grinding and abrasion; while automobile exhaust emissions and power plant stack gases are created during combustion. Of these it is combustion that accounts for the great majority of emissions, and it is the gases and particulate matter released when fuels are burned that have been the focus of most of the technical and legislative pollution control efforts.

In its simplest form, we can imagine the complete combustion of a pure hydrocarbon fuel such as methane (CH_4):

$$CH_4 + 2\,O_2 \rightarrow CO_2 + 2\,H_2O \tag{7.1}$$

The products of combustion are simple carbon dioxide (CO_2) and water (H_2O), neither of which had been considered an air pollutant until we realized that the accumulation of CO_2 in the atmosphere was enhancing the earth's natural greenhouse effect (as will be described in the next chapter).

If the temperature of combustion is not high enough, or there is not enough oxygen available, or if the fuel is not given enough time to burn completely, then the fuel will not be completely oxidized and some of the carbon will be released as carbon monoxide (CO) instead of CO_2. Also, some of the fuel will not be completely burned, so there will be emissions of various partially combusted hydrocarbons that we will represent by (HC). So we can write the following descriptive reaction to represent incomplete combustion of our pure hydrocarbon fuel, methane:

$$CH_4 + O_2 \rightarrow \text{mostly}\,(CO_2 + 2\,H_2O) + \text{traces of}\,[CO + (HC)] \tag{7.2}$$

Of course, most combustion takes place in air, not in a pure oxygen environment, and air is roughly 78 percent nitrogen (N_2) and 21 percent oxygen (O_2). When the temperature of combustion is high enough, some of that nitrogen reacts with the oxygen in air to form various nitrogen oxides (NO_x). Since this NO_x is formed when combustion temperatures are high, it is referred to as *thermal NO_x*.

$$\text{air}\,(N_2 + O_2) + \text{Heat} \rightarrow \text{Thermal}\,NO_x \tag{7.3}$$

So far we have assumed that the fuel being burned was a pure hydrocarbon such as methane. In reality, of course, most fuels have a number of other elements in them, such as nitrogen, sulfur, lead (in gasoline), and other unburnable materials called ash. Burning fuel with these "impurities" in them releases additional NO_x (called *fuel NO_x*), oxides of sulfur (SO_x), lead (Pb), more particulate matter, and ash.

Combining the effects of incomplete combustion, combustion in air, and combustion of fuels that are not pure hydrocarbons yields the following qualitative description of combustion:

$$\text{Fuel}\,(H, C, S, N, Pb, ash) + \text{air}\,(N_2 + O_2) \rightarrow$$
$$\text{Emissions}\,(CO_2, H_2O, CO, NO_x, SO_x, Pb, \text{particulates}) + \text{Ash} \tag{7.4}$$

Now let's add a simple representation of the photochemical reactions that produce ozone (O_3) and other constituents of photochemical smog. Hydrocarbons (HC) and other organic compounds that readily vaporize are called *volatile organic compounds* (VOCs). VOCs react with NO_x in the presence of sunlight to produce photochemical smog:

$$VOCs \ + \ NO_x \ + \ Sunlight \ \rightarrow \ Photochemical \ smog \ (O_3 \ + \ etc.) \quad (7.5)$$

To distinguish between the ozone that is formed near the ground by (7.5) from the ozone that exists in the stratosphere (next chapter), the designations *ground level ozone* and *stratospheric ozone* are sometimes used. As we shall see, ground level ozone is harmful to our health, while stratospheric ozone protects our health by shielding us from ultraviolet radiation from the sun.

Reactions (7.1) to (7.5) are greatly simplified, of course, but they do introduce the six principal players in urban air pollution: CO, NO_x, SO_x, Pb, O_3, and particulate matter (ash and unburned hydrocarbons).

Another way to approach emissions and controls of air pollutants is to categorize the sources as being *mobile* sources or *stationary* sources. Mobile sources include highway vehicles (automobiles and trucks) and other modes of transportation, including railroads, aircraft, farm vehicles, and boats and ships. Stationary sources are often categorized as stationary *fuel combustion*, which includes electric power plants and industrial energy systems; *industrial processes*, such as metals processing, petroleum refineries, and other chemical and allied product manufacturing; and *miscellaneous* sources. Emissions of the principal air pollutants following this categorization are illustrated in Figure 7.1.

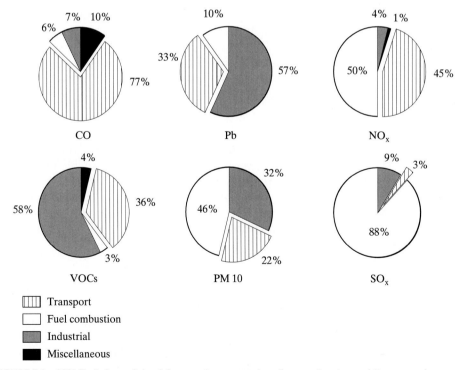

FIGURE 7.1 1993 Emissions of six of the most important air pollutants showing mobile sources (transport) and the breakdown of stationary sources. (Based on U.S. EPA, 1994b)

7.3 THE CLEAN AIR ACT

Initial efforts on the part of the U.S. Congress to address the nation's air pollution problem began with the passage of the Air Pollution Control Act of 1955. Although it provided funding only for research, and not control, it was an important milestone because it opened the door to federal participation in efforts to deal with air pollution. Up until that time, it had been thought to be a state and local problem. This was followed by a series of legislative actions by Congress that included the Clean Air Act Amendments of 1963, 1966, 1970, 1977, and 1990, all of which are sometimes lumped together and referred to as simply the *Clean Air Act* (CAA). In 1998, the Act will be due for reauthorization.

Much of the real structure to the Clean Air Act was established in the 1970 Amendments. In those amendments, the EPA was required to establish *National Ambient Air Quality Standards* (NAAQSs), and states were required to submit *State Implementation Plans* (SIPs) that would show how they would meet those standards. In addition, the Act required *New Source Performance Standards* (NSPSs) to be established that would limit emissions from certain specific types of industrial plants and from motor vehicles.

Air Quality and Emission Standards

The Clean Air Act requires the EPA to establish both air quality standards (NAAQS) and emission standards (NSPS), and it is important to keep in mind the fundamental difference between the two. Ambient air quality standards are acceptable *concentrations* of pollution in the atmosphere, while emission standards are allowable *rates* at which pollutants can be released from a source.

National Ambient Air Quality Standards have been established by EPA at two levels: *primary* and *secondary*. Primary standards are required to be set at levels that will protect public health and include an "adequate margin of safety," regardless of whether the standards are economically or technologically achievable. Primary standards must protect even the most sensitive individuals, including the elderly and those already suffering from respiratory and cardiopulmonary disorders. NAAQSs are, therefore, conceptually different from maximum contaminant levels (MCLs) that have been set for drinking water. Recall that the Safe Drinking Water Act requires the EPA to balance public health benefits with technological and economic feasibility in establishing drinking-water MCLs.

Secondary air quality standards are meant to be even more stringent than primary standards. Secondary standards are established to protect public welfare (e.g., structures, crops, animals, fabrics, etc.). Given the difficulty in achieving primary standards, secondary standards have played almost no role in air pollution control policy, and in fact they have usually been set at the same levels as primary standards.

National Ambient Air Quality Standards now exist for six *criteria* pollutants: carbon monoxide (CO), lead (Pb), nitrogen dioxide (NO_2), ground-level ozone (O_3), sulfur dioxide (SO_2), and particulate matter. The Clean Air Act requires that the list of criteria pollutants be reviewed periodically and that standards be adjusted according to the latest scientific information. Past reviews have modified both the list of pollutants and their acceptable concentrations. For example, the original particulate standard did not refer to size of particulates, but in 1987 the standard was modified to include only particulates with aerodynamic diameter less than a nominal 10 μm (PM 10), and in 1997 an additional category of fine particles with diameters less than or equal to 2.5 μm (PM 2.5) was added. Also in 1997 the ozone standard was tightened from 0.12 ppm to 0.08 ppm.

For a given region of the country to be in compliance with NAAQS, the concentrations cannot be exceeded more than once per calendar year. The law allows states to establish standards that are more stringent than the NAAQS, which California has done. Federal air quality standards along with California's standards are shown in Table 7.1.

For the gases in Table 7.1, the concentrations are expressed two ways—in parts per million by volume (ppm) as well as in mass per unit volume (μg/m^3 or mg/m^3). The volumetric units (ppm) are preferred since those are independent of pressure and temperature. The mass-per-volume concentrations assume a temperature of 25 °C and 1 atm of pressure. The conversion between units was discussed in Section 1.2, and the following example illustrates the procedure.

Besides establishing National Ambient Air Quality Standards, the Clean Air Act also requires the EPA to establish emission standards for mobile sources such as cars

TABLE 7.1   National Ambient Air Quality Standards and California State Standards

Pollutant	Averaging time	Federal primary	Federal secondary	California	Most relevant effects
Carbon monoxide (CO)	8 hr	9 ppm (10 mg/m^3)	None	9 ppm	Aggravation of angina pectoris; decreased exercise tolerance; possible risk to fetuses
	1 hr	35 ppm (40 mg/m^3)	None	20 ppm	
Nitrogen dioxide (NO$_2$)	Annual mean	0.053 ppm (100 μg/m^3)	Same	None	Aggravation of respiratory disease; atmospheric discoloration
	1 hr	None	None	0.25 ppm	
Ground level ozone (O$_3$)	8 hr	0.08 ppm (155 μg/m^3)	Same	0.09 ppm	Decreased pulmonary function; surrogate for eye irritation; materials and vegetation damage
Sulfur dioxide (SO$_2$)	Annual mean	0.03 ppm (80 μg/m^3)	None	None	Wheezing, shortness of breath, chest tightness; plant damage and odor
	24 hr	0.14 ppm (365 μg/m^3)	None	0.05 ppm	
	3hr	None	0.50 ppm	None	
	1hr	None	0.25 ppm		
PM 10	Annual mean	50 μg/m^3	Same	30 μg/m^3	Exacerbation of respiratory disease symptoms; excess deaths; visibility
	24 hr	150 μg/m^3	Same	50 μg/m^3	
PM 2.5[a]	Annual	15 μg/m^3	Same	None	
	24 hr	65 μg/m^3	Same	None	
Lead (Pb)	1 month	None	None	1.5 μg/m^3	Impaired blood formation; infant development.
	3 months	1.5 μg/m^3	Same	None	

[a] added in 1997

and trucks. The 1970 Amendments to the Clean Air Act gave the auto industry a five-year deadline to achieve a 90 percent reduction in emissions from new cars. At the time it was not even known whether such reductions were technologically possible, let alone how they could be implemented in such a short period of time. This "technology-forcing" legislation predictably led to numerous clashes between Congress and the automobile industry, and the standards were modified and delayed for many years. Later in this chapter the emission controls that eventually were implemented will be described in some detail.

The EPA is also required to establish emission standards for certain large stationary sources such as fossil-fuel-fired power plants, incinerators, Portland cement plants, nitric acid plants, petroleum refineries, sewage treatment plants, and smelters of various sorts. The methods of achieving the emission standards for stationary sources will be explored later in this chapter.

EXAMPLE 7.1.   Air Quality Standards Expressed in Volumetric Units

California's air quality standard for nitrogen dioxide (NO_2) is 470 $\mu g/m^3$ (at a temperature of 25 °C and 1 atmosphere of pressure). Express the concentration in ppm.

Solution   In Section 1.2 the ideal gas law was used to show that 1 mol of an ideal gas at 1 atm and 25 °C occupies a volume of 24.45 L (24.45×10^{-3} m^3). The molecular weight of NO_2 is

$$\text{mol wt} = 14 + 2 \times 16 = 46 \text{ g/mol}$$

so that

$$(NO_2) = \frac{24.45 \times 10^{-3}\,\text{m}^3/\text{mol} \times 470 \times 10^{-6}\,\text{g/m}^3}{46\,\text{g/mol}}$$

$$= 0.25 \times 10^{-6} = 0.25 \text{ ppm}$$

which agrees with Table 7.1.
Notice that parts per million by volume (ppm) is really a dimensionless volume fraction, independent of temperature and pressure.   ∎

The Clean Air Act Amendments of 1977

The goal of the 1970 Amendments was to attain clean air by 1975, as defined by the NAAQS, with allowable extensions in certain circumstances until 1977. For a number of reasons, only about one-third of the air quality control regions in the nation were meeting the standards by 1977. This forced Congress to readdress the problem through the Clean Air Act Amendments of 1977. Besides extending the deadlines, the 1977 Amendments had to deal with two important questions. First, what measures should be taken in *nonattainment areas* that were not meeting the standards? Second, should air quality in regions where the air is cleaner than the standards be allowed to degrade toward the standards, and if so, by how much?

For nonattainment areas, the 1970 Act appeared to prohibit any increase in emissions whatsoever, which would have eliminated industrial expansion and severely curtailed local economic growth. To counter this, the EPA adopted a policy of *emission offsets*. To receive a construction permit, a major new source of pollution in a nonattainment area must first find ways to reduce emissions from existing sources. The reductions, or offsets, must exceed the anticipated emissions from the new source. The net effect of this offset policy is that progress is made toward meeting air quality standards in spite of new emission sources being added to the airshed.

Offsets can be obtained in a number of ways. For example, emissions from existing sources in the area might be reduced by installing better emission controls on equipment that may or may not be owned by the permit seeker. In some cases, a permit seeker may simply buy out existing emission sources and shut them down. Emission offsets can be "banked" for future use, or they can be sold or traded to other companies for whatever the market will bear. In addition to offsets, new sources in nonattainment areas must use emission controls that yield the *lowest achievable emission rate* (LAER) for the particular process. LAER technology is based on the most stringent emission rate achieved in practice by similar sources, regardless of the economic cost or energy impacts.

The 1970 Amendments were not specific about regions that were cleaner than ambient standards required, and in fact appeared to allow air quality to deteriorate to those standards. The 1977 Amendments settled the issue of whether or not this would be allowed by establishing the concept of *prevention of significant deterioration* (PSD) in attainment areas. Attainment areas are put into one of three classes, and the amount of deterioration allowed is determined by the class. Class I areas include National Parks and Wilderness Areas, and almost no increase in pollution is allowed. At the other extreme, Class III areas are designated for development and allowable increments of new pollution are large. Everything else falls into Class II areas, where moderate deterioration in air quality is allowed. In PSD areas, *best available control technology* (BACT) is required on major new sources. BACT is less stringent than LAER, as it does allow consideration of economic, energy, and environmental impacts of the technology, but it can be more strict than allowed by NSPS.

In all PSD areas, the allowable increments of air quality degradation are constrained by the NAAQS. That is, in no circumstance would air quality be allowed to deteriorate to the point where the area is no longer in compliance with ambient air quality standards. To demonstrate compliance with these PSD increments and with air quality standards in general, mathematical models predicting ambient pollutant concentrations must be used for any proposed new source. Such models, which use meteorological and stack emission data to predict air quality impacts, will be described in Section 7.5.

The Clean Air Act Amendments of 1990

The Clean Air Act Amendments of 1990 significantly strengthened the government's efforts to assure healthful air in the United States and it broadened its scope to include control of pollutants that affect a global problem—stratospheric ozone depletion. Principal changes in the Act include the following:

- A new acid deposition control program (Title IV)
- New requirements for nonattainment areas (Title I)

- Tightened automobile emission standards and new fuel requirements (Title II)
- New toxic air pollution controls (Title I)
- Phase-out schedule for ozone-depleting substances (Title VI).

One of the most important shortcomings of the Clean Air Act before the 1990 Amendments was its inability to deal effectively with acid rain (or, more correctly, acid deposition). As will be described later, acid deposition results from emissions of sulfur dioxide (SO_2) that convert to sulfuric acid, and nitrogen oxides (NO_x) that become nitric acid. The goal of the Amendments is to cut annual SO_2 emissions to half of 1980 levels, establishing a cap of 8.9 million tons by 2000, and to cut NO_x emissions by 2 million tons. The NO_x reductions are to be achieved in the traditional way—that is, by tightening the emission standards for major stationary sources (sometimes referred to as the "command and control" approach), but the SO_2 reductions will result in large part from a new market-based approach.

In addition to specifying certain emission limits for SO_2, the EPA is administering a more flexible *allowance system*, in which one allowance authorizes the owner to emit one ton of SO_2. Large coal-fired power plants are not allowed to emit any more tons of SO_2 than the number of allowances they own. If insufficient allowances are owned to cover emissions, the owners are subject to an excess emissions penalty of $2000 per ton of SO_2. By controlling the number of allowances that the EPA issues each year, a cap is placed on emissions from these large sources. The intent is for these allowances to be bought and sold or banked in the same way that other commodities are traded. New sources that have no allowances would have to purchase allowances from existing sources or from annual EPA auctions. The idea, of course, is that major sources will find the least expensive ways to cut their emissions and then sell some of their allowances to others who cannot reduce their emissions as cheaply. The goal is a least-cost emission limitation that allows sources the flexibility they need to make the most cost-effective choices.

An even more innovative section of the acid deposition section of the 1990 Amendments includes the creation of a conservation and renewable energy reserve (*The Reserve*). The Reserve is a pool of 300,000 SO_2 allowances that are being given to eligible electric utilities as a reward for customer energy conservation programs and to utilities or independent power producers who build new renewable energy systems (such as wind or solar power). These bonus allowances are equivalent to emissions associated with 150 billion kWh of electricity, which will cut emissions of SO_2 by over 400,000 tons (U.S. EPA, 1994a). In addition, over 400,000 tons of NO_x and over 100 million tons of CO_2 reductions will result from this set aside (see Example 1.11 in Chapter 1).

The 1990 Amendments also address the slow progress being made in nonattainment areas. The Amendments establish a rating system for nonattainment areas based on the extent to which the NAAQSs are exceeded. A given area is designated as marginal, moderate, serious, severe, or extreme, and each category has its own deadlines and control requirements. One element of the new system is the amount of offsetting that is required for new sources. In *marginal* areas, industries must remove 10 percent more emissions from existing sources than the new emissions they intend to release. In *extreme* areas, that offset ratio is 1.5 to 1; that is, a new source must offset 1.5 times as

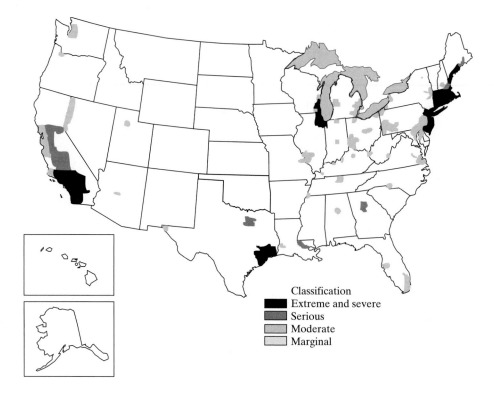

FIGURE 7.2   Areas designated nonattainment for 0.12 ppm ozone in 1994, showing the four severity classifications required by the 1990 Clean Air Act Amendments. (*Source*: U.S. EPA, 1994b)

much pollution as it is being permitted for. Nonattainment areas in the United States, including their severity classification, for ozone (O_3) are shown in Figure 7.2.

The 1990 Amendments also tighten the emission requirements for mobile sources, forcing 40 percent reductions of hydrocarbon emissions and 50 percent reductions for nitrogen oxide (NO_x) emissions by 1996. And for the first time, the Act now addresses the characteristics of the fuel that is burned in vehicles. In the nine worst ozone nonattainment areas (Los Angeles, San Diego, Houston, Baltimore, Philadelphia, New York, Hartford, Chicago, and Milwaukee) new, cleaner "reformulated" gasoline has been required since 1995.

Earlier versions of the Clean Air Act established *National Emission Standards for Hazardous Air Pollutants* (NESHAP), and through these standards a small list of pollutants (asbestos, benzene, beryllium, coke oven emissions, inorganic arsenic, mercury, radionuclides, and vinyl chloride) were controlled. The 1990 Amendments extend that list to include 189 pollutants listed in the legislation. The list can be changed by the EPA, and if there are objections to any additions the burden of proof is on the petitioner, who must show that the chemical may not reasonably be anticipated to cause any adverse human health or environmental effects. Emission standards for chemicals on the list are based on the Maximum Achievable Control Technology (MACT).

Finally, Title VI of the 1990 Amendments has been written to protect the stratospheric ozone layer by phasing out ozone-depleting substances such as chlorofluorocarbons (CFCs). The phase-out is mandated to be at least as strict as that required by the international treaty known as the Montreal Protocol. These chemicals and the Protocol itself will be more fully described in the next chapter.

7.4 THE POLLUTANT STANDARDS INDEX

The NAAQS forms the basis for an air pollution index, called the *Pollutant Standards Index* (PSI), that has been developed by the EPA and is used by all metropolitan areas in the United States to report to the public an overall assessment of a given day's air quality. The PSI integrates air quality data for five of the criteria pollutants into a single number that represents the worst daily air quality in an urban area (the only other criteria pollutant, lead, is not included since it does not have a short-term NAAQS). Descriptive terms ranging from good to hazardous are assigned to ranges of the PSI as shown in Table 7.2. A PSI value of 100 indicates that at least one pollutant just reached its ambient air quality standard on that day. When levels exceed 100, local officials may issue public health advisories and some industrial activities may be curtailed. A PSI of 200 results in an "Alert" and people with existing heart or lung disease are advised to stay indoors and reduce their physical activity. At 300, an air pollution "Warning" is issued and power plants, incinerators, and other industrial facilities may be required to limit operations. A PSI level of 400 would constitute an "Emergency" and would require a cessation of most industrial and commercial activity, plus a prohibition of almost all private use of motor vehicles. In the United States, PSI readings above 200 are quite unusual and values above 300 are extremely rare. That was not always the case. In the 1960s and 1970s, ozone concentrations in Los Angeles frequently exceeded 0.5 ppm, which corresponds to a PSI of over 400. By 1990 the highest PSI readings in the Los Angeles area were around 250.

The actual calculation of a day's PSI value is based on Table 7.3, which shows individual PSI numbers corresponding to various pollutant concentrations. Individual PSI "subindexes" are computed for each of the pollutants in the table using linear

TABLE 7.2.   Pollutant Standards Index values, descriptors, and general health effects

PSI Value	Descriptor	General health effects
0–50	Good	None for the general public
51–100	Moderate	Few or none for the general public
101–199	Unhealthful	Mild aggravation of symptoms among susceptible people, with irritation symptoms in the healthy population
200–299	Very unhealthful	Significant aggravation and decreased exercise tolerance in persons with heart or lung disease; widespread symptoms in the healthy population
≥ 300	Hazardous	Significant aggravation of symptoms in healthy persons; early onset of certain diseases; above 400, premature death of ill and elderly

Table 7.3.   Pollutant Standards Index (PSI) Breakpoints

Index	Designation	1 hr O_3 (ppm)	8 hr CO (ppm)	24 hr PM 10 ($\mu g/m^3$)	24 hr SO_2 (ppm)	1 hr NO_2 (ppm)
0	—	0	0	0	0	—[a]
50	—	0.06	4.5	50	0.03	—[a]
100	NAAQS	0.12[b]	9	150	0.14	—[a]
200	Alert	0.20	15	350	0.30	0.6
300	Warning	0.40	30	420	0.60	1.2
400	Emergency	0.50	40	500	0.80	1.6
500	Significant harm	0.60	50	600	1.00	2.0

[a]No index values reported at concentrations below the Alert level.
[b]does not yet reflect 1997 change in standard.
Source EPA, 1994b.

interpolation between the indicated breakpoints. The highest PSI subindex determines the overall PSI. A PSI calculation is provided in Example 7.2.

EXAMPLE 7.2.   Determining the PSI

Suppose on a given day the following maximum concentrations are measured:

1 hr O_3	0.18 ppm
8 hr CO	9 ppm
24 hr PM 10	130 $\mu g/m^3$
24 hr SO_2	0.12 ppm
1 hr NO_2	0.3

Find the PSI and indicate the descriptor that would be used to characterize the day's air quality.

Solution   Using Table 7.3, it can be seen that the ozone level (O_3) yields a subindex over 100; CO yields an subindex of 100; PM 10 and SO_2 are less than 100. There is no subindex for NO_2 since it is below the Alert level (0.6 ppm). The highest subindex therefore corresponds to O_3. To calculate the PSI, we must interpolate. An ozone concentration of 0.12 ppm yields a subindex of 100, while a value of 0.20 ppm corresponds to 200. By interpolation, the measured ozone concentration of 0.18 ppm yields a subindex of

$$\text{Subindex } O_3 = 100 + \frac{(0.18 - 0.12)}{(0.20 - 0.12)} \times (200 - 100) = 175 \text{ ppm}$$

The highest subindex is 175, so the PSI would be 175 and the air quality would be described as unhealthful.   ∎

The PSI is human health based and does not specifically take into account the damage air pollution can cause to animals, vegetation, and materials. It also does not take into account the possibility of synergistic effects associated with combinations of pollutants. For example, the combination of sulfur oxides and particulates is thought to be much more damaging to health than the sum of the individual effects, but the current version of the PSI does not account for that magnification.

The number of days that the PSI is above 100 is often used to describe progress that is being made in air quality. Figure 7.3 shows 10-year trend data from 1984 to 1993

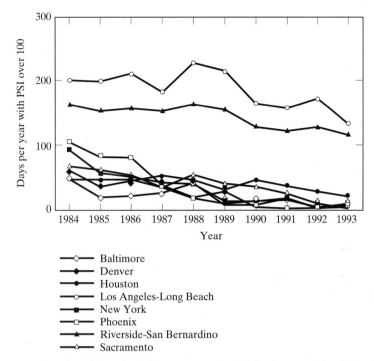

FIGURE 7.3   Annual number of days when PSI was greater than 100 for the most polluted American cities. Data cover only those sites in each metropolitan area that have complete data sets. (*Source*: U.S. EPA, 1994b)

for the most polluted American cities. In every case the number of days with PSI over 100 has decreased over time, with some cities, such as Sacramento, Denver, Baltimore, New York, and Phoenix, having cut their number of exceedances by over 70 percent. The worst cities in 1984, by a wide margin, were in southern California (Los Angeles–Long Beach and Riverside–San Bernardino). In 1993 those were still the most polluted cities (by this measure), with roughly one-third of their days having at least one pollutant exceeding the national ambient air quality standard. Almost always, that pollutant was ozone (O_3).

7.5 CRITERIA POLLUTANTS

Given the ongoing focus of the Clean Air Act, most of the monitoring of emissions, concentrations, and effects of air pollution has been directed toward the six criteria pollutants: ground level ozone (O_3), carbon monoxide (CO), sulfur dioxide (SO_2), small particulates (PM 10), nitrogen dioxide (NO_2), and lead (Pb). The original ambient air quality standards for these pollutants were based on extensive documentation assembled and published by the EPA in a multivolume set of *Air Quality Criteria* documents, from which the name *criteria pollutants* originated.

Figure 7.4 shows the progress that has been made in total emissions associated with these six criteria pollutants over the 25-year period from 1970 (the year the Clean

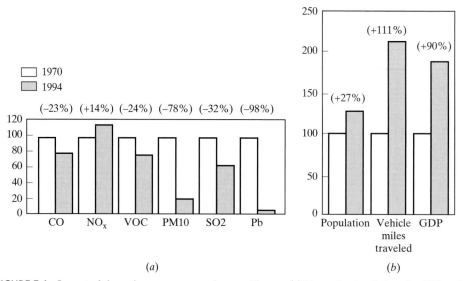

FIGURE 7.4   Impact of clean air programs over the past 25 years. (a) Normalized emissions for 1970 and 1994 (1970 equals 100). (b) Normalized indicators for growth in population, vehicle miles traveled and gross domestic product. (based on U.S. EPA, 1995)

Air Act initiated the NAAQS system) to 1994. These improvements occurred during a period of time when population grew by 27 percent, gross domestic product increased by 90 percent, and vehicle miles traveled more than doubled. Significant absolute decreases in emissions have been achieved in all categories except nitrogen oxides (NO$_x$), which have proven to be difficult to control. Notice that a category of pollutants known as volatile organic compounds (VOCs) is used as a surrogate for ozone (O$_3$) in these emission totals. Ozone is a *secondary* pollutant; that is, it is not actually emitted but rather is formed by reactions that take place in the atmosphere. Volatile organic compounds are inputs to the photochemical reactions that produce ozone, so they are used as indicators of the potential for ozone formation.

Despite the progress made in controlling emissions, in 1994 nearly one-fourth of the population of the United States lived in counties that did not meet national ambient air quality standards. As shown in Figure 7.5, ground-level ozone is the pollutant that more people encounter above standards than any other. Less than 6 percent of the population live in areas where any other pollutant exceeds standards.

Carbon Monoxide

More than half of the mass of all of the pollutant emissions in the United States is the colorless, odorless, tasteless, poisonous gas carbon monoxide (CO). It is produced when carbonaceous fuels are burned under less than ideal conditions. Incomplete combustion, yielding CO instead of CO$_2$, results when any of the following four variables are not kept sufficiently high: (1) oxygen supply, (2) combustion temperature, (3) gas residence time at high temperature, and (4) combustion chamber turbulence. These parameters are generally under much tighter control in stationary sources such as

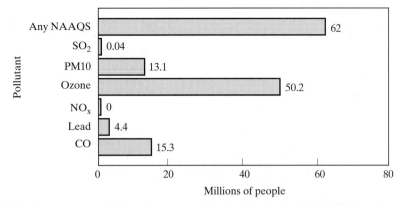

FIGURE 7.5   Number of people living in counties with air quality levels above the National Ambient Air Quality Standards in 1994 out of a total U.S. population of 260 million. (*Source*: U.S. EPA, 1995a).

power plants than in motor vehicles, and CO emissions are correspondingly less. For example, power plants that are designed and managed for maximum combustion efficiency produce less than 1/2 percent of all CO emissions in spite of the fact that they consume about 30 percent of our fossil fuel.

About 77 percent of total CO emissions are from the transportation sector, and almost all of the CO in urban areas comes from motor vehicles. Hourly atmospheric concentrations of CO over our cities often reflect city driving patterns: Peaks occur on weekdays during the morning and late afternoon rush hours, while on weekends there is typically but one lower peak in the late afternoon. Personal exposure to CO is determined by the proximity of motor vehicle traffic, with some occupational groups such as cab drivers, police, and parking lot attendants receiving far higher than average doses. Carbon monoxide emissions from motor vehicles decreased by 24 percent in the decade 1984–1993 despite a 33 percent increase in vehicle miles traveled. The impact of tightening emission controls, coupled with a drop in the fraction of older, more polluting cars still on the road, is clearly evident.

At levels of CO that occur in urban air, there are apparently no detrimental effects on materials or plants; those levels can, however, adversely affect human health. Carbon monoxide is an asphyxiant; it interferes with the blood's ability to carry oxygen from the lungs to the body's organs and tissues. When inhaled, it readily binds to hemoglobin in the bloodstream to form carboxyhemoglobin (COHb). Hemoglobin, in fact, has a much greater affinity for carbon monoxide than it does for oxygen, so even small amounts of CO can seriously reduce the amount of oxygen conveyed throughout the body. With the bloodstream carrying less oxygen, brain function is affected and heart rate increases in an attempt to offset the oxygen deficit.

The usual way to express the amount of carboxyhemoglobin in the blood is as a percentage of the saturation level, %COHb. The amount of COHb formed in the blood is related to the CO concentration and the length of time exposed, as suggested in the following (simplified from Ott and Mage, 1978):

$$\%\text{COHb} = \beta(1 - e^{-\gamma t})[\text{CO}] \tag{7.6}$$

where

$$
\begin{aligned}
\%\text{COHb} &= \text{carboxyhemoglobin as a percent of saturation} \\
[\text{CO}] &= \text{carbon monoxide concentration in ppm} \\
\gamma &= 0.402 \text{ hr}^{-1} \\
\beta &= 0.15\%/\text{ppm CO} \\
t &= \text{exposure time in hours}
\end{aligned}
$$

EXAMPLE 7.3   Federal Standard for CO

Estimate the %COHb expected for a 1-hour exposure to 35 ppm of CO (the federal standard).

Solution   From (7.6),

$$\%\text{COHb} = 0.15(\%/\text{ppm})[1 - \exp(-0.402/\text{hr} \times 1 \text{ hr})][35 \text{ ppm}] = 1.7 \text{ percent} \qquad \blacksquare$$

Physiological effects can be noted at small percentages of COHb, increasing in severity as the concentration increases. The elderly, the fetus, and individuals who suffer from cardiovascular disease— particularly those with angina pectoris (a heart condition characterized by chest pain)—or peripheral vascular disease are the most sensitive to COHb since the heart work must work harder in an attempt to offset the reduction in oxygen. Studies of patients with angina have shown an earlier than usual onset of pain during exercise when levels are as low as 2 percent COHb. In Example 7.3, it was calculated that an individual breathing CO at the federal ambient air quality standard of 35 ppm for one hour would be likely to reach 1.7 percent COHb. That is very close to the level at which health effects have been noted, and the federal standard has been criticized as a result. California's CO standard has been set lower (20 ppm) in an attempt to assure less than 2 percent COHb.

The reduction of oxygen in the bloodstream also affects the brain's ability to perceive and react. At 2.5 percent COHb, studies have shown an impairment in time-interval discrimination. (Subjects were less able to distinguish the duration of a tone signal.) Studies have also shown that at 5 percent, psychomotor response times may be affected and patients with heart disease experience increased physiological stress. At 5 to 17 percent, manual dexterity, ability to learn, and performance in complex sensorimotor tasks such as driving are diminished. At 10 percent headache is common, and at 30 percent COHb, most people will experience dizziness, headache, fatigue, and impaired judgment. Concentrations above 60 percent produce loss of consciousness and death if exposure continues (American Lung Association, 1994).

Carbon monoxide concentrations near busy roadways frequently range from 5 to 50 ppm, and measurements made on congested highways indicate that drivers can be exposed to CO concentrations of 100 ppm. Carbon monoxide is an important pollutant in indoor air as well. Cigarette smoke contains more than 400 ppm CO, and smokers frequently have COHb levels between 5 and 10 percent. Tobacco smoke in bars and restaurants often raises indoor CO levels to 20 to 30 ppm, which is close to the one-hour ambient standard (Wadden and Scheff, 1983). Fortunately, COHb is removed

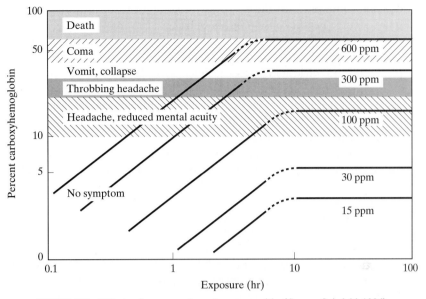

FIGURE 7.6   Effects of exposure to carbon monoxide. (*Source*: Seinfeld, 1986)

from the bloodstream when clean air is breathed. Healthy subjects clear about half of the CO from their blood in three to four hours, so adverse effects are usually temporary.

Figure 7.6 summarizes the relationship between exposure time and resulting percent carboxyhemoglobin, along with indications of some of the more severe impacts that can result. Notice the saturation effect for long exposures.

Oxides of Nitrogen

Although seven oxides of nitrogen are known to occur, NO, NO_2, NO_3, N_2O, N_2O_3, N_2O_4, and N_2O_5—the only two that are important air pollutants are nitric oxide (NO) and nitrogen dioxide (NO_2). As mentioned earlier, there are two sources of nitrogen oxides (or NO_x) when fossil fuels are burned. *Thermal NO_x* is created when nitrogen and oxygen in the combustion air are heated to a high enough temperature (about 1000 K) to oxidize the nitrogen. *Fuel NO_x* results from the oxidation of nitrogen compounds that are chemically bound in the fuel molecules themselves. Different fuels have different amounts of nitrogen in them, with natural gas having almost none and some coal having as much 3 percent by weight. Both thermal NO_x and fuel NO_x can be significant contributors to the total NO_x emissions, but fuel NO_x often is the dominant source.

Almost all NO_x emissions are in the form of NO, which is a colorless gas that has no known adverse health effects at concentrations found in the atmosphere. However, NO can oxidize to NO_2, and NO_2 can irritate the lungs, cause bronchitis and pneumonia, and lower resistance to respiratory infections. NO_x also can react with volatile organic compounds in the presence of sunlight to form photochemical oxidants that have adverse health consequences as well.

Nitrogen dioxide has other environmental consequences besides those directly associated with human health. It reacts with the hydroxyl radical (OH) in the atmosphere to form nitric acid (HNO_3), which corrodes metal surfaces and contributes to the acid rain problem. It also can cause damage to terrestrial plants and is a significant cause of eutrophication, especially in nitrogen-limited estuaries such as Chesapeake Bay. Nitrogen dioxide is responsible for the reddish-brown color in the smog that hovers over cities like Los Angeles.

Reductions in NO_x emissions have been harder to come by than reductions in other criteria pollutants. In fact, emissions have been modestly increasing for several decades, as was indicated in Figure 7.4. More recently, during the decade 1984 to 1993, NO_x emissions increased by about 1 percent. In view of the fact that transportation sources contribute roughly half of these emissions and vehicle miles traveled increased by one-third, this is still quite an accomplishment. As will be discussed further when mobile source controls are introduced, modifications to the combustion process that reduce emissions of carbon monoxide tend to make the NO_x problem worse, and vice versa. Controlling both at the same time has been a challenge.

Despite slowly rising emissions of NO_x, every monitoring location in the United States, including Los Angeles, met the NAAQS for NO_2 from 1992 through 1994 (U.S. EPA, 1995a).

Volatile Organic Compounds

This class of compounds includes unburnt hydrocarbons that are emitted from tailpipes and smoke stacks when fossil fuels are not completely combusted along with gaseous hydrocarbons that enter the atmosphere when solvents, fuels, and other organics evaporate. In addition, there are natural sources of reactive hydrocarbons such as deciduous trees and shrubs that emit isoprene, and conifers that emit pinene and limonene. In most urban areas, however, those natural sources provide only a small fraction of the hydrocarbons that exist in polluted air.

The transportation sector is responsible for about one-third of anthropogenic VOC emissions, and despite rapidly rising miles driven in motor vehicles, emissions decreased by 25 percent in the decade of 1984 to 1993. Reformulated gasolines that evaporate less easily and the decline in the percentage of dirtier, older vehicles on the road account for much of the drop. As emissions from motor vehicles become less significant, other sources, such as gasoline-powered lawnmowers, outboard motors, barbeque starter fluids, and oil-based paints, begin to look more important and are beginning to be regulated as well.

Industrial sources account for two-thirds of VOC emissions, with much of that again being caused by vaporization of hydrocarbons. Less than 2 percent of VOCs result from fossil-fuel combustion in power plants and industrial boilers.

Photochemical Smog And Ozone

When oxides of nitrogen, VOCs, and sunlight come together, they can initiate a complex set of reactions that produce a number of secondary pollutants known as photochemical oxidants. Ozone (O_3) is the most abundant of the photochemical oxidants, and it is the one for which an ambient air quality standard has been written. Although it is responsible for many of the undesirable properties of photochemical smog, from chest constriction and irritation of the mucous membrane in people to the cracking of rubber products and damage to vegetation, it is not itself a cause of the eye irritation that is our most common complaint about smog. Eye irritation is caused by other components of photochemical smog, principally formaldehyde (HCHO), peroxybenzoyl nitrate (PBzN), peroxyacetyl nitrate (PAN), and acrolein (CH_2CHCOH).

In the very simplest of terms, we can express the formation of photochemical smog as

$$\text{VOCs} + \text{NO}_x + \text{Sunlight} \rightarrow \text{Photochemical smog} \qquad (7.7)$$

The reaction in (7.7) only gives us the simplest overview. We can add a few details to give a sense of the key reactions involved, but a complete analysis is far beyond the scope of this book.

The $NO-NO_2-O_3$ Photochemical Reaction Sequence.   Consider some of the important reactions involving NO_x without the complications associated with the added hydrocarbons. We can begin with the formation of NO during combustion (for simplicity, we shall just show the thermal NO_x reaction):

$$N_2 + O_2 \rightarrow 2\,NO \qquad (7.8)$$

The nitric oxide thus emitted can oxidize to NO_2:

$$2\,NO + O_2 \rightarrow 2\,NO_2 \qquad (7.9)$$

If sunlight is available, a photon with the right amount of energy can decompose NO_2 in a process called *photolysis*.

$$NO_2 + h\nu \rightarrow NO + O \qquad (7.10)$$

where $h\nu$ represents a photon (with wavelength $\lambda < 0.39 \ \mu m$). The freed atomic oxygen (O) can then combine with diatomic oxygen (O_2) to form ozone (O_3):

$$O + O_2 + M \rightarrow O_3 + M \qquad (7.11)$$

where M represents a molecule (usually O_2 or N_2 since they are most abundant in air) whose presence is necessary to absorb excess energy from the reaction. Without M, the ozone would have too much energy to be stable, and it would dissociate back to O and O_2.

Ozone can then convert NO back to NO_2:

$$O_3 + NO \rightarrow NO_2 + O_2 \qquad (7.12)$$

Notice the general tendency for NO_2 to create O_3 (7.10) and (7.11), while NO tends to destroy O_3 (7.12). This set of reactions creates a cycle that is represented in Figure 7.7.

Figure 7.7, even without consideration of the effect of hydrocarbons, helps explain the sequence of stages through which atmospheric NO, NO_2, and O_3 progress on a typical smoggy day. The diagram suggests that we might expect NO concentrations to rise as early morning traffic emits its load of NO. Then, as the morning progresses, we would expect to see a drop in NO and a rise in NO_2 as NO gets converted

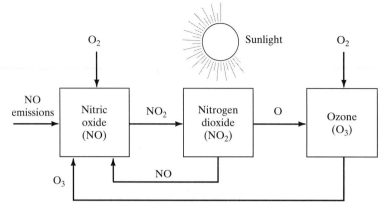

FIGURE 7.7 Simplified atmospheric nitrogen photolytic cycle.

to NO_2. As the sun's intensity increases toward noon, the rate of photolysis of NO_2 increases; thus NO_2 begins to drop while O_3 rises. Ozone is so effective in its reaction with NO (7.12) that as long as there is O_3 present, NO concentrations do not rise through the rest of the afternoon even though there may be new emissions.

The nitrogen dioxide photolytic cycle helps provide an explanation for the sequence: NO to NO_2 to O_3 observable in both laboratory smog chambers and in cities such as Los Angeles (see Figure 7.8). A careful analysis of the reactions, however, pre-

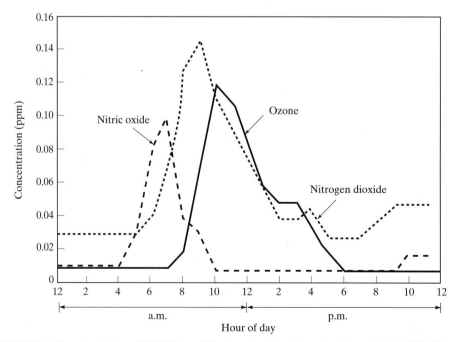

FIGURE 7.8 Diurnal variation of NO, NO_2, and O_3 concentrations in Los Angeles on July 19, 1965.(*Source*: U.S. HEW, 1970)

dicts O_3 concentrations that are much lower than those frequently found on smoggy days. If only the NO_2 photolytic cycle is involved, O_3 cannot accumulate in sufficient quantity to explain actual measured data. By introducing certain types of hydrocarbons into the cycle, however, the balance of production and destruction of O_3 can be upset, which allows more O_3 to accumulate, closing the gap between theoretical and actual levels.

Hydrocarbons and NO_x.   Expanding the nitrogen dioxide photolytic cycle to include hydrocarbons and other organics helps explain the increase in ozone above what would be predicted by the NO_x cycle alone. It also enables us to account for some of the other objectionable organic compounds found in photochemical smog.

The chemistry of photochemical smog is exceedingly complex. A multitude of organic chemicals are introduced to the atmosphere when fuels burn or volatiles evaporate, and many more are produced in the atmosphere as a result of chemical reactions. To help describe some of these reactions, let us begin with a very brief explanation of some of the nomenclature and notation used in organic chemistry, building on what was introduced in Chapter 2.

We begin by considering atoms or molecules with an odd number of electrons, called *free radicals*. Having an odd number of electrons means that one electron is not being used as a bonding electron to other atoms. For example, a carbon atom bonded to three hydrogens has one leftover electron. We can represent that molecule with the notation $CH_3{}^\bullet$, where the dot suggests the unpaired electron. Free radicals tend to be very reactive, and they are very important in the study of air pollution.

As described in Section 2.4, the *alkanes* are hydrocarbons in which each carbon forms single bonds with other atoms. The alkane series is the familiar sequence: methane (CH_4), ethane (C_2H_6), propane (C_3H_8), ... (C_nH_{2n+2}). If one of the hydrogens is removed from an alkane, the resulting free radical is called an *alkyl*. The alkyls then form a series beginning with methyl ($CH_3{}^\bullet$), ethyl ($C_2H_5{}^\bullet$), and so on. We could represent an alkyl with the general expression $C_nH_{2n+1}{}^\bullet$, but it is more convenient to call it simply $R^\bullet$.

Another basic chemical unit that comes up over and over again in the study of photochemical smog is a *carbonyl*, a carbon atom with a double bond joining it to an oxygen, as shown in the following diagram. A carbonyl with one bond shared with an alkyl, $R^\bullet$, and the other with a hydrogen atom forms an *aldehyde*. Aldehydes can thus be written as RCHO. The simplest aldehyde is *formaldehyde* (HCHO), which corresponds to $R^\bullet$ being just a single hydrogen atom. A more complex aldehyde is acrolein,

$$
\begin{array}{cccc}
\text{O} & \text{O} & \text{O} & \text{O} \\
\| & \| & \| & \| \\
-\,\text{C}\,- & \text{R}-\text{C}-\text{H} & \text{H}-\text{C}-\text{H} & \text{CH}_2=\text{CH}-\text{C}-\text{H} \\
\text{Carbonyl} & \text{Aldehyde} & \text{Formaldehyde} & \text{Acrolein}
\end{array}
$$

as shown. Both formaldehyde and acrolein are eye-irritating components of photochemical smog.

Another important key to understanding atmospheric organic chemistry is the hydroxyl radical $OH^{\bullet}$, which is formed when atomic oxygen reacts with water.

$$O \ + \ H_2O \ \rightarrow \ 2OH^{\bullet} \tag{7.13}$$

The OH radical is extremely reactive, and its atmospheric concentration is so low that it has been difficult to detect. Nevertheless, it plays a key role in many reactions, including the oxidations of NO_2 to nitric acid and CO to CO_2.

$$OH^{\bullet} \ + \ NO_2 \ \rightarrow \ HNO_3 \tag{7.14}$$

$$OH^{\bullet} \ + \ CO \ \rightarrow \ CO_2 \ + \ H^{\bullet} \tag{7.15}$$

In fact, the hydroxyl radical is responsible for initiating the oxidation of VOCs and most of the other important atmospheric pollutants. It is therefore crucial to the natural removal of pollution from the atmosphere.

It was mentioned previously that the NO_2 photolytic cycle, by itself, underpredicts the observed concentrations of O_3. As that cycle is described in (7.10) to (7.12), the availability of NO_2 affects the rate of production of O_3 while the availability of NO affects the rate of destruction of O_3. The balance of O_3 production and destruction can be upset if there are other reactions that will enhance the rate of conversion of NO to NO_2. Any reactions that will help convert NO to NO_2 will increase O_3 concentrations both by reducing the amount of NO available to destroy O_3 and increasing the amount of NO_2 available to create O_3.

The following three reactions provide one explanation for the way that hydrocarbons can enhance the rate of conversion of NO to NO_2 and hence increase O_3 concentrations. Starting with a hydrocarbon RH, we have

$$RH \ + \ OH^{\bullet} \ \rightarrow \ R^{\bullet} \ + \ H_2O \tag{7.16}$$

$$R^{\bullet} \ + \ O_2 \ \rightarrow \ RO_2^{\bullet} \tag{7.17}$$

$$RO_2^{\bullet} \ + \ NO \ \rightarrow \ RO^{\bullet} \ + \ NO_2 \tag{7.18}$$

These reactions are not complete, though they do show how hydrocarbons can help convert NO to NO_2 and thus increase O_3. As written, an already scarce hydroxyl $OH^{\bullet}$ is required to start the chain, and it appears to be destroyed in the process. Unless there is some way to rejuvenate that $OH^{\bullet}$, these reactions could not continue for long. The following pair of reactions shows one way that not only $OH^{\bullet}$ is regenerated, but also, in the process, how another NO is converted to NO_2. An aldehyde can also be formed from this pair of reactions:

$$RO^{\bullet} \ + \ O_2 \ \rightarrow \ HO_2^{\bullet} \ + \ R'CHO \tag{7.19}$$

$$HO_2^{\bullet} \ + \ NO \ \rightarrow \ NO_2 \ + \ OH^{\bullet} \tag{7.20}$$

where R' is the hydrocarbon that balances (7.19).

The net effect of reactions (7.16) through (7.20) is that one hydrocarbon molecule converts two molecules of NO to NO_2 and produces an aldehyde $R'CHO$. The removal of NO by these reactions slows the rate at which O_3 is removed, while the addition of NO_2 increases the rate at which it is produced, which allows higher levels of O_3 to accumulate in the air. These are summarized in Figure 7.9.

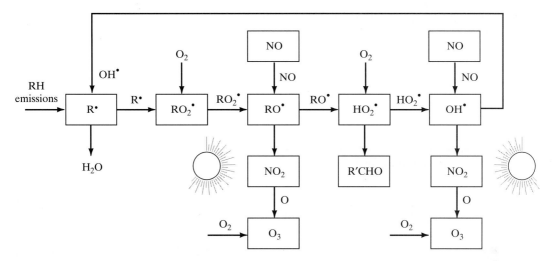

FIGURE 7.9 Showing one way that hydrocarbons can cause NO to convert to NO_2. Reducing NO slows the removal of O_3, while increasing NO_2 increases the production of O_3, so this cycle, when combined with Figure 7.7, helps account for elevated atmospheric O_3 levels.

EXAMPLE 7.4.   Ethane to Acetaldehyde

Suppose the hydrocarbon that begins reactions (7.16) through (7.20) is ethane (C_2H_6). Write the sequence of reactions.

Solution   The hydrocarbon RH that appears in (7.16) is ethane, C_2H_6, so the free radical $R^\bullet$ is $C_2H_5{}^\bullet$. Reaction (7.16) thus becomes

$$C_2H_6 \ + \ OH^\bullet \ \rightarrow \ C_2H_5{}^\bullet \ + \ H_2O$$

Reactions (7.17) to (7.19) are

$$C_2H_5{}^\bullet \ + \ O_2 \ \rightarrow \ C_2H_5O_2{}^\bullet$$
$$C_2H_5O_2{}^\bullet \ + \ NO \ \rightarrow \ C_2H_5O^\bullet \ + \ NO_2$$
$$C_2H_5O^\bullet \ + \ O_2 \ \rightarrow \ HO_2{}^\bullet \ + \ CH_3CHO$$

Notice that R' in reaction (7.19) is thus $CH_3{}^\bullet$. Finally, (7.20) is

$$HO_2{}^\bullet \ + \ NO \ \rightarrow \ NO_2 \ + \ OH^\bullet$$

The CH_3CHO produced is an aldehyde called *acetaldehyde*. As we shall see, it plays a role in the formation of the eye irritant peroxyacetyl nitrate (PAN). ◾

There are other ways that NO can be converted to NO_2. Carbon monoxide, for example, can do it too, in reactions that are similar to those given previously. Starting with CO and $OH^\bullet$, as in (7.15),

$$OH^\bullet \ + \ CO \ \rightarrow \ CO_2 \ + \ H^\bullet \tag{7.15}$$

The hydrogen atom then quickly combines with O_2 to form the hydroperoxyl radical, $HO_2{}^\bullet$:

$$H^\bullet + O_2 \rightarrow HO_2{}^\bullet \tag{7.21}$$

Now, going back to (7.20), NO is converted to NO_2:

$$HO_2{}^\bullet + NO \rightarrow NO_2 + OH^\bullet \tag{7.20}$$

So we see one way that another of our criteria pollutants, CO, can contribute to the photochemical smog problem. By increasing the rate at which NO is converted to NO_2, CO aids in the accumulation of O_3.

We can also extend these relationships to show the formation of another of the eye irritants, peroxyacetyl nitrate (PAN).

$$CH_3C(O)O_2NO_2$$

Peroxyacetyl nitrate (PAN)

$$
\begin{array}{ccc}
 & H & O \\
 & | & \| \\
H - & C - C & - O - O - NO_2 \\
 & | & \\
 & H & \\
\end{array}
$$

Acetaldehyde, CH_3CHO, which Example 7.4 indicated can be formed from ethane emissions, can react with $OH^\bullet$:

$$CH_3CHO + O_2 + OH^\bullet \rightarrow CH_3C(O)O_2{}^\bullet + H_2O \tag{7.22}$$

The resulting acetylperoxy radical can react with NO_2 to create PAN:

$$CH_3C(O)O_2{}^\bullet + NO_2 \rightarrow CH_3C(O)O_2NO_2 \tag{7.23}$$

It must be pointed out once again that the preceding reactions are only a very limited description of the complex chemistry going on each day in the atmosphere over our cities.

Sources and Effects of Photochemical Oxidants.   Since ozone and other photochemical oxidants are not primary pollutants, there is no direct way to specify emissions. An annual emissions inventory is kept by the EPA, however, for the precursors to photochemical oxidants—namely, nitrogen oxides and volatile organic compounds (VOCs). When all precursors are considered, especially if an adjustment is made for the sources' proximity to cities, the automobile is the dominant cause of photochemical smog. Conditions in Los Angeles, for example, are ideal for smog formation: The near total dependence on the automobile for transportation results in high hydrocarbon and NO_x emissions; there are long-lasting atmospheric inversions that restrict the vertical dispersion of pollutants; a ring of mountains nearly surrounds the city on three sides, reducing the horizontal dispersion; and there is an abundance of sunshine to power the photochemical reactions.

Photochemical smog is known to cause many annoying respiratory effects, such as coughing, shortness of breath, airway constriction, headache, chest tightness, and eye, nose, and throat irritation. These symptoms can be especially severe for asthmatics and others with impaired respiratory symptoms, but even healthy individuals who engage in strenuous exercise for relatively modest periods of time experience these symptoms at levels near the ambient air quality standard. Animal studies suggest that long-term exposures to ozone can lead to permanent scarring of lung tissue, loss of lung function, and reduced lung elasticity. As mentioned earlier, the Clean Air Act requires EPA to set air quality standards at levels that protect public health, without regard to cost. Accumulating evidence suggests the two-decades-old 0.12 ppm ozone standard did not adequately protect some individuals, especially children and those with asthma. In June, 1997, after considerable public debate, EPA strengthened the ozone standard to 0.08 ppm contending that 1 million fewer incidences of decreased lung function in children each year will result.

Ozone has been shown to cause damage to tree foliage and to reduce growth rates of certain sensitive tree species. It also reduces yields of major agricultural crops, such as corn, wheat, soybeans, and peanuts. Ozone alone is thought to be responsible for about 90 percent of all of the damage that air pollutants cause to agriculture, with a total economic cost that has been estimated at 6 to 7 percent of U.S. agricultural productivity (OTA, 1984).

Particulate matter

Atmospheric *particulate matter* consists of any dispersed matter, solid or liquid, in which the individual aggregates range from molecular clusters of 0.005 μm diameter to coarse particles up to about 100 μm (roughly the size of a human hair). As a category of criteria pollutant, particulate matter is extremely diverse and complex since size and chemical composition, as well as atmospheric concentration, are important characteristics.

A number of terms are used to categorize particulates, depending on their size and phase (liquid or solid). The most general term is *aerosol*, which applies to any tiny particles, liquid or solid, dispersed in the atmosphere. Solid particles are called *dusts* if they are caused by grinding or crushing operations. Solid particles are called *fumes* if they are formed when vapors condense. Liquid particles may be called *mist* or, more loosely, *fog*. *Smoke* and *soot* are terms used to describe particles composed primarily of carbon that result from incomplete combustion. *Smog* is a term that was derived from smoke and fog, originally referring to particulate matter, but now describing air pollution in general.

Although particles may have very irregular shapes, their size can be described by an equivalent *aerodynamic diameter* determined by comparing them with perfect spheres having the same settling velocity. The particles of most interest have aerodynamic diameters in the range of 0.1 μm to 10 μm (roughly the size of bacteria). Particles smaller than these undergo random (Brownian) motion and, through coagulation, generally grow to sizes larger than 0.1 μm. Particles larger than 10 μm settle quite quickly; a 10 μm particle, for example, has a settling velocity of approximately 20 cm/min.

We can use a fairly simple analysis to calculate the settling velocity of a spherical particle. When such a particle reaches its terminal velocity, the gravitational force pulling it down is balanced by a drag force that we can estimate. For particles that are smaller than about 30 μm, with density much greater than air, we can use a simplified version of Stoke's law to approximate the drag force:

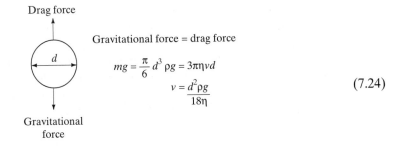

Drag force

Gravitational force = drag force

$$mg = \frac{\pi}{6} d^3 \rho g = 3\pi\eta v d$$

$$v = \frac{d^2 \rho g}{18\eta} \qquad (7.24)$$

Gravitational force

where

m = mass of the particle (g)
g = gravitational acceleration = 9.80 m/s^2
d = particle diameter (m)
ρ = particle density (g/m^3)
η = viscosity of air = 0.0172 g/m · s
v = settling velocity (m/s)

EXAMPLE 7.5   Settling velocity of a spherical particle

Find the settling velocity of a spherical droplet of water with diameter 2 μm, and estimate the residence time of such particles if they are uniformly distributed in the lower 1000 m of atmosphere and their removal rate is determined by how fast they settle in still air.

Solution   Using (7.24), with the density of water equal to 10^6 g/m^3, gives

$$v = \frac{d^2 \rho g}{18\eta} = \frac{(2 \times 10^{-6}\text{m})^2 \cdot (10^6 \text{g/m}^3) \cdot (9.8\text{m/s}^2)}{18 \times 0.0172\text{g/m} \cdot \text{s}}$$

$$v = 1.27 \times 10^{-4}\text{m/s}$$

which is about 0.5 m/hr.

We can use a simple box model to estimate the residence time τ of N particles uniformly distributed in a box of atmosphere with height h (m). We need to know the rate of removal of particles:

$$\text{Rate of removal of particles} = \frac{N(\text{particles})}{h(\text{m})} \cdot v(\text{m/s}) = \frac{Nv}{h} \text{ particles/s}$$

The residence time τ is the ratio of the number of particles in the box divided by the rate of removal of particles:

$$\tau = \frac{\text{Particles in box}}{\text{Rate of removal of particles}} = \frac{N}{(Nv/h)} = \frac{h}{v}$$

$$\tau = \frac{1000 \text{ m}}{1.27 \times 10^{-4} \text{ m/s}} = 7.9 \times 10^6 \text{ s} \approx 91 \text{ days}$$

In other words, small particles in the atmosphere do not settle very quickly at all and an estimate of residence time in terms of months is appropriate. ∎

The ability of the human respiratory system to defend itself against particulate matter is, to a large extent, determined by the size of the particles. To help understand these defense mechanisms, consider the illustration in Figure 7.10 . The upper respiratory system consists of the nasal cavity and the trachea, while the lower respiratory system consists of the bronchial tubes and the lungs themselves. Each bronchus divides over and over again into smaller and smaller branches, terminating with a large number of tiny air sacs called alveoli.

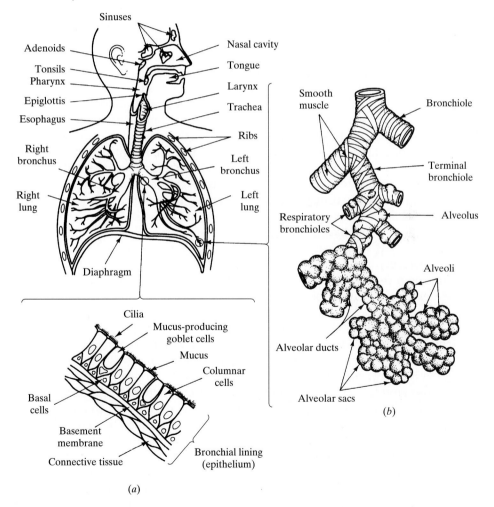

FIGURE 7.10   The human respiratory system. (a) The system as a whole and a cross section of the bronchial lining showing the cilia; (b) details of the lower respiratory system terminating in the alveoli. (*Source:* Williamson, 1973)

Large particles that enter the respiratory system can be trapped by the hairs and lining of the nose. Once captured, they can be driven out by a cough or sneeze. Smaller particles that make it into the tracheobronchial system can be captured by mucus, worked back to the throat by tiny hairlike *cilia*, and removed by swallowing or spitting. Particles larger than about 10 μm are effectively removed in the upper respiratory system by these defense mechanisms. Smaller particles, however, are often able to traverse the many turns and bends in the upper respiratory system without being captured on the mucous lining. These particles may make it into the lungs, but depending on their size, they may or may not be deposited there. Some particles are so small that they tend to follow the air stream into the lungs and then right back out again. Other submicron-size particles are deposited by Brownian motion. Roughly between 0.5 μm and 10 μm, particles may be small enough to reach the lung and large enough to be deposited there by sedimentation. Sedimentation in the lungs is most effective for particles between 2 μm and 4 μm.

The original NAAQS for particulates did not take size into account. Larger particles could dominate the mass per unit volume measure but be unimportant in terms of human health risk. In 1987, however, the PM 10 standard was introduced, and in 1997 the PM 2.5 standard was added. EPA contends that the new fine particle PM 2.5 standard will prevent approximately 15,000 premature deaths each year.

While it is probably the black soot we see emitted from buses and trucks or dark smoke pouring out of a power plant stack that first come to mind when thinking of particulate matter, such sources contribute a relatively small, but extremely important, portion of total particulate emissions. Other so-called fugitive sources, such as wildfires, wind-driven soil erosion, and dust from croplands, construction, mining activities, and the constant abrasion of paved and unpaved roadways account for roughly 95 percent of national particulate emissions. Many of those sources, however, are some distance from heavily populated urban centers, and the particle sizes tend to be large so their impact on human health is much less than the total tons of emissions would imply.

The PM 10 emissions tracked by the EPA are associated with fuel combustion (45 percent), industrial processing (33 percent), and transportation (22 percent). Emissions from these sources have decreased somewhat in the past few years in part due to better controls on diesel engines and restrictions on fireplace and woodstove use in homes.

The black smoke, or soot, we see emitted from diesel engines and smokestacks consists mostly of solid particles made up of vast numbers of carbon atoms fused together in benzene rings. While these particles themselves can irritate lungs, most often it is large organic molecules that stick to the surfaces of soot (adsorption) that are responsible for the most serious health effects. Of greatest concern is a class of benzenelike hydrocarbons called *polynuclear* (or polycyclic) *aromatic hydrocarbons* (PAHs). Polynuclear aromatic hydrocarbons consist of fused benzene rings, the simplest of which is napthalene, shown in Figure 7.11. PAHs are formed when carbon-containing materials are not completely oxidized during combustion. They are released as gases but quickly condense onto particles of soot. Common sources of PAHs include tobacco smoke, motor vehicle exhaust, the char on charcoal-broiled food, and smoke from wood and coal combustion. Probably the most dangerous of these PAHs is a substance called benzo[a]pyrene (BaP), which consists of five fused rings, as shown in Figure 7.11*c*. BaP is a Category A human carcinogen known to cause lung and kidney cancer.

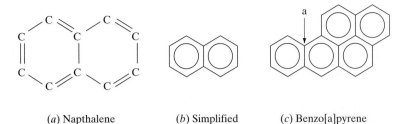

(*a*) Napthalene   (*b*) Simplified   (*c*) Benzo[a]pyrene

FIGURE 7.11   Polynuclear aromatic hydrocarbons consist of fused benzene rings. (*a*) Two fused benzene rings (napthalene) shown without their hydrogen bonds and (*b*) their simplified representation; (*c*) benzo[a]pyrene, showing the "a" connection point.

Many particulates, or aerosols, consist of sulfates (SO_4) that condense onto droplets of water. These sulfate aerosols are highly acidic and contribute to respiratory distress as will be described in the next section. They contribute to degradation of materials and are a major cause of reduced visibility. As will be described in the next chapter, by reflecting sunlight back toward space, sulfate aerosols offset some of the global warming being caused by accumulating greenhouse gases.

Particulates aggravate existing respiratory and cardiovascular disease and damage lung tissue, and some are carcinogenic. At concentrations near the national ambient standard, increases in the number of hospital visits for upper respiratory infections, cardiac disorders, bronchitis, asthma, pneumonia, and emphysema, as well as increased mortality rates, have been observed. A recent study estimated that hospital admissions for congestive heart failure (an inability of the heart to pump out all the blood that returns to it) increased 8 percent for each 100 $\mu m/m^3$ increase in PM 10 (Raloff, 1995). Evidence that tens of thousands of premature deaths are caused each year in the United States due to inhaled particles forced the EPA to reevaluate its particulate air quality standard, which resulted in the 1997 PM 2.5 standard.

Oxides of Sulfur

Almost 90 percent of the 22 million tons per year of anthropogenic sulfur oxide emissions are the result of fossil fuel combustion in stationary sources. Of that, almost 85 percent is emitted from electric utility power plants (16 million tons/yr). Only about 3 percent comes from highway vehicles. The only significant noncombustion sources of sulfur emissions are associated with petroleum refining, copper smelting, and cement manufacture. Total sulfur oxide emissions decreased by about 20 percent from 1970 to 1984, but since then the rate of decrease has been slowed considerably. The Clean Air Act Amendments of 1990 focus on emissions from coal-fired electric power plants, with the goal of cutting SO_2 emissions by 10 million tons compared to 1980 emissions.

All fossil fuels as they are extracted from the ground contain some sulfur. Coal, which has the most, typically contains from 1 to 6 percent sulfur. About half of that is organic sulfur that is chemically bound to the coal. The other half is simply physically trapped in the noncarbon portion of coal, and much of that half can be removed by pulverizing and washing the coal before combustion. The amount of sulfur in petroleum tends to be less than a few percent, and if it is refined almost all of that sulfur is removed during processing. Gasoline, for example, has much less than 1 ppm sulfur.

Natural gas as it leaves the wellhead contains a considerable amount of sulfur in the form of highly toxic hydrogen sulfide (H_2S), which must be removed before the gas can be used. Once natural gas is cleaned, however, it has negligible amounts of sulfur, which makes it a highly desirable replacement fuel for coal.

When these fuels are burned, the sulfur is released, mostly as sulfur dioxide (SO_2) but also with small amounts of sulfur trioxide (SO_3). Sulfur dioxide, once released, can convert to SO_3 in a series of reactions that, once again, involve a free radical such as $OH^\bullet$:

$$SO_2 + OH^\bullet \rightarrow HOSO_2^\bullet \tag{7.25}$$

$$HOSO_2^\bullet + O_2 \rightarrow SO_3 + HO_2^\bullet \tag{7.26}$$

The $HO_2^\bullet$ radical can then react with NO to return the initial $OH^\bullet$, as in (7.20). Sulfur trioxide reacts very quickly with H_2O to form sulfuric acid, which is the principal cause of acid rain.

$$SO_3 + H_2O \rightarrow H_2SO_4 \tag{7.27}$$

Sulfuric acid molecules rapidly become particles by either condensing on existing particles in the air or by merging with water vapor to form H_2O—H_2SO_4 droplets. Often a significant fraction of particulate matter in the atmosphere consists of such sulfate (SO_4) aerosols.

The transformation from SO_2 gas to sulfate particles (SO_4) is gradual, taking a matter of days. During that time, sulfur pollution may be deposited back onto the land or into water, either in the form of SO_2 or sulfate. In either form, sulfur pollution can be deposited by removal during precipitation (wet deposition) or by slow, continuous removal processes that occur without precipitation (dry deposition). Figure 7.12, suggests the effects of time and distance on the conversion and deposition of sulfur.

Figure 7.13 shows contours of pH for wet deposition in the United States and Canada. Recall from Chapter 2 that natural rainfall would have a pH value between 5 and 5.6, and anything less is loosely called "acid rain." A comparison between Figure 7.13 and Figure 5.25 shows that large areas of the eastern United States and Canada that unfortunately have the worst acid rain have lakes that are inherently the most sensitive to acidification. About two-thirds of U.S. coal consumption, the source of most sulfur emissions, is east of the Mississippi—a fact that correlates well with the acidity of rainfall in the eastern half of the United States, as shown in Figure 7.13.

Most sulfate particles in urban air have an effective size of less than 2 μm, with most of them being in the range of 0.2 to 0.9 μm. Their size is comparable to the wavelengths of visible light, and their presence greatly affects visibility. Their size also allows deep penetration into the respiratory system.

Sulfur dioxide is highly water soluble, much more so than any of the other criteria pollutants. As a result, when it is inhaled it is most likely to be absorbed in the moist passages of the upper respiratory tract, the nose and upper airways, where it does less long-term damage. Other gases, being less soluble, are more likely to reach the more critical terminal air sacs of the lungs. When sulfur is entrained in an aerosol, however, the aerodynamic properties of the particles themselves affect the area of deposition, and it is possible for sulfur oxides to reach far deeper into the lungs. The combination of particulate matter and sulfur oxides can then act synergistically, with the effects of both together being much more detrimental than either of them separately. In fact, in

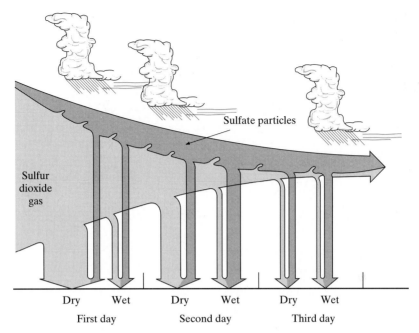

FIGURE 7.12   The effects of time and distance on conversion and deposition of sulfur pollution. (*Source*: OTA, 1984).

every major air pollution episode, the combination of sulfur oxides and particulates has been implicated as a cause of the excess mortality observed. The magnitude of the health risk posed by current levels of sulfates and other particulates has been estimated at 50,000 premature deaths (2 percent of total deaths) per year in the United States and Canada (OTA, 1984).

Sulfur oxides can damage trees, especially when trees are bathed in acid fog or clouds that tend to have very low pH levels. Acidification also damages plants by affecting their ability to extract nutrients from the soil. Nutrients are leached from soils more readily under acidic conditions, and low pH levels allow aluminum to solubilize, which interferes with the uptake of nutrients. Sulfurous pollutants can discolor paint, corrode metals, and cause organic fibers to weaken. Airborne sulfates significantly reduce visibility and discolor the atmosphere. Most of the visibility impairment in the Eastern United States is caused by sulfates, while in the West it is more often nitrogen oxides and dust.

Prolonged exposure to sulfates causes serious damage to building marble, limestone, and mortar, as the carbonates (e.g., limestone, $CaCO_3$) in these materials are replaced by sulfates. The reaction between limestone and sulfuric acid shows such a replacement:

$$CaCO_3 + H_2SO_4 \rightarrow CaSO_4 + CO_2 + H_2O \qquad (7.28)$$

The calcium sulfate (gypsum, $CaSO_4$) produced by this reaction is water soluble and easily washes away, leaving a pitted, eroded surface. Many of the world's historic buildings and statues are rapidly being degraded due to this exposure. It is common now in

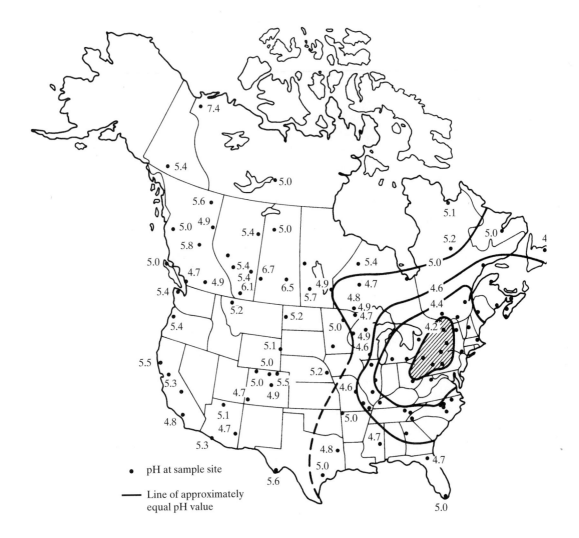

FIGURE 7.13   The pH of wet deposition in 1982 (precipitation-weighted annual average). (*Source*: Interagency Task Force on Acid Precipitation, 1983)

such monuments as the Acropolis in Greece for the original outdoor statuary to be moved into air conditioned museums, leaving plaster replicas in their place.

Lead

Most lead emissions in the past were from motor vehicles burning gasoline containing the antiknock additive tetraethyllead $Pb(C_2H_5)_4$. The antiknock properties of tetraethyllead were first discovered by Thomas Midgley and colleagues at the General

Motors Research Laboratory in 1921; by coincidence, in the same laboratory, Midgley also developed the first chlorofluorocarbons (Thomas, 1995).

In the United States (and just a handful of other developed countries), almost all lead emissions from gasoline have been eliminated so that by 1993, per capita emissions of lead in the United States were only 3 percent of the world average (Thomas, 1995). The decrease in the United States was originally motivated by The Clean Air Act Amendments of 1970, which dictated a 90 percent drop in CO, NO_x, and hydrocarbon emissions from automobiles. As will be described later in this chapter, the auto industry chose to use catalytic converters as their principal emission control system for those three pollutants. Catalytic converters, it turns out, are quickly rendered ineffective when exposed to lead, so beginning in the late 1970s many cars were designed to only burn unleaded fuels. By 1984 a good fraction of the motor vehicle fleet had these converters, and the use of leaded fuel dropped but was not entirely eliminated. To speed the reduction in lead emissions, in 1984 the EPA lowered the allowable lead content of leaded gasoline by over 90 percent (from 0.30 to 0.026 g/L). By the mid-1990's, almost all cars in the United States had catalytic converters and only about 1 percent of gasoline sold still had lead additives. The result has been a 95 percent reduction in lead emissions from motor vehicles since 1984 and an overall drop of 88 percent of all lead emissions, as shown in Figure 7.14. Unfortunately, in most of the rest of the world lead emissions remain very high.

With the nearly complete elimination of leaded gasoline in the United States, the remaining sources of ambient (outdoor) lead are mostly point sources rather than area-wide sources. Ambient lead levels tend to be high in the vicinity of industrial facilities such as metal smelters and plants that manufacture lead-acid batteries. The NAAQS for lead has been set at 1.5 mg/m^3 averaged over a three-month period. The only areas of the United States that exceed this standard are in the vicinity of Cleveland, Indianapolis, Memphis, Omaha, Philadelphia, and St. Louis. Figure 7.15 shows the maximum quarterly lead concentrations around significant point sources in the United States.

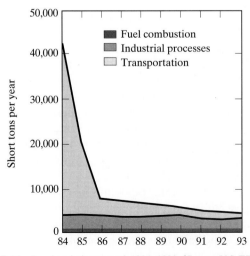

FIGURE 7.14   Lead emissions trend, 1984–1993. (*Source*:U.S. EPA, 1994)

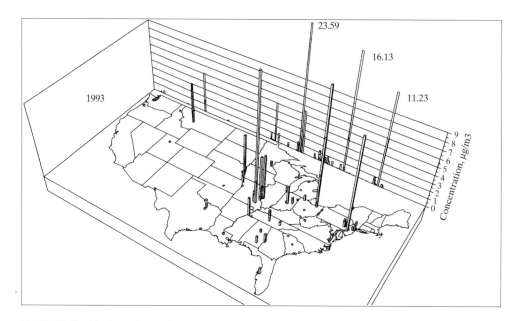

FIGURE 7.15   Maximum quarterly mean lead concentrations in the vicinity of lead point sources, 1993. (*Source*: U.S. EPA, 1994b)

Lead is emitted into the atmosphere primarily in the form of inorganic particulates. Much of this is removed from the atmosphere by settling in the immediate vicinity of the source. Airborne lead may affect human populations by direct inhalation, in which case people living nearest to major sources such as highways or metals processing plants are at greatest risk. Despite the elimination of most lead emissions from motor vehicles, the soil around highways is still heavily contaminated and it can become airborne when disturbed. It also can be tracked into homes, where it may end up imbedded in carpeting ready to become airborne once again. Wiping your feet on a doormat is one easy and effective way to help reduce that hazard.

Inhalation of lead can occur indoors as well as outdoors. A major indoor source of lead is chipped and flaking particles of lead-based paints that were commonly used in the past. Paint containing $Pb_3(CO_3)_2(OH)_2$ was widely used in white paint, and red paints with Pb_3O_4 are still used for outdoor protection of metal surfaces. Those substances are no longer allowed for paints used indoors, but they can be used for exterior surfaces. When those paints chip, peel, or are sanded, lead particles become dusts that are easily ingested or inhaled. Chips of leaded paint are somewhat sweet and are all too often eaten by children living in older homes.

While most human exposure to lead is from inhalation, it can also be ingested after airborne lead is deposited onto soil, water, and food crops such as leafy vegetables and fruits. It also can be ingested when water systems are contaminated. Lead used to be used to make water pipes and it was used until recently in solder to join copper water pipes, but those uses have been banned in new systems in the United States. Lead can still leach out of those older water systems, especially if the water is acidic or particularly soft, and for such systems running the water from a tap for a minute before drinking it is highly recommended.

Lead poisoning can cause aggressive, hostile, and destructive behavioral changes as well as learning disabilities, seizures, severe and permanent brain damage, and even death. Measurements in actual communities suggest that an increase in airborne lead concentration of 1 $\mu g/m^3$ (the NAAQS is 1.5 $\mu g/m^3$) results in an increase of about 1 to 2 μg per deciliter ($\mu g/dL$) in human blood levels. Blood levels of 10 to 15 $\mu g/dL$ are associated with reduced intelligence and detrimental effects on growth and development of children, leading the U.S. Public Health Service to label lead as the greatest environmental health threat to our children (Thomas, 1995; U.S. PHS, 1991). As lead was phased out of gasoline in the United States, the average concentration of lead in blood dropped from 16 to 3 $\mu g/dL$, which makes this one of the most successful of all environmental achievements.

7.6 TOXIC AIR POLLUTANTS

The Clean Air Act Amendments of 1990 require the EPA to speed up efforts to control toxic air pollutants. As defined in the Act, these are pollutants that "are known to be, or may reasonably be anticipated to be, carcinogenic, mutagenic, teratogenic, neurotoxic, which cause reproductive dysfunction, or which are acutely or chronically toxic." There are literally hundreds of such chemicals, including 189 compounds that are listed in the 1990 Amendments. The EPA is charged with setting emission standards that achieve "the maximum degree of reduction in emissions," taking into account cost and other non-air-quality factors, and risk-based health standards that will assure no more than a one-in-a-million risk for the most exposed individuals. In light of the years of effort that were needed to deal with the six criteria pollutants, the emission and air quality standard setting that will be required for hundreds of additional chemicals is daunting.

Even before the 1990 Amendments were passed, the EPA had initiated National Emission Standards for Hazardous Air Pollutants (NESHAP) under Section 112 of the Act, but by 1988 only asbestos, benzene, beryllium, coke-oven emissions, inorganic arsenic, mercury, radionuclides, and vinyl chloride were regulated. Another important step in the regulation of air toxics was initiated by the Emergency Planning and Community Right to Know Act of 1986 (EPCRA). EPCRA requires manufacturing facilities that use listed chemicals to submit annual reports to the EPA on their releases. These releases form a database called the Toxic Release Inventory (TRI), which has been in operation since 1987. The availability of the TRI to the public has undoubtedly had an impact on manufacturers who want to avoid negative publicity. In 1993 reported air emissions totaled 1.7 billion pounds of chemicals, which was a drop of 39 percent compared with 1988 levels (U.S. EPA, 1995b). A large fraction of these decreases were for ozone-depleting substances such as 1,1,1-trichloroethane and most of the chlorofluorocarbons.

7.7 AIR POLLUTION IN THE WORLD'S MEGACITIES

The world's population is becoming ever more urbanized and with that urbanization comes increasing environmental pressures, not the least of which is deterioration of air quality in many of the most rapidly growing cities. In 1970 there were only four cities in the world with populations over 10 million people (Tokyo, New York, Shanghai, and London); by 1990 there were 12; and by 2000 there will be 24 such cities. At that time fully half of the world's population will live in urban areas. The World Health Organization (WHO) and the United Nations Environment Programme (UNEP) have initiated a study of air quality in 20 megacities (cities with over 10 million people in 2000), with a focus on the same six criteria pollutants that are regulated in the United States' Clean Air Act, except that in most cities it is total suspended particulate matter (SPM) that is measured instead of PM 10. In many circumstances the data being gathered are difficult to obtain and of questionable quality, but a preliminary assessment is beginning to emerge that shows increasing numbers of people exposed to increasingly unhealthy concentrations of air contaminants, especially in developing countries.

In most megacities the principal source of air pollution is motor vehicles. In almost all cities in the developing world, leaded fuels are still being burned, a high percentage of the vehicles are diesel-powered trucks and buses with no emission controls, many streets are unpaved, and traffic congestion, which intensifies emissions, is over-

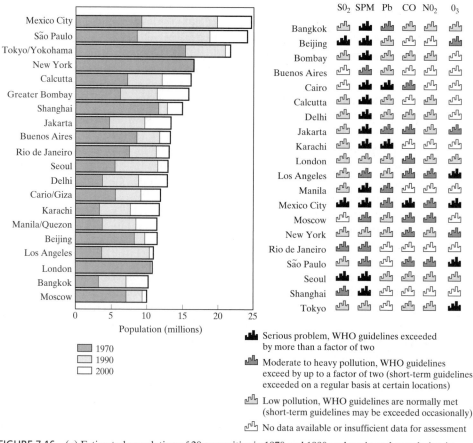

FIGURE 7.16 (*a*) Estimated population of 20 megacities in 1970 and 1990 and projected population in 2000; (*b*) overview of air quality. (*Source*: WHO, UNEP, 1992)

whelming. The resulting concentrations of Pb, CO, NO_x, O_3, and TSP are often many times higher than air quality guidelines defined by WHO. In addition, many countries have coal-fired power plants and other industrial facilities within city limits, and levels of SO_x, NO_x, and particulates are correspondingly high.

Figure 7.16 shows the populations and air quality in the 20 megacities studied by WHO/UNEP. High levels of SPM are the most prevalent air quality problem. Twelve of these 20 cities have serious problems with particulate matter, where *serious* is defined to mean that WHO air quality guidelines are exceeded by more than a factor of two. Beijing, Mexico City, and Seoul have serious SPM combined with serious SO_x problems, which is a lethal combination that leads to increased mortality and morbidity. In a number of areas, including parts of Beijing, Calcutta, Delhi, Shanghai and Seoul, combustion of coal and biomass fuels for cooking and heating leads to extremely high pollutant concentrations indoors, where many people, women especially, spend most of their time.

7.8 MOTOR VEHICLE EMISSIONS

As shown in Figure 7.1, the transportation sector in the United States accounts for a significant fraction of all of the criteria pollutants except for sulfur oxides. And most of those emissions are from motor vehicles. In large cities, motor vehicle emissions are especially noxious, in part because they tend to comprise an even higher fraction of total emissions and because those emissions are released right in the middle of crowded, urban populations. Figure 7.17 shows the intensity of those emissions in Los Angeles, a city known for its dependence on automobiles and, not unexpectedly, for its smog.

It is estimated that there are about 630 million vehicles in use around the world, with roughly 80 percent of those being located in Organization for Economic Cooperation and Development (OECD) countries. That number is expected to double within the next few decades, with much of that growth occurring in developing countries and in eastern Europe. That growth rate far outstrips the expected increases in both world population and in the population living in cities, as shown in Figure 7.18. Unless modern emission control systems are used, the burden of air pollution in cities located in the developing countries of the world will increase dramatically.

Emission Standards

The legislative history of auto emission controls began in California in 1959 with the adoption of state standards to control exhaust hydrocarbons (HC) and carbon monoxide (CO). These were supplemented in 1960 by standards to control emissions resulting from crankcase blowby. These standards, however, were not fully implemented until they were deemed technologically feasible in 1966. Federal standards for CO and HC began in 1968, and by 1970, the year in which major amendments to the Clean Air Act were enacted, the auto industry had reduced hydrocarbon emissions by almost three-fourths and carbon monoxide by about two thirds. At that time, no controls were required for NO_x, and to some extent, improvements in hydrocarbon and CO emissions were made at the expense of increased nitrogen oxide emissions.

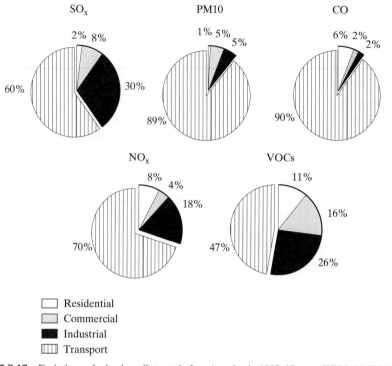

FIGURE 7.17 Emissions of criteria pollutants in Los Angeles in 1987. (*Source*: WHO, UNEP, 1992)

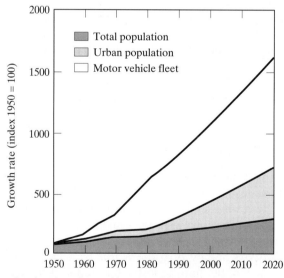

FIGURE 7.18 Estimated and projected rate of increase of the total world population, urban population, and number of motor vehicles, 1950–2020, indexed so that 100 equals the numbers in 1950. (*Source*: WHO, UNEP, 1992)

The Clean Air Act amendments of 1970 required that emissions from automobiles be reduced by 90 percent compared to levels already achieved by the 1970 model year. These standards were to be reached by 1975 for CO and HC, and by 1976 for NO_x, with a one-year delay allowed if industry could adequately prove that the technology was not available to meet them. The 1970 Amendments were unusual in that they were "technology forcing"—that is, they mandated specific emission levels before technology was available to meet the standards. The automobile industry successfully argued the case for the one-year delay, in spite of the fact that the 1973 Honda with a stratified charge engine was able to meet the 1975 emission requirements while getting 40 mpg.

Subsequently, in response to the oil embargo of 1973 and concerns for the impact of emission standards on fuel economy, the Federal Energy Supply and Coordination Act of 1974 rolled the standards back again to 1978–1979. Finally, the Clean Air Act amendments of 1977 delayed the standards once more, setting the date for HC and CO compliance at 1981. At the same time, the 1977 Amendments eased the standard for NO_x from 0.4 g/mile to 1 g/mile, calling the 0.4 g/mile a "research objective" despite the fact that California, which is allowed to set its own auto emission standards as long as they are more stringent than federal requirements, stayed with the original 0.4 g/mile standard for NO_x.

The 1990 Amendments to the Clean Air Act brought national standards to the level of California standards beginning with the 1996 models. The Act goes on to require another evaluation of emission standards in 1999 and, if warranted, the standards will be cut in half beginning with 2004 model year vehicles.

Table 7.4 indicates that vehicle emission regulations are expressed in terms of grams of pollutant per mile of driving. Since emissions vary considerably as driving conditions change, it has been necessary to define carefully a "standard" driving cycle and to base emission regulations on that cycle. The standard driving cycle used is based on an elaborate study of Los Angeles traffic patterns. It consists of a 7.5-mile simulated drive on a dynamometer, taking 22.8 minutes and including 18 stops. About 18 percent of the time is spent idling, as if waiting for traffic lights. Two tests are run, one that begins with a cold engine and one on an engine that has been completely warmed up. Then the results are averaged. This driving cycle is called the CVS-75 test, and it is shown in Figure 7.19.

Table 7.4   Federal Motor Vehicle Emission Standards (g/mile)

Vehicle	Year	CO	NO_x	NMHC[d]
Passenger cars[a]	1967	87.0	3.6	8.8
Passenger cars	1980	7.0	2.0	0.41
Passenger cars	1996	3.4[c]	0.4	0.25
Light-duty trucks (3751–5750 lb)	1996	4.4	0.7	0.32
Light-duty trucks (> 5750 lbs)	1997	5.0	1.1	0.39
Motorcycles[b]	1988	19.3	—	2.3

[a]prior to controls.
[b]California, displacement 700 cc or greater.
[c]10 g/mi at 20°F.
[d]NMHC stands for nonmethane hydrocarbons.

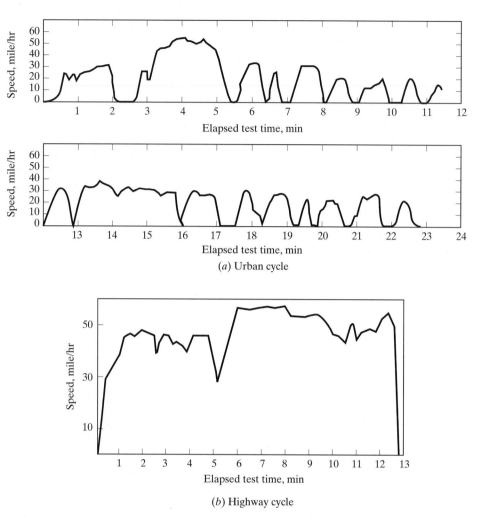

(a) Urban cycle

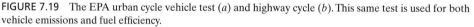

(b) Highway cycle

FIGURE 7.19   The EPA urban cycle vehicle test (a) and highway cycle (b). This same test is used for both vehicle emissions and fuel efficiency.

CAFE Fuel Economy Standards

The oil embargo of 1973 shifted Congress' attention from motor vehicle emissions to motor vehicle fuel efficiency. At the time, cars in the United States averaged about 14 miles per gallon (mpg). Corporate Average Fuel Economy (CAFE) standards were enacted in 1975 that required fuel efficiency for each manufacturer's fleet of new automobiles to average at least 27.5 mpg within a 10-year period. As further incentive, a gas guzzler tax was also initiated that subjects cars that achieve less than 22.5 mpg to a special sales tax ranging from $1000 to $7700.

The stop-and-go driving cycle shown in Figure 7.19 for emissions testing is the same cycle that is used to determine the "city" portion of fuel efficiency. A carbon balance performed on the gases collected from the tailpipe during the test is used to determine the amount of fuel consumed over the 7.5-mile simulated test drive. The "highway" portion of the fuel economy test simulates a 12.8-minute, 10-mile trip averaging 48 mph with no stops. The two tests are weighted using the assumption of 55 percent urban driving and 45 percent highway to produce the average fuel efficiency.

As memories of the oil crises of the 1970s faded and as cheap gasoline returned, the CAFE standard was relaxed to 26 mpg for 1986–1988 model years, but it was returned to 27.5 in 1990. As older, less efficient cars are replaced with newer models, not only are emissions for the average car on the road reduced, but the average fuel efficiency increases.

The average fuel economy of new passenger cars sold in the United States did rise to the required 27.5 mpg by 1986, but the overall average of new cars plus "light trucks," which includes pick-up trucks, minivans, and sport-utility vehicles (SUVs), has fallen short of that level. Those light trucks are only required to achieve 20.7 mpg, and as they increase in popularity they bring down the overall average for new vehicles. In fact the fleet average for new cars and light trucks peaked in 1986 at 26 mpg and fell to less than 25 mpg by 1995. In 1980 less than 20 percent of new vehicle sales were light trucks, but by 1995 they accounted for 40 percent of the domestic market. Figure 7.20 shows the impact of this rising popularity.

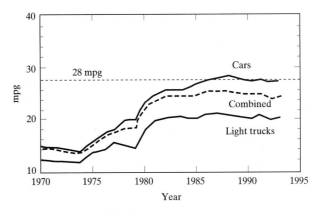

FIGURE 7.20   Fuel efficiency of new cars and light trucks and the composite average (dashed line). (*Source*: DeCicco and Ross, *An Updated Assesment of the Near-Term Potential for Improving Automotive Fuel Economy,* Reprinted with permission, © 1993, American Council for an Energy-Efficient Economy, Washington, DC 20036. All rights reserved.)

EXAMPLE 7.6   Average Fuel Efficiency

Suppose a fuel efficiency test indicates that a particular automobile gets 20 mpg in the city and 29 mpg on the highway. What would its CAFE fuel efficiency be?

If new cars get an average of 28 mpg and new light trucks get 20.2 mpg, what would be the fleet average if 40 percent of sales are light trucks?

Solution   The CAFE efficiency assumes 55 percent urban and 45 percent highway miles. If we want to imagine 1 mile of driving, the fuel efficiency would be

$$\text{mpg} = \frac{1 \text{ mile}}{\left(\dfrac{0.55 \text{ miles}}{20 \text{ miles/gal}} + \dfrac{0.45 \text{ miles}}{29 \text{ miles/gal}} \right)} = 23.2 \text{ mpg}$$

The fleet efficiency for 28-mpg cars and 20.2-mpg light trucks would be

$$mpg = 0.60 \times 28 \text{ mpg} + 0.4 \times 20.2 \text{ mpg} = 24.9 \text{ mpg}$$ ∎

Fuel efficiency is important for a number of reasons, not the least of which are national security issues related to our dependence on imported oil (approximately equal to the energy demands of our highway vehicles). While emissions of the conventional pollutants CO, HC, and NO_x are not necessarily directly related to fuel efficiency, emissions of the greenhouse gas, carbon dioxide (CO_2), certainly are.

The Conventional Otto Cycle Engine

The most common internal combustion engine is a four-stroke, spark-ignited piston engine invented around 1880 by a German engineer, Nicholas Otto. The operation of an Otto cycle, or four-stroke, engine is described in Figure 7.21. On the first, or *intake,* stroke, the descending piston draws in a mixture of fuel and air through the open intake valve. The *compression* stroke follows, in which the rising piston compresses the air-to-fuel mixture in the cylinder against the now closed intake and exhaust valves. As the piston approaches the top of the compression stroke, the spark plug fires, igniting the mixture. In the *power* stroke, the burning mixture expands and forces the piston down, which turns the crankshaft and delivers power to the drive train. In the fourth, or *exhaust,* stroke, the exhaust valve opens and the rising piston forces combustion products out of the cylinder, through the exhaust system, and into the air. Diesel engines operate with a similar cycle, but they compress the fuel more during the compression stroke, which makes the gases in the cylinder so hot that they ignite of their own accord without needing a spark plug.

The single most important factor in determining emissions from a four-stroke, internal combustion engine is the ratio of air-to-fuel in the mixture as it enters the cylinders during the intake stroke. To analyze that mixture ratio and its impact on emissions, let us begin with the stoichiometry of gasoline combustion. While modern gasolines are blends of various hydrocarbons, an average formulation can be represented as C_7H_{13}. We can write the following to represent its complete combustion in oxygen:

$$C_7H_{13} + 10.25 \, O_2 \rightarrow 7 \, CO_2 + 6.5 \, H_2O \tag{7.29}$$

If we want to show complete combustion in air, we can modify this reaction to account for the fact that about 3.76 mol of N_2 accompany every mole of O_2 in air. Thus, $10.25 \times 3.76 = 38.54$ mol of N_2 can be placed on each side of the reaction, yielding

$$C_7H_{13} + 10.25 \, O_2 + 38.54 \, N_2 \rightarrow 7 \, CO_2 + 6.5 \, H_2O + 38.54 \, N_2 \tag{7.30}$$

where any oxidization of nitrogen to nitrogen oxides has been neglected.

EXAMPLE 7.7   Stoichiometric Air-to-Fuel Ratio

Determine the ratio of air-to-fuel required for complete combustion of gasoline.

Solution   For each mole of gasoline, 10.25 mol of O_2 and 38.54 mol of N_2 are required. Using (7.30), we can determine the masses of each constituent as

$$1 \text{ mol } C_7H_{13} = 7 \times 12 + 13 \times 1 = 97 \text{ g}$$

$$10.25 \text{ mol } O_2 = 10.25 \times 2 \times 16 = 328 \text{ g}$$
$$38.54 \text{ mol } N_2 = 38.54 \times 2 \times 14 = 1079 \text{ g}$$

Considering air to be made up of only O_2 and N_2, the air-to-fuel ratio needed for complete oxidation of gasoline is

$$\frac{\text{Air}}{\text{Fuel}} = \frac{(328 + 1079)\text{g}}{97\text{g}} = 14.5$$

This is known as the *stoichiometric ratio* for gasoline.   ■

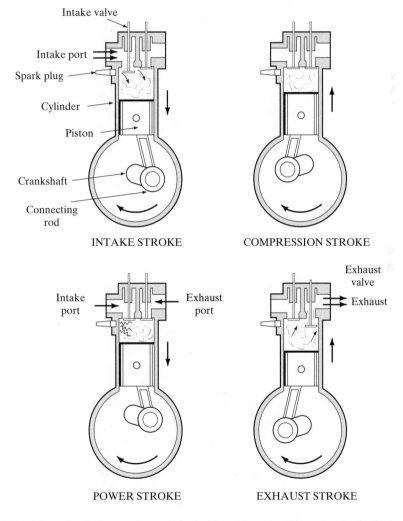

FIGURE 7.21   Schematic of a four-stroke, spark-ignited internal combustion engine. *Intake*: Intake valve open, piston motion sucks in fresh air/fuel charge. *Compression*: Both valves closed, air/fuel mixture is compressed by rising piston, spark ignites mixture near end of stroke. *Power*: Air/fuel mixture burns, increasing temperature and pressure, expansion of combustion gases drives pistons down. *Exhaust*: Exhaust valve opens, spent gases are pushed out of cylinder by rising piston. (*Source*: Powell and Brennan, 1988)

If the actual air-to-fuel mixture has less air than what the stoichiometric ratio indicates is necessary for complete combustion, the mixture is said to be *rich*. If more air is provided than is necessary, the mixture is *lean*. A rich mixture encourages production of carbon monoxide (CO) and unburned hydrocarbons (HC) since there is not enough oxygen for complete combustion. On the other hand, a lean mixture helps reduce CO and HC emissions unless the mixture becomes so lean that misfiring occurs. Production of NO_x is also affected by the air-to-fuel ratio. For rich mixtures, the lack of oxygen lowers the combustion temperature, reducing NO_x emissions. In the other direction, beyond a certain point lean mixtures may have enough excess air that the dilution lowers flame temperatures and reduces NO_x production.

Figure 7.22 shows the relationship between CO, HC, and NO_x emissions, and the air-to-fuel ratio. Also shown is an indication of how the air-to-fuel ratio affects both power delivered and fuel economy. As can be seen, maximum power is obtained for a slightly rich mixture, while maximum fuel economy occurs with slightly lean mixtures. Before catalytic converters became the conventional method of emission control, most cars were designed to run slightly rich for better power and smoothness. When the first automobile emission limitations were written into law in the 1960s, only CO and HC were controlled so manufacturers simply redesigned their engines to run on a less rich mixture. This simple change had the desired effect of reducing CO and HC, but at the

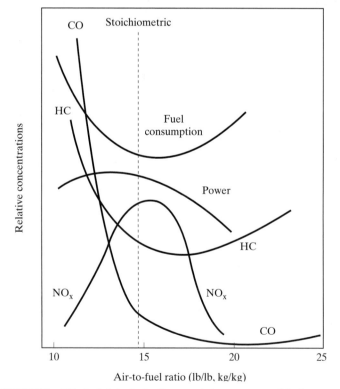

FIGURE 7.22 Effect of air-to-fuel ratio on emissions, power, and fuel economy.

same time NO_x emissions increased. Later, when the Clean Air Act Amendments of 1970 forced emission controls on all three pollutants, it was no longer possible to meet emission limitations by simply modifying the air-to-fuel ratio, and automotive engineers turned to the three-way catalytic converter.

The air-to-fuel ratio is, of course, not the only factor that influences the quantity of pollutants created during combustion. Other influencing factors include ignition timing, compression ratio, combustion chamber geometry, and, very important, whether the vehicle is idling, accelerating, cruising, or decelerating. In addition, not all of the pollutants created in the combustion chamber pass directly into the exhaust system. Some find their way around the piston during the power and compression strokes, into the crankcase. This *blowby,* as it is called, used to be vented from the crankcase to the atmosphere, but now it is recycled back into the engine air intake system to give it a second chance at being burned and released in the exhaust stream. The control method is called *positive crankcase ventilation* (PCV), and the main component is a crankcase ventilator valve that adjusts the rate of removal of blowby gases to match the changing air intake requirements of the engine, making sure that the desired air-to-fuel ratio is not upset by these added gases.

Two-stroke Engines

Two-stroke (or two-cycle) engines are quite similar to conventional four-stroke engines in that they use a spark to ignite an air/fuel mixture to deliver power to a piston connected to a crankshaft. The key difference is that every other stroke is a power stroke rather than every fourth stroke, as is the case with an Otto cycle engine. As suggested in Figure 7.23, the intake and compression strokes are combined into one as are the power and exhaust strokes. On the upstroke, the piston compresses an air/fuel mixture that is about to be ignited while it draws the next cycle's fuel into the crankcase. The fuel is fired by a spark plug, driving the piston down. On that downstroke, the air/fuel mixture in the crankcase is forced around the piston into the combustion chamber, forcing the exhaust gases out the exhaust port. By doubling the number of power strokes, the two-cycle engine has more power for its size than a conventional

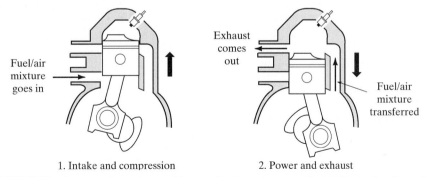

1. Intake and compression   2. Power and exhaust

FIGURE 7.23   A two-stroke engine combines the intake and compression strokes as the piston rises, and combines the power and exhaust strokes as the piston is driven downward. (*Source*: Powell and Brennan, 1992)

four-stroke engine. It also is simpler since there are no moving valves. The net is a smaller, lighter, less expensive engine that has been popular for use in motorcycles, lawnmowers, and outboard motors.

The problem with two-strokes is that they emit much more pollution than four-stroke engines. As exhaust gases are being pushed out of the combustion chamber, some of the fresh fuel entering the chamber is pushed out as well. Also, two-stroke engines need oil to be mixed with the gasoline in order to lubricate moving parts in the crankcase, and that leads to oily, smelly exhaust fumes as well as engine misfiring. While these engines constitute only a small fraction of the internal combustion engines in use, their emissions are disproportionately large. It has been estimated, for example, that a two-stroke outboard motor emits as much pollution in an hour as driving a late-model automobile 800 miles. In 1996, the EPA announced new rules that will eliminate conventional two-stroke outboard motors.

Diesel Engines

At this time, the only real competition to the basic spark-ignition, internal combustion engine for motor vehicles is the *diesel*. Diesel engines have no carburetor since fuel is injected directly into the cylinder, and there is no conventional ignition system with plugs, points, and condenser since the fuel ignites spontaneously during the compression stroke. Diesels have much higher compression ratios than conventional (Otto cycle) engines and, since they do not depend on spark ignition, they can run on very lean mixtures. Thus they are inherently more fuel efficient. The increased fuel efficiency of diesels was responsible for their momentary surge in popularity just after the oil crises of the 1970s.

Since diesels run with very lean mixtures, emissions of hydrocarbons and carbon monoxide are inherently very low. However, because high compression ratios create high temperatures, NO_x emissions are relatively high. Moreover, since the fuel is burned with so much excess oxygen, conventional catalysts that require a lack of oxygen to reduce NO_x are ineffective, and less efficient controls are used. In addition, diesels emit significant quantities of carbonaceous soot particles, and diesel engine exhaust has been labeled a probable human carcinogen. Filters have been proposed, but complex schemes would be required to clean the filter periodically of its accumulated particulate load. The difficulty in controlling NO_x and soot, along with the lack of interest in fuel economy after the crash in oil prices in the 1980s, has led to the virtual demise of the diesel in new automobiles.

Exhaust System Controls

During the exhaust stroke of an internal combustion engine, combustion gases are pushed through the exhaust manifold and out the tailpipe, and it is in this exhaust system that most of the control of automobile emissions now occurs. The most commonly used systems for treatment of exhaust gases are *thermal reactors, exhaust gas recirculation* (EGR) systems, and *catalytic converters.*

A thermal reactor is basically an afterburner that encourages the continued oxidation of CO and HC after these gases have left the combustion chamber. The reactor consists of a multipass, enlarged exhaust manifold with an external air source. Exhaust gases in the reactor are kept hot enough and enough oxygen is provided to allow combustion to continue outside of the engine itself, thus reducing CO and HC emissions. Usually, the carburetion system is designed to cause the engine to run rich in order to provide sufficient unburned fuel in the reactor to allow combustion to take place. This has the secondary effect of modestly reducing NO_x emissions, although it also increases fuel consumption since some fuel is not burned in the cylinders.

Some degree of control of NO_x can be achieved by recirculating a portion of the exhaust gas back into the incoming air/fuel mixture. This relatively inert gas that is added to the incoming mixture absorbs some of the heat generated during combustion without affecting the air-to-fuel ratio. The heat absorbed by the recirculated exhaust gas helps reduce the combustion temperature and, hence, helps decrease the production of NO_x. The coupling of exhaust gas recirculation with a thermal reactor, as shown in Figure 7.24, reduces emissions of all three pollutants, CO, HC, and NO_x, but at the expense of performance and fuel economy.

The approach most favored by automobile manufacturers to achieve the emission standards dictated by the Clean Air Act has been the three-way catalytic converter (three-way just means that it handles all three pollutants, CO, HC, and NO_x). A three-way converter is able to oxidize hydrocarbons and carbon monoxide to carbon dioxide while reducing NO_x to N_2 all in the same catalyst bed. These catalytic converters are very effective in controlling emissions, and they have the advantage of allowing the engine to operate at near stoichiometric conditions, where engine performance and efficiency are greatest. In fact, they *must* operate within a very narrow band of air-to-fuel ratios near the stoichiometric point or else their ability to reduce all three pollutants at once is severely compromised, as shown in Figure 7.25a. Maintaining that degree of

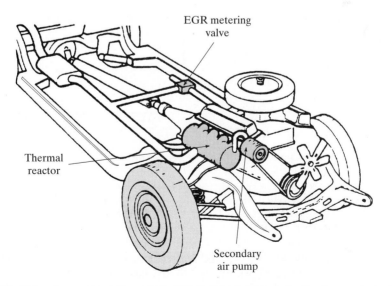

EGR metering valve

Thermal reactor

Secondary air pump

FIGURE 7.24   Exhaust gas recirculation reduces NO_x by lowering the combustion temperature, while the thermal reactor helps control CO and HC. (*Source*: Courtesy Gould Inc.)

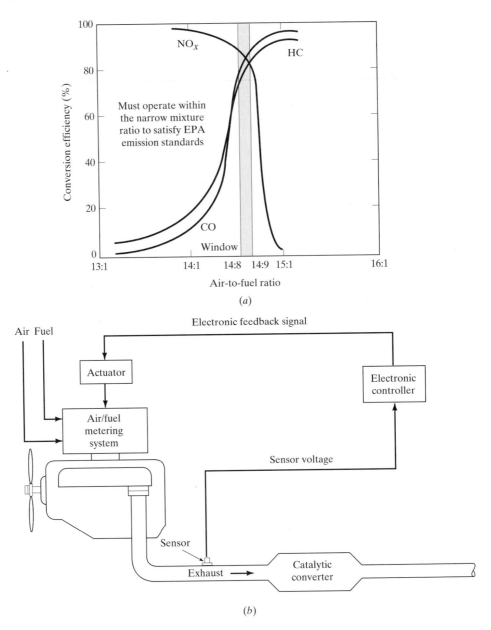

FIGURE 7.25 For a three-way catalyst to function correctly, the air-to-fuel ratio must be maintained within a very narrow band, as shown in (*a*). To maintain that ratio, a closed-loop control system monitors the composition of exhaust gases and sends corrective signals to the air-to-fuel metering system. (After Powell and Brennan, 1988)

control has required the development of precise electronic feedback control systems that monitor the composition of exhaust gas and feed that information to a microprocessor-controlled carburetor or fuel-injection system as shown in Figure 7.25b.

Catalysts are quickly destroyed if leaded fuels are burned, resulting in no pollution control at all. When leaded fuels were readily available, they were often cheaper than unleaded gasoline, and some motorists ruined their emission control systems by trying to use the cheaper leaded fuel. In an effort to decrease the likelihood of such misfueling, as well as to discourage tampering with emission control systems, areas with air quality problems have been required to institute periodic vehicle inspection and maintenance programs. Three-way converters are very effective once they are warmed up, but when they are cold, as well as when there are spurts of sudden acceleration or deceleration, they release excessive amounts of pollution that offset much of their perceived benefits. Those shortcomings are being addressed, and ultra-low-emission vehicles based on multiple catalytic converters will be on the market soon.

Cleaner Gasoline

Motor vehicle hydrocarbon emissions are not only the result of inefficient combustion, they are also caused by vaporization of the fuel itself. Vaporization rates are very dependent on temperature, so evaporation is especially high just after the engine is turned off, when the fans, cooling systems, and airflow due to vehicle movement are no longer operative. To help control these evaporative emissions automobiles are equipped with vapor-recovery systems, but these are not totally effective. A complementary approach is to reduce the volatility of the gasoline itself, but there are trade-offs. In cold, winter conditions, vehicles are harder to start unless fuels vaporize readily at low temperatures, but in the summer that higher volatility would lead to excessive evaporative emissions. Refineries then must formulate their fuels differently depending on location and season.

Another way to help reduce motor vehicle emissions is to provide more oxygen for combustion by incorporating oxygen-containing additives in gasoline. The most often used additives to create these "oxygenated" fuels are ethanol (C_2H_5OH) or MTBE, which stands for methyl tertiary butyl ether. The Clean Air Act Amendments of 1990 have been requiring oxygenated fuels in areas that have high levels of carbon monoxide pollution since 1992.

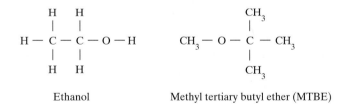

Ethanol   Methyl tertiary butyl ether (MTBE)

MTBE has certain advantages over ethanol: It has a higher octane number and it is not as volatile. On the other hand, MTBE may have health effects that have not yet been fully assessed. There have been complaints that MTBE causes nausea, dizziness, and headaches.

Another controversy over the formulation of gasoline has to do with the octane enhancers that have replaced tetraethyllead. Until recently, the antiknock properties of tetraethyllead have been achieved in the United States by increasing the proportion in gasoline of certain benzene derivatives such as toluene, $C_6H_5CH_3$, and xylene, $C_6H_4(CH_3)_2$. Along with benzene, these are collectively known as *BTX* (*benzene, toluene, xylene*). BTX boosts the octane rating of gasoline, but BTX hydrocarbons are more reactive than normal constituents of gasoline so they can increase formation of photochemical smog.

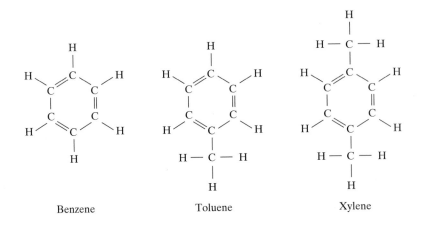

Benzene   Toluene   Xylene

There is another octane enhancer, methylcyclopentadienyl manganese tricarbonyl, better known as MMT, that was used in the United States until 1977, when health concerns led to its withdrawal. Combustion products of MMT contain toxic manganese, and the EPA opposed its use until health risks could be evaluated. The manufacturer of MMT argued that it had been used for many years in Canada without apparent health effects, and in 1995 a federal appeals court ruled in the manufacturer's favor and use of MMT began again in the United States in 1996.

The Clean Air Act Amendments of 1990 required *reformulated gasoline* to be sold year-round beginning in 1995 in areas where ozone nonattainment is rated "severe" (Baltimore, Chicago, Hartford, Houston, Milwaukee, New York, Philadelphia, and most of Southern California). Reformulated gasoline must have an oxygen content of at least 2 percent by weight and a maximum benzene content of 1 percent by volume (down from the normal 2 percent). In addition, high-volatility components of gasoline—aromatics, olefins, and sulfur—have been reduced. Diesel fuel has had its sulfur content cut drastically to 0.5 percent. Overall hydrocarbon and toxic emission reductions of 15 percent were required for reformulated gasoline in 1995, and at least 20 percent reductions in the year 2000.

Alternative Fuels

There are a number of alternatives to gasoline that are being investigated as possible fuels for the future. These include methanol, ethanol, compressed natural gas, propane, and hydrogen.

Methanol (CH_3OH), also known as wood alcohol, has long been the fuel of choice in high-performance racing cars. It has a much higher octane rating than gasoline, which enables engines to be designed with higher compression ratios for increased power. It burns with a lower flame temperature than does either gasoline or diesel fuel, so NO_x emissions are reduced. It burns more completely as well, so hydrocarbon and CO emissions are also reduced. Moreover, the hydrocarbons that are released are less photochemically reactive. As a result of these improved emission characteristics, the ozone-producing potential of methanol is perhaps half that of gasoline (California Energy Commission, 1989).

Methanol is not without its problems. Its emissions are much higher in formaldehyde (HCHO), an eye irritant and a suspected carcinogen. The low volatility of pure methanol makes it difficult to start engines in cold conditions, and it is highly toxic. It burns without a visible flame, which can lead to especially dangerous conditions in an accident involving fire. Moreover, its energy content is only about half that of gasoline, so either much larger fuel tanks would be required or the driving range would be considerably reduced.

Some of the problems of pure methanol can be overcome by using a blend of methanol and gasoline. A mixture of 85 percent methanol and 15 percent gasoline, called M85, eliminates the cold start problem and yields a visible flame as well. It also tastes much worse than pure methanol, which should discourage ingestion. The blend has an octane rating of 102 (versus 87–92 for gasoline) and a volumetric energy content of a little more than half that of gasoline. While the increase in combustion efficiency does slightly offset the lower energy density, the driving range for M85 is still only about 60 percent of that for gasoline. To help ease the transition to methanol, automobile manufacturers have introduced *flexible fuel vehicles* (FFVs) designed to run on M85, gasoline, or any combination of both. Eventually, if methanol becomes widely available, it would be possible to take advantage of the increased octane level with more efficient engines dedicated to M85 use.

With regard to emissions of the greenhouse gas, carbon dioxide, methanol derived from natural gas produces slightly less CO_2 than gasoline. On the other hand, if methanol is produced from coal, total CO_2 emissions released during coal processing and methanol combustion are double that of gasoline (Moyer, 1989).

Ethanol (C_2H_5OH), also called grain alcohol, can be produced by fermentation of all sorts of plant crops as well as from materials such as wood or paper wastes. In Brazil large amounts of ethanol are fermented from sugar cane, and ethanol supplies roughly one-fourth of that country's vehicle fuel. In the United States ethanol, produced mostly from corn, is blended in a 10 percent ethanol, 90 percent gasoline mixture and sold as "gasohol." Ethanol fuels yield lower carbon monoxide and reactive hydrocarbon emissions than gasoline, but that advantage may be offset by increased fuel vaporization. Since ethanol can be made from renewable crops, carbon dioxide emissions for the complete fuel cycle can be reduced (carbon removed from the atmosphere during plant photosynthesis is returned when the resulting fuel is burned). This may not be the case if fossil fuels are used to provide the heat needed to distill the ethanol during its manufacture.

Compressed natural gas (CNG) is already used in fleets of trucks, delivery vans, and buses, especially in Canada and New Zealand. CNG is a very clean fuel, and very low emissions of reactive hydrocarbons, carbon monoxide, particulates, and toxics are possible. NO_x emissions tend to be higher than from gasoline-fueled vehicles, but only because NO_x emission-control systems are not as developed as those designed for gasoline. Refueling is rather inconvenient; heavy fuel tanks are required to contain the highly pressurized gas, and at 2400 pounds per square inch of pressure it takes five times as much volume to store an equivalent amount of energy compared with gasoline. Methane is a much more potent greenhouse gas than carbon dioxide, so leakage could make methane worse for global warming than gasoline.

Liquefied petroleum gas (95 percent propane) is another fuel that has been used for vehicle fleets such as school buses, trucks, fork lifts, taxis, and farm vehicles. Some vehicles have been designed with dual tanks so that either gasoline or LPG can be burned, but the added flexibility of fuel sources is offset some by the added space, weight, and complications of such systems. Compared with gasoline, carbon monoxide emissions are somewhat lower; hydrocarbon emissions are roughly the same, but they are less reactive; NO_x emissions tend to be higher, but could be lower if appropriate control systems are developed; and carbon dioxide may be lower or higher depending on whether the LPG is derived from natural gas or petroleum.

In the eyes of some, hydrogen could be the ultimate fuel for many purposes, including vehicles. It can be pressurized and stored as a compressed gas, but even at 8000 psi it would take nearly five times the volume to store the same energy as gasoline. A more promising method of storage is in the form of metal hydrides, in which hydrogen gas is absorbed into the crystalline structure of a metal such as titanium or magnesium. Metal hydrides have been used to store hydrogen for use in experimental, fuel-cell-powered buses. Burning hydrogen releases energy without carbon monoxide, hydrocarbons, or toxic emissions. Thermal NO_x can be produced, however, so it is not entirely "clean." It should be realized that hydrogen is fundamentally different from the other fuels just considered in that it is an energy *carrier*, much like electricity, rather than a primary fuel. It must be produced using energy from some other source, such as natural gas, coal, or perhaps solar energy, so the pollution associated with the entire hydrogen fuel cycle is very dependent on the characteristics of the primary fuel.

One old idea that continues to intrigue many in the engineering community is the electric vehicle. From an air quality perspective, a fleet of electric cars would shift pollution from the city, where vehicles are concentrated, to the countryside, where power plants that would generate electricity for the vehicles would tend to be located. Not only would the location of emissions shift, but the type of emissions would change as well. If the power plants burn fossil fuels, then HC and CO emissions would be reduced, NO_x could still be a problem, and particulate and SO_x emissions could greatly increase. Photochemical smog problems in the city could be eased, but acid rain and other industrial smog problems could be exacerbated. Of course, to the extent that electric power production might shift from fossil fuels, especially coal, to renewable energy sources or nuclear power, there would be an obvious air quality advantage.

Electric vehicles are entering the marketplace now, and how well the public will accept their limitations will be known soon. Their limited range, high price, and recharging inconvenience will deter some, but their lack of emissions and their quietness clearly will attract a dedicated following.

A Car for the Future: Hybrids

As Figure 7.20 attests, the fuel efficiency of new vehicles has reached a plateau. Electric vehicles are emerging, but their limitations are significant. Some believe that the best of both worlds can be combined in a hybrid vehicle with an onboard, fuel-fired system that generates electricity, coupled to an electric-motor-based drive system. The onboard "engine" could be as conventional as a small internal combustion engine that spins a generator, or as exotic as a fuel cell that converts fuel to electricity using electrochemical reactions. Fuel cells are described in a later section of this chapter on stationary source emissions.

Figure 7.26 shows one version of the major components needed in a hybrid vehicle. In this case a small internal combustion engine spins a generator that delivers electricity to a modest battery bank. The engine is designed to supply only the average amount of power needed for driving, which allows it to be optimized for a very limited combination of requirements. When bursts of energy are required to accelerate or climb hills, the peaking power needed comes from an energy storage system, in this case batteries, but some designs use flywheels for storage. The storage system would be designed to allow the car to operate as an emission-free electric vehicle for short trips, as well as to handle short periods of high-power demands. The power to the wheels would come from an electric drive motor. An added advantage of electric drive is the ability to do regenerative braking; that is, when the brakes are applied the electric motor becomes an electric generator that recharges the batteries with the energy that would normally be dissipated by brake pads. Thus, the energy storage device is charged partly from the onboard electric power plant and partly from regenerative braking.

If the body of a hybrid vehicle would be built using synthetic composites instead of steel, it could be very lightweight as well as strong. There is an interesting positive feedback that can be initiated by reducing the weight of a vehicle. A lighter vehicle can use a smaller, lighter onboard power system and still have the zip that drivers seem to demand. A lighter vehicle would also have better fuel efficiency, so the size and weight

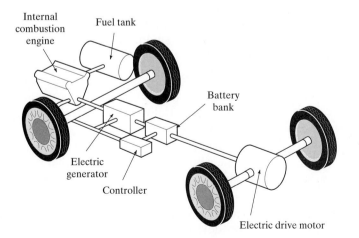

FIGURE 7.26   A hybrid vehicle uses electricity generated onboard to power an electric drive system. (*Source*: McCosh, 1994)

of the fuel tank could be smaller. Having a lighter body, engine, and fuel tank allows the structural portions of the vehicle to be lightened as well, and so on. Amory Lovins, a proponent of hybrid vehicles, asserts that hybrids made with composites could have fuel efficiencies on the order of 300 miles per gallon with emissions that would be a small fraction of today's vehicles (Lovins and Lovins, 1995).

7.9 STATIONARY SOURCES

Nontransportation, fossil-fuel combustion is responsible for almost 90 percent of the SO_x and half of the NO_x and PM 10 emitted in the United states. Most of that is released at electric power plants, and most of the power plant emissions result from the combustion of coal. Both the regulatory and technological approaches to reducing those emissions are very different from those we just described for motor vehicles. Although most stationary source emissions are caused by combustion of fossil fuels, other processes, such as evaporation of volatile organic substances, grinding, and forest fires, can be important as well. Our focus, however, will be on combustion.

Since most air pollutants are produced during combustion, one of the most important, but most overlooked, approaches to reducing emissions is simply to reduce the consumption of fossil fuels. There are three broad approaches that can be taken to reduce fossil fuel consumption: (1) Increase the conversion efficiency from fuel to energy, (2) increase the efficiency with which energy is used, and (3) substitute other less polluting energy sources for fossil fuels. Progress has been made on all three fronts. The conversion efficiency from fuel to electricity for power plants has improved eightfold from the 5 percent efficiency of Edison's original plants a century ago to nearly 40 percent for a modern power plant. Encouraging gains have been made in improving the energy efficiency of end uses, and some estimates suggest that improvements in electric motors and motor controls, better lighting systems, more efficient manufacturing processes, and so forth can cut per capita energy demands in half. The third approach is to reduce our reliance on fossil fuels by increasing power production from solar, wind, hydroelectric, geothermal, and perhaps nuclear sources.

To the extent that fossil fuels continue to be used, there are three general approaches that can be used to reduce emissions:

1. *Precombustion controls* reduce the emission potential of the fuel itself. Examples include switching to fuels with less sulfur or nitrogen content in power plants. In some cases fossil fuels can be physically or chemically treated to remove some of the sulfur or nitrogen before combustion.

2. *Combustion controls* reduce emissions by improving the combustion process itself. Examples include new burners in power plants that reduce NO_x emissions and new fluidized bed boilers that reduce both NO_x and SO_x.

3. *Postcombustion controls* capture emissions after they have been formed but before they are released to the air. On power plants, these may be combinations of particulate collection devices and flue-gas desulfurization techniques, used after combustion but before the exhaust stack.

Coal-fired Power Plants

Coal-fired power plants emit great quantities of sulfur oxides, nitrogen oxides, and particulates. Most early emission controls were designed to reduce particulates using post-combustion equipment such as baghouses and electrostatic precipitators. More recently, especially as a result of increased awareness of the seriousness of global acid deposition, attention has shifted somewhat toward control of nitrogen and sulfur oxides using redesigned combustors for NO_x reductions and scrubbers for SO_x control.

Before going into any of the details on these and other emission control techniques, it is useful to introduce a typical conventional coal-fired power plant, as shown in Figure 7.27. In this plant, coal that has been crushed in a pulverizer is burned to make steam in a boiler for the turbine-generator system. The steam is then condensed, in this case using a cooling tower to dissipate waste heat to the atmosphere, and the condensate is then pumped back into the boiler. The flue gas from the boiler is sent to an electrostatic precipitator, which adds a charge to the particulates in the gas stream so that they can be attracted to electrodes that collect them. Next, a wet scrubber sprays a limestone slurry over the flue gas, precipitating the sulfur and removing it in a sludge of calcium sulfite or calcium sulfate, which then must be treated and disposed of.

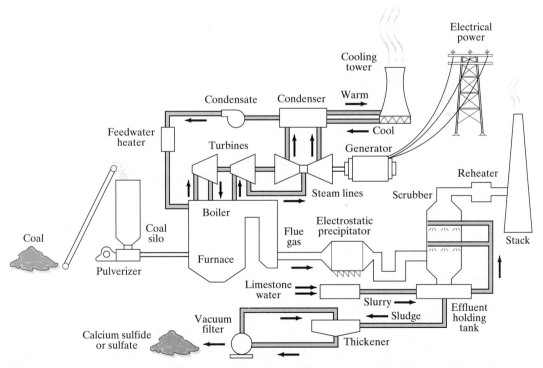

FIGURE 7.27.   Typical modern coal-fired power plant using an electrostatic precipitator for particulate control and a limestone-based SO_2 scrubber. A cooling tower is shown for thermal pollution control.

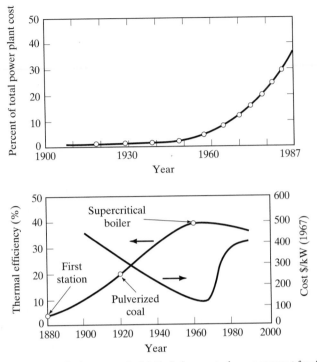

FIGURE 7.28 The impact of pollution controls. (*a*) Emission controls now account for 40 percent of the cost of building a coal-fired power plant. (*b*) Coal plant efficiency rose until the 1960s, when emission controls began to be required. About half of the jump in cost shown in (*b*) is for emission controls, and about half is due to the increased cost of siting, licensing, and lengthened construction time. (*Source*: Yeager, 1991)

Figure 7.28 shows the impact that pollution controls have had on the cost of a new power plant. Up until the 1960s, coal-fired power plants were dirty, and not much was being done to control those emissions. As Figure 7.28*a* shows, however, that picture has changed dramatically and now close to 40 percent of the cost of building a new coal plant is money spent to pay for pollution controls. Not only have existing levels of control been expensive in monetary terms, they also use up about 5 percent of the power generated by the plant, reducing the plant's overall efficiency. Figure 7.28*b* shows the decline in coal plant efficiency after emission controls were required.

Precombustion Controls

One of the two precombustion control techniques used on coal-fired power plants to control sulfur emissions is *fuel switching*. As the name suggests, it involves either substituting low sulfur coal for fuel with higher sulfur content or, perhaps, blending the two. Fuel switching can reduce emissions by anywhere from 30 to 90 percent, depending on the sulfur content of the fuel currently being burned. Often fuel switching can be a temporary measure taken only during periods of time when atmospheric conditions are particularly adverse.

The sulfur content of coal is sometimes expressed as a percent, or, quite often, in terms of mass of SO_2 emissions per unit of heat energy delivered. Power plant coal typically has sulfur content between 0.2 to 5.5 percent (by weight) corresponding to an uncontrolled SO_2 emission rate of about 0.4 to 10 lb/MBtu (0.17 to 4.3 g/MJ). New Source Performance Standards (NSPS) for coal plants restrict emissions to no more than 1.2 lb/MBtu (0.5 g/MJ). Coal that can meet the NSPS without controls is sometimes referred to as "compliance" coal or, more loosely, as "low-sulfur" coal.

In round numbers, just under half of U.S. coal reserves are in the eastern half of the United States (principally Illinois, West Virginia, Kentucky, Pennsylvania, and Ohio), and just over half are in the western United States (Montana, Wyoming, Colorado, and Texas). While that distribution is nearly even, the location of low-sulfur coal reserves is very unevenly divided. Almost all of our low sulfur coal, about 85 percent, is located in the western states (the only notable exceptions are small reserves in West Virginia and Kentucky). But two-thirds of our coal consumption and three-fourths of coal production occurs east of the Mississippi (OTA, 1984). Increasing reliance on fuel switching, then, would greatly increase the cost to transport the coal from mine to power plant. It would also significantly affect the economies of a number of states if eastern mines decrease production while western ones increase theirs. In addition, western coal often has different ash, moisture, and energy content, which may require modifications to existing power plants to allow fuel switching. Such costs, however, are relatively minor compared to the increased cost of low-sulfur fuel.

The other precombustion approach currently used to reduce sulfur emissions is coal cleaning. Sulfur in coal is either bound into organic coal molecules themselves, in which case precleaning would require chemical or biological treatment (at the research stage now), or it can be in the form of inorganic pyrite (FeS_2), which can readily be separated using physical treatment. Pyrite has a specific gravity that is 3.6 times greater than coal, and that difference allows various coal "washing" steps to separate the two. Such physical cleaning not only reduces the sulfur content of coal, but it also reduces the ash content, increases the energy per unit weight of fuel (which reduces coal transportation and pulverization costs), and creates more uniform coal characteristics that can increase boiler efficiency. These benefits can offset much of the cost of coal cleaning. It has been estimated that this fairly simple coal cleaning can reduce sulfur emissions by about 10 percent.

Fluidized-bed Combustion

Fluidized-bed combustion (FBC) is one of the most promising clean coal technologies. In an FBC boiler, crushed coal mixed with limestone is held in suspension (fluidized) by fast rising air injected from the bottom of the bed. Sulfur oxides formed during combustion react with the limestone ($CaCO_3$) to form solid calcium sulfate ($CaSO_4$), which falls to the bottom of the furnace and is removed. Sulfur removal rates can be higher than 90 percent.

In an FBC boiler, the hot, fluidized particles are in direct contact with the boiler tubes. This enables much of the heat to be transferred to the boiler tubes by conduction, which is much more efficient than convection and radiation heat transfer occurring in conventional boilers. The increase in heat transfer efficiency enables the boilers to operate at about half the temperature of conventional boilers (800 °C versus 1600 °C), greatly reducing the formation of NO_x. In addition to efficient SO_x and NO_x control, FBC boilers are less sensitive to variations in coal quality. Coal with higher ash content can be burned without fouling heat exchange surfaces since the lower combustion temperature is below the melting point of ash. In fact, fluidized-bed combustors can efficiently burn most any solid fuel, including cow manure and tree bark.

Utility-scale FBC applications are just beginning, the first unit having gone into operation in 1986. The future of this technology looks promising, however, and it is likely to become an important technology as time goes on.

Integrated Gasification Combined Cycle

Another promising technology, called integrated gasification combined cycle (IGCC), offers the combination of increased combustion efficiency with reduced emissions (Figure 7.29). In the first stage of an IGCC a coal-water slurry is brought into contact with steam to form a fuel gas consisting of carbon monoxide (CO) and hydrogen (H_2):

$$C \; + \; H_2O \; \rightarrow \; CO \; + \; H_2 \tag{7.31}$$

The fuel gas is cleaned up, removing most of the particulate and sulfur, and then burned in a high-efficiency gas turbine that spins a generator. Waste heat from the gas turbine runs a conventional steam cycle, generating more electricity. Despite energy losses in the gasification process, the combination of gas turbine and steam turbine offsets those losses, and the overall efficiency of fuel to electricity can be near 40 percent.

Low NO_x Combustion

Recall that nitrogen oxides are formed partly by the oxidation of nitrogen in the fuel itself (fuel NO_x) and partly by the oxidation of nitrogen in the combustion air (thermal NO_x). Coal-fired power plants, which are responsible for about one-fourth of total NO_x emissions in the United States, emit roughly twice as much fuel NO_x as thermal NO_x.

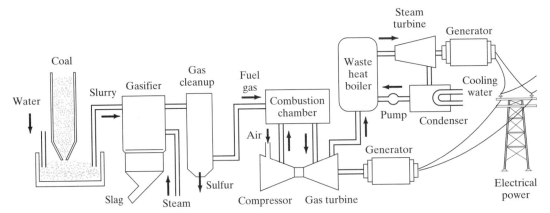

FIGURE 7.29   Integrated gasification combined cycle. Gasified coal fuels an efficient gas turbine; waste heat from the gas turbine powers a steam cycle.

Modifications to the combustion processes described in this subsection are designed to reduce both sources of NO_x. In one technique, called *low excess air*, the amount of air made available for combustion is carefully controlled at the minimum amount required for complete combustion. Low excess air technology can be retrofitted onto some boilers at a modest cost, yielding from 15 to 50 percent lower NO_x emissions.

There is a new, second-generation *low NO_x burner* technology that promises even greater NO_x removal efficiencies and can be retrofitted onto more existing furnaces. Low NO_x burners employ a staged combustion process that delays mixing the fuel and air in the boiler. In the first stage of combustion, the fuel starts burning in an air-starved environment, causing the fuel-bound nitrogen to be released as nitrogen gas, N_2, rather than NO_x. The following stage introduces more air to allow complete combustion of the fuel to take place. Potential NO_x reductions of 45 to 60 percent seem likely.

Another combustion modification incorporates a staged burner for NO_x control combined with limestone injection for SO_2 control. This *limestone injection multistage burner* (LIMB) technology is still under development but looks promising.

Flue Gas Desulfurization (Scrubbers)

Flue gas desulfurization (FSD) technologies can be categorized as being either *wet* or *dry* depending on the phase in which the main reactions occur, and as either *throwaway* or *regenerative*, depending on whether or not the sulfur from the flue gas is discarded or recovered in a usable form. Most scrubbers operating in the United States use wet, throwaway processes.

In most wet scrubbers, finely pulverized limestone ($CaCO_3$) is mixed with water to create a slurry that is sprayed into the flue gases. The flue gas SO_2 is absorbed by the slurry, producing a calcium sulfite ($CaSO_3$) or a calcium sulfate ($CaSO_4$) precipitate. The precipitate is removed from the scrubber as a sludge. Though the chemical reactions between SO_2 and limestone involve a number of steps, an overall relationship resulting in the production of inert calcium sulfite dihydrate is

$$CaCO_3 \; + \; SO_2 \; + \; 2\,H_2O \;\; \rightarrow \;\; CaSO_3 \cdot 2\,H_2O \; + \; CO_2 \qquad (7.32)$$

About 90 percent of the SO_2 can be captured from the flue gas using limestone in wet scrubbers.

Wet scrubbers sometimes use lime (CaO) instead of limestone in the slurry, and the following overall reaction applies:

$$CaO \; + \; SO_2 \; + \; 2\,H_2O \;\; \rightarrow \;\; CaSO_3 \cdot 2\,H_2O \qquad (7.33)$$

Lime slurries can achieve greater SO_2 removal efficiencies, up to 95 percent. However, lime is more expensive than limestone, so it is not widely used. Dry scrubbers must use lime, and that increased cost is one reason for their relative lack of use.

Although wet scrubbers can capture very high fractions of flue gas SO_2, they have been accepted by the utilities with some reluctance. They are expensive, costing on the order of $200 million for a large power plant. If they are installed on older plants, with less remaining lifetime, their capital costs must be amortized over a shorter period of time, and annual revenue requirements increase accordingly. Scrubbers also reduce the net energy delivered to the transmission lines. The energy to run scrubber pumps, fans, and flue gas reheat systems requires close to 5 percent of the total power produced by the plant. Scrubbers are also subject to corrosion, scaling, and plugging problems, which may reduce overall power plant reliability.

Scrubbers also use large amounts of water and create similarly large volumes of sludge that has the consistency of toothpaste. A large, 1000-MW plant burning 3 percent sulfur coal can produce enough sludge each year to cover a square mile of land to a depth of over 1 ft (Shepard, 1988). Sludge treatment often involves oxidation of calcium sulfite to calcium sulfate (which precipitates easier), thickening, and vacuum filtration. Calcium sulfate (gypsum) can be reused in the construction industry.

Particulate Control

There are a number of gas cleaning devices that can be used to remove particulates. The most appropriate device for a given source will depend on such factors as particle size, concentration, corrosivity, toxicity, volumetric flow rate, required collection efficiency, allowable pressure drops, and costs.

For relatively large particles, the most commonly used control device is the centrifugal, or *cyclone*, collector. As shown in Figure 7.30, particle-laden gas enters tangentially near the top of the cyclone. As the gas spins in the cylindrical shell, centrifugal forces cause the particles to collide with the outer walls, and then gravity causes them to slide down into a hopper at the bottom. The spiraling gases then exit the collector from the top. Efficiencies of cyclones can be above 90 percent for particles larger than 5 μm, but that efficiency drops off rapidly for the small particle sizes that are of greatest concern for human health. While they are not efficient enough to meet emission standards, they are relatively inexpensive and maintenance free, which makes them ideal as precleaners for more expensive, and critical, final control devices, such as baghouses and electrostatic precipitators.

To collect really small particles, either *baghouses* or *electrostatic precipitators* are used, with the majority of utility power plants using precipitators. Figure 7.31 shows one configuration for an electrostatic precipitator consisting of vertical wires placed between parallel collector plates. The plates are grounded and the wires are charged up to a very high (negative) voltage of perhaps 100,000 V. The intense electric field created near the wires causes a corona discharge, ionizing gas molecules in the air stream. The negative ions and free electrons thus created move toward the grounded plates, and along the way some attach themselves to passing particulate matter. The particles now carry a charge, which causes them to move under the influence of the electric field to a grounded collecting surface. They are removed from the collection electrode either by gravitational forces, by rapping, or by flushing the collecting plate with liquids. Actual electrostatic precipitators, such as the one shown in Figure 7.32, may have hundreds of parallel plates, with total collection areas measured in the tens of thousands of square meters.

Electrostatic precipitators can easily remove more than 98 percent of the particles passing through them, including particles of submicron size. Some have efficiencies even greater than 99.9 percent . They can handle large flue gas flow rates with little pressure drop, and they have relatively low operation and maintenance costs. They are quite versatile, operating on both solid and liquid particles. Precipitators are very efficient, but they are expensive and take a lot of space. Area requirements increase nonlinearly

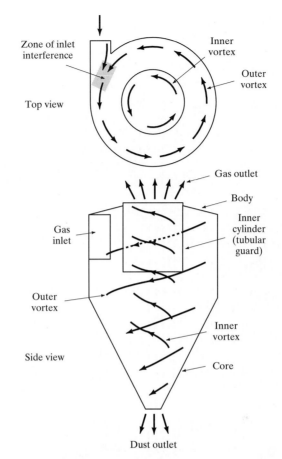

FIGURE 7.30   Conventional reverse-flow cyclone. (*Source*: U.S. HEW, 1969)

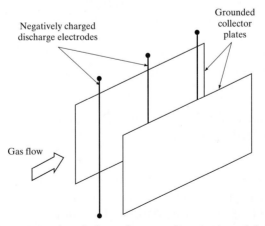

FIGURE 7.31   Schematic representation of a flat surface-type electrostatic precipitator. Particles in the gas stream acquire negative charge as they pass through the corona and are then attracted to the grounded collecting plates.

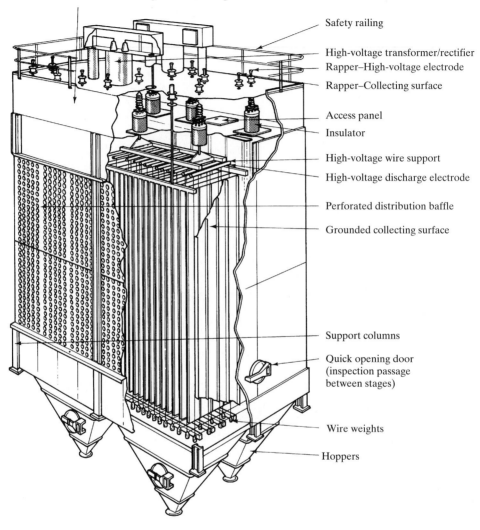

Penthouse enclosing insulators and gas seals

Safety railing

High-voltage transformer/rectifier
Rapper–High-voltage electrode
Rapper–Collecting surface

Access panel
Insulator

High-voltage wire support
High-voltage discharge electrode

Perforated distribution baffle

Grounded collecting surface

Support columns

Quick opening door
(inspection passage
between stages)

Wire weights

Hoppers

FIGURE 7.32   Cutaway view of a flat surface-type electrostatic precipitator. (*Source*: U.S. HEW, 1969)

with collection efficiency, in accordance with the following relationship, known as the Deutsch-Anderson equation:

$$\eta = 1 - e^{-wA/Q} \tag{7.34}$$

where η is the fractional collection efficiency, A is the total area of collection plates, Q is the volumetric flow rate of gas through the precipitator, and w is a parameter known as the effective drift velocity. The effective drift velocity is the terminal speed at which particles approach the collection plate under the influence of the electric field. It is determined from either pilot studies or previous experience with similar units. Any consistent set of units can be used for these quantities.

EXAMPLE 7.8   Electrostatic Precipitator Area

An electrostatic precipitator with 6000 m² of collector plate area is 97 percent efficient in treating 200 m³/s of flue gas from a 200-MW power plant. How large would the plate area have to be to increase the efficiency to 98 percent and 99 percent?

Solution   Rearranging (7.34) to solve for the drift velocity w gives

$$w = -\frac{Q}{A}\ln(1 - \eta) = -\frac{200 \text{ m}^3/\text{s}}{6000 \text{ m}^2}\ln(1 - 0.97)$$

$$= 0.117 \text{ m/s}$$

To achieve 98 percent efficiency, the area required would be

$$A_{98} = -\frac{Q}{w}\ln(1 - \eta) = -\frac{200 \text{ m}^3/\text{s}}{0.117 \text{ m/s}}\ln(1 - 0.98) = 6690 \text{ m}^2$$

To achieve 99 percent, the area required would be

$$A_{99} = -\frac{200 \text{ m}^3/\text{s}}{0.117 \text{ m/s}}\ln(1 - 0.99) = 7880 \text{ m}^2$$

That is about the area of two football fields! As these calculations suggest, the additional collector area required to achieve incremental improvements in collection efficiency goes up rapidly. To increase from 97 to 98 percent required 690 m² of added area, while the next 1 percent increment requires 1190 m². ∎

The major competition that electrostatic precipitators have for efficient collection of small particles is *fabric filtration*. Dust-bearing gases are passed through fabric filter bags, which are suspended upside-down in a large chamber, called a baghouse, as shown in Figure 7.33. A baghouse may contain thousands of bags that are often distributed among several compartments. This allows individual compartments to be cleaned while others remain in operation.

Part of the filtration is accomplished by the fabric itself, but a more significant part of the filtration is caused by the dust itself that accumulates on the inside of the bags. Efficiencies approach 100 percent removal of particles as small as 1 μm; substantial quantities of particles as small as 0.01 μm are also removed. They have certain disadvantages, however. As is the case for precipitators, baghouses are large and expensive. They can be harmed by corrosive chemicals in the flue gases, and they cannot operate in moist environments. There is also some potential for fires or explosions if the dust is combustible. The popularity of baghouse filters is rising, however, and they now rival precipitators in total industrial sales.

Fuel Cells

Fuel cells are devices that generate electricity from fuels without combustion and without most of the emissions associated with combustion. A fuel cell is basically electrolysis of water played in reverse. That is, instead of passing an electrical current through water to liberate oxygen (O_2) and hydrogen (H_2), in a fuel cell, an electrochemical reaction between hydrogen and oxygen produces electricity and water.

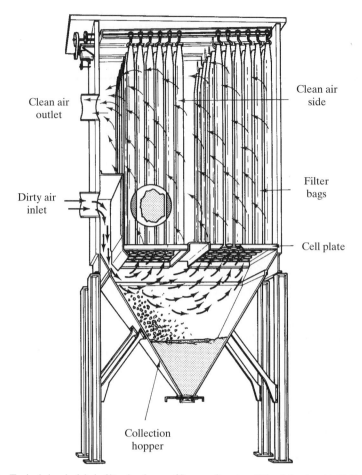

Clean air outlet

Clean air side

Dirty air inlet

Filter bags

Cell plate

Collection hopper

FIGURE 7.33 Typical simple fabric filter baghouse. (*Source*: Courtesy Wheelabrator Air Pollution Control)

Figure 7.34 shows a basic fuel cell system. The source of hydrogen for a fuel cell can be any of a number of hydrocarbon fuels, including natural gas, methanol, gasoline, and coal gas, or it may be hydrogen gas obtained by electrolysis of water. If it is a hydrocarbon fuel, it is first cleaned of any sulfur, and then it is mixed with steam and passed over a catalyst in the fuel reformer, producing a hydrogen rich gas that is sent to the anode of the fuel cell. At the same time, oxygen from air is sent to the cathode. As hydrogen is drawn through the anode, electrons are stripped off and flow to the external electrical load. The hydrogen ions pass through an electrolyte (or ion-permeable membrane) to the cathode, where they combine with oxygen and electrons to produce water. The electrons flowing through the load deliver electrical power in the form of direct current (dc). For some applications dc power is acceptable, but in a power station there would be additional equipment to convert the dc to ac with the proper frequency and voltage.

Fuel cells have a number of desirable characteristics. They are quiet, they emit almost no pollution, they are modular (so they can be small enough to power an auto-

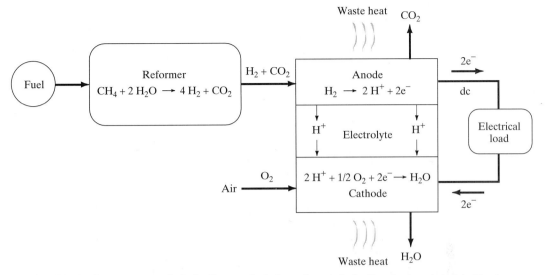

FIGURE 7.34   Basic components of a fuel cell system include a reformer, fuel cell, and external electrical load.

mobile or large enough to provide utility-scale power), and they have the potential to be more efficient than electricity generated by burning fuels (40 to 60 percent efficiencies are possible). Given these benign characteristics, fuel cells can be sited in locations where no conventional power plant would be allowed, in which case the waste heat from the fuel cell can often be used for on-site needs. For example, a number of buildings in Japan and in the United States are now using fuel cells to generate electricity on site, while the waste heat provides domestic hot water and heat for absorption-cycle air conditioning. By doing on-site *cogeneration* (simultaneous electricity and heating), the overall efficiency of fuel cells can climb toward 80 or 85 percent.

A number of different electrolytes have been used in fuel cells, and fuel cells are named accordingly. The four types that have received the most attention to date are proton-exchange membrane (PEM), phosphoric acid (PA), molten carbonate (MC), and solid oxide (SO) fuel cells. Some of their key characteristics, such as efficiency, operating temperature, and potential applications, are shown in Table 7.5. Fuel cells, such as PEM cells, that operate at lower temperatures can be started more quickly

Table 7.5   Fuel Cell Characteristics

Fuel Cell Type	Operating temp (°C)	Efficiency (percent)	Features
Proton-exchange membrane	80	≈ 60	Electric vehicles; anything that uses batteries
Phosphoric acid	200	≈ 40–45	Nearing commercialization; used in buildings
Molten carbonate	600–700	≈ 45–60	Wide source of fuels; power plants
Solid oxide	1000	≈ 50–60	Power plants; steam cogeneration

Source: NEDO, 1994.

Table 7.6   Emissions from the 200-kW, PC-25 phosphoric acid
fuel cell compared with a modern gas turbine

Pollutant	PC-25 Fuel Cell (g/MWh)	Modern clean gas turbine (g/MWh)
NO_x	7.2	105
CO	10.5	77
SO_x	0	2.3
VOC	0	123

Source: SMUD, 1993.

since they do not need as much warmup time. For small appliances and for vehicles
that is advantageous. For electric power plants, on the other hand, startup times are less
important and higher temperatures more easily allow cogeneration of electricity and
useful heat, raising the overall efficiency. Solid oxide and molten carbonate cells are
likely to be used in that way.

Air pollutant emissions for a PA fuel cell system compared with a modern gas
turbine are shown in Table 7.6.

7.10 AIR POLLUTION AND METEOROLOGY

Obviously, air quality at a given site varies tremendously from day to day even though
emissions may remain relatively constant. The determining factors have to do with the
weather: how strong are the winds, in what direction are they blowing, what is the temperature profile, how much sunlight is available, and how long has it been since the last
strong winds or precipitation were able to clear the air. Air quality is dependent on the
dynamics of the atmosphere, the study of which is called *meteorology*.

Adiabatic Lapse Rate

The ease with which pollutants can disperse vertically into the atmosphere is largely
determined by the rate of change of air temperature with altitude. For some temperature profiles the air is *stable*; that is, air at a given altitude has physical forces acting on
it that make it want to remain at that elevation. Stable air discourages the dispersion
and dilution of pollutants. For other temperature profiles, the air is unstable. In this
case rapid vertical mixing takes place that encourages pollutant dispersal and increases
air quality. Obviously, vertical stability of the atmosphere is an important factor that
helps determine the ability of the atmosphere to dilute emissions; hence, it is crucial to
air quality.

Let us investigate the relationship between atmospheric stability and temperature. It is useful to imagine a "parcel" of air being made up of a number of air molecules with an imaginary boundary around them. If this parcel of air moves upward in
the atmosphere, it will experience less pressure, causing it to expand and cool. On the
other hand, if it moves downward, more pressure will compress the air and its temperature will increase. This heating or cooling of a gas as it is compressed or expanded

should be a familiar concept. Pumping up a bicycle tire, for example, warms the valve on the tire; releasing the contents of a pressurized spray can allows the contents to expand and cool, cooling your finger on the button as well.

As a starting point, we need a relationship for the rate of change of temperature of a parcel of air as it moves up or down in the atmosphere. As it moves, we can imagine its temperature, pressure, and volume changing, and we might imagine its surroundings adding or subtracting energy from the parcel. If we make small changes in these quantities and apply both the ideal gas law and the first law of thermodynamics, it is relatively straightforward to derive the following expression (see Problem 7.12):

$$dQ = C_p \, dT - V \, dP \qquad (7.35)$$

where

dQ = heat added to the parcel per unit mass (J/kg)
C_p = specific heat of air at constant pressure; that is, the amount of heat required to raise the temperature of 1 kg of air by $1\,^\circ\text{C}$ while holding its pressure constant (= 1005 J/kg-K)
dT = incremental temperature change (K)
V = volume per unit mass (m^3/kg)
dP = incremental pressure change in the parcel (Pa)

Let us make the accurate assumption that as the parcel moves, there is no heat transferred across its boundary, that is, this process is *adiabatic*. This means that $dQ = 0$, so we can rearrange (7.35) as

$$\frac{dT}{dP} = \frac{V}{C_p} \qquad (7.36)$$

Equation (7.36) gives us an indication of how atmospheric temperature would change with air pressure, but what we are really interested in is how it changes with altitude. To do that we need to know how pressure and altitude are related.

Consider a static column of air with cross section A, as shown in Figure 7.35. A horizontal slice of air in that column of thickness dz and density ρ, will have mass $\rho A \, dz$. If the pressure at the top of the slice due to the weight of air above it is $P(z + dz)$, then

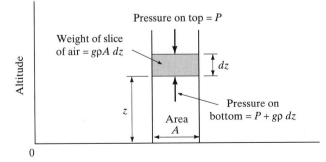

FIGURE 7.35   A column of air in static equilibrium used to determine the relationship between air pressure and altitude.

the pressure at the bottom of the slice, $P(z)$, will be $P(z + dz)$ plus the added weight per unit area of the slice itself:

$$P(z) = P(z + dz) + \frac{g\rho A \, dz}{A}$$

where g is the gravitational constant. We can write the incremental pressure dP for an incremental change in elevation, dz as

$$dP = P(z + dz) - P(z) = -g\rho \, dz \tag{7.37}$$

Expressing the rate of change in temperature with altitude as a product, and substituting in (7.36) and (7.37), gives

$$\frac{dT}{dz} = \frac{dT}{dP} \cdot \frac{dP}{dz} = \left(\frac{V}{C_p}\right)(-g\rho) \tag{7.38}$$

However, since V is volume per unit mass and ρ is mass per unit volume, the product $V\rho = 1$, and (7.38) simplifies to

$$\frac{dT}{dz} = \frac{-g}{C_p} \tag{7.39}$$

The negative sign indicates that temperature decreases with increasing altitude. Substituting the constant $g = 9.806$ m/s^2, and the constant-volume specific heat of dry air at room temperature $Cp = 1005$ J/kg-K, into (7.39) yields

$$\frac{dT}{dz} = \frac{-9.806 \text{ m/s}^2}{1005 \text{ J/kg-K}} \times \frac{1 \text{ J}}{\text{kg-m}^2/\text{s}^2} = -0.00976 \text{ K/m} \tag{7.40}$$

This is a very important result. When its sign is changed to keep things simple, $-dT/dz$ is given a special name: the *dry adiabatic lapse rate* Γ_d.

$$\Gamma_d = -\frac{dT}{dz} = 9.76\,^{\circ}\text{C/km} = 5.4\,^{\circ}\text{F}/1000 \text{ ft} \tag{7.41}$$

Equation (7.41) tells us that moving a parcel of dry air up or down will cause its temperature to change by 9.76 °C/km, or roughly 10 °C/km. This temperature profile will be used as a reference against which actual ambient air temperature gradients will be compared. As we shall see, if the actual air temperature decreases faster with increasing elevation than the adiabatic lapse rate, the air will be unstable and rapid mixing and dilution of pollutants will occur. Conversely, if the actual air temperature drops more slowly than the adiabatic lapse rate, the air will be stable and air pollutants will concentrate.

Equation (7.41) was derived assuming that our parcel of air could be treated as an ideal gas that could be moved about in the atmosphere without any heat transfer between it and its surroundings. Both assumptions are very good. It was also assumed that the air was dry, but this may or may not be such a good assumption. If air has some water vapor in it, C_p changes slightly from the value assumed, but not enough to warrant a correction. On the other hand, if enough water vapor is present that condensation occurs when the parcel is raised and cooled, latent heat will be released. The added heat means a saturated air parcel will not cool as rapidly as a dry one. Unlike the dry adiabatic rate, the *saturated adiabatic lapse rate* (Γ_s) is not a constant since the amount of moisture that air can hold before condensation begins is a function of temperature. A reasonable average value of the moist adiabatic lapse rate in the troposphere is about 6 °C/km. Figure 7.36 shows the dry and saturated adiabatic rates.

EXAMPLE 7.9   Air Conditioning for a High-altitude Aircraft?

An aircraft flying at an altitude of 9 km (30,000 ft) draws in fresh air at −40 °C for cabin ventilation. If that fresh air is compressed to the pressure at sea level, would the air need to be heated or cooled if it is to be delivered to the cabin at 20 °C?

Solution   As the air is compressed, it warms up. As it warms up, it is even easier for the air to hold whatever moisture it may have had, so there is no condensation to worry about and the dry adiabatic lapse rate can be used. At 10 °C per km, compression will raise the air temperature by $10 \times 9 = 90$ °C, making it $-40 + 90$ °C $= 50$ °C (122 °F). It needs to be air conditioned!   ∎

Atmospheric Stability

Our interest in lapse rates is based on the need to understand atmospheric stability since it is such a crucial factor in the atmosphere's ability to dilute pollution. There are a number of factors, such as windspeeds, sunlight, and geographical features, that cause the actual *ambient lapse rate* (Γ) in the real atmosphere to differ from the 1 °C/100 m dry adiabatic lapse rate just calculated. Differences between the ambient lapse rate and the adiabatic lapse rate determine the stability of the atmosphere.

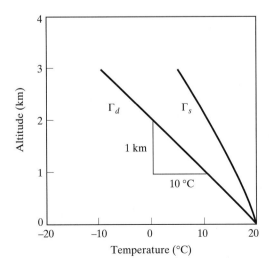

FIGURE 7.36   The dry adiabatic lapse rate Γ_d is a constant 10 °C/km, but the saturated adiabatic lapse rate Γ_s varies with temperature. In the troposphere, Γ_s is approximately 6 °C/km.

Consider a parcel of air at some given altitude. It has the same temperature and pressure as the air surrounding it. Our test for atmospheric stability will be based on the following mental experiment. If we imagine raising the parcel of air slightly, it will experience less atmospheric pressure, so it will expand. Because it will have done work on its environment (by expanding), the internal energy in the parcel will be reduced so its temperature will drop. Assuming that the parcel is raised fast enough to be able to ignore any heat transfer between the surrounding air and the parcel, the cooling will follow the adiabatic lapse rate. After raising the parcel, note its temperature, and compare it with the temperature of the surrounding air. If the parcel, at this higher elevation, is now colder than its surroundings, it will be denser than the surrounding air and will want to sink back down again. That is, whatever it was that caused the parcel to start to move upward will immediately be opposed by conditions that make the parcel want to go back down again. The atmosphere is said to be *stable*. If, however, raising the parcel causes its temperature to be greater than the surrounding air, it will be less dense than the surrounding air and it will experience buoyancy forces that will encourage it to keep moving upward. The original motion upward will be reinforced and the parcel will continue to climb. This is an *unstable* atmosphere.

Consider Figure 7.37*a*, which shows an ambient temperature profile for air that cools more rapidly with altitude than the dry adiabatic lapse rate. In this case the ambient air temperature is said to be *superadiabatic*. Imagine a 20 °C parcel of air at 1 km to be just like the air surrounding it. If that parcel is raised to 2 km, it will cool adiabatically to 10 °C. The 10 °C parcel of air at 2 km is now warmer than the surrounding air (0 °C in the figure), so it is less dense and more buoyant and wants to keep rising

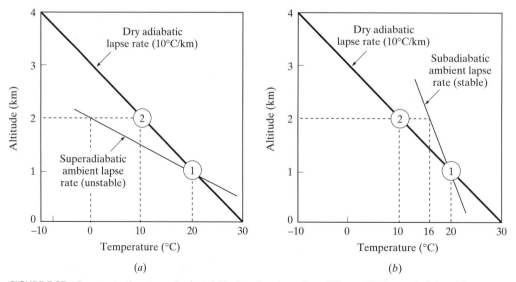

FIGURE 7.37 Demonstrating atmospheric stability in a dry atmosphere. When a 20 °C parcel of air at 1 km (position 1) moves up to 2 km (position 2), its temperature drops to 10 °C (following the dry adiabatic lapse rate). In (*a*) the parcel of air at 2 km is warmer than the surrounding ambient air so the parcel keeps rising (unstable). In (*b*) the parcel at 2 km is colder than ambient, so it sinks back down (stable).

("warm air rises"). In other words, shoving the parcel of air up some in a superadiabatic atmosphere creates forces that want to keep it moving upward. Conversely, a parcel of air at 1 km that is nudged downward becomes colder than the surrounding air and so it keeps sinking. It does not matter whether the parcel of air is nudged up or down; in either case, it keeps on going so the atmosphere is said to be unstable.

In Figure 7.37b, the ambient temperature profile is drawn for the subadiabatic case in which the ambient temperature cools less rapidly than the adiabatic lapse rate. If we once again imagine a parcel of air at 1 km and 20 °C that is for one reason or another nudged upward, it will find itself colder than the surrounding air. At its new elevation, it experiences the same pressure as the air around it, but it is colder, so it will be denser and will sink back down. Conversely, a parcel starting at 1 km and 20 °C that starts moving downward will get warmer and less dense than its surroundings, so buoyancy forces will push it back up again. In other words, anything that starts a parcel of air moving up or down will cause the parcel to experience forces that will want it to return to its original altitude. The subadiabatic atmospheric profile is therefore stable.

If the ambient lapse rate is equal to the adiabatic lapse rate, moving the parcel upward or downward results in its temperature changing by the same amount as its surroundings. In any new position, it experiences no forces that either make it continue its motion or make it want to return to its original elevation. The parcel likes where it was, and it likes its new position too. Such an atmospheric is said to be *neutrally stable*.

Figure 7.37 was drawn for a dry atmosphere; that is, as a parcel moves upward the air remains unsaturated, there is no condensation of moisture, and it cools at 10 °C/km. If condensation does occur, the adiabatic lapse rate drops to around 6 °C/km, and that slope determines the condition of neutral stability. Figure 7.38 shows a summary of how lapse rates determine atmospheric stability. If the ambient lapse rate shows cooling at a faster rate than the dry adiabatic lapse rate, the atmosphere is *absolutely unstable*. The air will always want to move to some new altitude so vertical dispersion of pollution is enhanced. For ambient temperatures that cool less rapidly than the saturated adiabatic lapse rate, the atmosphere is *absolutely stable*. That air wants to stay where it is. For ambient lapse rates between the dry and wet adiabatic lapse rates, the

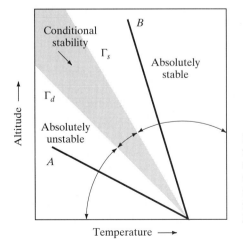

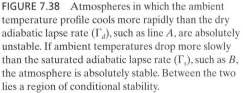

FIGURE 7.38   Atmospheres in which the ambient temperature profile cools more rapidly than the dry adiabatic lapse rate (Γ_d), such as line A, are absolutely unstable. If ambient temperatures drop more slowly than the saturated adiabatic lapse rate (Γ_s), such as B, the atmosphere is absolutely stable. Between the two lies a region of conditional stability.

atmosphere may be stable or it may be unstable; we cannot tell without knowing the actual ambient temperature profile and the actual adiabatic profile. That region is labeled as having *conditional stability*.

An extreme case of a subadiabatic lapse rate is one in which ambient temperatures increase with altitude. Such *temperature inversions* yield a very stable air mass, and pollution tends to stay trapped there. In Chapter 8, it will be noted that the heating caused by absorption of incoming solar energy by oxygen and ozone in the upper atmosphere creates a stratospheric temperature inversion. That inversion causes the stratosphere to be extremely stable, often trapping pollutants for many years.

Temperature Inversions

Temperature inversions represent the extreme case of atmospheric stability, creating a virtual lid on the upward movement of pollution. There are several causes of inversions, but the two that are the most important from an air quality standpoint are *radiation inversions* and *subsidence inversions*. Radiation inversions are caused by nocturnal cooling of the earth's surface, especially on clear winter nights. The second, subsidence inversions, are the result of the compressive heating of descending air masses in high-pressure zones. There are other, less important, causes of inversions such as *frontal* inversions. A frontal inversion is created when a cold air mass passes under a warm air mass, but these are short lived and tend to be accompanied by precipitation which cleanses the air. There are also inversions associated with geographical features of the landscape. Warm air passing over a cold body of water, for example, creates an inversion. There are also inversions in valleys when cold air rolls down the canyons at night under warmer air that might exist aloft.

Radiation Inversions.   The surface of the earth cools down at night by radiating energy toward space. On a cloudy night, the earth's radiation tends to be absorbed by water vapor, which in turn reradiates some of that energy back to the ground. On a clear night, however, the surface more readily radiates energy to space, and thus ground cooling occurs much more rapidly. As the ground cools, the temperature of the air in contact with the ground also drops. As is often the case on clear winter nights, the temperature of this air just above the ground becomes colder than the air above it, creating an inversion. Radiation inversions begin to form at about dusk. As the evening progresses, the inversion extends to a higher and higher elevation, reaching perhaps a few hundred meters before the morning sun warms the ground again, breaking up the inversion. Figure 7.39 shows the development of a radiation inversion through the night, followed by the erosion of the inversion that takes place the next day.

Radiation inversions occur close to the ground, mostly during the winter, and last for only a matter of hours. They often begin at about the time traffic builds up in the early evening, which traps auto exhaust at ground level and causes elevated concentrations of pollution for commuters. Without sunlight, photochemical reactions cannot take place, so the biggest problem is usually the accumulation of carbon monoxide (CO). In the morning, as the sun warms the ground and the inversion begins to break up, pollutants that have been trapped in the stable air mass are suddenly brought back to earth in a process known as *fumigation*. Fumigation can cause short-lived, high concentrations of pollution at ground level.

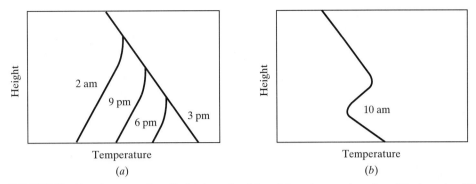

FIGURE 7.39   Development of a radiation inversion (*a*), and the subsequent erosion of the inversion (*b*). The times are representative only. The breakup of the inversion in the morning leads to a process called fumigation.

Radiation inversions are important in another context besides air pollution. Fruit growers in places like California have long known that their crops are in greatest danger of frost damage on winter nights when the skies are clear and a radiation inversion sets in. Since the air even a few meters up is warmer than the air at crop level, one way to help protect sensitive crops on such nights is simply to mix the air with large motor-driven fans.

Subsidence Inversions.   While radiation inversions are mostly a short-lived, ground level, wintertime phenomenon, the other important cause of inversions, subsidence, creates the opposite characteristics. Subsidence inversions may last for months on end, occur at higher elevations, and are more common in summer than winter.

Subsidence inversions are associated with high-pressure weather systems, known as *anticyclones*. Air in the middle of a high-pressure zone is descending, while on the edges it is rising. Air near the ground moves outward from the center, while air aloft moves toward the center from the edges. The result is a massive vertical circulation system. As air in the center of the system falls, it experiences greater pressure and is compressed and heated. If its temperature at elevation z_1 is T_1, then as it falls to elevation z_2 it will be heated adiabatically to $T_2 = T_1 + \Gamma_d(z_1 - z_2)$ as shown in Figure 7.40. As is often the case, this compressive heating warms the descending air to a higher temperature than the air below, whose temperature is dictated primarily by conditions on the ground.

Since subsiding air is getting warmer, it is more and more able to hold moisture as it descends. Without sources of new moisture, its relative humidity drops and there is little chance that clouds will form. The result is that high-pressure zones create clear, dry weather with lots of sunshine during the day and clear skies at night. Clear skies during the day allow solar warming of the earth's surface, which often creates superadiabatic conditions under the inversion. The result of subsidence and surface warming is an inversion located anywhere from several hundred meters above the surface to several thousand meters that lasts as long as the high-pressure weather system persists. At night, the surface can cool quickly by radiation, which may result in a radiation inversion, located under the subsidence inversion as shown in Figure 7.41.

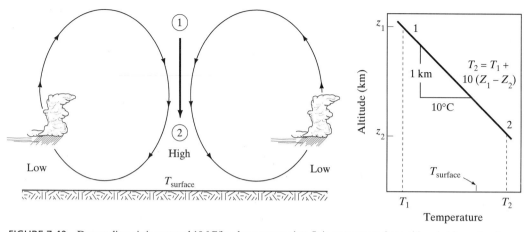

FIGURE 7.40   Descending air is warmed 10 °C/km by compression. It is common to have this subsidence heating create warmer conditions aloft than near the surface, which causes a temperature inversion. Clear skies are typical in a region of high-pressure, while cloudiness and precipitation are common in areas with low atmospheric pressure.

Some anticyclones and their accompanying subsidence inversions drift across the continents (west to east in northern midlatitudes) so that at any given spot they may appear to come and go with some frequency. On the other hand, other anticyclones are semipermanent in nature and can cause subsidence inversions to last for months at a time. These semipermanent highs are the result of the general atmospheric circulation patterns shown in Figure 7.42.

At the equinox, the equator is directly under the sun, and the air there will be heated, become buoyant, and rise. As that air approaches the top of the troposphere (roughly 10 to 12 km), it begins to turn, some heading north and some south. An eighteenth-century meteorologist, George Hadley, postulated that the air would continue to the poles before descending. In actuality, it descends at a latitude of about 30° and then returns

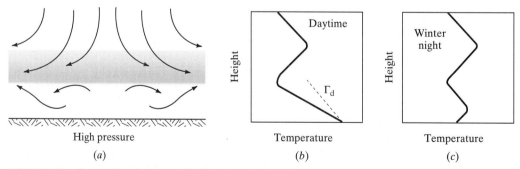

FIGURE 7.41   Descending air causes subsidence inversions (*a*). During the day, air under the inversion may be unstable due to solar warming of the surface (*b*). Radiation inversions may form under the subsidence inversion when nights are clear, especially in winter (*c*).

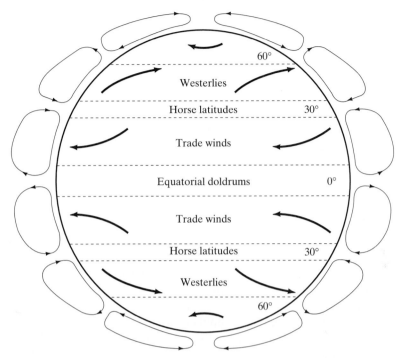

FIGURE 7.42 Idealized general air circulation patterns, drawn at the equinox. High-pressure zones are common around 30° latitude, producing clear skies and subsidence inversions.

to the equator, forming what is known as a *Hadley cell*. In similar fashion, though not as distinct, there are other cells, linked as in a chain, between 30° and 60° latitude and between 60° and the poles. The descending air at 30° creates a persistent high-pressure zone with corresponding lack of clouds or rainfall that is a contributing factor to the creation of the world's great deserts. The deserts of southern California, and the southwestern United States, the Sahara, the Chilean, the Kalahari in South Africa, and the great deserts in Australia are all located at roughly 30° latitude for this reason. Conversely, rising air near the equator and near 60° latitude tends to be moist after passing over oceans. As it cools, the moisture condenses, causing clouds and rainfall. For this reason, some of the wettest regions of the world are in those bands of latitude.

The air that moves along the surface of the earth close to a Hadley cell is affected by Coriolis forces, causing it to curve toward the west if it is moving toward the equator or toward the east if it is moving toward the poles. The resulting winds between roughly 30° and 60° are known as the *westerlies*, and between the 30° and the equator they are the *trade winds*. Near the equator, there is little wind since the air is mostly rising. That band is called the *equatorial doldrums*. Similarly, the surface air is relatively calm around 30° latitude, forming a band called the *horse latitudes*. (Apparently, early explorers were sometimes forced to lighten the load by throwing horses overboard to avoid being becalmed as they sailed to the New World.)

All of these important bands of latitudes, the doldrums, horse latitudes, high-pressure and low-pressure zones, move up and down the globe as the seasons change. Figure 7.42 shows global circulation patterns at an equinox, when the sun is directly over the equator. In the northern hemisphere summer, the sun moves northward, as do the persistent high-and low-pressure zones associated with these bands. In the northern-hemisphere winter, the sun is directly overhead somewhere in the southern hemisphere and the bands move southward. There is, for example, a massive high-pressure zone off the coast of California that moves into place over Los Angeles (latitude 34°) and San Francisco (latitude 38°) in the spring and remains there until late fall. That is the principal reason for California's sunny climate, as well as its smog. Clear skies assure plenty of sunlight to power photochemical reactions, the lack of rainfall eliminates that atmospheric cleansing mechanism, and prolonged subsidence inversions concentrate the pollutants. In the case of Los Angeles, there is also a ring of mountains around the city that tends to keep winds from blowing the smog away.

The global circulation patterns of Figure 7.42 are, of course, idealized. The interactions between sea and land, the effects of storms and other periodic disturbances, and geographical features such as high mountain ranges all make this model useful only on a macroscale. Even a simplified global model, however, helps explain a number of significant features of the world's climate and some aspects of regional air pollution problems.

Atmospheric Stability and Mixing Depth

Atmospheric stability is important because it determines the ability of pollutants to disperse vertically into the atmosphere. We already know that layers of the atmosphere have stability characteristics that depend on their local lapse rate in comparison with adiabatic lapse rates. That picture is incomplete, however, and we must introduce another conceptual experiment before we can be sure of the consequences of local lapse rates.

Consider the temperature profile in Figure 7.43a. For the moment, let us assume that the air is dry so we do not have to worry about the difference between dry and saturated lapse rates. If we look only at the lapse rates, the layer of air from ground up to 0.5 km is superadiabatic and hence we would label it "unstable"; the layer from 0.5 to 1.5 km is subadiabatic, so it is labeled "stable"; and the layer above 1.5 km has a lapse rate equal to the dry adiabatic lapse rate 10 °C/km, so it is labeled "neutral." We call these "local" stability classifications.

Now consider three parcels of air, labeled 1, 2, and 3, in Figure 7.43b. If for some reason one of these parcels starts to move up or down its temperature will change by 10 °C/km. After imagining that the parcel has moved, compare its temperature with the temperature of the surrounding air at the new altitude. If the parcel is warmer, it will be buoyant and will rise; if it is colder than the surrounding air, it will be denser and will sink. With this in mind, consider parcel 1. If for some reason parcel 1 starts to move slightly upward, it will cool at the dry adiabatic lapse rate of 10 °C/km, making it warmer than the air surrounding it, so the parcel will continue rising on its own. If par-

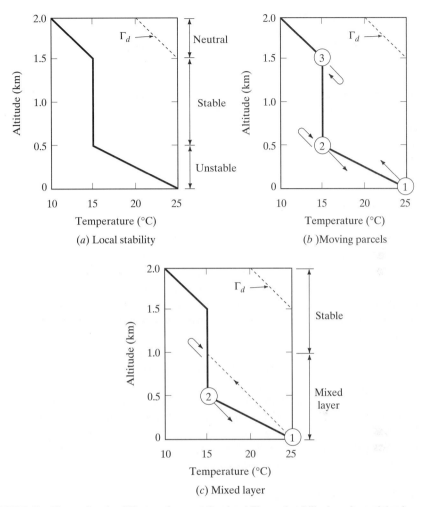

FIGURE 7.43   Illustrating the difference between local stability and stability based on mixing layers (*a*) Local stability determined by lapse rates; (*b*) parcels 1 and 2 will move; (*c*) parcel 1 will rise to 1km, creating turbulence and mixing from ground level to 1 km.

cel 2 starts to rise, however, it will find itself colder than the surrounding air, so it will sink back to where it started from. If it sinks below its initial level, it will be colder than surrounding air and will continue to sink. If parcel 3 starts down, it will be warmer than the surrounding air and will rise back to where it was.

So how far up will parcel 1 go? It will keep rising as long as it is warmer than the surrounding air. As shown in 7.43*c*, it will rise to 1 km and then stop. If it went higher than 1 km, it would find itself colder than its surroundings and would sink back to 1 km. In other words, air from ground level to 1 km will actually experience turbulent mixing despite the "stable" lapse rate above 0.5 km.

The process illustrated in Figure 7.43 helps provide a sense of the ability of the atmosphere to provide dilution of air pollutants. By projecting the ground temperature upward at the dry adiabatic lapse rate until it crosses the actual ambient temperature profile, a region of turbulent mixing is identified. The altitude of the top of that mixed layer is called the *mixing depth* (or sometimes the *mixing height*).

The product of mixing depth and average windspeed within the mixing depth is sometimes used as an indicator of the atmosphere's dispersive capability. This product is known as the *ventilation coefficient* (m²/s). Values of ventilation coefficient less than about 6000 m²/s are considered indicative of high air pollution potential (Portelli and Lewis, 1987).

Saturated Adiabatic Lapse Rate and Cloud Formation

In determining the mixing depth, it is assumed that air rising from ground level cools at the dry adiabatic lapse rate. But as air cools, it is less and less able to hold the moisture that it contains. At a temperature called the *dew point*, the air is saturated and any further cooling will cause liquid cloud droplets to form. The latent heat given off when this condensation occurs causes the parcel of rising air to cool less rapidly. The slope of the temperature profile then switches from the dry adiabatic lapse rate to the saturated adiabatic lapse rate.

To determine the elevation at which a rising parcel of air reaches its dew point temperature, we start by imagining a parcel of air at ground level. Project its temperature as it rises and cools using the dry adiabatic lapse rate of 10 °C/km. If the dew point temperature were fixed, the problem would be simple: Just project the temperature until it reaches the dew point. This determination is complicated, however, by the fact that the dew point temperature changes as the air parcel rises. An adequate assumption that can be used to handle this complication is to estimate the drop in dew point at 2°C/km, as illustrated in the following example.

EXAMPLE 7.10   Formation of Clouds

Suppose air at ground level is 38 °C and has a dew point of 30 °C. The atmosphere is unstable and there are rising parcels of air. Estimate the altitude at which clouds begin to form if the dew point drops by 2 °C/km.

Solution   At ground level the air is unsaturated (its temperature is above the dew point), so as a parcel rises it will cool following the dry adiabatic lapse rate slope of 10 °C/km. At ground level the dew point is 30 °C and it drops by 2 °C/km. We need to know at what altitude the parcel temperature equals the dew point temperature. Letting h represent altitude,

$$\text{Parcel temperature} = \text{Dew point temperature}$$

$$38\,°C - 10(°C/km) \times h\,(km) = 30\,°C - 2(°C/km) \times h\,(km)$$

$$h = 1.0\,km = 1000\,m$$

A plot of the parcel and dew point temperatures shows this crossover point (Figure 7.44). As the air rises beyond 1 km, both the lapse rate and dew point drop at the same rate of about 6 °C/km.   ∎

Smokestack Plumes and Adiabatic Lapse Rates

The atmospheric temperature profile affects the dispersion of pollutants from a smokestack, as shown in Figure 7.45. If a smokestack were to emit pollutants into a neutrally stable atmosphere, we might expect the plume to be relatively symmetrical, as shown in Figure 7.45*a*. The term used to describe this plume is *coning*. In Figure 7.45*b*, the atmosphere is very unstable and there is rapid vertical air movement, both up and down, producing a *looping* plume. In Figure 7.45*c*, a *fanning* plume results when a stable atmosphere greatly restricts the dispersion of the plume in the vertical direction, although it still spreads horizontally. In Figure 7.45*d*, when a stack is under an inversion layer, emissions move downward much more easily than upward. The resulting *fumigation* can lead to greatly elevated downwind, ground level concentrations.

When the stack is above an inversion layer, as in Figure 7.45*e*, mixing in the upward direction is uninhibited, but downward motion is greatly restricted by the inversion's stable air. Such *lofting* helps keep the pollution high above the ground, reducing exposure to people living downwind. In fact, a common approach to air pollution control in the past has been to build taller and taller stacks to emit pollutants above inversions. An unfortunate consequence of this approach, however, has been that pollutants released from tall stacks are able to travel great distances and may cause unexpected effects, such as acid deposition, hundreds of miles from the source.

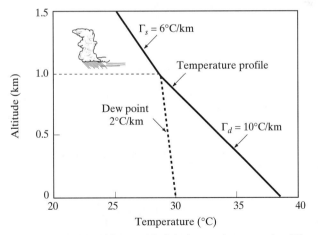

FIGURE 7.44 Temperature profile for a rising parcel of air that reaches saturation. Where the curves of dew point and dry adiabatic lapse rate cross, clouds form and further ascension proceeds at the saturated adiabatic lapse rate. (Numbers correspond to Example 7.10.)

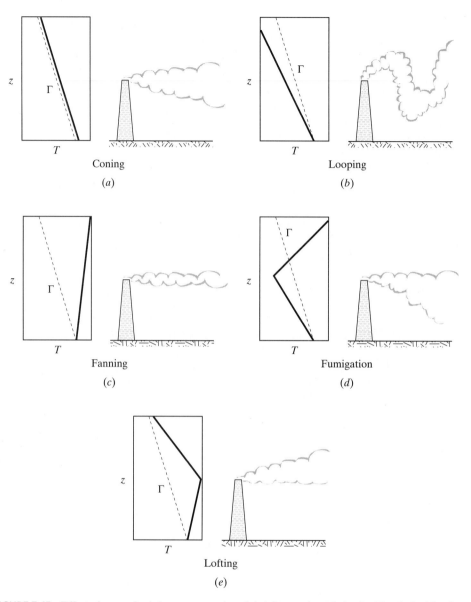

FIGURE 7.45 Effect of atmospheric lapse rates and stack heights on plume behavior. The dashed line is the dry adiabatic lapse rate for reference.

7.11 THE POINT-SOURCE GAUSSIAN PLUME MODEL

The Clean Air Act specifies new source emission standards, and it specifies ambient air quality standards. The connecting link between the two is the atmosphere. How do pollutants behave once they have been emitted, and how may we predict their concentrations in the atmosphere? How can the limitations imposed by a Prevention of Significant Deterioration policy be shown to be satisfied for new sources in an attainment area. How can we predict the improvement in air quality that must be achieved when new sources are proposed for a nonattainment area? To help answer questions like these, computer models that use such information as predicted emissions, smokestack heights, wind data, atmospheric temperature profiles, ambient temperatures, solar insolation, and local terrain features have been developed and will be discussed in this section.

At the heart of most every computer program that attempts to relate emissions to air quality is the assumption that the time-averaged pollutant concentration downwind from a source can be modeled using a normal, or Gaussian, distribution curve (for a brief discussion of the Gaussian curve, see Section 3.3). The basic Gaussian dispersion model applies to a single *point source*, such as a smokestack, but it can be modified to account for *line sources* (such as emissions from motor vehicles along a highway) or *area* sources (these can be modeled as a large number of point sources).

To begin, consider just a single point source such as that shown in Figure 7.46. The coordinate system has been set up to show a cross section of the plume, with z representing the vertical direction and x the distance directly downwind from the source. If we were to observe the plume at any particular instant, it might have some irregular shape, such as the outline of the looping plume shown. A few minutes later, however, the plume might have an entirely different boundary. If we were to set up a camera and leave the shutter open for a while, we could imagine getting a photograph of a time-averaged plume envelope such as that shown in Figure 7.46.

Since stack emissions have some initial upward velocity and buoyancy, it might be some distance downwind before the plume envelope might begin to look symmetri-

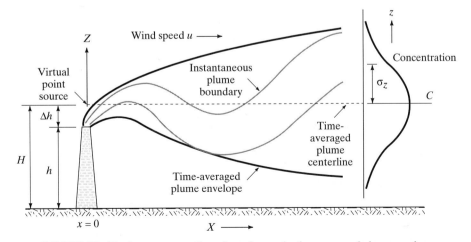

FIGURE 7.46   The instantaneous plume boundary and a time-averaged plume envelope.

cal about a centerline. The centerline would be somewhat above the actual stack height. The highest concentration of pollution would be along this centerline, with decreasing concentrations as the distance from this centerline increases. The Gaussian plume model assumes that the pollutant concentration follows a normal distribution about this centerline in both the vertical plane, as shown in the figure, and in the horizontal direction, not shown. It also treats emissions as if they came from a virtual point source along the plume centerline, at an *effective stack height H.*

The Gaussian point-source dispersion equation relates average, steady-state pollutant concentrations to the source strength, wind speed, effective stack height, and atmospheric conditions. Its form can be derived from basic considerations involving gaseous diffusion in three-dimensional space. The derivation, however, is beyond the scope of this book (see, for example, Wark and Warner, 1981). It is important to note the following assumptions, which are incorporated into the analysis:

- The rate of emissions from the source is constant.
- The windspeed is constant both in time and with elevation.
- The pollutant is conservative, that is, it is not lost by decay, chemical reaction, or deposition. When it hits the ground, none is absorbed and all is reflected.
- The terrain is relatively flat, open country.

The three-dimensional coordinate system established in Figure 7.47 has the stack at the origin, with distance directly downwind given by x, distance off the downwind axis specified by y, and elevation given by z. Since our concern is going to be only with

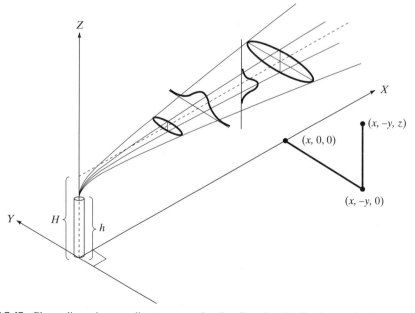

FIGURE 7.47 Plume dispersion coordinate system, showing Gaussian distributions in the horizontal and vertical directions. (*Source*: Turner, 1970)

receptors (people and ecosystems) at ground level, the form of the Gaussian plume equation given here is less general than it can be, and applies only for $z = 0$:

$$C(x, y) = \frac{Q}{\pi\, u_\mathrm{H}\sigma_y\sigma_z} \exp\left(\frac{-H^2}{2\sigma_z^2}\right) \exp\left(\frac{-y^2}{2\sigma_y^2}\right) \tag{7.42}$$

where

$C(x, y)$ = concentration at ground-level at the point (x, y), $\mu g/m^3$

x = distance directly downwind, m

y = horizontal distance from the plume centerline, m

Q = emission rate of pollutants, $\mu g/s$

H = effective stack height, m ($H = h + \Delta h$, where h = actual stack height and Δh = plume rise)

u_H = average windspeed at the effective height of the stack, m/s

σ_y = horizontal dispersion coeffecient (standard deviation), m

σ_z = vertical dispersion coefficient (standard deviation), m

Before we get into the details of (7.42), several features are worth noting. Ground-level pollution concentration is directly proportional to the source strength Q, so it is easy to determine how much source reduction is necessary to achieve a desired decrease in downwind concentration. The units of Q, by the way, have been given in micrograms so concentrations will be in the usual micrograms per cubic meter ($\mu g/m^3$). The ground level pollution decreases when taller stacks are used, although the relationship is not linear. Also note that there is no explicit relationship between emission rate Q and downwind distance x. The downwind distance will be brought into the equation when σ_y and σ_z are evaluated.

Equation (7.42) suggests that downwind concentration appears to be inversely proportional to the windspeed, which is what might be expected intuitively. In actuality, the inverse relationship is slightly modified by the dependence of plume rise Δh on windspeed. Higher windspeeds reduce the effective height of the stack H, which keeps ground level pollution from dropping quite as much a simple inverse relationship would imply.

Finally, and most important, although the Gaussian plume equation is based on both theory and actual measured data, it is still, at best, a crude model. Predictions based on the model should be assumed to be accurate to within perhaps ±50 percent. In spite of that uncertainty, however, it is still very useful since it has wide acceptance, is easy to use, and allows comparison between the estimates made by different modelers in varying situations.

Windspeed Changes with Elevation

The windspeed to be used in (7.42), u_H, is the windspeed at the effective stack height. Usually windspeed is measured with an anemometer that is set up at a height of 10 m above the ground, so we need some way to relate windspeed at the anemometer height with windspeed at the effective stack height. The following power law expression is frequently used for elevations less than a few hundred meters above the ground:

$$\left(\frac{u_{\mathrm{H}}}{u_{\mathrm{a}}}\right) = \left(\frac{H}{z_{\mathrm{a}}}\right)^p \tag{7.43}$$

where

u_{H} = windspeed at the elevation H
u_{a} = windspeed at the anemometer height
H = effective height of the plume
z_{a} = anemometer height above ground
p = a dimensionless parameter that depends on surface roughness and atmospheric stability

Table 7.7 gives values for p that are recommended by the EPA when there are rough surfaces in the vicinity of the anemometer (Peterson, 1978). For smooth terrain such as flat fields or near bodies of water, the values of p given in Table 7.7 should be multiplied by 0.6. The *stability class* indicators in the table will be further clarified in Table 7.8, when we compute the characteristics of smokestack plumes.

Table 7.7   Wind Profile Exponent p, for Rough Terrain[a]

Stability Class	Description	Exponent p
A	Very unstable	0.15
B	Moderately unstable	0.15
C	Slightly unstable	0.20
D	Neutral	0.25
E	Slightly stable	0.40
F	Stable	0.60

[a] For smooth terrain, multiply p by 0.6; see Table 7.8 for further descriptions of the stability classifications used here (Peterson, 1978).

EXAMPLE 7.11   Windspeed at the Stack

Suppose an anemometer at a height of 10 m above ground measures the windspeed at 2.5 m/s. Estimate the windspeed at an elevation of 300 m in rough terrain if the atmosphere is slightly unstable.

Solution   From Table 7.7 we find that the wind profile exponent for a slightly unstable atmosphere (stability class C) is 0.20. Rearranging (7.43) gives

$$u_{\mathrm{H}} = u_{\mathrm{a}}\left(\frac{H}{z_{\mathrm{a}}}\right)^p = 2.5 \cdot \left(\frac{300}{10}\right)^{0.2} = 4.9 \text{ m/s} \qquad \blacksquare$$

Table 7.8 Atmospheric Stability Classifications

Surface wind speed[a] (m/s)	Day solar insolation			Night cloudiness[e]	
	Strong[b]	Moderate[c]	Slight[d]	Cloudy ($\geq 4/8$)	Clear ($\leq 3/8$)
< 2	A	A–B[f]	B	E	F
2–3	A–B	B	C	E	F
3–5	B	B–C	C	D	E
5–6	C	C–D	D	D	D
> 6	C	D	D	D	D

[a]Surface wind speed is measured at 10 m above the ground.
[b]Corresponds to clear summer day with sun higher than 60° above the horizon.
[c]Corresponds to a summer day with a few broken clouds, of a clear day with sun 35–60° above the horizon.
[d]Corresponds to a fall afternoon, or a cloudy summer day, or clear summer day with the sun 15–35° above the horizon.
[e]Cloudiness is defined as the fraction of sky covered my clouds.
[f]For A–B, B–C, or C–D conditions, average the values obtained for each.
Note: A, Very unstable; B, moderately unstable; C, slightly unstable; D, neutral; E, slightly stable; F, stable. Regardless of windspeed, class D should be assumed for overcast conditions, day or night.
Source: Turner (1970).

The Gaussian Dispersion Coefficients

The two dispersion coefficients in (7.42), σ_y and σ_z, need explanation. These are really just the standard deviations of the horizontal and vertical Gaussian distributions, respectively (about 68 percent of the area under a Gaussian curve is within $\pm 1 \, \sigma$ of the mean value). Smaller values for a dispersion coefficient mean the Gaussian curve is narrower, with a higher peak, while larger values mean the opposite. The further downwind we go from the source, the larger these coefficients become. This causes the Gaussian curves to spread further and further. These coefficients are not only a function of downwind distance, but they also depend, in a complex way, on atmospheric stability.

The most common procedure for estimating the dispersion coefficients was introduced by Pasquill (1961), modified by Gifford (1961), and adopted by the U.S. Public Health Service (Turner, 1970); it is presented here as Figure 7.48. The parameters A through F in Figure 7.48 represent stability classifications based on qualitative descriptions of prevailing environmental conditions. Table 7.8 describes these parameters. For example, a clear summer day, with the sun higher than 60° above the horizon and windspeeds less than 2 m/s (at an elevation of 10 m), creates a *very unstable* atmosphere with stability classification A. The opposite extreme is classification F, which is labeled *stable* and corresponds to a clear night (less than 3/8 of the sky covered by clouds) with winds less than 3 m/s.

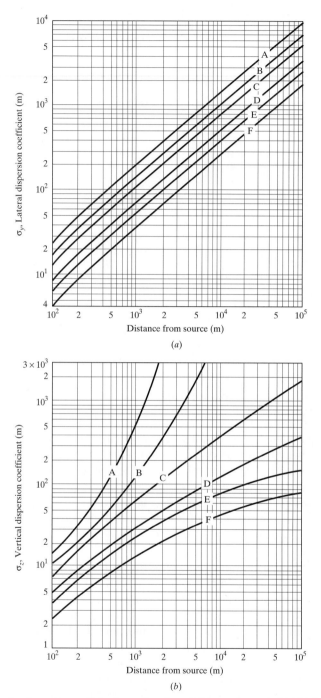

FIGURE 7.48   Gaussian dispersion coefficients as a function of distance downwind (*a*) horizontal coefficient, σ_y; (*b*) vertical coefficient σ_z. (*Source*: Turner, 1970).

It is often easier to use a computer to work with (7.42), in which case the graphical presentation of dispersion coefficients in Figure 7.48 is inconvenient. A reasonable fit to those graphs can be obtained using the following equations (Martin, 1976):

$$\sigma_y = ax^{0.894} \qquad (7.44)$$

and

$$\sigma_z = cx^d + f \qquad (7.45)$$

where the constants a, c, d, and f are given in Table 7.9 for each stability classification. The downwind distance x must be expressed in kilometers to yield σ_y and σ_z in meters. For convenience, a few values for the dispersion coefficients calculated using (7.44) and (7.45) are listed in Table 7.10.

Table 7.9   Values of the constants a, c, d, and f for use in (7.44) and (7.45)

| Stability | a | $x \le 1$ km | | | | $x \ge 1$ km | | |
		c	d	f		c	d	f
A	213	440.8	1.941	9.27		459.7	2.094	−9.6
B	156	106.6	1.149	3.3		108.2	1.098	2.0
C	104	61.0	0.911	0		61.0	0.911	0
D	68	33.2	0.725	−1.7		44.5	0.516	−13.0
E	50.5	22.8	0.678	−1.3		55.4	0.305	−34.0
F	34	14.35	0.740	−0.35		62.6	0.180	−48.6

Note: The computed values of σ will be in meters when x is given in kilometers.
Source: Martin (1976).

Table 7.10   Dispersion coefficients (m) for selected distances downwind (km), computed with (7.44) and (7.45)

| Distance x (km) | Stability class and σ_y | | | | | | Stability class and σ_z | | | | | |
	A	B	C	D	E	F	A	B	C	D	E	F
0.2	51	37	25	16	12	8	29	20	14	9	6	4
0.4	94	69	46	30	22	15	84	40	26	15	11	7
0.6	135	99	66	43	32	22	173	63	38	21	15	9
0.8	174	128	85	56	41	28	295	86	50	27	18	12
1	213	156	104	68	50	34	450	110	61	31	22	14
2	396	290	193	126	94	63	1953	234	115	51	34	22
4	736	539	359	235	174	117		498	216	78	51	32
8	1367	1001	667	436	324	218		1063	406	117	70	42
16	2540	1860	1240	811	602	405		2274	763	173	95	55
20	3101	2271	1514	990	735	495		2904	934	196	104	59

Downwind Ground level Concentration

The ground level concentration of pollution directly downwind of the stack is of interest since pollution will be highest along that axis. With $y = 0$, (7.42) simplifies to

$$C(x, 0) = \frac{Q}{\pi\, u_H \sigma_y \sigma_z}\, \exp\left(\frac{-H^2}{2\sigma_z^2}\right) \tag{7.46}$$

The following example illustrates the use of the Gaussian plume equation.

EXAMPLE 7.12   A Power Plant Plume

A 40 percent efficient, 1000-MW (10^6 kW) coal-fired power plant emits SO_2 at the legally allowable rate of 0.6 lb SO_2 per million Btu of heat into the plant. The stack has an effective height of 300 m. An anemometer on a 10-m pole measures 2.5 m/s of wind, and it is a cloudy summer day. Predict the ground level concentration of SO_2 4 km directly downwind.

Solution   We know the output power of the plant, 10^6 kW, but emission standards are written in terms of input power. Using the conversion 3412 Btu = 1 kWh, the input power is

$$\text{Input power} = \frac{\text{Output power}}{\text{Efficiency}} = \frac{1.0 \times 10^6\,\text{kW}}{0.40} \times \frac{3412\,\text{Btu}}{\text{kWh}} = 8530 \times 10^6\,\text{Btu/h}$$

So the SO_2 emission rate would be

$$Q = \frac{8530 \times 10^6\,\text{Btu}}{\text{h}} \times \frac{0.6\,\text{lb } SO_2}{10^6\,\text{Btu}} \times \frac{\text{kg}}{2.2\,\text{lb}} \times \frac{10^9\,\mu\text{g}}{\text{kg}} \times \frac{\text{h}}{3600\,\text{s}} = 6.47 \times 10^8\,\mu\text{g } SO_2/\text{s}$$

For the anemometer windspeed and solar conditions given, Table 7.8 indicates that the appropriate stability classification is C. Note that Table 7.8 uses the windspeed as measured at the standard 10-m anemometer height. The windspeed in the Gaussian plume equation, however, requires that we estimate the wind at the effective stack height. We did that in Example 7.11 and found it to be 4.9 m/s at 300 m.

At 4 km downwind, Table 7.10 indicates that the dispersion coefficients are $\sigma_y = 359$ m and $\sigma_z = 216$ m. Plugging these into (7.46) gives

$$C(4, 0) = \frac{6.47 \times 10^8\,\mu\text{g/s}}{\pi \times 4.9\,\text{m/s} \times 359\,\text{m} \times 216\,\text{m}}\, \exp\left[\frac{-(300)^2}{2 \times (216)^2}\right]$$

$$= 206 \mu\text{g/m}^3 \qquad \blacksquare$$

So the power plant in Example 7.12 would add 206 μg/m^3 to whatever SO_2 level is already there from other sources. For perspective, let us compare these emissions to ambient air quality standards. Table 7.1 indicates that the annual average SO_2 concentration must be less than 80 μg/m3. This power plant by itself would cause pollution to greatly exceed that standard if these atmospheric conditions prevailed over the full year. The 24-hour SO_2 standard is 365 μg/m^3, so under the conditions stated, this plant would not violate that standard, but it would make it difficult to have much in the way of other sources in the area.

Example 7.12 opens the door to many interesting questions. How does the concentration vary as distance downwind changes? What would be the effect of changes in windspeed or stability classification? How would we utilize statistical data on windspeed, wind direction, and atmospheric conditions to be sure that all air quality standards will be met? If they will not be met, what are the alternatives? Some examples come to mind. We might raise the stack height (adding to the acid deposition problem); we might increase the efficiency of the scrubber to clean flue gases, or we might use coal with a lower sulfur content. Perhaps energy conservation efforts might reduce the size of the power plant required to meet projected needs. Clearly, to do a proper siting analysis for a new source, such as the power plant in Example 7.12, would require a complex study. Though we will not carry out such calculations here, we do have the crucial starting point for such a study—namely, the Gaussian plume model .

Peak Downwind Concentration

Plotting (7.46) by hand is tedious, especially if we want to do a sensitivity analysis to see how the results change with changing stack heights and atmospheric conditions. It is much easier to work this out with a computer, and it is especially simple if a spreadsheet program with graphics capability is used. Using Example 7.12 as a base case, the effect on downwind concentration of changes in effective stack height has been plotted in Figure 7.49a. The effect of changing stability class, while keeping stack height constant at 300 m, is shown in Figure 7.49b.

The downwind concentration of pollution is quite sensitive to changes in effective stack height, as can be seen in Figure 7.49a. Raising the effective stack height from 250 to 300 m reduces the peak concentration by more than half. That ability of tall stacks to cut ground level pollution significantly has led to some tremendously tall stacks. In fact, the tallest stack in the world is for a smelter in Subdury, Ontario, and it is as tall as the Empire State Building in New York (380 m). The effective stack height would be even higher.

The impact of changing stability classification shown in Figure 7.49b is perhaps unexpected. The highest peak downwind concentration occurs when the atmosphere is very unstable rather than stable. The turbulence in an unstable atmosphere brings the looping plume to earth very quickly, resulting in high peak values near the stack. Downwind, however, concentrations drop off very quickly. Having a high peak concentration near the stack may be a satisfactory situation as long as any populations or ecosystems that might be damaged by the pollution are more than a few kilometers away.

The neutral atmosphere shown in Figure 7.49b, on the other hand, causes a relatively low peak concentration. The diagram of a fanning plume in Figure 7.45c may help to understand that conclusion. The plume fans out horizontally, but not much reaches the ground until the plume is some distance downwind. As shown in Figure 7.49b, the neutral atmosphere is worse than the unstable atmospheres beyond about 14 km. It should be noted, however, that the concentrations shown in Figure 7.49 do not account for the fact that effective stack height depends on the stability classification. If that complication is included, the high peaks shown for unstable atmospheres are reduced modestly.

An obvious question is, how can the peak downwind concentration be predicted from (7.46)? Unfortunately, it is not possible to derive a mathematical solution. One way to predict the peak is to simply plot curves of the sort shown in Figure 7.49 using a computer. With the ready availability of personal computers and spreadsheet programs, that has become an easy enough way to deal with the problem. For hand calculations, however, this approach would be far too tedious. Turner (1970) has derived curves that use the stability classification and effective stack height as parameters to find the distance downwind to the maximum concentration (x_{max}). The same process yields a normalized concentration $\left(\frac{Cu_H}{Q}\right)_{max}$, from which the maximum concentration can be found using the following:

$$C_{max} = \frac{Q}{u_H}\left(\frac{Cu_H}{Q}\right)_{max} \tag{7.47}$$

The curves are presented in Figure 7.50, and the following example illustrates their use.

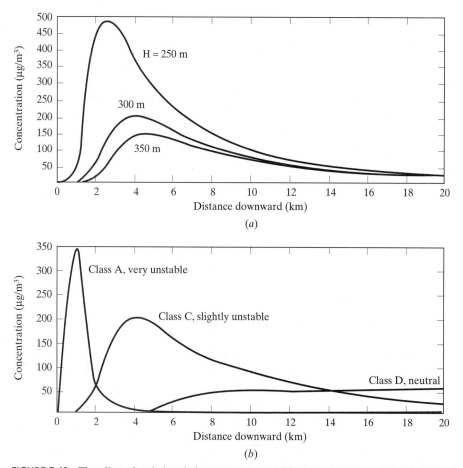

FIGURE 7.49 The effect of variations in key parameters on SO_2 plume for the coal plant in Example 7.12. (a) Impact of changes in the effective stack height for a constant stability classification, and (b) effect of stability classification given a constant effective stack height.

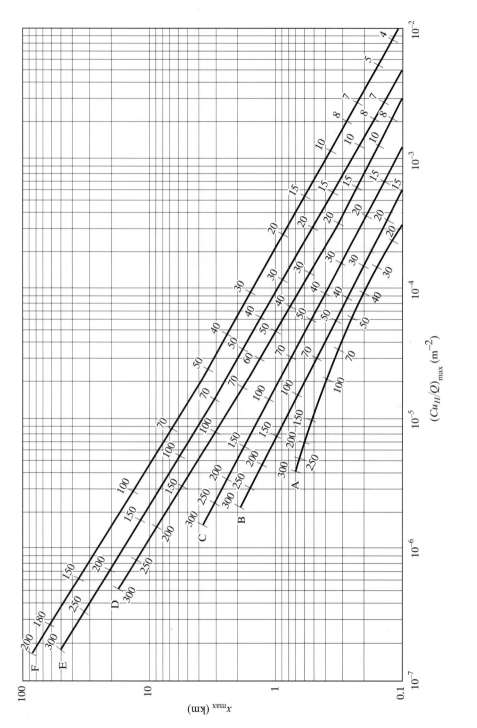

FIGURE 7.50 To determine the downwind concentration peak, enter the graph at the appropriate stability classification and effective stack height (numbers on the graph in meters) and then move across to find the distance to the peak, and down to find a parameter from which the peak concentration can be found (Turner, 1970).

EXAMPLE 7.13   Peak Downwind Concentration

For the 1000 MW, coal-fired power plant of Example 7.12, use Figure 7.50 to determine the distance downwind to reach the maximum SO_2 concentration. Then find that concentration.

Solution   The stability classification is C and the effective stack height is 300 m. From Figure 7.50 the distance downwind, x_{max}, is about 4 km (which agrees with Figure 7.49). Note the scales on Figure 7.50 are logarithmic, so be careful when reading values.

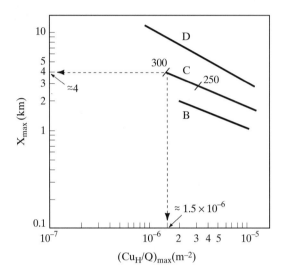

Reading down from the point on the figure corresponding to stability class C and effective stack height $H = 300$ m, $(Cu_H/Q)_{max}$ looks to be about 1.5×10^{-6} m^{-2}.

Windspeed at the effective height of the stack was found in Example 7.11 to be $u_H = 4.9$ m/s, and Q was 6.47×10^8 μg/s, so that

$$C_{max} = \frac{Q}{u_H} \left(\frac{C u_H}{Q} \right)_{max}$$

$$= \frac{6.47 \times 10^8 \ \mu g/s}{4.9 \ m/s} \times \frac{1.5 \times 10^{-6}}{m^2} = 198 \mu g/m^3$$

which is fairly close to the 206 μg/m^3 calculated at 4 km in Example 7.12. It is quite difficult to read Figure 7.50, so high accuracy should not be expected.   ∎

Plume Rise

So far, we have dealt only with effective stack height H in the calculations. The difference between the actual stack height h and the effective height H is called the plume rise Δh. Plume rise is caused by a combination of factors, the most important ones being the bouyancy and momentum of the exhaust gases and the stability of the atmosphere itself. Bouyancy results when exhaust gases are hotter than the ambient air or when the molecular weight of the exhaust is lower than that of air (or a combination of both factors). Momentum is caused by the mass and velocity of the gases as they leave the stack.

A number of techniques have been proposed in the literature for dealing with plume rise, and they tend to yield very different results. The EPA recommends a model based on work by Briggs (1972), and so we will use it. Plume rise depends on momentum and bouyancy. The bouyancy assumed in this analysis is due to the temperature of the stack gases being higher than the surrounding ambient, but one could also include differences in molecular weight of the exhaust gas versus air.

The following plume rise equation can be used for *stable* conditions (stability categories E and F):

$$\Delta h = 2.6 \left(\frac{F}{u_h S} \right)^{1/3} \tag{7.48}$$

The quantity F is called the *bouyancy flux parameter* (m^4/s^3); it can be used for all stability conditions.

$$F = g r^2 v_s \left(1 - \frac{T_a}{T_s} \right) \tag{7.49}$$

where

$$
\begin{aligned}
\Delta h &= \text{plume rise, m} \\
g &= \text{gravitational acceleration, } 9.8 \text{ m/s}^2 \\
r &= \text{inside radius of the stack, m} \\
u_h &= \text{windspeed at the height of the stack, m/s} \\
v_s &= \text{stack gas exit velocity, m/s} \\
T_s &= \text{stack gas temperature, K} \\
T_a &= \text{ambient temperature, K}
\end{aligned}
$$

The quantity S is a *stability parameter* with units of s^{-2} given by

$$S = \frac{g}{T_a} \left(\frac{\Delta T_a}{\Delta z} + 0.01 \,°\text{C/m} \right) \tag{7.50}$$

The quantity $\Delta T_a / \Delta z$ represents the actual rate of change of ambient temperature with altitude in °C/m (note that a positive value means temperature is increasing with altitude).

For *neutral* or *unstable* conditions in the atmosphere (stability categories A–D), the following equation can be used to estimate plume rise:

$$\Delta h = \frac{1.6 F^{1/3} x_f^{2/3}}{u_h} \tag{7.51}$$

where

$$x_f = \text{distance downwind to point of final plume rise, m}$$

Since (7.51) is used when conditions are neutral or unstable, it may be difficult to define the distance downwind at which the plume centerline stops rising. The following is sometimes used:

$$\text{Use } x_f = 120F^{0.4} \quad \text{if } F \geq 55 \text{ m}^4/\text{s}^3$$
$$x_f = 50F^{5/8} \quad \text{if } F < 55 \text{ m}^4/\text{s}^3$$

EXAMPLE 7.14   Plume Rise

A large power plant has a 250-m stack with inside radius 2 m. The exit velocity of the stack gases is estimated at 15 m/s, at a temperature of 140 °C (413 K). Ambient temperature is 25 °C (298 K) and winds at stack height are estimated to be 5 m/s. Estimate the effective height of the stack if (a) the atmosphere is stable with temperature increasing at the rate of 2 °C/km, (b) the atmosphere is slightly unstable, Class C.

Solution   First, find the bouyancy parameter F from (7.49):

$$F = gr^2 v_s \left(1 - \frac{T_a}{T_s} \right)$$

$$= 9.8 \text{ m/s}^2 \times (2\text{m})^2 \times 15 \text{ m/s} \times \left(1 - \frac{298}{413} \right) = 164 \text{ m}^4/\text{s}^3$$

(a) With the atmosphere stable, we need to use (7.48) and (7.50):

$$S = \frac{g}{T_a} \left(\frac{\Delta T_a}{\Delta z} + 0.01 °\text{C/m} \right)$$

$$= \frac{9.8 \text{ m/s}^2}{298\text{K}} (0.002 + 0.01)\text{K/m} = 0.0004/\text{s}^2$$

$$\Delta h = 2.6 \left(\frac{F}{u_h S} \right)^{1/3}$$

$$= 2.6 \left(\frac{164 \text{ m}^4/\text{s}^3}{5 \text{ m/s} \times 0.0004/\text{s}^2} \right) = 113 \text{ m}$$

So the effective stack height is $H = h + \Delta h = 250 + 113 = 363$ m.
(b) With an unstable atmosphere, Class C, we need to use (7.51). Since $F > 55$ m^4/s^3, the distance downwind to the point of final plume rise that should be used is

$$x_f = 120F^{0.4} = 120 \times (164)^{0.4} = 923 \text{ m}$$

$$\Delta h = \frac{1.6F^{1/3} \times x_f^{2/3}}{u_h} = \frac{1.6(164)^{1/3}(923)^{2/3}}{5} = 166 \text{ m}$$

and the effective stack height is $H = 250 + 166 = 413$ m.   ∎

Downwind Concentration under a Temperature Inversion

The Gaussian plume equation, as presented thus far, applies to an atmosphere in which the temperature profile is a simple straight line. If, as is often the case, there is an inversion above the effective stack height, then the basic Gaussian equation must be modified to account for the fact that the vertical dispersion of pollutants is limited by the inversion.

   If the pollutants are assumed to reflect off the inversion layer, just as they were assumed to reflect off the ground in the basic Gaussian equation, then an estimate of the concentration at any point downwind would require an analysis of these multiple reflections. That complexity can be avoided if we are willing to restrict our predictions of plume concentration to distances far enough downwind that the summation of these multiple reflections converges into a closed-form solution. Beyond that distance, the air is considered to be completely mixed under the inversion, with uniform concentrations from ground level to the bottom of the inversion layer.

   Turner (1970) suggests the following modified Gaussian equation to estimate ground level concentrations downwind under an inversion. It is derived based on the assumption that the downwind distance from the source is at least twice the distance to the point where the plume first interacts with the inversion layer:

$$C(x, 0) = \frac{Q}{(2\pi)^{1/2}u_H\sigma_y L} \qquad \text{for } x \geq 2X_L \qquad (7.52)$$

where L = elevation of the bottom of the inversion layer (m)

$\quad X_L$ = the distance downwind where the plume first encounters the inversion layer

   Notice that (7.52) is applicable only for distances $x \geq 2 X_L$ (see Figure 7.51). That distance, X_L, occurs at the point where the vertical dispersion coefficient, σ_z, is equal to

$$\sigma_z = 0.47(L - H) \quad \text{at } x = X_L \qquad (7.53)$$

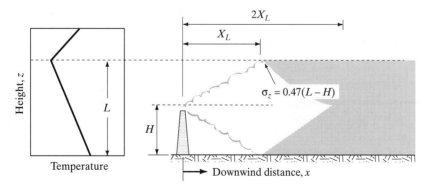

FIGURE 7.51 Plume dispersion under an elevated inversion. Equation (7.52) applies for distances greater than $2X_L$ downwind, where X_L occurs at the point where $\sigma_z = 0.47(L-H)$.

Once σ_z is found from (7.53), the distance X_L can be estimated using Figure 7.48b, or using (7.45) and Table 7.9.

For distances $x \le X_L$, the standard Gaussian plume equation (7.42) can be used to estimate downwind concentrations. For $X_L \ge 2X_L$, (7.52) applies. For distances between X_L and $2X_L$, concentrations can be estimated by interpolating between the values computed for $x = X_L$ and $x = 2X_L$.

EXAMPLE 7.15   Concentration Under an Inversion Aloft

Consider a stack with effective height 100 m, emitting SO_2 at the rate of 2×10^8 μg/s. Windspeed at 10-m is 5.5 m/s and at 100-m it is 10 m/s. It is a clear summer day with the sun nearly overhead and an inversion layer starting at 300 m. Estimate the groundlevel SO_2 concentration at a distance downwind twice that where reflection begins to occur from the inversion.

Solution   Equation (7.53) gives σ_z at the distance X_L,

$$\sigma_z = 0.47 \, (L\text{-}H) = 0.47 \times (300 - 100) = 94 \text{ m}$$

To find X_L we can use Figure 7.48b, but first we need the stability classification below the inversion. From Table 7.8, a clear summer day with 5.5 m/s windspeed corresponds to stability class C. Entering Figure 7.48b on the vertical axis at $\sigma_z = 94$ m, then going across to the Class C line and dropping down to the horizontal axis, leads to an estimate for X_L of about 1600 m. (Note the difficulty in reading this value from the figure; we could also use (7.45) with Table 7.9 to obtain an estimate analytically).

To find the concentration at a distance $x = 2X_L = 2 \times 1600$ m $= 3200$ m, we can use (7.52), but first we need to estimate σ_y at that point. Using Figure 7.48(a) at $x = 3200$ m, and Class C, we can estimate σ_y to be about 300 m (we could have used Eq. 7.44 with Table 7.9). Using (7.52), the concentration directly downwind at groundlevel is

$$C(x, 0) = \frac{Q}{(2\pi)^{1/2} u_H \sigma_y L}$$

$$= \frac{2 \times 10^8 \, \mu\text{g/s}}{(2\pi)^{1/2} 10\text{m/s} \times 300 \text{ m} \times 300\text{m}} = 90 \, \mu\text{g/m}^3 \qquad \blacksquare$$

A Line-Source Dispersion Model

In some circumstances it is appropriate to model sources distributed along a line as if they formed a continuously emitting, infinite line source. Examples of line sources that might be modeled this way include motor vehicles traveling along a straight section of highway, agricultural burning along the edge of a field, or a line of industrial sources on the banks of a river. For simplicity, we will consider only the case of an infinite-length source at ground level, with winds blowing perpendicular to the line as shown in Figure 7.52.

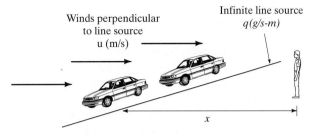

FIGURE 7.52   Showing the geometry of a line-source, such as a straight section of highway, and the receptor located x distance from the source.

Under these specialized circumstances, the ground level concentration of pollution at distance x from the line source can be described by the following:

$$C(x) = \frac{2q}{\sqrt{2\pi}\,\sigma_z\,u} \qquad (7.54)$$

where

q = emission rate per unit of distance along the line (g/m-s).

EXAMPLE 7.16   CO Near a Freeway

The federal emission standard for CO from new vehicles is 3.4 g/mile. Suppose a highway has 10 vehicles per second passing a given spot, each emitting 3.4 g/mile of CO. If the wind is perpendicular to the highway and blowing at 5 mph (2.2 m/s) on an overcast day, estimate the ground level CO concentration 100 m from the freeway.

Solution   We need to estimate the CO emission rate per meter of freeway:

$$q = 10 \text{ vehicles/s} \times 3.4 \text{ g/veh-mile} \times 1 \text{ mile}/1609 \text{ m} = 0.021 \text{ g/s-m}$$

and we need the vertical dispersion coefficient σ_z. Since this is an overcast day, the footnotes of Table 7.8 indicate that we should use stability class D. Checking Table 7.10 for σ_z does not help since it does not give values for such short distances. So, with (7.45) and values of c, d, and f from Table 7.9 we can compute σ_z:

$$\sigma_z = cx^d + f = 33.2(0.1)^{0.725} - 1.7 = 4.6 \text{ m}$$

Substituting these values into (7.54) gives

$$C(0.1 \text{ km}) = \frac{2q}{\sqrt{2\pi}\,\sigma_z u} = \frac{2 \times 0.021 \text{ g/m} \cdot \text{s} \times 10^3 \text{ mg/g}}{(2\pi)^{1/2} \times 4.6 \text{ m} \times 2.2 \text{ m/s}} = 1.7 \text{ mg/m}^3$$

The eight-hour CO air quality standard is 10 mg/m^3, so this estimate is well within the air quality standard.   ∎

Area-Source Models

For distributed sources, there are a number of approaches that can be taken to estimate pollutant concentrations. If there are a modest number of point sources, it is reasonable to use the point-source Gaussian plume equation for each source to predict its individual contribution. Then, by superposition, find the total concentration at a given location by summing the individual contributions. Multiple use of the Gaussian line-source equation is another approach. By dividing an area into a series of parallel strips, and then treating each strip as a line-source, the total concentration on any strip can be estimated.

A much simpler, more intuitive approach can be taken to estimate pollutant concentrations over an area (such as a city) by using the box model concepts introduced in Section 1.3. Consider the air shed over an urban area to be represented by a rectangular box, such as is shown in Figure 7.53, with base dimensions L and W and height H. The box is oriented so that wind, with speed u, is normal to one side of the box. The height of the box is determined by atmospheric conditions, and we could consider it to be just the mixing depth. Emissions per unit area will be represented by q_s (g/m²-s).

Consider the air blowing into the box on the upwind side to have pollutant concentration C_{in} and, for simplicity, assume that no pollution is lost from the box along the sides parallel to the wind or from the top. We will also assume that the pollutants are rapidly and completely mixed in the box, creating a uniform average concentration C. Finally, we will treat the pollutants as if they are conservative (that is, they do not react, decay, or fall out of the air stream). All of these restrictions can be modified in more sophisticated versions of a box model.

Working with pollutant mass, the amount of pollution in the box is the volume of the box times the concentration, $LWHC$. The rate at which air is entering and leaving the box is the area of either end times the windspeed, WHu, so the rate at which pollution is entering the box is $WHuC_{in}$. The rate that it leaves the box is $WHuC$. If we assume the pollutant is conservative, then we can then write the following mass balance for the box:

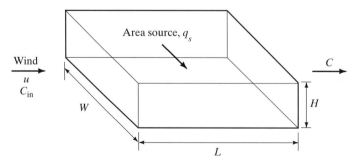

FIGURE 7.53   Box model for an air shed over a city. Emissions per unit area are given by q_s, pollutants are assumed to be uniformly mixed in the box with concentration C, while upwind of the box the concentration is C_{in}.

$$\begin{pmatrix} \text{Rate of change of} \\ \text{pollution in the box} \end{pmatrix} = \begin{pmatrix} \text{Rate of pollution} \\ \text{entering the box} \end{pmatrix} - \begin{pmatrix} \text{Rate of pollution} \\ \text{leaving the box} \end{pmatrix}$$

or

$$LWH\frac{dC}{dt} = q_s LW + WHuC_{\text{in}} - WHuC \tag{7.55}$$

where

C = pollutant concentration in the airshed, mg/m^3
C_{in} = concentration in the incoming air, mg/m^3
q_s = emission rate per unit area, mg/m^2-s
H = mixing height, m
L = length of airshed, m
W = width of airshed, m
u = average windspeed against one edge of the box, m/s

The steady-state solution to (7.55) can be obtained by simply setting $dC/dt = 0$, so that

$$C(\infty) = \frac{q_s L}{uH} + C_{\text{in}} \tag{7.56}$$

which looks reasonable. If the air entering the box is clean, the steady-state concentration is proportional to the emission rate and inversely proportional to the ventilation coefficient (the product of mixing depth and windspeed). If it is not clean, we just add the effect of the incoming concentration. We can also solve (7.55) to obtain the time-dependent increase in pollution above the city. Letting C(0) be the concentration in the airshed above the city (the box) at time t = 0, the solution becomes

$$C(t) = \left(\frac{q_s L}{uH} + C_{\text{in}}\right)(1 - e^{-ut/L}) + C(0)e^{-ut/L} \tag{7.57}$$

If we assume that the incoming wind blows no pollution into the box and if the initial concentration in the box is zero, then (7.57) simplifies to

$$C(t) = \frac{q_s L}{uH}(1 - e^{-ut/L}) \tag{7.58}$$

When $t = L/u$, the exponential function becomes e^{-1} and the concentration reaches about 63 percent of its final value. That value of time has various names. It is called the *time constant*, the *ventilation time*, the *residence time*, or the *e-folding time*.

EXAMPLE 7.17   Evening Rush Hour Traffic

Suppose within a square city, 15 km on a side, there are 200,000 cars on the road, each being driven 30 km between 4 P.M. and 6 P.M., and each emitting 3 g/km of CO. It is a clear winter evening with a radiation inversion that restricts mixing to 20.0 m. The wind is bringing clean air at a steady rate of 1.0 m/s along an edge of the city. Use a box model to estimate the CO concentration at 6 P.M. if there was no CO in the air at 4 P.M. and the only source of CO is cars. Assume that CO is conservative and that there is complete and instantaneous mixing in the box.

Solution   The emissions per m², q_s, would be

$$q_s = \frac{200{,}000 \text{ cars} \times 30 \text{ km/car} \times 3 \text{ g/km}}{(15 \times 10^3 \text{ m})^2 \times 3600 \text{ s/hr} \times 2 \text{ hr,}} = 1.1 \times 10^{-5} \text{ g/s} - \text{m}^2$$

Using (7.58), the concentration after two hours (7200 s) would be

$$C(t) = \frac{q_s L}{uH} \left(1 - e^{-ut/L}\right)$$

$$C(2 \text{ hr}) = \frac{1.1 \times 10^{-5} \text{ g/sm}^2 \times 15 \times 10^3 \text{ m}}{1.0 \text{ m/s} \times 20.0 \text{ m}} \left[1 - \exp\left(\frac{-1.0 \text{ m/s} \times 7200 \text{ s}}{15{,}000 \text{ m}}\right)\right]$$

$$= 3.2 \times 10^{-3} \text{ g/m}^3 = 3.2 \text{ mg/m}^3$$

which is considerably below both the one-hour NAAQS for CO of 40 mg/m³ and the eight-hour standard of 10 mg/m³. Any CO that was already in the air at 4 P.M. would, of course, increase this estimate. The time constant, $L/u = 15{,}000\text{m}/(1 \text{ m/s}) = 15{,}000 \text{ s} = 4.2$ hours, suggests that in these two hours the concentration is well below what it would become if these conditions were to continue.   ∎

7.12 INDOOR AIR QUALITY

This chapter began with the rather startling assertion that, in terms of human exposure to respirable particulate matter, a 2 percent reduction in emissions of *environmental tobacco smoke* (ETS) would be equivalent to eliminating all the coal-fired power plants in the United States (Smith, 1993b). The key, of course, goes back to the theme of Chapter 5, that health effects are the result of exposure, not emissions. We are exposed to chemicals and particulates in the air we breathe, and the air we breathe is mostly inside of buildings where we spend almost all of our time.

Some of the indoor pollutants are the same ones that we are now familiar with from our study of ambient air quality. For example, combustion that takes place inside of homes and other buildings to cook, heat water, and provide space heating and cooling can produce elevated levels of carbon monoxide, nitrogen oxides, hydrocarbons, and respirable particulates. Cigarette smoke emits carbon monoxide, benzene, acrolein and other aldehydes, and particulates, as well as about 4000 other chemicals. Some photocopying machines emit ozone. Building materials such as particleboard, plywood, urea-formaldehyde foam insulation, and various adhesives emit formaldehyde. Chipped and peeling paint containing lead becomes airborne toxic dust. A long list of volatile organic compounds are emitted from household cleaning products, paints, carpeting, and a variety of other chemicals we use in our homes.

Some pollutants are somewhat unique to the indoor environment, such as asbestos used for fireproofing and insulation; radon gas, which seeps out of the soil and collects in houses; and biological pollutants, such as house dust mites, fungi, and other microorganisms. Many pollutants, such as cigarette smoke and radon gas, if they are emitted outdoors, have plenty of dilution air so people tend not to be exposed to hazardous levels of contamination. Indoors, however, these pollutants can be concentrated, leading all too often to harmful exposure levels.

Air pollution in homes is largely under our control, but in offices and other workplaces we depend on the builder's choice of materials used in construction and the proper design and maintenance of building air filtration and ventilation systems. It is not at all unusual these days for occupants of new and remodeled buildings to experience a bewildering array of symptoms that characterize what is now called *sick-building syndrome*. Sneezing or coughing, watery eyes, headaches, eye, nose, and throat irritation, dry or itchy skin, nausea and dizziness, fatigue, difficulty in concentrating, and general malaise arise without any clearly identifiable casues. Studies suggest that nearly one-quarter of U.S. office workers perceive air quality problems in their work environments and 20 percent believe their work is impaired by reactions to indoor pollution (Kreiss, 1990). Despite the number of complaints about sick-building syndrome, surprisingly little research has been done on the phenomenon and it remains poorly understood.

Table 7.11 summarizes some of the sources and exposure guidelines for a number of pollutants that are commonly found in indoor environments.

TABLE 7.11.   Sources and Exposure Guidelines of Indoor Air Contaminants

Pollutant and indoor sources	Guidelines, average concentrations
Asbestos and other fibrous aerosols	
Friable asbestos: fireproofing, thermal and acoustic insulation, decoration. Hard asbestos: vinyl floor and cement products	0.2 fibers/mL for fibers longer than 5 μm
Carbon monoxide	
Kerosene and gas space heaters, gas stoves, wood stoves, fireplaces, smoking.	10 mg/m^3 for 8 hr, 40 mg/m^3 for 1 hr
Formaldehyde	
Particleboard, paneling, plywood, carpets, ceiling tile, urea-formaldehyde foam insulation, other construction materials.	120 μg/m^3
Inhalable particulate matter	
Smoking, vacuuming, wood stoves, fireplaces.	55–110 μg/m^3 annual. 150–350 μg/m^3 for 24 hr
Nitrogen dioxide	
Kerosene and gas space heaters, gas stoves.	100 μg/m^3 annual
Ozone	
Photocopying machines, electrostatic air cleaners.	235 μg/m^3/hr once a year
Radon and radon progeny	
Diffusion from soil, groundwater, building materials.	0.01 working levels annual
Sulfur dioxide	
Kerosene space heaters.	80 μg/m^3 annual, 365 μg/m^3 24 hr
Volatile organics	
Cooking, smoking, room deodorizers, cleaning sprays, paints, varnishes, solvents, carpets, furniture, draperies.	None available

Source: Nagda et al. (1987).

Environmental Tobacco Smoke

One category of indoor air pollution stands out from all others: tobacco smoke. It is estimated that about 26 percent of the U.S. adult population are smokers, who consume more than 500 billion cigarettes annually. Tobacco smoke contains over 4000 chemicals, including more than 40 that are known to cause cancer in humans or animals, and many are strong respiratory irritants. Smoking is thought to be responsible for between 80 and 90 percent of the lung cancer deaths in the United States. The combined death rate from lung cancer, emphyzema, and cardiovascular disease attributable to smoking is estimated to be around 400,000 per year. Unfortunately, it is not only smokers who suffer the ill effects of tobacco.

First some terminology. Smokers inhale what is referred to as *mainstream* smoke. *Sidestream* smoke, emitted from smoldering cigarettes, mixed with smoke exhaled by smokers is known as *environmental tobacco smoke* (ETS), or *secondhand smoke*. Breathing air with ETS is called *involuntary* or *passive smoking*.

In early 1993, the EPA released a report (*Respiratory Health Effects of Passive Smoking: Lung Cancer and Other Disorders*) that concluded that ETS causes lung cancer in adult nonsmokers and has serious effects on the respiratory system of children. ETS has been listed as a Group A, known human carcinogen. The EPA concludes that secondhand smoke is causing an estimated 3000 lung cancer deaths annually in the United States as well as 150,000 to 300,000 cases of pneumonia or bronchitis among children under 18 months of age each year. It worsens the condition of up to 1 million asthmatic children. Smokers have nearly 10 times as much of the carcinogen benzene in their blood as nonsmokers, and a pregnant smoker undoubtedly passes benzene to her developing fetus. That correlates with the fact that children of smokers die of leukemia at several times the rate of children of nonsmokers. An even more recent study concludes that the incidence rate of sudden infant death syndrome increases as exposure to tobacco smoke increases.

Asbestos

Another somewhat special indoor air quality problem can be caused by asbestos-containing materials. Asbestos used to be a common building material found in structural fireproofing, heating-system insulation, floor and ceiling tiles, and roofing felts and shingles. It has also been used in consumer products such as fireplace gloves, ironing board covers, and certain hair dryers.

As some asbestos-containing materials age, or if they are physically damaged in some way during their use, microscopic fibers may be dispersed into the indoor air environment. Inhalation of these fibers can lead to a number of life-threatening diseases, including asbestosis, lung cancer, and mesothelioma. Asbestos fibers have long been known to be human carcinogens; in fact, asbestos was one of the first substances categorized as a hazardous air pollutant under Section 112 of the Clean Air Act. It was identified by the EPA as a hazardous pollutant in 1971, and in 1973 the Asbestos National Emissions Standards for Hazardous Air Pollutants (NESHAP) were initiated. An especially deadly combination is asbestos exposure along with tobacco smoking, which elevates the lung cancer risk by approximately fivefold.

There are two broad categories of asbestos containing materials: friable and non-friable. Friable materials are those that can be crumbled, pulverized, or reduced to powder by hand pressure. Obviously, these are the most dangerous since asbestos fibers are easily released into the environment. Nonfriable materials, such as vinyl asbestos floor tiles and asphalt roofing products, do not usually require special handling even during demolition and renovation of buildings. Friable substances, on the other hand, are subject to NESHAP during such work. Since disturbing asbestos that is in good condition may create more of a risk than simply sealing it in place, it is often best not to attempt to remove asbestos-containing materials in a building unless there is known to be a problem or major construction work is to ensue.

Radon

One of the most publicized indoor air pollution problems is caused by the accumulation of radon gas in some homes in some parts of the country. Radon gas and its radioactive daughters are known carcinogens and may be the second leading cause of lung cancer, after smoking.

Recall from Section 2.5 that radon is a radioactive gas that is part of a natural decay chain beginning with uranium and ending with lead. A simplified description of the sequence, along with half-lives and alpha, gamma, and beta radiation emitted, is shown in Figure 7.54. Radon gas formed in pore spaces between mineral grains in soil can work its way to the surface, where it can enter buildings through the floor. Radon itself is inert, but its short-lived decay products—polonium, lead, and bismuth—are chemically active and easily become attached to inhaled particles that can lodge in the lungs. In fact, it is the alpha-emitting polonium, formed as radon decays, that causes the greatest lung damage.

Radon can also be emitted from some earth-derived building materials, such as brick, concrete, and tiles. It can be captured in groundwater, to be released when that water is aerated, such as during showers (the radon risk in water is from inhalation of the released gas, not from drinking the water itself). Radon has also been detected in some natural gas supplies, so modest amounts may be released during cooking. By far the most important source of radon, however, is through soils under homes.

The units of measurement used in describing radon are somewhat unusual and are summarized in Table 7.12. The decay rate of 1 gram of radium, 3.7×10^{10} disintegrations per second, is the origin of the *Curie*. A more intuitive unit is the *Becquerel*, which corresponds to one radioactive decay per second. The *working-level-month* (WLM) is based on the exposure a miner received working 173 hours per month. Indoor concentrations of radon have traditionally been expressed in *picocuries per liter* of air (pCi/L). It is estimated that the average house in the United States has 1.25 pCi/L, and around 6 million homes (6 percent of the total) have levels above 4 pCi/L, which is the EPA's level of concern. Retrofitting those 6 million U.S. homes with radon control measures would cost on the order of $50 billion.

Estimates of the lung cancer risk associated with inhalation of radon and its progeny are based in large part on epidemiologic studies of uranium miners who have been exposed to high radon levels and for whom the incidence rate of lung cancer is

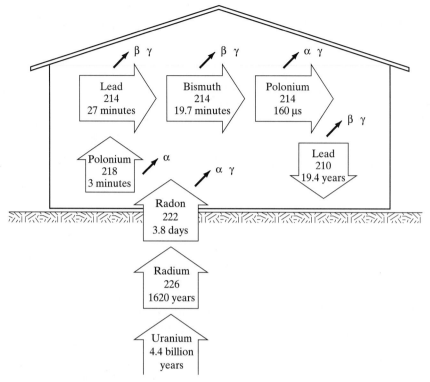

FIGURE 7.54   Simplified uranium decay series, with half-lives and emissions. Radon gas that seeps out of soils can decay inside of buildings.

TABLE 7.12   Radon measurement units

Curie	1 Ci	=	3.7×10^{10} radioactive decays per second (1 g radium)
Picocurie	1 pCi	=	2.2 radioactive decays per minute
Becquerel	1 Bq	=	1 radioactive decay per second
Picocurie per liter	1 pCi/L	=	37 Bq/m^3
EPA criterion	4 pCi/L	=	150 Bq/m^3 (0.0000028 parts per trillion of radon)
Working level[a]	1 WL	$\approx$	200 pCi/L = 7400 Bq/m^3 of radon
Working-level month	1 WLM	=	1 working-level of exposure for 173 hours

[a] One WL is defined as 100 pCi/L of radon in equilibrium with its progeny.

much higher than that of the general population. Translating data from miners, however, to risks in residences introduces large uncertainties. Working miners, for example, would be expected to have higher breathing rates than people resting at home, they would be exposed to much greater concentrations of particulate matter along with the radon, and they have much higher exposure to radon. A number of more recent studies have tried to find evidence of elevated lung cancer rates at the comparatively low radon levels found in homes. In the largest study of its kind, a Swedish report (Pershagen, 1994) showed progressively higher lung cancer rates as home radon exposure increased. At exposure levels of 3.8 to 10.8 picocuries per liter (pCi/L), lung cancer risks increased 30 percent over those who lived in homes with less than 1.4 pCi. Above 10.8 pCi/L, the incidence rate was 80 percent higher.

Much of the danger associated with radon is caused by radon daughters, which attach themselves to small particulate matter that is inhaled deep into the lungs. The combination of cigarette smoke and radon is, as a result, an especially deadly combination. According to data summarized in Table 7.13, approximately 16,000 lung cancer deaths per year in the United States are thought to be attributable to radon, but only about 500, or 3 percent, of this mortality rate is estimated to occur among individuals who have never smoked. If these data are true, well over 90 percent of the lung-cancer risk associated with radon could be controlled by eliminating smoking without any change in radon concentrations (Nazaroff and Teichman, 1990).

Table 7.13   Estimated Annual Lung Cancer Deaths in the United States Attributable to Radon Exposure, 1986 data.

Smoking History	Population (millions)	Lung cancer deaths	
		All causes	Radon attributable
Never smoked	145	5,000	500
Former smoker	43	57,000	6,400
Current, light smoker[a]	38	37,600	4,500
Current, heavy smoker	14	30,800	4,200
Total	241	130,400	15,700

[a]Less than 25 cigarettes per day.
Source: Nazaroff and Teichman (1990).

Homes built over radon-rich, high-permeability soils are most vulnerable to elevated radon levels. The emanation rate from soil seems to range from about 0.1 pCi/m^2-s to over 100 pCi/m^2-s, with a value of 1 pCi/m^2-s being fairly typical. This large variation in emission rates is the principal cause of variations in indoor radon concentrations measured across the country. When emission rates are low, as is the case for most of the United States, homes can be tightly constructed for energy efficiency without concern for elevated levels of radon. In areas where soil radon emissions are high, tight building construction techniques can elevate indoor radon concentrations. Smokers, of course, would be most affected.

The EPA has attempted to evaluate the radon potential across the United States and presents their results in three categories. Based on soil and home measurements, areas in Zone 1 are predicted to have indoor radon levels that are likely to exceed 4 pCi/L. The EPA has been considering recommending preventative radon control measures in new construction in these areas. Areas of Zone 2 are between 2 and 4 pCi/L, while Zone 3 areas are below 2 pCi/L. A map of these zones is shown in Figure 7.55. This mapping is only a very rough first approximation to the extent of the radon problem. Large variations in radon are often found from one neighborhood to another in the same city, so decisions about remediation measures should only be made after thorough testing of individual homes.

Available techniques to help reduce indoor radon concentrations depend somewhat on what type of floor construction has been employed. Many houses in the United States. are built over basements that may or may not be heated; other homes, especially in the West, are built over crawl spaces or on concrete slabs. With every type of construction, mitigation begins with efforts to seal any and all cracks and openings between the floor and the soil beneath. For houses with basements or crawl spaces, increased ventilation of those areas, either by natural or mechanical means can be very effective. If the basement is heated, a heat recovery ventilator may be called for to avoid wasting excessive amounts of energy. One of the most widely used radon reduction techniques uses subslab suction, as shown in Figure 7.56. Radon exhaust pipes pass through the floor into the aggregate below. Fans suck the radon gas out of the aggregate and exhaust it into the air above the house.

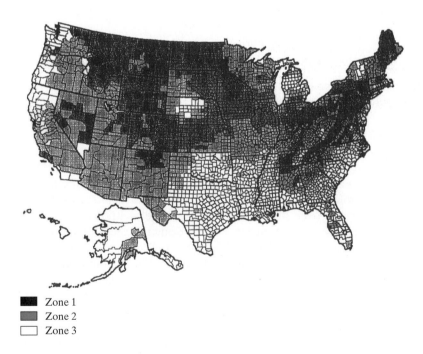

■■■ Zone 1
▒▒▒ Zone 2
▢ Zone 3

FIGURE 7.55   The EPA's map of radon potential zones. Zone 1 has predicted indoor radon above 4 pCi/L; Zone 2 is between 2 and 4 pCi/L; Zone 3 is less than 2 pCi/L.

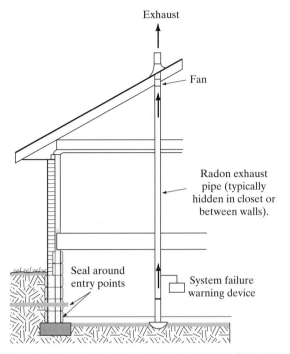

FIGURE 7.56   A subslab suction radon mitigation system. (U. S. EPA, 1993)

Exposure Assessment

Toxic chemicals affect human health when people are exposed to them. Perhaps it is because this simple notion is so obvious that its importance is so often overlooked. To be exposed to pollutants, pollutants have to be in the air we breathe, the water we drink, or the food we eat, so it makes sense to approach health risk assessment by evaluating those pathways. Such an approach, called the *total exposure assessment methodology* (TEAM), was launched by the EPA's Office of Research and Development in 1979, and considerable data of actual exposures have now been collected (Wallace, 1993, 1995).

From the perspective of air pollution, the starting point for a total environmental assessment is to determine the amount of time spent indoors, in vehicles, and outdoors. Large-scale field studies based on carefully kept diaries suggest that Americans are inside of buildings 89 percent of the time, in vehicles 6 percent, and outdoors 74 minutes per day, or 5 percent of the time (Figure 7.57). Perhaps surprisingly, Californians, who might think they are outdoors much more than national averages would suggest, are only slightly different, 6 percent outdoors rather than 5 (Jenkins, 1992). Also shown in Figure 7.57 are estimates of the time spent indoors in the less developed countries. As countries develop, a greater fraction of their populations live in urban areas and spend more time indoors.

The importance of the amount of time spent indoors is dramatized by studies of human exposure to a number of especially important air pollutants. These TEAM

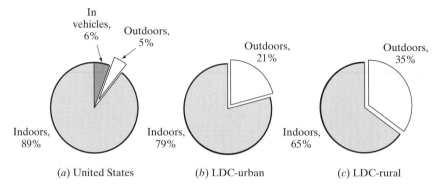

FIGURE 7.57 (*a*) In the United States, 89 percent of our time is spent indoors (Robinson et al., 1991). Percentage of time spent indoors and outdoors in less developed countries, (*b*) urban and (*c*) rural (Smith, 1993a).

studies depend on actual exposure measurements made with personal air quality monitors that accompany participants during their normal daily activities. Exposure measurements are supplemented with blood gas analyses to verify the actual uptake of pollutants. One such study (Smith, 1993b) estimates that per capita emissions of particulates from coal-fired power plants in the United States are about 1.6 kg/person, while emissions of environmental tobacco smoke are roughly 0.050 kg/person. That is, coal plant particulate emissions are about 30 times higher than ETS emissions. But coal plant emissions are released into a large reservoir, the atmosphere, while almost all of ETS particulates are released indoors, where there is less dilution and where people spend most of their time. In terms of particulate matter inhaled, it has been estimated that human exposure to ETS is 60 times the exposure from coal-fired power plants even though ETS constitutes only 3 percent of emissions. These data, summarized in Figure 7.58, tell us that major reductions in coal plant emissions would probably have relatively little overall effect on public health—certainly not nearly as much as very modest reductions in ETS might provide.

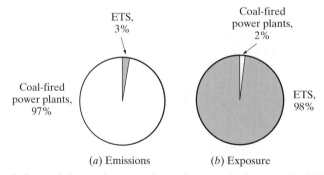

FIGURE 7.58 Particulate emissions and exposure for environmental tobacco smoke (ETS) and U.S. coal-fired power plants. Of the total, (*a*) ETS is 3 percent of emissions, but (*b*) 98 percent of exposure. (Based on Smith, 1993a)

Similar conclusions can be reached for the Class A human carcinogen, benzene. Combustion of gasoline accounts for 82 percent of benzene emissions in the United States, while industrial sources account for most of the rest, 14 percent. Cigarette smoke is responsible for only 0.1 percent of all benzene emissions, but on the average 40 percent of all benzene exposure is direct inhalation by smokers, and 5 percent is exposure to ETS. In comparison, all industrial sources combined are responsible for less benzene exposure (3 percent) than passive smoking (5 percent). Figure 7.59 shows these data (Wallace, 1995).

Even in communities located next to industrial facilities, indoor exposure dominates. One such study focused on the large petroleum storage and loading facility at the Alyeska Marine Terminal located just 3 miles from a residential community in Valdez, Alaska. The Marine Terminal accounted for 11 percent of benzene exposure, while personal activities such as smoking, driving, consumer products used in homes, and gasoline vapors in attached garages accounted for the remaining 89 percent (Yokum et al., 1991). Another major study in Northern New Jersey, which has one of the greatest concentrations of oil refineries and chemical plants in the nation, measured exposure to 25 different VOCs. No difference in exposure was seen between residents near the plants and farther away (Wallace, 1995).

Cigarettes emit not only benzene, but a host of other toxic chemicals, including styrene, toluene, xylenes, ethylbenzene, formaldehyde, and 1,3-butadiene. Newly dry-cleaned clothes emit tetrachloroethylene for days after putting them in your closet. Moth repellents and spray room air fresheners emit *p*-dichlorobenzene, which causes cancer in rats and mice. Hot showers, clothes, and dishwashing are major sources of airborne chloroform, another carcinogen. Cooking adds polynuclear aromatic hydro-

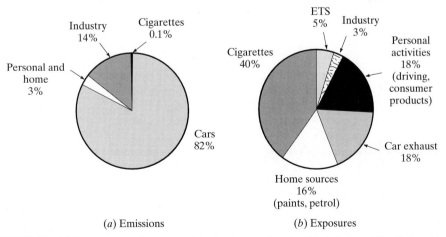

(*a*) Emissions   (*b*) Exposures

FIGURE 7.59   (*a*) Benzene emissions are mainly due to car exhaust and to industry, while all cigarettes smoked in the United States provide only 0.1 percent. (*b*) Benzene exposures are overwhelmingly due to cigarettes, mostly from active smoking. ETS contributes more exposure than all industrial sources combined. (*Source:* Wallace, 1995)

carbons (PAHs) to indoor air. Pesticides stored in garages and homes are major sources of many other carcinogens found indoors. Pesticides that are used outdoors, and lead particles that settled onto the soil, are tracked into homes on dirt from shoes. All told, TEAM studies of human exposure to 33 VOCs and pesticides in the United States show that indoor exposures account for about 85 percent of total cancer risk from these chemicals (Figure 7.60).

Clearly, if we want to reduce the hazards of air pollution, the place to start is at home. Simple things like using a floor mat, or removing shoes at the entrance, can reduce the amount of outside pollution that is brought into a home on our shoes. Dry cleaning clothes less often, or at least airing them out for a day or two before putting them in your closet, will help. Exhaust fans over ranges can reduce particulate and PAH exposure, and bathroom fans can help clear out chloroform from chlorinated shower water. Be sure carbon monoxide and other combustion products formed in water and space heaters are properly vented, and certainly do not try to heat a house with a charcoal grill or an unvented kerosene heater. Be sure paints, cleaning fluids, and other sources of VOCs are sealed properly before storage, and provide plenty of ventilation when you use them. If you smoke, do so outside if at all possible. Simple, no-cost measures such as these can significantly reduce your exposure to air pollution.

In the less developed countries of the world, indoor exposure to hazardous air pollutants is much more severe. Traditional methods of cooking often involve "dirty" fuels such as coal, wood, animal dung, charcoal, and kerosene, and these fuels are often burned in homes or cooking huts without chimneys or proper ventilation. Women and their infant children are often exposed to extremely high concentrations of particulates and other products of combustion for lengthy periods of time. One result is elevated levels of acute respiratory infections (ARI) in children that a number of studies correlate with indoor exposure to smoke. Acute respiratory infections, such as pneumonia, are the main cause of childhood deaths worldwide, killing approximately 4.3 million children per year. That is 30 percent more deaths than are caused by the

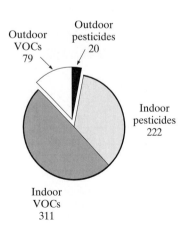

FIGURE 7.60 TEAM studies of 33 carcinogens commonly found in homes show an upper-bound cancer risk of about 5×10^{-4}. This is similar in magnitude to risk from radon and passive smoking. Numbers are lifetime cancer risks per million people. (*Source:* Wallace, 1995)

number two killer, diarrhea. A number of studies show two to six times the incidence rate of ARI among children living in homes where cooking is done on an open biomass stove, compared to those using more modern fuels, such as LPG, with chimneys. Similar elevated incidence rates for lung cancer have been observed in urban homes in China that burn coal for cooking or heating compared with homes that use natural gas (Smith, 1993a). The enormity of the indoor air quality problem in developing countries is truly staggering.

Infiltration, Ventilation, and Air Quality

Just as with outdoor air, the amount of air available to dilute pollutants is an important indicator of likely contaminant concentrations. Indoor air can be exchanged with outdoor air by any combination of three mechanisms: infiltration, natural ventilation, and forced ventilation. *Infiltration* is the term used to describe the natural air exchange that occurs between a building and its environment when doors and windows are closed. That is, it is leakage that occurs through various cracks and holes that exist in the building envelope. *Natural ventilation* is the air exchange that occurs when windows or doors are purposely opened to increase air circulation, while *forced ventilation* occurs when mechanical air handling systems induce air exchange using fans or blowers.

Large amounts of energy are lost when conditioned air (heated or cooled) that leaks out of buildings is replaced by outside air that must be mechanically heated or cooled to maintain desired interior temperatures. It is not uncommon, for example, for one-third of a home's space heating and cooling energy requirements to be caused by unwanted leakage of air (that is, infiltration). Nationwide, about 10 percent of total U.S. energy consumption is accounted for by infiltration, which translates into tens of billions of dollars worth of wasted energy each year. Since infiltration is quite easily and cheaply controlled, it is not surprising that tightening buildings has become a popular way to help save energy. Unfortunately, in the process of saving energy we may exacerbate indoor air quality problems unless we simultaneously reduce the sources of pollution.

Air leaks in and out of buildings through numerous cracks and openings in the building envelope. The obvious cracks around windows and doors are the usual ones we try to plug with caulk and weatherstripping, but there are many less obvious, but potentially more important, leakage areas, such as those created when plumbing, ducts, and electrical wiring penetrate walls, ceilings, and floors; fireplaces without dampers; ceiling holes created around recessed light fixtures, attic access hatches and any other bypasses created between heated spaces and the attic; gaps where foundations are connected to walls; exhaust vents in bathrooms and kitchens; and, perhaps most important, leaky ductwork in homes with forced-air heating systems. Figure 7.61 shows some of these infiltration sites.

Infiltration is driven by pressure differences between the inside of the building and the outdoor air. These pressure differences can be caused by wind, or by inside-to-outside temperature differences. Wind blowing against a building creates higher pressure on one side of the building than the other, inducing infiltration through cracks and other openings in the walls. Temperature-induced infiltration (usually referred to as

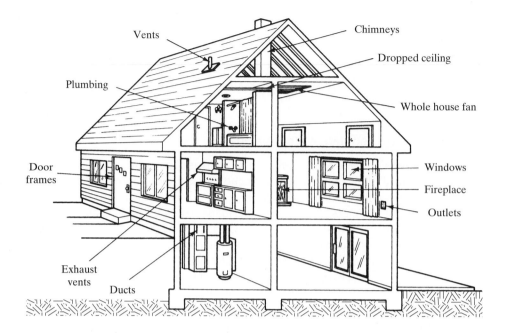

FIGURE 7.61 Infiltration sites in homes. (*Source:* Diamond and Grimsrud, 1984)

the *stack effect*) is influenced less by holes in the walls than by various openings in the floors and ceilings. In the winter, warm air in a building wants to rise, exiting through breaks in the ceiling and drawing in colder air through floor openings. Thus, infiltration rates are influenced not only by how fast the wind is blowing and how great the temperature difference is between inside and out, but also by the locations of the leaks in the building envelope. Greater leakage areas in the floor and ceiling encourage stack-driven infiltration, while leakage areas in vertical surfaces encourage wind-driven infiltration.

Moreover, while it is usually assumed that increasing the infiltration rate will enhance indoor air quality, that may not be the case in one important circumstance—namely, for radon that is emitted from the soil under a building. For radon, wind-driven infiltration helps reduce indoor concentrations by allowing radon-free fresh air to blow into the building. Stack-driven infiltration, which draws air through the floor, may actually encourage new radon to enter the building, negating the cleaning that infiltration usually causes. Figure 7.62 illustrates these important differences.

Infiltration rates may be expressed in units such as m^3/hr or cubic feet per minute (cfm), but more often the units are given in air changes per hour (ach). The air exchange rate in air changes per hour is simply the number of times per hour that a volume of air equal to the volume of space in the house is exchanged with outside air. Typical average infiltration rates in American homes range from about 0.5 ach to 1 ach, with newer houses being more likely to have rates at the lower end of the scale

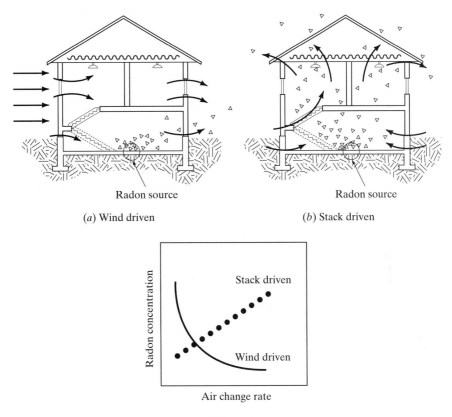

(a) Wind driven   (b) Stack driven

(c) Impact

FIGURE 7.62   Wind-driven infiltration helps reduce radon, while stack-driven infiltration may cause it to increase. (*Source:* Reece, 1988)

while older homes are closer to the top end. Some very poorly built houses have rates as high as 3–4 ach. Carefully constructed new homes today can easily be built to achieve infiltration rates that are as low as 0.1 ach by using continuous plastic sheet "vapor barriers" in the walls along with careful application of foam sealants and caulks to seal cracks and holes. At such low infiltration rates, moisture and pollutant build-up can be serious enough to require extra ventilation, and the trick is to get that ventilation without throwing away the heat that the outgoing stale air contains.

One way to get extra ventilation with minimal heat loss is with a mechanical *heat-recovery ventilator* (HRV), in which the warm, outgoing stale air transfers much of its heat to the cold, fresh air being drawn into the house. Another simpler and cheaper approach is to provide mechanical ventilation systems that can be used intermittently in the immediate vicinity of concentrated sources of pollants. Exhaust fans in bathrooms and range hoods over gas stoves, for example, can greatly reduce indoor pollution, and by using them only as necessary, heat losses can be modest.

An Indoor Air Quality Model

It is straightforward to apply the box model concepts developed earlier to the problem of indoor air quality. The simple model we will use treats the building as a single, well-mixed box, with sources and sinks for the pollutants in question. If necessary, the simple model can be expanded to include several boxes, each characterized by uniform pollutant concentrations. A two-box model, for example, is sometimes used for radon estimates, where one box is used to model radon concentrations within the living space of a dwelling and the other models the air space beneath the house.

Consider the simple, one-box model of a building shown in Figure 7.63. There are sources of pollution within the building that can be characterized by various emission rates. In addition, ambient air entering the building may bring new sources of pollution, which adds to whatever may be generated inside. Those pollutants may be removed from the building by infiltration or ventilation, or they may be nonconservative and decay with time. In addition, if there is a mechanical air cleaning system, some pollutants may be removed as indoor air is passed through the cleaning system and returned. To help keep the model simple, we will ignore such mechanical filtration.

A basic mass balance for pollution in the building, assuming well-mixed conditions, is

$$\begin{pmatrix} \text{Rate of increase} \\ \text{in the box} \end{pmatrix} = \begin{pmatrix} \text{Rate of pollution} \\ \text{entering the box} \end{pmatrix} - \begin{pmatrix} \text{Rate of pollution} \\ \text{leaving the box} \end{pmatrix} - \begin{pmatrix} \text{Rate of decay} \\ \text{in the box} \end{pmatrix}$$

$$V \frac{dC}{dt} = (S + C_a IV) - CIV - KCV \qquad (7.59)$$

where

$$V = \text{volume of conditioned space in building } (\text{m}^3/\text{air change})$$
$$I = \text{air exchange rate (air changes per hour, ach)}$$

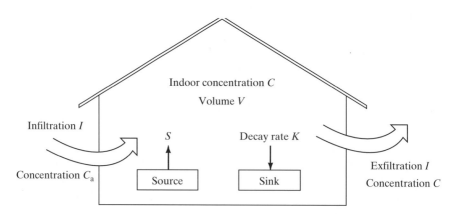

FIGURE 7.63   Box model for indoor air pollution.

S = source emission rate (mg/h)
C = indoor concentration (mg/m^3)
C_a = ambient concentration (mg/m^3)
K = pollutant decay rate of reactivity (1/hr)

An easy way to find the steady-state solution is to set $dC/dt = 0$ to yield

$$C(\infty) = \frac{(S/V) + C_a I}{I + K} \tag{7.60}$$

A general solution is

$$C(t) = \left[\frac{(S/V) + C_a I}{I + K} \right][1 - e^{-(I+K)t}] + C(0)e^{-(I+K)t} \tag{7.61}$$

where $C(0)$ = the initial concentration in the building.

Some of the pollutants that we might want to model, such as CO and NO, can be treated as if they were conservative—that is, they do not decay with time or have significant reactivities, and thus $K = 0$. Equation (7.61) simplifies considerably for the basic condition where there is no ambient concentration of the pollutant ($C_a = 0$), there is no initial concentration of pollution in the building, $C(0) = 0$, and the pollutant is conservative ($K = 0$). Under these circumstances,

$$C(t) = \left(\frac{S}{IV} \right)(1 - e^{-It}) \tag{7.62}$$

Examples of emission rates for various sources are given in Table 7.14, and estimates of appropriate decay rates for common indoor pollutants are given in Table 7.15.

Table 7.14   Some Measured Pollutant Emission Rates for Various Sources

Source	Pollutant Emission Rate, (mg/hr)			
	CO	NO$_x$[a]	SO$_2$	HCHO
Gas range				
Oven	1900	52	0.9	23
One top burner	1840	83	1.5	16
Kerosene heater[b]				
Convective	71	122	—	1.1
Radiant	590	15	—	4.0
One cigarette				
(sidestream smoke)[c] (mg)	86	0.05	—	1.44

[a]NO$_x$ reported as N.
[b]New portable heaters, warmup emissions not included, SO$_2$ not measured.
[c]*Source:* National Research Council (1981).
Sources: Traynor et al. (1981, 1982).

Table 7.15 Decay constants or reactivities K

Pollutant	K (1/hr)
CO	0.0
NO	0.0
NO_x (as N)	0.15
HCHO	0.4
SO_2	0.23
Particles ($< 0.5\ \mu$m)	0.48
Radon	7.6×10^{-3}

Source: Traynor et al. (1981).

EXAMPLE 7.18   A Portable Kerosene Heater

An unvented, portable, radiant heater, fueled with kerosene, is tested under controlled laboratory conditions. After running the heater for two hours in a test chamber with a 46.0 m³ volume and an infiltration rate of 0.25 air-changes per hour (ach), the concentration of carbon monoxide (CO) reaches 20 mg/m³. Initial CO in the lab is zero, and the ambient CO level is negligible throughout the run. Treating CO as a conservative pollutant, find the rate at which the heater emits CO. If the heater were to be used in a small home to heat 120 m³ of space having 0.4 ach, predict the steady-state concentration.

Solution   Rearranging (7.62) to solve for the heater's emission rate S gives

$$S = \frac{IVC(t)}{(1 - e^{-It})} = \frac{0.25\text{ ac/hr} \times 46.0\text{ m}^3/\text{ac} \times 20\text{ mg/m}^3}{1 - e^{-0.25/\text{hr} \times 2\text{ hr}}} = 585\text{ mg/hr}$$

In the small home, the steady-state concentration using (7.60) would be

$$C(\infty) = \frac{S}{IV} = \frac{585\text{ mg/hr}}{0.4\text{ ac/hr} \times 120\text{ m}^3/\text{ac}} = 12.2\text{ mg/m}^3$$

which exceeds the eight-hour ambient standard of 10 mg/m³ for CO.   ∎

Portable kerosene heaters, such as the one introduced in Example 7.18, became quite popular soon after the energy crises of the 1970s, and they continue to be used indoors despite warning labels. There are two types of kerosene heaters: radiant heaters, which heat objects in a direct line-of-sight by radiation heat transfer, and convective heaters, which warm up the air, which, in turn, heats objects in the room by convection. Radiant heaters operate at a lower temperature so their NO_x emissions are reduced, but their CO emissions are higher than the hotter running convective types. Both types of kerosene heaters have high enough emission rates to cause concern if used in enclosed spaces such as a home or trailer.

The following example shows how we can estimate the radon concentration in a home, using this simple indoor air pollution model.

EXAMPLE 7.19   Indoor Radon Concentration

Suppose the soil under a single-story house emits 1.0 pCi/m²-s of radon gas. As a worst case, assume that all of this gas finds its way through the floor and into the house. The house has 250 m² of floor space, an average ceiling height of 2.6 m, and an air change rate of 0.9 ach. Estimate the steady-state concentration of radon in the house, assuming that the ambient concentration is negligible.

Solution We can find the steady-state concentration using (7.60). The decay rate of radon is given in Table 7.15 as $K = 7.6 \times 10^{-3}$/hr, or we could have calculated that value from (3.7), which shows the relationship between K and half-life:

$$C(\infty) = \frac{(S/V) + C_a I}{I + K}$$

$$= \frac{\left(\dfrac{1 \text{ pCi/m}^2\text{s} \times 3600 \text{ s/hr} \times 250 \text{ m}^2}{250 \text{ m}^2 \times 2.6 \text{ m}} \right)}{0.9/\text{hr} + 7.6 \times 10^{-3}/\text{hr}} = 1.5 \times 10^3 \text{ pCi/m}^3 = 1.5 \text{ pCi/L}$$

which is fairly typical of homes in the United States.   ∎

PROBLEMS

7.1 Convert the following (eight-hour) indoor air quality standards established by the U.S. Occupational Safety and Health Administration (OSHA) from ppm to mg/m³ (at 25 °C and 1 atm), or vice versa.

   (a) Carbon dioxide (CO_2), 5000 ppm

   (b) Formaldehyde (HCHO), 3.6 mg/m³

   (c) Nitric oxide (NO), 25 ppm

7.2 Consider a new 38 percent efficient 600-MW power plant burning 9000 Btu/lb coal containing 1 percent sulfur. If a 70 percent efficient scrubber is used, what would be the emission rate of sulfur (lb/hr)?

7.3 In the power plant in Problem 7.2, if all of the sulfur oxidizes to SO_2, how many pounds per hour of SO_2 would be released if the plant is equipped with a 90 percent efficient scrubber? How many pounds of SO_2 per kilowatt-hour of electricity generated would be released?

7.4 A new coal-fired power plant has been built using a sulfur emission control system that is 70 percent efficient. If all of the sulfur oxidizes to SO_2 and if the emissions of SO_2 are limited to 0.6 lb SO_2 per million Btu of heat into the power plant, what maximum percent-sulfur content can the fuel have?

   (a) If 15,000 Btu/lb coal is burned?

   (b) If 9000 Btu/lb coal is burned?

7.5 Compliance coal releases no more than 1.2 lb of SO_2 per 10^6 Btu of heat released, without controls. What maximum percentage sulfur could 12,000 Btu/lb compliance coal contain if all of the sulfur oxidizes to SO_2 during combustion?

7.6 What PSI, and what air quality description, should be reported for the air pollution on the days given?

Pollutant	Day 1	Day 2	Day 3
O_3, 1 hr (ppm)	0.15	0.18	0.12
CO, 8 hr (ppm)	12	9	14
PM 10, 24 hr ($\mu g/m^3$)	150	350	90
SO_2, 24 hr	0.12	0.28	0.14
NO_2, 1 hr	0.4	0.3	0.5

7.7 The OSHA standard for worker exposure to 8 hours of CO is 50 ppm. What percentage COHb would result from this exposure?

7.8 An indoor "tractor pull" competition resulted in a CO concentration of 436 ppm. What percent COHb would result for a spectator who is exposed to one hour at that level? How long would it take to reach 10% COHb, which is a level at which most people will experience dizziness and headache?

7.9 What hydrocarbon, RH, reacting with the $OH^\bullet$ radical in (7.16), would produce formaldehyde, HCHO, in (7.19)?

7.10 Suppose propene, $CH_2 = CH\!-\!CH_3$, is the hydrocarbon (RH) that reacts with the hydroxyl radical $OH^\bullet$ in reaction (7.16). Write the set of chemical reactions that end up with an aldehyde. What is the final aldehyde?

7.11 In 1989, 0.39×10^{12} g of particulates were released when 685 million (2000 lb) tons of coal were burned in power plants that produced 1400 billion kWh of electricity. Assume that the average heat content of the coal is 10,000 Btu/lb. What must have been the average efficiency (heat to electricity) of these coal plants? How much particulate matter would hvae been released if all of the plants met the New Source Perfomance Standards that limit particulate emissions to 0.03 lb per 10^6 Btu of heat?

7.12 In this chapter, the following expression was used in the derivation of the dry adiabatic lapse rate: $dQ = C_p\, dT - V\, dP$ (Eq. 7.35). Derive that expression starting with a statement of the first law of thermodynamics: $dQ = dU + dW$, where $dU = C_v\, dT$ is the change in internal energy when an amount of heat, dQ, is added to the gas, raising its temperature by dT and causing it to expand and do work $dW = P\, dV$. C_v is the specific heat at constant volume. Then use the ideal gas law, $PV = nRT$, where n is moles (a constant) and R is the gas constant, along with the definition of the derivative of a product—$d(PV) = P\, dV + V\, dP$ to find another expression for dQ. Finally, using the definition $C_p = (dQ/dT)$ with the pressure held constant, show that $C_p = C_v + nR$, and you're about there.

7.13 Suppose the following atmospheric altitude versus temperature data have been collected.

Altitude (m)	Temp (°C)
0	20
100	18
200	16
300	15
400	16
500	17
600	18

(a) What would be the mixing depth?

(b) How high would you expect a plume to rise if it is emitted at 21 °C from a 100-m stack if it rises at the dry adiabatic lapse rate? Would you expect the plume to be looping, coning, fanning, or fumigating?

7.14 For the temperature profile given in Problem 7.13, if the daytime surface temperature is 22 °C and a weather station anemometer at 10 m height shows winds averaging 4 m/s, what would be the ventilation coefficient? Assume Stability Class C and use the wind at the height halfway to the mixing depth.

7.15 A tall stack and a nearby short stack have plumes as shown in Figure P7.15. Which atmospheric temperature profile would be most likely to cause that pair of plumes? The dotted lines represent the dry adiabatic lapse rate.

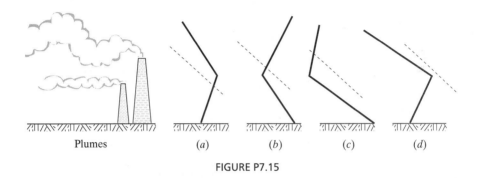

FIGURE P7.15

7.16 Suppose 30 °C air at ground level has a dew point temperature of 14 °C. The atmosphere is unstable and there are rising parcels of air.

(a) Estimate the altitude at which clouds begin to form if the dew point drops by 2 °C/km.

(b) If air continures to rise, estimate its temperature at an elevation of 3 km. (Assume dry adiabatic lapse rate cooling until clouds form, followed by cooling at the wet adiabatic lapse rate as air continues to rise).

(c) If an air parcel that started at ground level, which rose to 3 km, then falls back to ground level, what would its temperature be if the movement was done adiabatically?

7.17 Find the settling velocity of a 20-μm diameter particle with density 1500 kg/m³. If this particle had been hurled to a height of 8000 m during a volcanic eruption, estimate the time required to reach the ground (sea level). You many assume that the viscosity of air does not change enought to worry about. If winds average 10 m/s, how far away would it blow on its way down?

7.18 Find the residence time in the atmosphere for a 10-μm particle with unit density (i.e., the density of water, 10^6 g/m³) at 1000 m elevation.

7.19 Equation (7.24) is valid for Reynolds numbers much less than 1:

$$\text{Re} = \frac{\rho_{air} \, dv}{\eta}$$

where ρ_{air} is air density (1.29×10^3 g/m³), d is the particle diameter, v is its velocity, and η

is viscosity. Find the settling velocity and Reynolds numbers for particles having density of water droplets, with diameters

(a) 1 μm

(b) 10 μm

(c) 20 μm

7.20 A point-source Gaussian plume model for a power plant uses 50 m as the effective stack height (see Figure P7.20). The night is overcast. (*Note*: This is not the same as "cloudy" in this model; check the footnotes in Table 7.8). Your concern is with ground- level pollution at two locations, A and B, which are 1.2 and 1.4 km directly downwind from the stack.

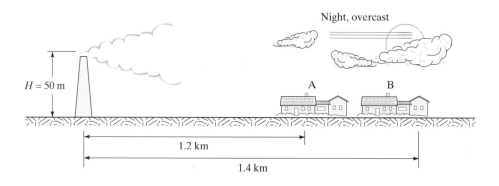

FIGURE P7.20

(a) At what distance will the maximum concentration of pollution occur? Which loca- tion (A or B) would have the higher level of pollution?

(b) Suppose the sky clears up and the windspeed stays less than 5 m/s. Will the location downwind at which the maximum concentration occurs move? If so, will it move closer to the stack or further away from the stack? (*H* remains 50 m.)

(c) Under the new conditions in part b, which house would experience the most pollution?

7.21 Suppose a bonfire emits CO at the rate of 20 g/s on a clear night when the wind is blow- ing at 2 m/s. If the effective stackheight at the fire is 6 m, (a) what would you expect the ground-level CO concentration to be at 400 m downwind? (b) Estimate the maximum ground-level concentration.

7.22 A coal-fired power plant with effective stack height of 100 m emits 1.2 g/s of SO_2 per megawatt of power delivered. If winds are assumed to be 4 m/s at that height and just over 3 m/s at 10 m, how big could the plant be (MW) without having the ground level SO_2 exceed 365 μg/m³? (First decide which stability classification leads to the worst conditions.)

7.23 A 35 percent efficient coal-fired power plant with effective height of 100 m emits SO_2 at a the rate of 0.6 lb/10⁶ Btu into the plant. If winds are assumed to be 4 m/s at the stack height and just over 3 m/s at 10 m, how large could the plant be (MW) without having the ground level SO_2 exceed 365 μg/m³?

7.24 A stack emitting 80 g/s of NO has an effective stack height of 100 m. The windspeed is 4 m/s at 10 m, and it is a clear summer day with the sun nearly overhead. Estimate the ground level NO concentration

(a) Directly downwind at a distance of 2 km.

(b) At the point downwind where NO is a maximum.

(c) At a point located 2 km downwind and 0.1 km off the downwind axis.

7.25 For stability class C, the ratio of σ_y/σ_z is essentially a constant, independent of distance x. Assuming it is a constant, take the derivative of (7.46) and

(a) Show that the distance downwind from a stakc at which the maximum concentration occurs corresponds to the point where $\sigma_z = H/\sqrt{2} = 0.707\ H$.

(b) Show that the maximum concentration is

$$C_{\max} = \frac{Q}{\pi\sigma_y\,\sigma_z ue} = \frac{0.117\,Q}{\sigma_y\sigma_z u}$$

(c) Show that $C_{\max}$ is inversely proportional to H^2.

7.26 The world's tallest stack is on a copper smelter in Sudbury, Ontario. It stands 380 m high and has an inner diameter at the top of 15.2 m. If 130 °C gases exit the stack at 20 m/s while the ambient temperature is 10 °C and the winds at stack height are 8 m/s, use the Briggs model to estimate the effective stack height. Assume a slightly unstable atmosphere, Class C.

7.27 Repeat Problem 7.26 for a stable, isothermal, atmosphere (no temperature change with altitude).

7.28 A power plant has a 100-m stack with an inside radius of 1 m. The exhaust gases leave the stack with an exit velocity of 10 m/s at a temperature of 120 °C. Ambient temperature is 6°C, winds at the effective stack height are estimated to be 5 m/s, surface windspeed is 3 m/s, and it is a cloudy summer day. Estimate the effective height of this stack.

7.29 A 200-MW power plant has a 100-m stack with radius 2.5 m, flue gas exit velocity 13.5 m/s, and gas exit temperature 145 °C. Ambient temperature is 15 °C, windspeed at thet stack is 5 m/s, and the atmosphere is stable, Class E, with a lapse rate of 5 °C/km. If it emits 300 g/s of SO_2, estimate the concentration at ground level at a distance of 16 km directly downwind.

7.30 A source emits 20 g/s of some pollutant from a stack with effective height 50 m in winds that average 5 m/s. On a single graph, sketch the downwind concentration for stability classifications A, C, and F, using Figure 7.50 to identify the peak concentration and distance.

7.31 A source emits 20 g/s of some pollutant in winds that average 5 m/s, on a Class C day. Use Figure 7.50 to find the peak concentrations for effective stack heights of 50, 100, and 200 m. Note whether the concentration is roughly proportional to $(1/H^2)$, as Problem 7.25 suggests.

7.32 A paper plant is being proposed for a location 1 km upwind from a town. It will emit 40 g/s of hydrogen sulfide, which has an odor threshold of about 0.1 mg/m^3. Winds at the stack may vary from 4 to 10 m/s blowing toward the town. What minimum stack height should be used to assure concentrations are nor more than 0.1 times the odor threshold at the near edge of town on a Class B day? To be conservative, the stack will be designed assuming no plume rise. If the town extends beyond the 1-km distance, will any buildings experience higher concentrations than a residence at the boundary under these conditions?

7.33 A stack with effective height of 45 m emits SO_2 at the rate of 150 g/s. Winds are estimated at 5 m/s at the stack height, the stability class is C, and there is an inversion at 100 m. Estimate the ground-level concentration at the point where reflections begin to occur from the inversion and at a point twice that distance downwind.

7.34 A point source with effective stack height of 50 m emits 80 g/s of SO_2 on a clear summer day with surface winds at 4 m/s. Winds at 50 m are 5 m/s. An inversion layer starts at an elevation of 250 m.

(a) Estimate the ground-level SO_2 concentration at a distance of 4 km downwind from the stack.

(b) If there had been no inversion layer, estimate the concentration 4 km downwind.

7.35 A long line of burning agricultural waste emits 0.3 g/m-s of particulate matter on a clear fall afternoon with winds blowing 3 m/s perpendicular to the line. Estimate the ground-level particulate concentration 400 m downwind from the line.

7.36 A freeway has 10,000 vehicles per hour passing a house 200 m away. Each car emits an average of 1.5 g/mi of NO_x, and winds are blowing at 2 m/s across the freeway towards the house. Estimate the NO_x concentration at the house on a clear summer day near noon (assuming that NO_x is chemically stable).

7.37 Consider an area-source box model for air pollution above a peninsula of land (see Figure P7.37). The length of the box is 15 km, its width is 80 km, and a radiation inversion restricts mixing to 15 m. Wind is blowing clean air into the long dimension of the box at 0.5 m/s. Between 4 and 6 P.M. there area 250,000 vehicles on the road, each being driven 40 km and each emitting 4 g/km of CO.

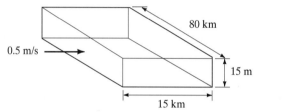

0.5 m/s
80 km
15 m
15 km
FIGURE P7.37

(a) Find the average rate of CO emissions during this two-hour period (g CO/s per m^2 of land).

(b) Estimate the concentration of CO at 6 P.M. if there was no CO in the air at 4 P.M. Assume that CO is conservative and that there is instantaneous and complete mixing in the box.

(c) If the windspeed is zero, use (7.55) to derive a relationship between CO and time and use it to find the CO over the peninsula at 6 P.M.

7.38 Consider a box model for an air shed over a city 1×10^5 m on a side, with a mixing depth of 1200 m. Winds with no SO_2 blow at 4 m/s against one side of the box. SO_2 is emitted in the box at the rate of 20 kg/s. If SO_2 is considered to be conservative, estimate the steady-state concentration in the air shed.

7.39 With the same air shed and ambient conditions as given in Problem 7.38, assume that the emissions occur only on week days. If emissions stop at 5 P.M. on Friday, estimate the SO_2 concentration at midnight. If they start again on Monday at 8 A.M., what would the concentration be by 5 P.M.

7.40 If steady-state conditions have been reached for the city in Problem 7.38 and then the wind drops to 2 m/s, estimate the concentration of SO_2 two hours later.

7.41 If the wind blowing in to the air shed in Problem 7.38 has 5 $\mu g/m^3$ of SO_2 in it, and the SO_2 concentration in the air shed at 8 A.M. Monday is 10 $\mu g/m^3$, estimate the concentration at noon, assuming that emission are still 20 kg/s.

7.42 With the same air shed and ambient conditions as given in Problem 7.38, if SO_2 is not conservative and in fact has a reactivity of 0.23/hr, estimate its steady-state concentration over the city.

7.43 Consider use of a tracer gas to determine the air exchange rate in a room. By injecting a stable gas into the room and then monitoring the decay in concentration with time, we can estimate I (ach). The governing equation is

$$C = C_0 e^{-It}$$

Taking the log of both sides gives: $\ln C = (\ln C_0) - It$

Thus if you plot $\ln C$ vs. time, you should get a straight line with negative slope equal to the infiltration rate I (see Figure P7.43).

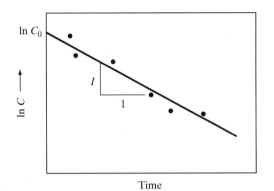

FIGURE P7.43

Suppose you gather the following data:

Time (hr)	Concentration (ppm)
0	10.0
0.5	8.0
1.0	6.0
1.5	5.0
2.0	3.3

Plot $\ln C$ versus time; find the slope and thus the infiltration rate I (ach).

7.44 A single-story home with infiltration rate of 0.5 ach has 200 m² of floor space and a total volume of 500 m³. If 0.6 pCi/m²-s of radon is emitted from the soil and enters the house, estimate the steady-state indoor radon concentration.

7.45 If the house in Problem 7.44 were to have been built as a two-story house with 100 m² on each floor and the same total volume, what would the estimated radon concentration be?

7.46 If an individual lived for 30 years in the house given in Problem 7.44, estimate the cancer risk from the resulting radon exposure. Use radon potency data from Problems 4.10 and 4.12 and standard exposure factors for residential conditions given in Table 4.10.

7.47 Consider a "tight" 300 m³ home with 0.2 ach infiltration rate. The only source of CO in the home is the gas range, and the ambient concentration of CO is always zero. Suppose there is no CO in the home at 6 P.M., but then the oven and two burners are on for one hour. Assume that the air is well mixed in the house, and estimate the CO concentration in the home at 7 P.M. and again at 10 P.M.

7.48 A convective kerosene heater is tested in a well-mixed 27 m³ chamber having an air exchange rate of 0.39 ach. After one hour of operation, the NO concentration reached 4.7 ppm. Treating NO as a conservative pollutant,

(a) Estimate the NO source strength of the heater (mg/hr).

(b) Estimate the NO concentration that would be expected in the lab one hour after turning off the heater.

(c) If this heater were to be used in the home described in Problem 7.47, what steady-state concentration of NO would you expect to be caused by the heater?

7.49 Consider a 100-MW (100,000 kilowatt), 33.3 percent efficient coal-fired power plant that operates at the equivalent of full power 70 percent of the time and no power 30 percent of the time (i.e., its capacity factor is 0.70).

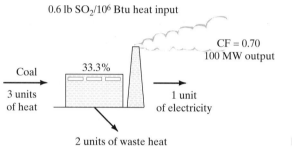

FIGURE P7.49

(a) How much electricity (kWhr/yr) would the plant produce each year?

(b) How may Btu of heat per year would be needed to generate that much electricity? (*Note*: At 33.3 percent efficiency, 3 kWhr of heat, at 3412 Btu/kWhr, are needed to produce 1 kWhr of electricity, as shown in Figure P7.49).

(c) Suppose this coal plant has been emitting 0.6 lb of SO_2 per million Btu of heat input, and suppose it has enough SO_2 allowances for it to continue to do so. If the power company decides to shut down this coal plant and replaces it with a natural-gas-fired plant that emits no SO_2, how many (2000-lb) tons of SO_2 emissions would be saved in one year? If SO_2 allowances are trading at $1500 each, how much could this power company receive by selling one year's worth of allowances?

REFERENCES

American Lung Association, American Medical Association, U.S. Consumer Product Safety Commission, U.S. Environmental Protection Agency, 1994, *Indoor Air Pollution: An Introduction for Health Professionals,* Environmental Protection Agency, Washington, DC.

Briggs, G. A., 1972, Dispersion of Chimney Plumes in Neutral and Stable Surroundings, *Atmospheric Environment,* 6(1).

California Energy Commission, 1989, *News and Comment, CEC Quarterly Newsletter,* No. 23, Fall.

DeCicco, J., and M. Ross, 1993, *An Updated Assessment of the Near-Term Potential for Improving Automotive Fuel Economy,* American Council for an Energy-Efficient Economy, Washington, DC.

Diamond, R. C., and D. T. Grimsrud, 1984, *Manual on Indoor Air Quality,* prepared by Lawrence Berkeley Laboratory for the Electric Power Research Institute, Berkeley, CA.

Gifford, F. A., 1961, Uses of routine meteorological observations for estimating atmospheric dispersion, *Nuclear Safety,* 2(4).

Interagency Taskforce on Acid Precipitation, 1983, *Annual Report 1983 to the President and Congress,* Washington, DC.

Kreiss, K., 1990, The sick building syndrome: Where is the epidemiologic basis?, *American Journal of Public Health,* 80:1172–1173.

Lovins, A., and H. Lovins, 1995, Reinventing the wheels. *The Atlantic Monthly,* January.

Martin, D. O., 1976, The change of concentration standard deviation with distance, *Journal of the Air Pollution Control Association,* 26(2).

McCosh, D., 1994, Emerging technologies for the supercar, *Popular Science,* June, 95–101.

Moyer, C. B., 1989, *Global Warming Benefits of Alternative Motor Fuels,* Acurex Corporation, Mountain View, CA.

Nagda, N. L., H. E. Rector, and M. D. Koontz, 1987, *Guidelines for Monitoring Indoor Air Quality,* Hemisphere, Washington, DC.

National Research Council, 1981, *Indoor Pollutants,* National Academy Press, Washington, DC.

Nazaroff, W. W., and K. Teichman, 1990, Indoor radon, exploring U.S. federal policy for controlling human exposures, *Environmental Science and Technology,* 24:774–782.

NEDO. 1994, *NEDO Creates New Energy,* New Energy and Industrial Technology Development Organization, Tokyo.

OTA, 1984, *Acid Rain and Transported Air Pollutants, Implications for Public Policy,* Office of Technology Assessment, Washington, DC.

Pasquill, F., 1961, The estimation of the dispersion of windborne material, *Meteorological Magazine,* 90:1063.

Perkins, H. C., 1974, *Air Pollution,* McGraw-Hill, New York.

Pershagen, G., G. Akerblom, O. Axelson, B. Clavensjo, L. Damber, G. Desai, A. Enflo, F. Lagarde, H. Mellamder, and M. Svartengren, 1994, Residential radon exposure and lung cancer in Sweden, *New England Journal of Medicine,* Vol 330, No. 3, Jan. 20, pp 159–164.

Peterson, W. B., 1978, *User's Guide for PAL—A Guassian Plume Algorithm for Point, Area, and Line Resources,* U.S. Environmental Protection Agency, Research Triangle Park, NC.

Portelli, R. V., and P. J. Lewis, 1987, Meteorology, in *Atmospheric Pollution,* E.E. Pickett (ed.), Hemisphere, Washington DC.

Powell, J. D., and R. P. Brennan, 1988, *The Automobile, Technology and Society,* Prentice Hall, Englewood Cliffs, NJ.

Powell, J. D., and R. P. Brennan, 1992, *Automotive Technology,* Stanford University, Stanford, CA.

Raloff, J., 1995, Heart-y risks from breathing fine dust, *Science News,* Vol. 148, July 1.

Reece, N. S., 1988, Indoor air quality, *Homebuilding & Remodeling Resource,* November/December.

Robinson, J. P., J. Thomas, and J. V. Behar, 1991, Time spent in activities, locations, and microenvironments: a California-National comparison. Report under Contract No. 69-01-7324, Delivery Order 12, Exposure Assessment Research Division, National Exposure Research Center, U.S. Environmental Protection Agency, Las Vegas, NV.

Seinfeld, J. H., 1986, *Atmospheric Chemistry and Physics of Air Pollution,* Wiley Interscience, New York.

Shepard, M., 1988, Coal technologies for a new age, *EPRI Journal,* Electric Power Research Institute, Jan/Feb.

Smith, K. R., 1993a, Fuel combustion, air pollution exposure, and health: The situation in the developing countries, *Annual Review of Energy and Environment,* 1993, Annual Reviews Inc., Palo Alto, CA 18:529–566.

Smith, K. R., 1993b, taking the true measure of air pollution, *EPA Journal,* US Environmental Protection Agency, October–December, pp. 6–8.

SMUD, 1993, *Fuel Cell Power Plants,* Sacramento Municipal Utility District, Sacramento, CA.

Thomas, V. M., 1995, The elimination of lead in gasoline, *Annual Review of Energy and the Environment,* Annual Reviews, Inc., Palo Alto.

Traynor, G. W., D. W. Anthon, and C. D. Hollowell, 1981, *Technique for Determining Pollutant Emissions from a Gas-Fired Range,* Lawrence Berkeley Laboratory, LBL-9522, December.

Traynor, G. W., J.R. Allen, M. G. Apte, J. F. Dillworth, J. R. Girman, C. D. Hollowell, and J. F. Koonce, Jr., 1982, *Indoor Air Pollution from Portable Kerosene-Fired Space Heaters, Wood-Burning Stoves, and Wood-Burning Furnaces,* Lawrence Berkeley Laboratory, LBL-14027, March.

Turner, D. B., 1970, *Workbook of Atmospheric Dispersion Estimates,* U.S. Environmental Protection Agency, Washington, DC.

U.S. EPA, 1993, *Radon, The Threat with a Simple Solution,* EPA 402-K-93-008, Environmental Protection Agency, Washington, DC.

U.S. EPA, 1994a, *Acid Rain Program: Conservation and Renewable Energy Reserve Update,* EPA 430-R-94-010, Environmental Protection Agency, Washington, DC.

U.S. EPA, 1994b, *National Air Quality and Emissions Trends Report, 1993,* EPA-454/R-94-026, Environmental Protection Agency, Washington, DC.

U.S. EPA, 1995a, *Air Quality Trends,* EPA-454/F-95-003, Environmental Protection Agency, Washington, DC.

U.S. EPA, 1995b, *1993 Air Toxic Release Inventory, Public Data Release,* EPA 745-R-95-019, Environmental Protection Agency, Washington, DC.

U.S. HEW, 1969, *Control Techniques for Particulate Air Pollutants,* National Air Pollution Control Administration, Washington, DC.

U.S. HEW, 1970, *Air Quality Criteria for Carbon Monoxide,* AP-62, National Air Pollution Control Administration, Washington, DC.

U.S. PHS, 1991, *Preventing Lead Poisoning in Young Children,* U.S. Department of Health and Human Services, Washington, DC.

Wadden, R. A., and P. A. Scheff, 1983, *Indoor Air Pollution, Characterization, Prediction, and Control,* Wiley-Interscience, New York.

Wallace, L., 1993, The TEAM studies, *EPA Journal,* U.S. Environmental Protection Agency, October–December, 23–24.

Wallace, L., 1995, Human exposure to environmental pollutants: A decade of experience, *Clinical and Experimental Allergy,* 25:4–9.

Wark, K., and C. F. Warner, 1981, *Air Pollution, Its Origin and Control,* Harper & Row, New York.

Williamson, S. J., 1973, *Fundamentals of Air Pollution,* Addison-Wesley, Reading, MA.

WHO, UNEP, 1992, *Urban Air Pollution in Megacities of the World,* World Health Organization and the United Nations Environment Programme, Blackwell Publishers, Oxford, UK.

Yeager, K. E., 1991, Powering the second electrical century, in *Energy and the Environment in the 21st Century,* J. W. Tester (ed.), MIT Press, Boston.

Yokum, J. E., D. R. Murray, and R. Mikkelsen, 1991, Valdez Air Health Study—Personal, Indoor, Outdoor, and Tracer Monitoring, Paper No. 91–172.12, 84th Annual Meeting of the Air and Waste Management Association, Vancouver, BC.

CHAPTER 8

Global Atmospheric Change

The balance of evidence suggests a discernible human influence on global climate. —Intergovernmental Panel on Climate Change, 1996

Nobody makes a greater mistake than he who thinks he knows nothing because he knows so little! —Edmund Burke

8.1 INTRODUCTION

The assertion of the first of the preceding quotations—that recent increases in global temperatures are unlikely to be entirely natural in origin—is both a bold and cautious statement. It acknowledges the uncertainties that have plagued the discussion of climate change and it stops short of saying what fraction of the observed changes are the result of human activities, but it clearly indicates that the preponderance of evidence suggests that we have begun to change the earth's climate. The quote is taken from the 1996 report of the Intergovernmental Panel on Climate Change (IPCC), and it represents the opinion of hundreds of the world's best scientists in the field from over 60 countries.

The IPCC was established in 1988 by the World Meteorological Organization (WMO) and the United Nations Environment Program (UNEP) to (1) assess available scientific information on climate change, (2) assess the environmental and socioeconomic impacts of climate change, and (3) formulate appropriate response strategies (IPPC, 1995). The IPCC provided the technical documentation that led to the United Nations Framework Convention on Climate Change (UNFCCC), which was signed by 150 nations at the UN Conference on Environment and Development in Rio de Janeiro in 1992. The UNFCCC is a climate treaty that lays the groundwork for nations to stabilize greenhouse gas concentrations in the atmosphere at a level that would prevent dangerous interference with the climate system. It took effect on March 21, 1994 after 50 countries ratified the treaty.

Another effort by WMO and UNEP to organize the scientific community around the issue of stratospheric ozone depletion led to the signing, in 1987, of the Montreal Protocol on Substances that Deplete the Ozone Layer. The Montreal Protocol, and subsequent Amendments framed in London (1990), Copenhagen (1992), and Vienna (1995), has been extraordinarily important both in terms of its success in reducing emissions of ozone-depleting substances and its clear demonstration that nations can come together to address global environmental problems.

Much of this chapter is based on the multivolume scientific assessment reports on climate change and ozone depletion that have been written as part of the WMO and UNEP efforts.

8.2 THE ATMOSPHERE OF EARTH

While the atmosphere is made up almost entirely of nitrogen and oxygen, other gases and particles existing in very small concentrations determine to a large extent the habitability of our planet. In this chapter we will focus on several of these other gases, including carbon dioxide (CO_2), nitrous oxide (N_2O), methane (CH_4), and ozone (O_3), as well as a category of human-made gases called *halocarbons* that includes chlorofluorocarbons (CFCs), hydrochlorofluorocarbons (HCFCs), hydrofluorocarbons (HFCs), carbon tetrachloride (CCl_4), methylchloroform (CH_3CCl_3), and halons. The two problems of greenhouse effect enhancement, leading to global climate change, and stratospheric ozone depletion (which increases our exposure to life-threatening ultraviolet radiation), are linked to changes in these trace gases and are the subject of this chapter.

When the earth was formed, some 4.6 billion years ago, it probably had an atmosphere made up of helium and compounds of hydrogen, such as molecular hydrogen, methane, and ammonia. That early atmosphere is thought to have escaped into space, after which our current atmosphere slowly began to form. Through volcanic activity, gases such as carbon dioxide, water vapor, and various compounds of nitrogen and sulfur were released over time. Molecular oxygen (O_2) eventually began to form both as a result of photodissociation of water vapor and by photosynthesis by plants that were evolving underwater where life was protected from the sun's intense, biologically damaging ultraviolet radiation. As photosynthesis gradually increased atmospheric oxygen levels, more and more ozone (O_3) was formed in the atmosphere. It is thought that the absorption of incoming ultraviolet radiation by that ozone provided the protection necessary for life to begin to emerge onto the land.

TABLE 8.1   Composition of Clean, Dry Air (fraction by volume in troposphere, 1994)

Constituent	Formula	Percent by volume	Parts per million
Nitrogen	N_2	78.08	780,800
Oxygen	O_2	20.95	209,500
Argon	Ar	0.93	9300
Carbon dioxide	CO_2	0.035	358
Neon	Ne	0.0018	18
Helium	He	0.0005	5.2
Methane	CH_4	0.00017	1.7
Krypton	Kr	0.00011	1.1
Nitrous oxide	N_2O	0.00003	0.3
Hydrogen	H_2	0.00005	0.5
Ozone	O_3	0.000004	0.04

Table 8.1 shows the composition of the earth's atmosphere as it exists now, expressed in volumetric fractions (see Section 1.2 for a reminder of the difference between gaseous concentrations expressed by volume and by mass). The values given are for "clean," dry air and do not include the relatively small but extremely important amounts of water vapor and particulate matter. While most of the values in the table are essentially unchanging, that is not the case for the principal greenhouse gases carbon dioxide (CO_2), methane (CH_4), and nitrous oxide (N_2O), which are rising.

It is convenient to think of the atmosphere as being divided into various horizontal layers, each characterized by the slope of its temperature profile. Starting at the earth's surface, these layers are called the *troposphere, stratosphere, mesosphere,* and *thermosphere.* The troposphere and mesosphere are characterized by decreasing temperatures with altitude, while the stratosphere and thermosphere show increasing temperatures. The transition altitudes separating these layers are called the *tropopause, stratopause,* and *mesopause.* Obviously, the conditions in the true atmosphere of earth vary with time and location, but a useful, idealized temperature profile, known as the *U.S. Standard Atmosphere,* provides a convenient starting point for atmospheric studies (Figure 8.1).

More than 80 percent of the mass of the atmosphere and virtually all of the water vapor, clouds, and precipitation occur in the troposphere. At midlatitudes, the troposphere extends up to 10 or 12 km (about the altitude of a typical airline flight). At the poles it may be only about 5 or 6 km, while at the equator it is about 18 km. In the troposphere, temperatures typically decrease at 5 to 7 °C per km, which is essentially the wet adiabatic lapse rate corresponding to the rate of change of temperature as water-saturated air rises (see Section 7.10). The troposphere is usually a very turbulent place; that is, there are strong vertical air movements that lead to rapid and complete mixing. This mixing is good for air quality since it rapidly disperses pollutants.

Above the troposphere is a stable layer of very dry air called the stratosphere. Pollutants that find their way into the stratosphere may remain there for many years before they eventually drift back into the troposphere, where they can be more easily diluted and ultimately removed by settling or precipitation. In the stratosphere, short-wavelength ultraviolet energy is absorbed by ozone O_3 and oxygen O_2, causing the air

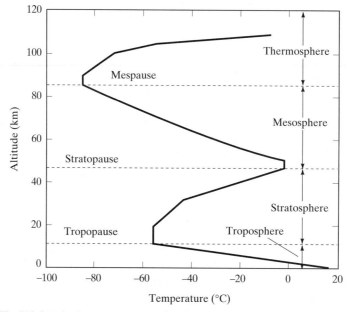

FIGURE 8.1.   The U.S. Standard Atmosphere, showing the four major layers.

to be heated. The resulting temperature inversion is what causes the stratosphere to be so stable. The troposphere and stratosphere combined account for about 99.9 percent of the mass of the atmosphere. Together they extend only about 50 km above the surface of the earth, a distance equal to less than 1 percent of the earth's radius.

Beyond the stratosphere lies the mesosphere, another layer where air mixes fairly readily, and above that the thermosphere. The heating of the thermosphere is due to the absorption of solar energy by atomic oxygen. Within the thermosphere is a relatively dense band of charged particles, called the *ionosphere*. (Before satellites, the ionosphere was especially important to worldwide communications because of its ability to reflect radio waves back to earth.)

8.3 GLOBAL TEMPERATURE

The common definition of climate suggests that it is the prevailing or average weather of a place as determined by temperature and other meteorological conditions over a period of years. Average temperature, then, is but one measure of climate, and many others, including precipitation, winds, glaciation, and frequency of extreme events such as typhoons and hurricanes, are also important parameters. Long-term variations in average temperature, however, are the single most important attribute of climate change.

Climatologists have used a number of clues to piece together past global temperatures, including evidence gathered from historical documents, tree rings, changes in ice volume and sea level, fossil pollen analysis, and geologic observations related to glacial movements. One of the most fruitful approaches involves a well-established correlation between the world's ice cover and the concentration of the heavy isotope of oxygen, ^{18}O, in sea-floor sediments. When water evaporates from the oceans it contains a mix of two isotopes of oxygen, ^{18}O and ^{16}O. Being slightly heavier, ^{18}O does not evaporate as easily, and when it does, it condenses and falls as precipitation somewhat sooner than water vapor made up of ^{16}O. Thus, precipitation over the oceans tends to be slightly richer in ^{18}O than precipitation that must travel further to reach polar ice sheets. Similarly, precipitation that forms glaciers and ice sheets is relatively depleted of ^{18}O. As the world's ice volume increases, it selectively removes ^{16}O from the hydrologic cycle and concentrates the remaining ^{18}O in the decreasing volume of the oceans. Hence, marine organisms that build their shells out of calcium carbonate in seawater will have a higher ratio of ^{18}O to ^{16}O in their shells when it is cold and more of the world's water is locked up in glaciers and ice. By dating marine sediments extracted from deep-sea cores and observing the ratio of the two oxygen isotopes in their carbonates, a historic record of the volume of ice storage on the earth can be created. From that it is possible to estimate the temperature.

The ratio of the two oxygen isotopes is described in the literature using the notation $\delta^{18}O$. Usually what is of interest is the shift in the isotopic ratio from some reference or standard. Also, since the mass ratio of the two isotopes shifts only slightly under climatic temperature changes, it is conventional practice to multiply the changes by 1000, giving the following:

$$\delta^{18}O(^0/_{00}) = \left[\frac{(^{18}O/^{16}O) \text{ sample} - (^{18}O/^{16}O) \text{ standard}}{(^{18}O/^{16}O) \text{ standard}} \right] \times 10^3 \qquad (8.1)$$

Using this notation, negative numbers represent a decrease in ^{18}O below the standard, while positive numbers correspond to an increase. A similar notation exists for the ratio of the heavier isotope of hydrogen, deuterium, to ordinary hydrogen and is represented by δD.

EXAMPLE 8.1   Interpreting $\delta^{18}O$ in ocean sediments

Find $\delta^{18}O$ for sediments having $(^{18}O/^{16}O) = 0.00201$ if 0.00200 is the standard.

Solution

$$\delta^{18}O(^0/_{00}) = \left[\frac{0.00201 - 0.00200}{0.00200} \right] \times 10^3 = 5$$

Since it is positive, the sample corresponds to a colder climate. ∎

The $\delta^{18}O$ record in ocean sediments and glacial ice cores provides a thermometer that tracks the surface air temperature at the time the ice or sediments were formed. To avoid confusion, it is worth noting that the isotope record in ocean sediments moves in the opposite direction to the isotope record in glacial ice. That is, while warmer temperatures cause *decreases* in $\delta^{18}O$ in the oceans and sediments, warmer temperatures cause *increases* in $\delta^{18}O$ (and δD) in that year's layer of glacial ice. Thus, for example, in

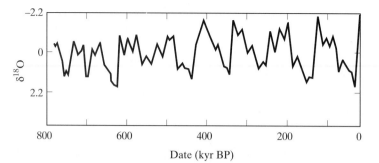

FIGURE 8.2 Oxygen isotope record from deep-sea cores, drawn so that peaks correspond to warmer, interglacial periods and valleys correspond to glacial conditions. The notation kyr BP means thousands of years before the present. (*Source:* Imbrie, 1984)

order for the plot of $\delta^{18}O$ in deep-sea cores to show interglacial periods (warmer temperatures) at the top of the graph and glacial conditions near the bottom, it was necessary to flip the vertical axis of the isotope record in Figure 8.2.

Analysis of ice cores taken in Greenland and the Antarctic has provided a remarkable picture of the climate and composition of the atmosphere over many thousands of years of the earth's history. Isotopic analysis of glacial ice provides the temperature record, while analysis of the composition of the air bubbles sealed off in the ice as it was formed gives a corresponding record of the concentrations of various atmospheric gases. In one of the most significant scientific studies of past climates, an ice core 2083 m long was recovered by the Soviets at Vostok in East Antarctica. Careful analysis of the hydrogen isotope ratios and gaseous concentrations in this Vostok ice core have provided a continuous 160,000-year record. Subsequent ice core studies at Vostok have extended the data back 220,000 years (Jouzel et al., 1993). Figure 8.3 shows the remarkable correlation between atmospheric carbon dioxide and methane concentrations and Antarctic surface temperature from the Vostok core. During glacial periods, the greenhouse gases CO_2 and CH_4 are low; during the warmer interglacial period, they are high. What is less clear, however, is whether CO_2 and CH_4 changes caused temperature to change, or vice versa.

In the opposite hemisphere, scientists have drilled through the 3-km-thick glacial cap in Greenland and produced climate data on the past 110,000 years (Dansgaard, and Oeschger, 1989). The Greenland data show cycles of gradual cooling, followed by rapid warming, during the last glacial period. These cycles, which have a period of between 500 and 2000 years, are called *Dansgaard-Oeschger* events after their discoverers. They seem to be linked to another phenomenon, called *Heinrich* events, in which it is thought that enormous flotillas of icebergs periodically drifted across the North Atlantic; their melting deposited unusual layers of sediments on the sea floor that were used to identify and date these events. One of the most startling observations from Greenland ice cores is the speed with which large, regional temperature jumps can occur. During the last glacial period, 10,000 to 110,000 years ago, the temperature of Greenland showed jumps of 5 to 7 °C within just a decade or two. Global temperatures, on the other hand, do not seem to have changed nearly that rapidly in the past. Over

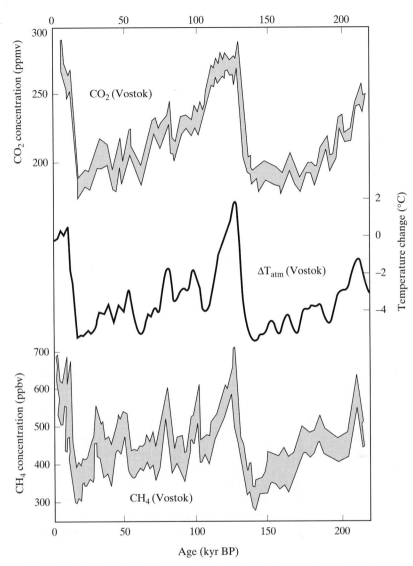

FIGURE 8.3.   Carbon dioxide and methane concentrations (ppmv) and Antarctic temperature change (°C) over the past 220,000 years. Temperatures are referenced to current Vostok surface temperature. (*Source:* IPCC, 1995)

the past 10,000 years (the present interglacial period), it seems unlikely that global mean temperatures have ever changed faster than 1 °C per century (IPCC, 1995).

By combining many types of climate data, a record of midlatitude air temperatures has been reconstructed, some of which is summarized in Figure 8.4. A striking feature of this figure is how modest the temperature differences have been in the past 850,000 years, between the warmest interglacials, such as we are in now, and the coldest glacial periods, when ice covered much of North America. A rise in temperature of only a few degrees Celsius would make the earth warmer than it has been in the last

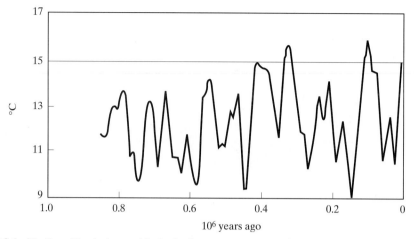

FIGURE 8.4 Northern Hemisphere, midlatitude air temperatures over the past one million years. (*Source:* Clarke, W. C., ed., *Carbon Dioxide Review 1982.* Copyright © 1982 by Oxford University Press. Reprinted by permission of Oxford University Press.)

million years, but it would take an increase of closer to 10 °C for temperatures to reach what they were during the age of dinosaurs, roughly 65 to 200 million years ago.

In the more recent past, the global mean temperature record is based on actual measurements that have been taken since 1856. During that time, the mean temperature has increased by about 0.3 to 0.6 °C, with about 0.2 to 0.3 °C of that occurring over the last 40 years (IPCC, 1995). The five-year period from 1990 through 1994 was the warmest on record, in spite of a several-year period of cooling that is attributed to the increased reflection of sunlight off the atmosphere caused by the eruption of Mount Pinatubo in the Philippines in June 1991. The warmest year on record as of this writing was 1995, followed by 1990, 1991, and 1994. Figure 8.5 shows the year-by-year global mean temperature along with a smoothed curve to see the trends more easily. The temperatures are expressed as a deviation, or anomaly, from a fixed reference temperature.

Orbital Variations and Sunspots

The history and future of the earth's orbit around the sun can be calculated precisely. The connection between orbital variations and climate were first proposed in the 1930s by an astronomer, Milutin Milankovitch, and the orbital cycles are now referred to as *Milankovitch* oscillations. Changes in the orbit affect the amount of sunlight striking the earth as well as the distribution of sunlight both geographically and seasonally. Those variations are thought to be influential in the timing of the coming and going of ice ages and interglacial periods.

There are three primary orbital cycles. The shape of the earth's orbit oscillates from elliptical to more nearly circular with a period of 100,000 years (*eccentricity*). The earth's tilt angle with respect to its orbit fluctuates from 21.5° to 24.5° with a period of

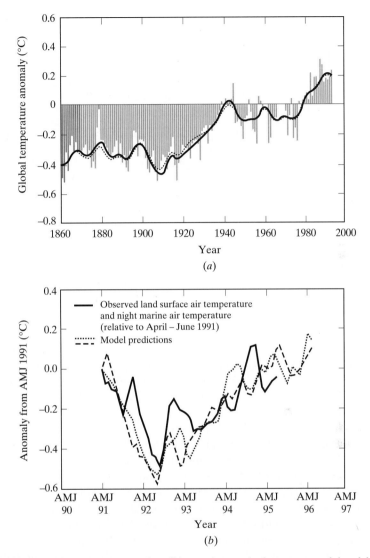

FIGURE 8.5 (*a*) Global temperature anomaly; solid curve is smoothed to suppress subdecadal variations, dashed corresponds to estimates from IPCC (1992). (*b*) Two model predictions and observed changes after the June 1991 eruption of Mount Pinatubo; lines are running three-month means from April through June 1991. (*Source:* IPCC, 1996).

41,000 years (*obliquity*). Finally, there is a 23,000-year period associated with the precession, or wobble, of the earth's spin axis (*precession*). This precession determines where in the earth's orbit a given hemisphere's summer occurs. Figure 8.6 illustrates these variations.

Careful analysis of the historical record of global temperatures does show a primary cycle between glacial episodes of about 100,000 years, mixed with secondary oscillations with periods of 23,000 years and 41,000 years that match the Milankovitch theory reasonably well. Although these orbital variations only change the total annual

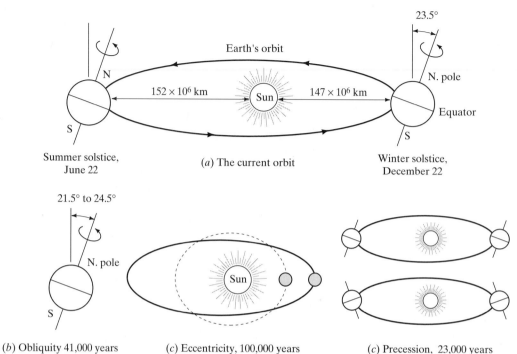

FIGURE 8.6   Orbital variations affect the timing of ice ages. (*a*) The current orbit; (*b*) The tilt angle variation, with period 41,000 years; (*c*) eccentricity variation, with period 100,000 years; (*d*) precession, with period 23,000 yrs.

dosage of sunlight by about 0.1 percent, the impacts on seasons and the resulting patterns of oceanic and atmospheric heat distribution around the globe are thought to be significant enough to trigger major climate changes.

Another factor that affects the amount of solar radiation reaching the top of the earth's atmosphere is variations in the intensity of radiation emitted from the sun itself. For example, the sun spins on its axis, making one complete rotation every 27 days. There are darker and brighter areas on the solar surface that cause variations of up to 0.2 percent in the amount of solar radiation reaching the earth over that 27-day period. Those variations occur so rapidly, however, that they are not thought to be particularly important in determining average global climate.

Of greater importance is an 11-year cycle of sunspots that were first described by an amateur astronomer, Heinrich Schwabe, in 1843 (Figure 8.7). During peak periods of magnetic activity on the sun, the surface has large numbers of cooler, darker regions, called *sunspots,* that in essence block solar radiation, accompanied by other regions, called *faculae,* that are brighter than the surrounding surface. The net effect of sunspots that dim the sun, and faculae that brighten it, is an increase in solar intensity during periods of increased numbers of sunspots. The last two peaks in these *Schwabe cycles* occurred in 1979–1981 and 1989–1991, and the next one is due in the year 2000. The variation in solar radiation reaching the earth as a result of these cycles is roughly 0.1 percent, which is thought to be enough to change the earth's temperature by

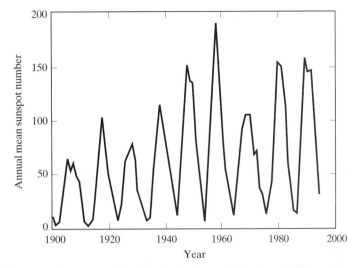

FIGURE 8.7   Mean annual sunspot number showing the 11-year Schwabe cycle. (*Source:* Lean and Rind, 1996)

roughly 0.2 °C. Sunspot activity helps provide an explanation for the continuous up and down jiggling of the earth's temperature, which complicates the problem of deciding whether a true global warming "signal" is emerging from the "noise."

A Simple Global Temperature Model

Measurements of historical global temperatures show that our planet has maintained its average temperature within a limited range, but within those bounds it seems to be continuously changing. If we hope to predict future impacts of anthropogenic changes in our environment, we need to develop mathematical models that explain the past. Mathematical modeling of global climate and predicting the impacts of changes in key environmental parameters is an extremely important but difficult task. Such models range from very simple back-of-the-envelope calculations to complex, three-dimensional *general circulation models* (GCMs), which attempt to predict climate on a regional, seasonal, and annual basis. The most sophisticated of these models can take weeks to run on a supercomputer, yet they must still be considered primitive. In comparison, our treatment here is just the briefest of introductions.

The simplest starting point for modeling climate begins with models that focus on factors influencing the single variable *temperature*. Obviously, other factors, such as precipitation patterns, winds and storms, ocean currents, soil moisture, sea ice, glacial cover, and so forth, are exceedingly important, but they are more difficult to approach with simple models. Even beginning with just temperature as the single quantity of interest, we could try to find out how temperature varies in all four dimensions (latitude, longitude, altitude, time), which is the domain of very complex general circulation models. At the other end of the complexity scale is a simple zero-dimensional model, in which a single average global temperature is obtained that is not a function of location or time. The following is such a model.

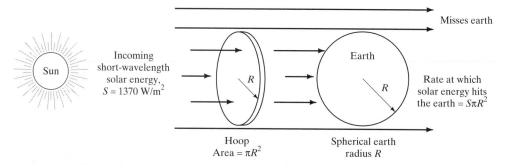

FIGURE 8.8.   Solar energy passing through a "hoop" with the same radius as that of the earth hits the earth. Radiation that misses the hoop also misses the earth.

The basic zero-dimensional energy balance model equates solar energy absorbed by the earth with the energy that the earth radiates back to space. Radiation from the sun arrives just outside the earth's atmosphere with an average annual intensity, called the *solar constant, S,* currently equal to about 1370 W/m². A simple way to calculate the total rate at which energy hits the earth is to note that all of the flux passing through a "hoop" having radius equal to that of the earth, and placed normal to the incoming radiation, strikes the earth's surface. From Figure 8.8, we can write

$$\text{Rate at which solar energy strikes the earth} = S\pi R^2 \text{(watts)} \qquad (8.2)$$

where

$S =$ the solar constant, taken to be 1370 W/m²
$R =$ the radius of the earth (m)

Some of the incoming solar energy that hits the earth is reflected back into space, as shown in Figure 8.9. Such reflected energy is not absorbed by the earth or its atmosphere and does not contribute to their heating. The fraction of incoming solar radiation that is reflected is called the *albedo,* and for the earth, the global annual mean value is now estimated to be about 31 percent. What is not reflected is absorbed, which leads to the following expressions:

$$\text{Energy reflected by earth} = S\pi R^2 \alpha \qquad (8.3)$$
$$\text{Energy absorbed by earth} = S\pi R^2 (1 - \alpha) \qquad (8.4)$$

where

$\alpha =$ the earth's albedo, taken to be 0.31.

On the other side of the energy balance equation is the rate at which earth sends energy back to space. Since there can be no heat transfer from the earth to space by conduction or convection, the only way for the earth to get rid of energy is by radiation. Recall from Section 1.4 that every object radiates energy at a rate that is proportional to its surface area times its absolute temperature raised to the fourth power (Eq. 1.42). For this model, we will assume that the earth is a blackbody; that is, it radiates as much as

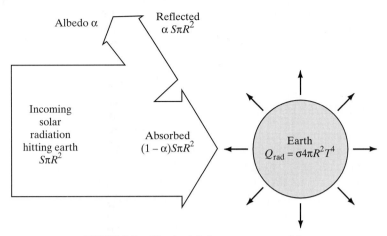

FIGURE 8.9.   Simple global temperature model.

any object with the same temperature and area possibly can (emissivity = 1). We also assume that it is isothermal; that is, the temperature is the same everywhere on the planet. Since the area of a spherical object is $4\pi R^2$, we can write

$$\text{Energy radiated back to space by earth} = \sigma 4\pi R^2 T_e^4 \qquad (8.5)$$

where

$$\sigma = \text{Stefan-Boltzmann constant} = 5.67 \times 10^{-8} \text{ W/m}^2\text{-K}^4$$

and

$$T_e = \text{earth's "effective" blackbody temperature (kelvins)}$$

If we go on to assume steady-state conditions—that is, the earth's temperature is not changing with time—we can equate the rate at which energy from the sun is absorbed (8.4), with the rate at which energy is radiated from earth back to space (8.5):

$$S\pi R^2(1 - \alpha) = \sigma 4\pi R^2 T_e^4 \qquad (8.6)$$

Solving for T_e,

$$T_e = \left[\frac{S(1 - \alpha)}{4\sigma} \right]^{1/4} \qquad (8.7)$$

Substituting appropriate values into (8.7) yields

$$T_e = \left[\frac{1370 \text{ W/m}^2(1 - 0.31)}{4 \times 5.67 \times 10^{-8} \text{ W/m}^2\text{-K}^4} \right]^{1/4} = 254 \text{ K} = -19\,^\circ\text{C}$$

Notice the conversion from kelvins to Celsius ($^\circ\text{C} = \text{K} - 273$). The actual value of the earth's average surface temperature is about 288 K (15 °C). While we are off by only 12 percent, which might seem a modest error, in terms of life on earth the 254 K ($-19\,^\circ$C) estimate for T_e is terribly wrong. We need to find an explanation for why the earth is (fortunately) not that cold. The key factor that makes our model differ so much from reality is that it does not account for interactions between the atmosphere and the radiation that is emitted from the earth's surface. That is, it does not include the greenhouse effect.

8.4 THE GREENHOUSE EFFECT

The surface of the earth is 34 °C higher than what is predicted by (8.7). To understand the reason for the higher temperature, it is helpful to begin by recalling the discussion in Chapter 1 concerning the relationship between the spectrum of wavelengths radiated by an object and its temperature. Wien's displacement rule (Eq. 1.43), repeated here, gives the wavelength at which a blackbody spectrum peaks as a function of its absolute temperature:

$$\lambda_{max}(\mu m) = \frac{2898}{T(K)} \qquad (8.8)$$

The sun can be represented as a blackbody with temperature 5800 K, so its spectrum peaks at 0.5 μm. The earth, at 288 K, has its peak at 10.1 μm. Figure 8.10 shows these two spectra. Recall from Chapter 1 that the area under these curves, between any two wavelengths, is the total radiant energy in that range of frequencies. For example, the area under the curve of incoming solar radiation just outside the atmosphere is the total solar radiant flux; that is, it is the solar constant, 1370 W/m^2. Notice that nearly all

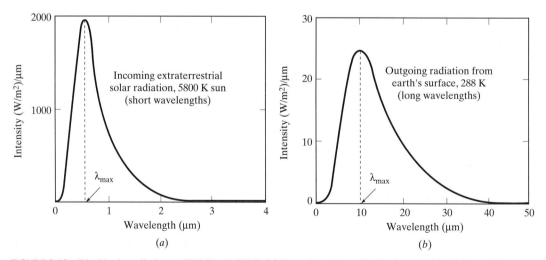

FIGURE 8.10   Blackbody radiation at 5800 K and 288 K. (*a*) Incoming solar radiation just outside of the earth's atmosphere. (*b*) Radiation from the earth's surface at 288 K.

the incoming solar energy as it arrives just outside the earth's atmosphere has wavelengths less than 3 μm, while the outgoing energy radiated by the earth has essentially all of its energy in wavelengths greater than 3 μm. With so little overlap, it is convenient to speak of solar energy as being *short-wavelength* radiation, while energy radiated from the earth's surface is *long-wavelength,* or *thermal,* radiation. The infrared (IR) portion of the spectrum begins at about 0.7 μm and extends out to 100 μm, so some of the incoming solar radiation and all of the outgoing thermal radiation is IR.

As radiant energy attempts to pass through the atmosphere, it is affected by various gases and aerosols in the air. Those atmospheric constituents can either let the radiant energy pass through unaffected, they can scatter the energy by reflection, or they can stop it by absorption. The key phenomenon of interest here is the ability of gases to absorb radiant energy. As the atoms in gaseous molecules vibrate toward and away from each other (vibrational energy) or rotate around each other (rotational energy), they absorb and radiate energy in specific wavelengths. When the frequency of these molecular oscillations is close to the frequency of the passing radiant energy, the molecule can absorb that energy. This absorption occurs over a rather limited range of frequencies, not just at the oscillatory frequency of the molecule, and results in an absorptivity spectrum, which is a plot of the fraction of incoming radiant energy that is absorbed as a function of wavelength.

Figure 8.11 shows the absorption spectra for the key gases of concern in this chapter, along with their effect on incoming solar radiation and outgoing infrared radiation emitted by the earth's surface. Most of the long-wavelength energy radiated by the earth is absorbed by a combination of radiatively active gases, most important of which are water vapor (H_2O), carbon dioxide (CO_2), methane (CH_4), nitrous oxide (N_2O), molecular oxygen (O_2) and ozone (O_3). Water vapor, which is by far the most important greenhouse gas, strongly absorbs thermal radiation with wavelengths less than 8 μm and greater than 18 μm. Carbon dioxide shows a strong absorption band centered at 15 μm, as well as bands centered at 2.7 μm and 4.3 μm. Between 7 μm and 12 μm there is a relatively clear sky for outgoing thermal radiation, referred to as the *atmospheric radiative window.* Radiation in those wavelengths easily passes through the atmosphere, except for a small but important absorption band between 9.5 μm and 10.6 μm associated with ozone (O_3).

Notice in Figure 8.11 that essentially all of the incoming solar radiation with wavelengths less than 0.3 μm (ultraviolet) is absorbed by oxygen and ozone. This absorption of ultraviolet occurs in the stratosphere, shielding the earth's surface from harmful ultraviolet radiation. Later in this chapter we will return to this important phenomenon, in our discussion of the stratospheric ozone reduction being caused by halocarbons.

Radiatively active gases that absorb wavelengths longer than 4 μm are called greenhouse gases. This absorption heats the atmosphere, which, in turn, radiates energy back to the earth as well as out to space, as shown in Figure 8.12. These greenhouse gases act as a thermal blanket around the globe, raising the earth's surface temperature beyond the equivalent temperature calculated earlier. The importance of water vapor as a greenhouse gas is quite evident on clear nights, when the earth cools much more rapidly than it does on cloudy nights. It is also interesting to note that the

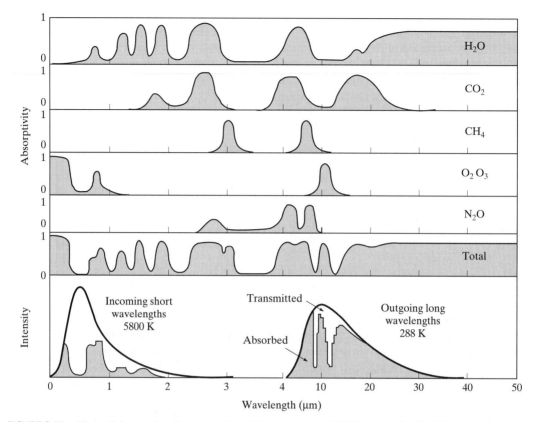

FIGURE 8.11 Absorptivity as a function of wavelength for water vapor (H_2O), carbon dioxide (CO_2), methane (CH_4), oxygen and ozone (O_2, O_3), and nitrous oxide (N_2O), and the total absorptivity of the atmosphere. Shown here are the spectra for incoming solar energy and outgoing thermal energy from the 288 K surface of the earth. Note the wavelength scale change at 4 μm.

term *greenhouse effect* is based on the concept of a conventional greenhouse with glass acting much like the aforementioned gases. Glass, which easily transmits short-wavelength solar energy into the greenhouse, absorbs almost all of the longer wavelengths radiated by the greenhouse interior. This radiation trapping is partly responsible for the elevated temperatures inside the greenhouse, although much of the effect is simply due to the reduction in convective cooling of the interior space caused by the enclosure. The elevated interior temperature of your car after it has been parked in the sun is another common example of the greenhouse effect.

If the earth did not already have a greenhouse effect, its temperature would be 254 K, as predicted by (8.7). That is, the planet would have an average temperature of $-19\ °C$, or about $-2\ °F$. In fact, one way to quantify the magnitude of the greenhouse effect is to compare the effective temperature T_e given in (8.7) with the actual surface temperature T_s

$$\text{Magnitude of greenhouse effect} = T_s - T_e \qquad (8.9)$$

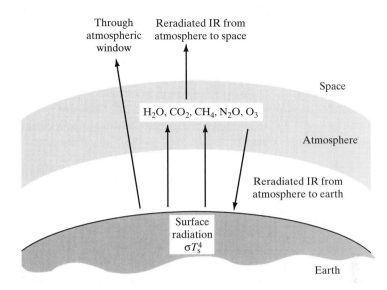

FIGURE 8.12   Most of the long-wavelength energy radiated from the earth's surface is absorbed by greenhouse gases in the atmosphere, but some passes directly through the atmospheric window. The atmosphere, in turn, radiates energy to space and back to the earth.

Thus, since the actual temperature of the earth is 288 K and its effective temperature is 254 K, we can say that the greenhouse effect adds 34 °C of warming to the surface of the earth.

In Table 8.2, this notion is applied to Venus and Mars. Mars has almost no greenhouse effect. Even though the atmosphere of Mars is almost entirely carbon dioxide, there is so little atmosphere that the greenhouse effect is barely apparent. The atmospheric pressure on Venus, on the other hand, is nearly 100 times that of earth, and its atmosphere is 97 percent CO_2. The greenhouse effect on Venus is correspondingly very pronounced. It is interesting to note that without the greenhouse effect, the greater albedo of Venus would make it cooler than the earth in spite of its closer proximity to the sun.

TABLE 8.2.   Application of the simple model (8.7) to compute effective temperatures, compared with actual surface temperatures

Planet	Distance to sun (10^6 km)	Atmosphere pressure (atm)	Solar constant S(W/m^2)	Albedo α (%)	Effective temperature T_e (K)	Surface temperature T_s (K)	Greenhouse warming (°C)
Venus	108	90	2620	76	229	750	521
Earth	150	1	1370	31	254	288	34
Mars	228	0.006	589	25	210	218	8

Note: Mars, with little atmosphere, shows almost no greenhouse effect, while on Venus it is quite pronounced.
Source: Hoffert, 1992

8.5 GLOBAL ENERGY BALANCE

Let us add some quantitative information to the simple greenhouse diagram presented in Figure 8.12. As suggested there, it is convenient to consider the earth, its atmosphere, and outer space as three separate regions. We will normalize energy flows between these regions by expressing them in terms of rates per unit of surface area of the earth. For example, (8.2) indicates that the total amount of solar radiation striking the earth is $S\pi R^2$. Distributed over the entire surface of the earth, the average incoming solar radiation is equal to

$$\frac{\text{Incoming solar radiation}}{\text{Surface area of earth}} = \frac{S\pi R^2}{4\pi R^2} = \frac{S}{4} = \frac{1370 \text{ W/m}^2}{4} = 342 \text{ W/m}^2 \quad (8.10)$$

Since the albedo is 31 percent, the amount of incoming radiation reflected back into space per square meter of the earth's surface is

$$\frac{\text{Solar energy reflected}}{\text{Surface area of earth}} = \frac{S\pi R^2 \alpha}{4\pi R^2} = \frac{S}{4}\alpha \quad (8.11)$$

$$= 342 \text{ W/m}^2 \times 0.31 = 107 \text{ W/m}^2$$

Of this 107 W/m^2, it is estimated that 77 W/m^2 are reflected off the atmosphere itself, while the remaining 30 W/m^2 are reflected off the earth's surface. The solar radiation that is not reflected is absorbed by the earth and its atmosphere. Calling that absorbed energy Q_{abs} (again, with units of watts per square meter of surface) gives

$$\frac{\text{Solar radiation absorbed}}{\text{Surface area of earth}} = Q_{abs} = \frac{S\pi R^2(1-\alpha)}{4\pi R^2} = \frac{S}{4}(1-\alpha) \quad (8.12)$$

$$= 342 \text{ W/m}^2 \times (1 - 0.31) = 235 \text{ W/m}^2$$

Of that 235 W/m^2, 67 W/m^2 are absorbed by the atmosphere and the remaining 168 W/m^2 are absorbed by the surface of the earth.

If we assume that global temperatures are unchanging with time, then the rate at which the earth and its atmosphere receive energy from space must equal the rate at which energy is being returned to space. The 107 W/m^2 of reflected energy is already balanced; that is, 107 W/m^2 hits the earth/atmosphere and 107 W/m^2 is reflected back into space so we can ignore that component for now. The earth and its atmosphere absorb the remaining 235 W/m^2, so the same amount must be radiated back into space. If the earth's surface were at 254 K, it would radiate 235 W/m^2, which is just enough to balance the incoming energy. We know, however, that greenhouse gases would absorb most of that outgoing 235 W/m^2, so the required energy balance would not be realized. Therefore, to force enough energy through the atmosphere to create the necessary balance, the temperature of the earth's surface must be higher than 254 K.

If we treat the earth as a blackbody radiator, we can use (8.5) to estimate the rate at which energy is radiated from the earth's surface toward the atmosphere. With the surface of the earth at 288 K, it will radiate the following amount per unit of surface area.

$$\frac{\text{Energy radiated by surface}}{\text{Surface area of earth}} = \frac{\sigma 4\pi R^2 T_s^4}{4\pi R^2} = \sigma T_s^4 \quad (8.13)$$

$$= 5.67 \times 10^{-8} \text{ W/m}^2\text{K}^4 \times (288 \text{ K})^4 = 390 \text{ W/m}^2$$

Of that 390 W/m^2, only 40 W/m^2 passes directly through the atmosphere, mostly through the atmospheric radiative window between 7 and 12 μm. The remaining 350 W/m^2 is absorbed by greenhouse gases in the atmosphere. The atmosphere then radiates 324 W/m^2 back to the surface.

There is also heat transfer from the surface to the atmosphere by convective heating and by evaporation and condensation of water. Convection transfers 24 W/m^2 to the atmosphere, while condensation of water vapor provides 78 W/m^2 of latent heat (see Section 1.4).

All of these energy flows are shown in Figure 8.13. If this model is internally self-consistent, the rate of energy gain should equal the rate of energy loss in each of the three regions: space, the atmosphere, and the earth's surface. Consider the following checks:

$$\text{Rate of energy gain} = \text{Rate of energy loss?}$$

Earth's surface	$168 + 324 + 30 = 78 + 24 + 30 + 390$	(checks)
Atmosphere	$67 + 78 + 24 + 350 = 165 + 30 + 324$	(checks)
Space	$107 + 165 + 30 + 40 = 342$	(checks)

So the model shows the necessary balances.

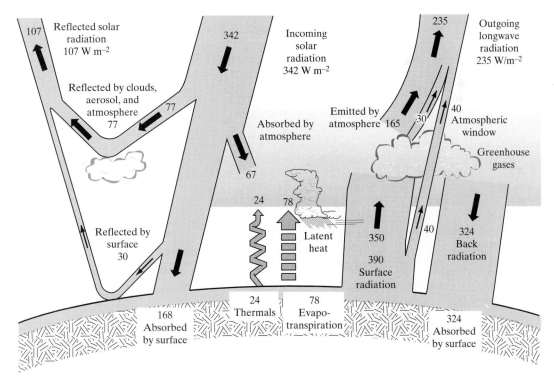

FIGURE 8.13.   Global average energy flows between space, the atmosphere, and the earth's surface. Units are watts per square meter of surface area. Values given are from Kiehl and Trenberth (1996) as they appear in IPCC, 1996.

8.6 RADIATIVE FORCING OF CLIMATE CHANGE

Thus far, the greenhouse effect has been described as a natural phenomenon that is responsible for earth having an average surface temperature 34 °C warmer (288 K vs. 254 K) than it would have if it did not have radiatively active gases in the atmosphere. As is now well known, anthropogenic sources of a number of gases and aerosols are affecting the greenhouse effect, leading us into a future of uncertain global climate.

In the model shown in Figure 8.13, the incoming solar energy absorbed by the earth and its atmosphere is 235 W/m², which is balanced by the 235 W/m² of outgoing longwave radiation. If, for some reason, an additional amount of energy is added to the incoming energy, then the balance will be temporarily upset. Over time, however, the climate system will adjust to that change by either increasing or decreasing the surface temperature of earth until a balance is once again attained.

Mathematically, we can represent that process as follows. In Figure 8.14, the incoming energy absorbed and outgoing energy being radiated are shown as they exist at the top of the troposphere—that is, at the tropopause. Initially, the balanced system has energy absorbed equal to energy radiated:

$$Q_{abs} = Q_{rad} \tag{8.14}$$

When the system is perturbed by adding *radiative forcing*, $\Delta F (\mathrm{W/m^2})$, to the incoming absorbed energy, a new equilibrium will eventually be established so that

$$Q_{abs} + \Delta Q_{abs} + \Delta F = Q_{rad} + \Delta Q_{rad} \tag{8.15}$$

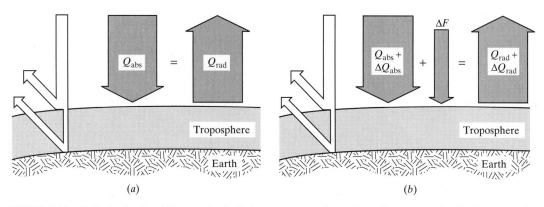

FIGURE 8.14 Radiative forcing, ΔF, perturbs the balance between incoming solar energy absorbed, Q_{abs}, and outgoing radiant energy, Q_{rad}. (*a*) The balanced system before perturbation; (*b*) the balanced system after radiative forcing is added.

where the deltas refer to changes in the quantity in question. Subtracting (8.14) from (8.15) gives

$$\Delta F = \Delta Q_{rad} - \Delta Q_{abs} \tag{8.16}$$

Determining the radiative forcing associated with changes in greenhouse gas concentrations, changes in aerosols, changes in albedo, and changes in solar constant has been one of the most important research areas for atmospheric scientists.

Climate Sensitivity Parameter

The key question, of course, is how much the surface temperature changes, (ΔT_s), for a given change in radiative forcing ΔF. The quantity that links the two is called the *climate sensitivity parameter,* λ, which can be defined by the following relationship:

$$\Delta T_s = \lambda \Delta F \tag{8.17}$$

If we use (8.16) in (8.17), we can get another expression for λ:

$$\lambda = \frac{\Delta T_s}{\Delta F} = \frac{\Delta T_s}{\Delta Q_{rad} - \Delta Q_{abs}} = \left(\frac{\Delta Q_{rad}}{\Delta T_s} - \frac{\Delta Q_{abs}}{\Delta T_s} \right)^{-1} \tag{8.18}$$

To be a little more precise, lambda can be defined by the following:

$$\lambda = \frac{dT_s}{dF} = \left(\frac{\partial Q_{rad}}{\partial T_s} - \frac{\partial Q_{abs}}{\partial T_s} \right)^{-1} \tag{8.19}$$

Equation (8.18) shows that the climate sensitivity factor depends on two components: (1) how much the outgoing radiant energy at the top of the atmosphere changes as the surface temperature changes, and (2) the change of incoming energy that is absorbed as surface temperature changes. The latter component (the change of absorbed energy) allows us to enter some estimate of how the albedo might change with temperature.

A great deal of work has gone into trying to determine appropriate values for this climate sensitivity parameter. One approach is based on IR radiation data obtained from satellites. By correlating monthly surface temperatures with infrared emissions from the top of the atmosphere, it has been noted that a linear relationship between the two works quite well for modest temperature changes. The following is one such relationship (Warren and Schneider, 1979), and an application of it is given in Example 8.2:

$$Q_{rad}(W/m^2) = 1.83\, T_s\,(°C) + 209 \tag{8.20}$$

EXAMPLE 8.2   Temperature change for a doubling of CO_2

It is believed that doubling the atmospheric concentration of CO_2 causes a radiative forcing of 4.35 W/m². If the earth's albedo does not change, estimate the climate sensitivity factor λ and use it to estimate the eventual change in the surface temperature of earth needed to balance incoming and outgoing radiation.

Solution   By thinking in terms of a derivative, (8.20) allows us to write

$$\frac{\Delta Q_{rad}}{\Delta T_s} = 1.83$$

Since the albedo is assumed not to change,

$$\frac{\Delta Q_{abs}}{\Delta T_s} = 0$$

So a first cut at the climate sensitivity factor gives us

$$\lambda = \left(\frac{\Delta Q_{rad}}{\Delta T_s} - \frac{\Delta Q_{abs}}{\Delta T_s} \right)^{-1} = (1.83)^{-1} = 0.55 \, \frac{°C}{W/m^2}$$

We have been told that the doubling of CO_2 creates a radiative forcing of 4.35 W/m², so, using (8.17), our estimated change in surface temperature would be

$$\Delta T_s = \lambda \, \Delta F = 0.55(°C/W/m^2) \times 4.35 \, W/m^2 = 2.4 \, °C \qquad ■$$

Estimates of the climate sensitivity factor λ are still very difficult to make. One way to appreciate the uncertainty in λ is to note that the IPCC (1996) estimate of the surface temperature change associated with a doubling of CO_2 is between 1.5 and 4.5 °C with the best estimate being about 2.5 °C. If the radiative forcing for CO_2 given in the preceding example is accurate, and if the range of temperature change is as predicted by the IPCC, the value of λ would be somewhere between 0.34 to 1.03, with the best estimate being 0.57 (very close to the value found in Example 8.2).

This very large range of uncertainties in λ is due, in large part, to the difficulty in modeling feedback factors, which are still not well understood. For example, it is clear that an increase in carbon dioxide concentration causes more radiative forcing, which would increase the earth's surface temperature. As the temperature goes up, it is reasonable to expect more evaporation, which would increase the water vapor content of the atmosphere. Since water vapor is an important greenhouse gas, one might expect that its increase would add to the original warming, making it even warmer (positive feedback). On the other hand, more water vapor could mean more cloudiness, and that could cause an increase in the albedo. Increasing the albedo reduces the solar energy reaching the earth and would tend to offset the original warming (negative feedback). This water-vapor/cloud feedback process is one of the most important, yet least understood, problems in climate modeling.

EXAMPLE 8.3   Changing the albedo

Suppose glacial melting causes the earth's albedo to change from 0.31 to 0.30. Estimate the resulting radiative forcing. If the climate sensitivity factor is somewhere between 0.34 and 1.03 °C $W^{-1}m^2$, estimate the change in surface temperature.

Solution   In this case, the radiative forcing is caused by a change in energy absorbed. From (8.12) we have

$$\Delta F = \Delta[\frac{S}{4}(1 - \alpha)] = 342 \, W/m^2 \cdot (1 - 0.31) - 342 \, W/m^2(1 - 0.30) = 3.42 \, W/m^2$$

This radiative forcing would result in a surface temperature change between

$$\Delta T_s = \lambda \, \Delta F = 0.34 \, \frac{°C}{W/m^2} \times 3.42 \, W/m^2 = 1.16 \, °C$$

and

$$\Delta T_s = \lambda \, \Delta F = 1.03 \, \frac{°C}{W/m^2} \times 3.42 \, W/m^2 = 3.52 \, °C$$

∎

Equilibrium Temperature and Realized Temperature

The surface temperatures dealt with thus far, T_s and ΔT_s are known as *equilibrium temperatures*—that is, they are temperatures that would eventually be reached for a given greenhouse gas forcing. In reality, of course, the temperature of the earth will not adjust instantaneously to changes in radiative forcing. A considerable period of time is needed to warm the upper layer of the oceans and the surface of the land, which means the actual *realized* surface temperature of the earth will lag somewhat behind the equilibrium temperature.

Figure 8.15 illustrates the concepts of equilibrium (T_{eq}) and realized temperature (T_r) corresponding to a radiative forcing that increases linearly with time. At time t_1, the earth has already warmed to T_r and even if there is no further increase in radiative forcing, it is *committed* to a temperature increase of $(T_{eq} - T_r)$. For the assumed linear increase in radiative forcing, there is a time lag of $(t_2 - t_1)$ years before that temperature commitment is realized. Coupled atmosphere-ocean general circulation models indicate that the realized temperature is roughly 60 to 85 percent of the equilibrium temperature.

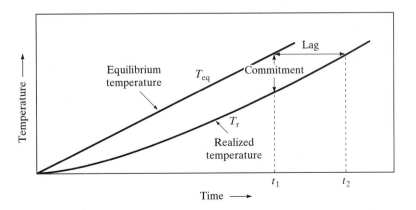

FIGURE 8.15   Illustrating the difference between equilibrium temperature and realized temperature for a linear increase in greenhouse gas forcing. At time t_1, the earth is committed to a temperature increase of $(T_{eq} - T_r)$ that will take $(t_2 - t_1)$ years to achieve.

8.7 RADIATIVE FORCINGS SINCE PREINDUSTRIAL TIMES

The concept of radiative forcing of climate change can be applied to the accumulation of greenhouse gases in the atmosphere, changes in aerosols from natural and anthropogenic sources, ozone depletion in the stratosphere, photochemically produced ozone accumulation in the troposphere, and the natural variability in solar intensity reaching earth's outer atmosphere. Both positive and negative forcings are possible. Positive forcings contribute to global warming, while negative forcings tend to cool the earth.

Gases and particulate matter added to the atmosphere can exert both *direct* and *indirect* radiative forcing effects. Direct forcing is caused by substances in the atmosphere that have actually been emitted from some source. Indirect forcings are those that occur when those substances go on to cause other atmospheric changes that affect radiative properties of the atmosphere. For example, aerosols have a direct effect on forcing when they absorb or reflect sunlight. Aerosols can also cause an indirect effect when they induce changes in the albedo of clouds. Halocarbons provide another example of direct and indirect effects. The direct effect of halocarbons is an increase in radiative forcing as these gases absorb longwave radiation from the earth. They also, however, cause an indirect effect by destroying ozone in the stratosphere. Recall from Figure 8.11 that ozone absorbs in the middle of the atmospheric window, so ozone destruction opens the window and allows the earth to cool more readily. The direct effect of halocarbons therefore contributes to global warming, while the indirect effect of destroying ozone works in the opposite direction to help to cool the planet.

Table 8.3 presents a summary of current estimates of the radiative forcings caused by the direct and indirect effects of greenhouse gases, aerosols, and solar radiation. The following sections will describe each of these components in greater detail.

TABLE 8.3 Globally averaged radiative forcing due to changes in greenhouse gases, aerosols, and solar intensity from 1850 to the present

	Radiative forcing, ΔF (W/m^2)	Confidence level	Comments and uncertainty estimates
DIRECT GREENHOUSE			
Carbon dioxide, CO_2	1.56	High	CO_2 from 278 to 356 ppm
Methane, CH_4	0.47	High	CH_4 from 700 to 1714 ppb
Halocarbons	0.28	High	Mostly CFC-11, -12, -113
Nitrous oxide, N_2O	0.14	High	N_2O from 275 to 311 ppb
Total	2.45	High	Range 2.1 to 2.8 W/m^2
INDIRECT GREENHOUSE			
Stratospheric ozone, O_3	−0.1	Low	Range −0.05 to −0.2 W/m^2
Tropospheric ozone, O_3	0.4	Low	Range 0.2 to 0.6 W/m^2
DIRECT AEROSOLS			
Sulfate SO_4	−0.4	Low	Range −0.2 to −0.8 W/m^2
Fossil fuel soot	0.1	Very low	Range 0.03 to 0.3 W/m^2
Biomass burning	−0.2	Very low	Range −0.07 to −0.6
Total Direct	−0.5	Very low	Range −0.2 to −1.0
INDIRECT AEROSOLS	0 to −1.5	Very low	Impact on cloud formation
SOLAR (since 1850)	0.3	Very low	Range 0.1 to 0.5 W/m^2

Source: Based on IPCC, 1996.

Direct Forcing by Greenhouse Gases

The principal greenhouse gases listed in Table 8.3 are carbon dioxide (CO_2), methane (CH_4), nitrous oxide (N_2O), and a category of carbon-based gases called halocarbons that contain atoms of carbon plus fluorine, chlorine, and/or bromine. All of these gases are well mixed in the atmosphere, and their radiative forcings are well understood.

Figure 8.16 shows the relative importance of these principal greenhouse gases in terms of the changes in their radiative forcings from 1850 to the present. Of the total 2.45 W/m^2 of forcing since the preindustrial period, carbon dioxide accounts for 64 percent, methane 19 percent, halocarbons 11 percent, and nitrous oxide 6 percent. The contribution of halocarbons is oversimplified in this figure because their indirect cooling effect associated with ozone destruction is not included.

Carbon Dioxide (CO_2)

Carbon dioxide has been recognized for its importance as a greenhouse gas for over a century. Arrhenius (1896) is usually credited with the first calculations on global temperature as a function of atmospheric CO_2 content, and his results are not that far from those obtained today. It is the gas that has received the most attention in discussions on the greenhouse effect since it does account for almost two-thirds of the current radiative forcing.

The first continuous, precise, and direct measurements of atmospheric carbon dioxide began in 1957 at the South Pole, and 1958 at Mauna Loa, Hawaii. At that time, the concentration was around 315 ppm and growing at a little under 1 ppm per year. By 1994 it had reached 358 ppm and was climbing at about 1.6 ppm per year (Figure 8.17). Anthropogenic carbon dioxide emissions are located predominantly in the northern hemisphere, but the mixing of northern and southern hemisphere air is sufficiently complete that the concentrations of CO_2 are only a few parts per million higher at Mauna Loa than at the South Pole.

The oscillations shown in the Mauna Loa data plotted in Figure 8.17 are caused by seasonal variations in the rates of photosynthesis and respiration. During photosynthesis, carbon is transferred from the air into plant material (indicated by the carbohydrate, glucose, in the following reaction.) During spring and summer, when plants are growing their fastest, atmospheric CO_2 levels drop, tending to reach their lowest point in the northern hemisphere in about October.

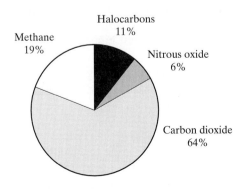

FIGURE 8.16   Direct radiative forcing due to changes in greenhouse gases from pre-industrial times to the present day as presented in Table 8.3. The total direct forcing is estimated to be 2.45 W/m^2. The overall forcing of halocarbons is actually less than 11 percent when the indirect effects are included.

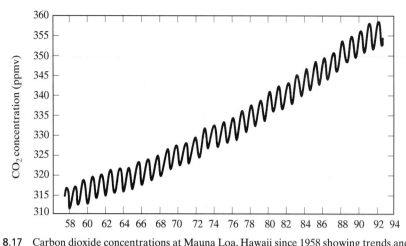

FIGURE 8.17 Carbon dioxide concentrations at Mauna Loa, Hawaii since 1958 showing trends and seasonal cycle. (*Source:* IPCC, 1995)

$$6\,CO_2 + 6\,H_2O \xrightarrow{\text{Sunlight, chlorophyll}} C_6H_{12}O_6 + 6\,O_2 \qquad (8.21)$$

Reversing the preceding reaction yields the equation describing respiration, which is the process that living things use to obtain energy. During respiration, complex organic molecules are broken down, returning carbon to the atmosphere. When the rate of respiration exceeds the rate of photosynthesis, as tends to occur in the fall and winter seasons, there is a net replacement of carbon into the atmosphere that results in peak concentrations in the northern hemisphere around May.

$$C_6H_{12}O_6 + 6\,O_2 \longrightarrow 6\,CO_2 + 6\,H_2O + \text{Energy} \qquad (8.22)$$

Carbon thus moves continually from the atmosphere into the food chain during photosynthesis and returns to the atmosphere during respiration.

Atmospheric carbon dioxide concentrations inferred from Antarctica ice cores over the past 1000 years, combined with more recent direct measurements made at Mauna Loa, are shown in Figure 8.18. Over most of that time period the concentration of carbon dioxide hovered at close to 280 ppm, and that is the value that is commonly used as a reference point for comparison with current readings and future projections. Carbon dioxide concentrations are now almost 30 percent higher than they were just before the Industrial Revolution.

Carbon dioxide is clearly the most important greenhouse gas, but further discussion of its characteristics will be postponed in this chapter until the other gases and aerosols that are affecting climate can be introduced.

Methane (CH₄)

The accumulation of methane in the atmosphere accounts for 0.47 W/m² of radiative forcing, which is 19 percent of the total direct greenhouse gas forcings. It has an

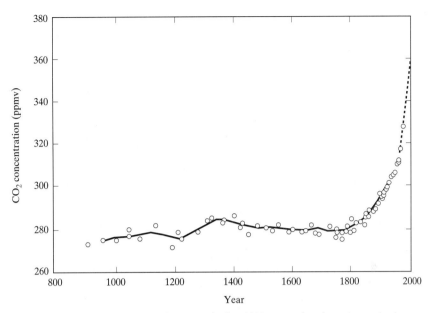

FIGURE 8.18 Carbon dioxide concentrations over the last 1000 years taken from Antarctica ice cores and Mauna Loa. The smooth curve is based on a 100-year running mean. (*Source:* IPCC, 1995)

absorption band centered at 7.7 μm, on the edge of the atmospheric window. As shown in Figure 8.19, the concentration of methane in the atmosphere was approximately 700 parts per billion (ppb) for many hundreds of years before it began its rapid climb in the 1800s. In 1992 it reached 1714 ppb, which is an increase of almost 250 percent over preindustrial levels.

Methane is another naturally occurring gas that is increasing in concentration as a result of human activities. It is produced by bacterial fermentation under anaerobic conditions, such as occurs in swamps, marshes, rice paddies, landfills, and in the digestive tracts of ruminants. It is also released during the production, transportation, and consumption of fossil fuels. Natural sources of methane, which include wetlands, termites, and oceans, release on the order of 160 million tonnes per year, while anthropogenic sources account for approximately 375 million tonnes (IPCC, 1995). Almost half of the anthropogenic emissions are the result of human food production (Figure 8.20). Livestock, including cattle, sheep, and buffalo, belch some 80 million tonnes of methane per year; rice paddies release 60 million; and clearing and burning of biomass in part to prepare land for grazing and crops account for another 40 million. Over one-fourth of anthropogenic emissions, some 100 million tonnes per year, are associated with fossil fuel use. As food and energy production rises to meet the increasing demands of a growing population, methane emissions will continue to be a significant fraction of total radiative forcing.

Methane is removed from the atmosphere primarily through reactions with the hydroxyl radical (OH), as the following reaction suggests:

$$CH_4 + OH + 9\,O_2 \longrightarrow CO_2 + 0.5\,H_2 + 2\,H_2O + 5\,O_3 \qquad (8.23)$$

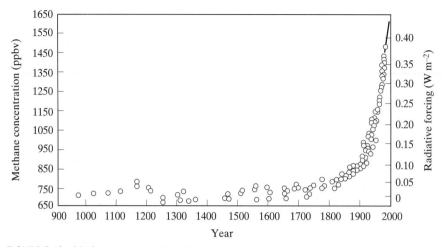

FIGURE 8.19 Methane concentrations from Antarctic ice cores combined with more recent direct measurements (smooth line). Radiative forcing is also shown. (*Source:* IPCC, 1995)

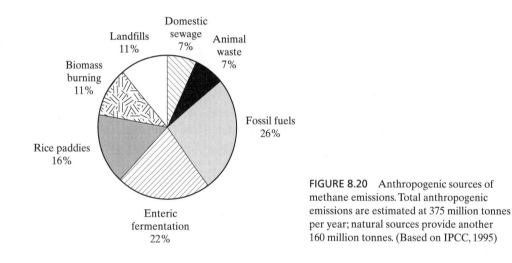

FIGURE 8.20 Anthropogenic sources of methane emissions. Total anthropogenic emissions are estimated at 375 million tonnes per year; natural sources provide another 160 million tonnes. (Based on IPCC, 1995)

Methane is, of course, a greenhouse gas so it has direct effects on radiative forcing. There are, however, a number of indirect effects that provide additional radiative forcing, which reaction (8.23) helps explain. First, when methane reacts with hydroxyl (OH), the concentration of OH decreases. With less OH available, the rate of removal of CH_4 slows down, lengthening the atmospheric lifetime of the remaining methane. With a longer lifetime for CH_4 in the atmosphere, it continues to absorb infrared for a longer time, increasing its global warming potential. The second indirect effect that (8.23) implies is that as methane reacts with hydroxyl it produces water vapor. When this reaction occurs in the troposphere, the increase in water vapor is insignificant, but in the stratosphere it becomes quite important. Finally, (8.23) indicates that the destruction of methane produces an increase in ozone, which is itself a greenhouse gas. The sum of all of these indirect effects greatly increases the impact that methane has on radiative forcing.

There is some concern for the possibility that global warming could free large amounts of methane currently frozen in permafrost in the far northern regions of the world and could allow anaerobic decomposition of organic matter also frozen in permafrost, producing more methane. This is another example of a positive feedback loop. Warming leading to increased releases of the greenhouse gas, methane, could reinforce the original warming.

Nitrous Oxide (N_2O)

Nitrous oxide ("laughing gas") is another naturally occurring greenhouse gas that has been increasing in concentration due to human activities. The current atmospheric concentration is about 312 ppb, which is a 13 percent increase over the preindustrial concentration of 275 ppb. The current growth rate of N_2O is estimated at about 0.6 ppb, or 0.2 percent, per year. Figure 8.21 shows the recent history of concentration and corresponding radiative forcing. At 0.14 W/m^2 it accounts for 6 percent of the industrial period radiative forcing.

Nitrous oxide is released into the atmosphere mostly during the nitrification portion of the nitrogen cycle:

$$NH_4^+ \quad \rightarrow \quad N_2 \quad \rightarrow \quad N_2O \quad \rightarrow \quad NO_2^- \quad \rightarrow \quad NO_3^-$$

Ammonium	Molecular	Nitrous	Nitrite	Nitrate
ion	nitrogen	oxide	ion	ion

It is estimated that natural sources of N_2O deliver 9 million tonnes of nitrogen per year to the atmosphere, with most of that coming from oceans and wet forest soils. Anthropogenic sources contribute about 40 percent of total N_2O emissions, or 5.7 million tonnes per year (IPCC, 1995). Almost two-thirds of anthropogenic emissions are

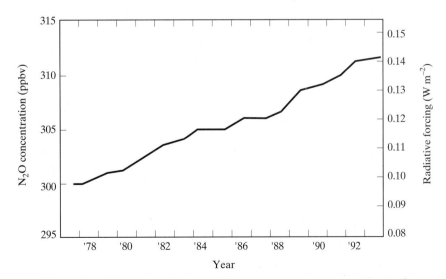

FIGURE 8.21   Increasing concentration of nitrous oxide and corresponding radiative forcing. (*Source:* IPCC, 1995)

the result of tropical agriculture. Newly cleared forest land that is converted to grassland produces significant N_2O emissions for a number of years, and nitrogen fertilizers on cropland provide additional releases. Other sources include three-way catalytic converters on cars, combustion of fuels containing nitrogen, and a variety of industrial processes such as the production of nylon.

Apparently, there are no significant tropospheric sinks for N_2O and it is only slowly degraded in the stratosphere by photolysis. As a result, it has a long atmospheric lifetime, estimated at about 120 years, which means perturbations in the natural cycle will have long-lasting repercussions. Nitrous oxide has an absorption band at 7.8 μm that is associated with a stretching of the bonds, and another at 8.6 μm associated with bending of the bond angle. The band at 7.8 μm is on the shoulder of the atmospheric window and the band at 8.6 μm is right in the window, so N_2O is a very potent greenhouse gas.

Halocarbons

Halocarbons are carbon-based molecules that have chlorine, fluorine, or bromine in them. The carbon-to-fluorine bonds in halocarbons oscillate, and hence absorb, at wavelengths around 9 μm, and other bond stretching and bending in halocarbons also occurs with frequencies in the atmospheric window, so these molecules are all potent greenhouse gases. They are environmentally important not only because they contribute to global warming, but also because chlorine and bromine atoms that find their way into the stratosphere have the ability to destroy ozone catalytically, as will be described later in this chapter.

Subcategories of halocarbons include *chlorofluorocarbons* (CFCs), which have only carbon, fluorine, and chlorine, but no hydrogen; *hydrochlorofluorocarbons* (HCFCs), which are like CFCs but do contain hydrogen; *hydrofluorocarbons* (HFCs), which contain no chlorine; and *halons,* which are carbon-based molecules containing bromine along with fluorine and perhaps chlorine. These halocarbon gases differ from all of the other radiatively active gases in that they do not occur naturally and their presence in the atmosphere is due entirely to human activities. Other important halocarbons include carbon tetrachloride (CCl_4), methyl chloroform (CH_3CCl_3), and methyl bromide (CH_3Br). With so many categories it may help to show some examples, which Table 8.4 does.

Chlorofluorocarbons (CFCs) are nontoxic, nonflammable, nonreactive, and they are not water soluble, which led to the belief when they were first invented that they were truly benign chemicals. But because they are inert and do not dissolve easily in water, they are not destroyed by chemical reactions or removed from the troposphere by rain. That means they have long atmospheric lifetimes, as Table 8.4 suggests. The only known removal mechanism is photolysis by short-wavelength solar radiation, which occurs after the molecules drift into the stratosphere. It is the chlorine freed during this process that can go on to destroy stratospheric ozone. Similarly, halons, which contain bromine, have no tropospheric sinks and their only removal mechanism is also photochemical decomposition in the stratosphere, which releases the ozone-depleting bromine.

TABLE 8.4   Examples of halocarbons

Formula	Designation	Atmospheric lifetime (yrs)
Chlorofluorocarbons (CFCs)		
$CFCl_3$	CFC–11	50
CF_2Cl_2	CFC–12	102
$CF_2ClCFCl_2$	CFC–113	85
Hydrochlorofluorocarbons (HCFCs)		
CHF_2Cl	HCFC–22	12.1
CH_3CFCl_2	HCFC–141b	9.4
$CF_3CF_2CHCl_2$	HCFC–225ca	2.1
Hydrofluorocarbons (HFCs)		
CF_3CH_2F	HFC–134a	14.6
CH_3CF_3	HFC–143a	48.3
Halons		
CF_3Br	H–1301	65
CF_2ClBr	H–1211	20
Other		
CH_3CCl_3	Methyl chloroform	4.9
CCl_4	Carbon tetrachloride	42
CF_4	Perfluoromethane	50,000

Source: IPCC, 1996.

Hydrochlorofluorocarbons (HCFCs) are being introduced as replacements for CFCs. By adding hydrogen to the molecules, they are no longer chemically inert, which means chemical reactions can destroy them in the troposphere before they have a chance to drift into the stratosphere. Notice the much shorter atmospheric lifetimes for the HCFCs listed in Table 8.4. HCFCs are only temporary replacements for CFCs, however, since they do still have some potential to deplete the ozone layer and they are still potent greenhouse gases.

Hydrofluorocarbons (HFCs) have no chlorine at all, so they are even better than HCFCs in terms of stratospheric ozone protection. The hydrofluorocarbon CH_2FCF_3 (HFC 134a) is quickly becoming the refrigerant of choice for automobile air conditioners and refrigeration equipment. Even this chemical, however, has a sizable atmospheric lifetime (14.6 years), which contributes to its rather significant global warming potential.

Halons have the ozone-destroying element bromine in them. They are very stable molecules with no tropospheric sinks, so they only release that bromine when they eventually drift into the stratosphere and are broken apart by photolysis. The primary use of halons is in fire extinguishers. They are nontoxic and leave no residue when sprayed onto fires, so they are ideal for use in confined spaces containing critical equipment, such as computers.

Halocarbon Numbering Systems.   In Table 8.4 most of the halocarbons are identified by a simple numerical designation as well as a chemical formula. The CFCs, HCFCs, and HFCs are referred to using a number system developed years ago by

DuPont. For example, trichlorofluoromethane, $CFCl_3$ is CFC–11, and the hydrochloro-fluorocarbon CHF_2Cl is HCFC–22. To determine the chemical formula from a fluoro-carbon number, begin by adding "90" to the number and then interpret the three-digit result as follows: The leftmost digit is the number of carbon atoms, the middle digit is the number of hydrogen atoms, and the right digit is the number or fluorines.

To determine the number of chlorine atoms, begin by visualizing molecules in which each carbon atom forms four single bonds to other atoms (if the other atoms are all hydrogens, this is the familiar alkane series: methane, ethane, propane, etc.):

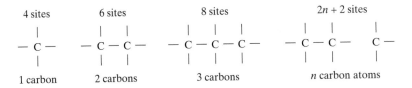

A single carbon atom has four sites to fill, two carbon atoms have six sites, and so forth. All of the sites will be occupied by hydrogen, fluorine, or chlorine. So to find the number of chlorines, just subtract the number of hydrogens and fluorines from the total available sites. Each vacant bonding site not taken up by fluorine or hydrogen is occupied by chlorine.

For example, to figure out what CFC–12 is, add 90 to 12, giving 102. So there is one carbon, no hydrogens, and two fluorines. With one carbon there are four sites available, two of which are taken by fluorine, leaving two for chlorine.

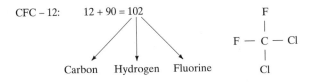

Thus, CFC–12 is CF_2Cl_2.

The halons also have a number system, but this one is not so complicated. Halons are given a four-digit designation, with the leftmost digit being the number of carbons. The second is fluorines, the third is chlorines, and the fourth is bromine. For example, H–1211 is CF_2ClBr.

EXAMPLE 8.4   Halocarbon numbering

   a. What is the chemical composition of CFC–115?

   b. What is the CFC number for CCl_2FCClF_2 ?

   c. What is H–2402?

Solution:

   a. CFC–115: Adding 90 to 115 gives 205. Thus, a molecule contains two carbons, no hydrogens, and five fluorines. Two carbons have six bonding sites, five of which are taken by fluorine. The remaining site is taken by chlorine. The chemical formula would therefore be C_2F_5Cl (or CF_3CF_2Cl).

b. $C_2H_2F_4$ has two carbons, two hydrogens, and four fluorine atoms. Subtracting 90 from 224 gives 134. There is no chlorine so this halocarbon is a hydrofluorocarbon, HFC–134. Notice that we cannot tell whether this is CHF_2CHF_2 or CH_2FCF_3. To distinguish one isomer from another, a letter designation is added. For example, CHF_2CHF_2 is HFC–134, and CH_2FCF_3 is HFC–134a.

c. H–2402 is $C_2F_4Br_2$.   ∎

Refrigerants.   Chlorofluorocarbons have many unusual properties that have led to their widespread use. For example, they are easily liquefied under pressure, and when that pressure is released they evaporate and produce very cold temperatures. In fact, they were originally developed to satisfy the need for a nontoxic, nonflammable, efficient refrigerant for home refrigerators. Before they were introduced in the early 1930s, the most common refrigerants were ammonia, carbon dioxide, isobutane, methyl chloride, methylene chloride, and sulfur dioxide. All had significant disadvantages. They were either toxic, noxious, highly flammable, or required high operating pressures, which necessitated heavy, bulky equipment. From those perspectives, CFCs are far superior to any of the gases that they replaced. As a side note, fluorocarbons used in refrigeration equipment are often referred to with a "refrigerant" number. For example, CFC–12 is often called "refrigerant–12" or, more simply, R–12. The DuPont trade name Freon™ has also been used, so occasionally it is still called F–12.

Until recently, most refrigerators, freezers, and automobile air conditioners used CFC–12 as the refrigerant, and large building air conditioning systems tended to use CFC–11. CFC refrigerants do not wear out, so as long as they are sealed into equipment the total emission rate can be small. Automobile air conditioners, however, tend to develop leaks that necessitate periodic recharging of their systems. In the recent past, when car air conditioners were serviced, the old refrigerant was usually vented to the atmosphere rather than being captured and recycled, which compounded the loss rate. As a result, automobile air conditioners used to contribute on the order of 20 percent of all emissions of CFCs in the United States. The Montreal Protocol and subsequent changes in the Clean Air Act have changed that picture significantly. Production and importation of CFCs ended in 1996, and air conditioners on new cars now tend to use HFC–134a, which contains no chlorine. Only licensed facilities that use CFC recycling equipment can service older automobile air conditioners.

Aerosol Propellants

When CFCs were first hypothesized as a danger to the ozone layer by Molina and Rowland (1974), over half of worldwide emissions were from aerosol propellants for products such as deodorants, hair spray, and spray paint cans. At that time the United States alone was using over 200,000 tonnes per year of CFC–11 and CFC–12 in aerosols. The EPA responded rather quickly to the threat, and acting under the Toxic Substances Control Act, it banned the use of CFCs in nonessential aerosol propellant applications beginning in 1979. Norway, Sweden, and Canada adopted similar restrictions, but the rest of the world lagged far behind until recently. The Montreal Protocol has led to total bans in most of the developed world. Replacements for CFCs in aerosols include isobutane, propane, and carbon dioxide. In some applications, simple pumps or "roll on" systems have replaced the propellants.

Foamed Plastics. The second most common use for CFCs in general, and the most common use for CFC–11 in particular, has been in the manufacture of various rigid and flexible plastic foams found in everything from seat cushions to hamburger "clamshell" containers to building insulation. When liquid CFCs are allowed to vaporize in plastic, they create the tiny bubbles that make the plastic foamy.

Sheets of rigid "closed-cell" urethane or isocyanurate foams are used primarily as thermal insulation in buildings and refrigeration equipment. In such applications the CFCs, which are poor thermal conductors, trapped in foam cells reduce the heat transfer capabilities of the product. CFCs are also used to manufacture nonurethane, rigid foams such as extruded polystyrene (Dow's trade name is Styrofoam) used extensively for egg cartons and food service trays, and expanded polystyrene foam, which is used to make drinking cups. Since CFCs are trapped in the holes in closed-cell foams, they are only slowly released into the atmosphere as the material ages or is eventually crushed.

Replacements for these foams are possible. Fiberglass insulation, though thermally not as effective per unit of thickness, contains no CFCs and can be used to replace rigid foams used in the construction industry. Various cardboards and other paper products can be used to replace many of the polystyrene applications in the food industry. Finally, the CFC foaming agent itself can be modified or replaced. HCFC–22, which is less damaging to the ozone layer, is sometimes used, as is pentane or methylene chloride (CH_2Cl_2).

Flexible foams, which are used in furniture, automobile seats, and packaging, have cells that are open to the atmosphere ("open cell"). Hence, CFC release is almost immediate. These foams are made using carbon dioxide as the primary blowing agent, but the CO_2 is often augmented with CFC–11 or methylene chloride.

Radiative Forcing of Halocarbons

The total direct radiative forcing associated with halocarbons since the preindustrial period is given in Table 8.3 as 0.28 W/m², which is 11 percent of the direct forcings of the greenhouse gases (CO_2, CH_4, N_2O, and halocarbons). Of that amount, CFC–11 is responsible for about 0.06 W/m², CFC–12 contributes 0.14 W/m², and most of the remaining 0.08 W/m² is associated with CCl_4, HCFC–22, and CFC–113. These data are for the direct radiative forcing, which is positive, and do not include the negative forcing that results from the indirect greenhouse effect of halocarbons. Since halocarbons induce a loss of ozone in the lower stratosphere, and ozone is a greenhouse gas, the net forcing of halocarbons is reduced by about 0.1 W/m².

The good news on the halocarbon front is the impact that the Montreal Protocol and subsequent agreements on ozone-depleting substances are beginning to have. As will be described in more detail later in this chapter, these agreements call for a phaseout of the production of CFCs and halons. Production in the developed countries supposedly stopped in 1995, while the developing countries have been given a few extra

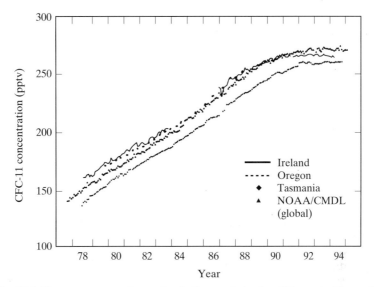

FIGURE 8.22   CFC–11 measurements taken at sites in Oregon, Ireland, and Tasmania. (*Source:* IPCC, 1996)

years to comply. Figure 8.22 shows the impact of the Montreal Protocol on the atmospheric concentration of one of the most important CFCs, CFC–11. The concentration of CFC–11 in the atmosphere stopped rising in the 1990s as a result of the production bans, but the long atmospheric lifetime of CFC–11 (50 years) means it will take many decades for significant decreases to occur.

Ozone (O_3)

Ozone has a strong absorption band at 9 μm, right in the middle of the atmospheric window (Figure 8.11), so clearly it is a greenhouse gas of importance. It has, however, proven to be a difficult one to understand. It is not a "well-mixed" gas in the atmosphere in the way all of the greenhouse gases described up to this point are. Not only does its concentration vary from place to place around the globe, but its effect on climate depends on its vertical distribution as well.

As was discussed in Chapter 7, ozone in the troposphere is formed by photochemical reactions involving relatively short-lived precursor gases, including NO_x, non-methane hydrocarbons, and CO. It is the principal gas in photochemical smog. And since smog is associated with the major industrialized areas of the world, it is not surprising that tropospheric ozone concentrations are higher in the northern hemisphere than in the southern hemisphere. Those concentrations also vary seasonally as sunnier summer months energize the formation of ozone. Considerable uncertainty in the total radiative forcing of tropospheric ozone still exists, and the IPCC (1996) tentatively puts the forcing at between 0.2 and 0.6 W/m^2.

In the stratosphere, ozone concentrations are decreasing as a result of the attacks by chlorine and bromine released by UV-exposed CFCs and halons. These stratospheric ozone losses also vary by geographic location and by season, as the annual appearance of the ozone hole over Antarctica in September and October certainly demonstrates. It has been estimated that the loss of stratospheric ozone has a globally averaged negative forcing of about -0.1 W/m^2 with a factor of 2 uncertainty. Since this ozone loss is an indirect result of our use of CFCs and halons, this negative radiative forcing has tended to offset some of the positive forcing caused by those halocarbon emissions. As emissions of CFCs and halons into the atmosphere are curtailed, it is expected that ozone will begin to recover in the early part of the twenty-first century, so this negative forcing will diminish.

Aerosols

Suspensions of particles having an effective diameter of less than 10 μm are called aerosols. Some particles enter the atmosphere as solids (e.g., soil dust) and others are formed in the atmosphere when gases such as sulfur dioxide condense into liquid particles such as sulfates. Combustion of fossil fuels and biomass burning are the principal anthropogenic sources of aerosols.

As suggested in Figure 8.23, aerosols affect the earth's energy balance in three ways: (1) They can reflect incoming solar radiation back into space, which increases the earth's albedo; (2) They can provide cloud condensation nuclei, which increases cloud reflectivity and cloud lifetime, and those also increase albedo; and (3) carbonaceous particles, such as soot from fossil-fuel combustion, can increase the atmospheric absorption of incoming solar energy. Increasing the albedo by enhancing atmospheric and cloud reflection yields negative radiative forcing that is counteracted to some extent by the positive forcing due to solar absorption by soot. The enhanced reflectivity of the atmosphere caused by some particulates and the increased solar absorption

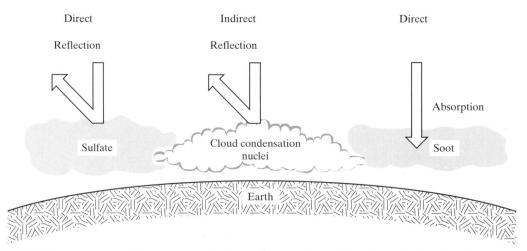

FIGURE 8.23 Aerosols provide negative radiative forcing by reflecting sunlight and by enhancing cloud formation, which also increases reflection. Carbonaceous particles (soot) can provide negative forcing by absorbing sunlight. Aerosol absorption and reflection are direct effects; cloud formation and reflection is an indirect effect.

caused by others are *direct* effects of aerosols, while the impact of aerosols on cloud formation and lifetime is referred to as an *indirect* effect.

Aerosols have characteristics that make them considerably different from the well-mixed, long-lived greenhouse gases CO_2, CH_4, N_2O, and halocarbons. For one, they have atmospheric lifetimes that are measured in days in the troposphere, and a few years in the stratosphere, while greenhouse gases have lifetimes measured in decades. The aerosols in the troposphere today are particles that entered the atmosphere within the last few days. That means aerosol concentrations can vary considerably from week to week and from place to place. That short lifetime also implies that the radiative impacts of aerosols are likely to be regional phenomena, centered around the industrialized areas of the world, whereas greenhouse gas forcings are much more uniformly distributed around the globe. In fact, in some heavily industrialized parts of the world, the cooling caused by aerosols can be greater than the warming due to greenhouse gases. That does not necessarily mean, however, that such an area would experience cooling since regional temperatures are so affected by other factors.

Natural sources of aerosols include wind-blown soil dust, evaporation of sea-spray droplets, and volcanic eruptions. While the natural emission rates from these and other sources far exceed anthropogenic emissions, most of these particulates are so large that they quickly drop out of the troposphere. Smaller particles, especially ones that reach the stratosphere, are much more important in terms of their impact on climate. Volcanic eruptions are a particularly important source of data for aerosol modeling since they are discrete events that can be monitored and evaluated. Such eruptions can emit enormous quantities of sulfur dioxide (SO_2), which is a precursor to sulfate aerosols, solid particles (ash), water vapor, and carbon dioxide.

There have been two major volcanic eruptions in particular that have been well studied: the El Chichon eruption in 1982 in Mexico, and the June 1991 eruption of Mount Pinatubo in the Philippines. One measure of the impact of such eruptions is the reduction in sunlight that passes through the atmosphere, called the *optical depth*. Changes in optical depth provides a surrogate for the rate at which aerosols are removed from the atmosphere. As shown in Figure 8.24a, El Chichon affected the aerosol optical depth for only about one year, while Mount Pinatubo, which injected much more SO_2 into the relatively stable stratosphere, affected the atmosphere for more than two years. The radiative forcing caused by Mount Pinatubo has been calculated to be -4 W/m^2 one year after the eruption, but by the end of the second year the forcing had decayed to about -1 W/m^2. These negative radiative forcings are comparable to the $+2.45$ W/m^2 forcing due to greenhouse gases, and the drop in global surface temperature just after Mount Pinatubo was consistent with what was predicted by general circulation models (Figure 8.5b).

While acknowledging the difficulties in estimating global average radiative forcings for aerosols, it is still important to get some quantitative sense of their impact. The IPCC (1996) suggests that the direct forcing due to reflection and absorption since the preindustrial period is approximately -0.5 W/m^2 with a factor of 2, while the indirect forcing associated with cloud impacts is an additional 0 to -1.5 W/m^2. A central value of -0.8 W/m^2 is tentatively suggested, which would make the total direct and indirect aerosol forcing to be on the order of -1.3 W/m^2. Compare that with the estimated $+2.45$ W/m^2 of forcing associated with the well-mixed greenhouse gases CO_2, CH_4,

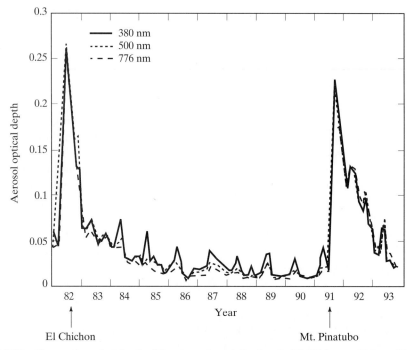

FIGURE 8.24   Changes in optical depth of the atmosphere following the El Chichon and Mount Pinatubo eruptions. (*Source:* IPCC, 1995)

N_2O, and halocarbons. If these estimates are valid, it would mean that global cooling due to aerosols is currently offsetting around half of the global warming caused by greenhouse gases.

Recent estimates of the impact of aerosols on global temperature have helped bring computer simulations into closer agreement with actual temperature measurements. Previously, when greenhouse gases alone were modeled, simulated temperatures were much higher than they should have been. By including the cooling effect of aerosols, however, there is now much better correlation between simulation and reality, as Figure 8.25 shows.

Combined Forcings

Figure 8.26 shows a summary of estimated radiative forcings from 1850 to 1990 caused by greenhouse gases; stratospheric and tropospheric ozone; sulfates, soot, and biomass aerosols; the indirect effect of aerosols on clouds; and the estimated increase in solar intensity that has occurred in this period. An indication of the confidence of the estimates is given along with uncertainty bars. The confidence level for most of the estimates is low, or very low, except for the well-understood direct greenhouse gas forcings of CO_2, CH_4, N_2O, and halocarbons.

It is, of course, very tempting to simply add the best estimates of positive forcings and subtract the best estimates of negative forcings to arrive at some indication of the

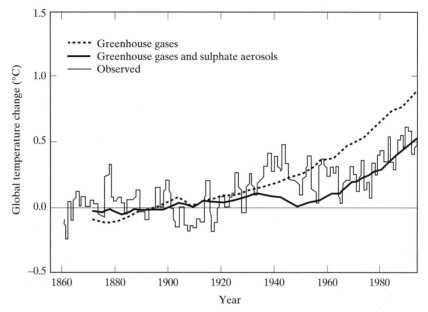

FIGURE 8.25 Simulated global annual mean warming from 1860 to 1990 for greenhouse gases alone (dashed curve) and for greenhouse gases plus sulfate aerosols (solid curve) compared with measured values. (*Source:* IPCC, 1996)

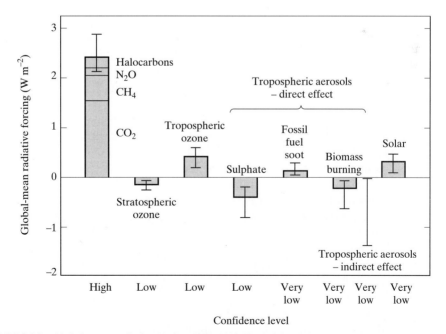

FIGURE 8.26 Global-mean radiative forcing, 1850–1990. The height of the rectangular bar represents a midrange estimate, while the error bars show an estimate of the uncertainty range. (*Source:* IPCC, 1996)

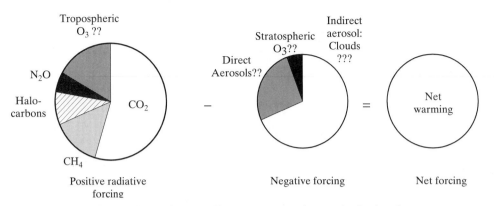

Figure 8.27   Positive radiative forcings are offset to some extent by negative forcings, but current uncertainties make it difficult to quantify this diagram.

net effect. Unfortunately, too many questions remain unanswered to justify combining the radiative forcings of the individual components into a single numerical sum, but perhaps a nonquantitative visual presentation can serve some purpose. Figure 8.27 purposely leaves off the numbers while trying to give a sense of the net impacts.

8.8 RADIATIVE FORCING SATURATION EFFECTS

The radiative forcing of an incremental amount of some greenhouse gas depends on the wavelengths the gas tends to absorb and how well existing concentrations are already absorbing those wavelengths. The first factor has to do with whether or not the absorption band falls in the $7~\mu m$ to $12~\mu m$ atmospheric radiative window. A greenhouse gas that has its absorption band within the window is much more potent than one that does not. The second factor is based on the notion that absorption bands can approach saturation, which means that each new addition to the atmosphere has less and less impact.

The fraction of the radiation of any wavelength that is absorbed is related to the number of greenhouse gas molecules that the photons encounter. The opportunities for radiation to be absorbed are therefore a function of the concentration of the greenhouse gas and the distance that the radiation must travel as it passes through the atmosphere. This suggests that a relationship between radiative forcing and greenhouse gas concentration would be nonlinear and might look something like Figure 8.28. For gases that exist in small concentrations, such as the halocarbons, absorption is directly proportional to concentration. Doubling the concentration doubles the absorption. As concentrations increase, absorption becomes more and more nonlinear. For example, CH_4 and N_2O are somewhere in the middle of the graph, and absorption is roughly proportional to the square root of concentration. Finally, when concentration is already high, as is the case for CO_2, absorption tends to increase as the logarithm of concentration.

If we express the radiative forcing as a function of concentration as follows

$$F = f(C)$$

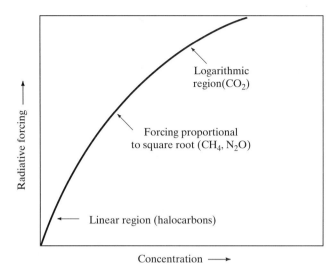

FIGURE 8.28 Radiative forcing for greenhouse gases. At low concentrations forcing is linear, but as concentration increases saturation effects become more evident.

where C is the concentration, then, as shown in Figure 8.29, an incremental change in concentration from C_0 to C will yield an incremental change in radiative forcing of

$$\Delta F = f(C) - f(C_0) \tag{8.24}$$

For the three regions specified in Figure 8.28, changes in concentration will yield the following changes in forcing:

$$\text{Linear region (halocarbons):} \quad \Delta F = k_1(C - C_0) \tag{8.25}$$

$$\text{Square root region (CH}_4\text{, N}_2\text{O):} \quad \Delta F = k_2(\sqrt{C} - \sqrt{C_0}) \tag{8.26}$$

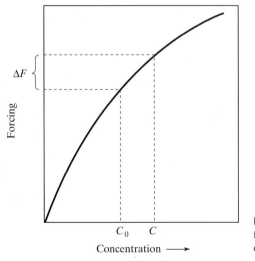

FIGURE 8.29 Illustrating the change in radiative forcing associated with a change in greenhouse gas concentration.

$$\text{Logarithmic region } (CO_2): \quad \Delta F = k_3(\ln C - \ln C_0) = k_3 \ln\left(\frac{C}{C_0}\right) \quad (8.27)$$

where the coefficients k_1, k_2, and k_3 are coefficients that need to be determined for each gas.

Equations (8.25) and (8.26) assume that the CH_4 and N_2O forcings are independent of each other. That is, changes in the concentration of one do not affect the forcing function of the other. That would be the case if their absorption bands did not overlap. In reality, as shown in Figure 8.11, the absorption bands of CH_4 and N_2O do overlap, so increasing the concentration of one of these gases increases the saturation effect on the other. More careful analysis (IPCC, 1990) does include an "overlap" term, but the impact is modest and will be ignored here (see Problem 8.19).

While those coefficients can be derived from fundamental principles, we can work backward from published studies to figure them out, as the following example suggests.

EXAMPLE 8.5   Radiative forcing functions for principal greenhouse gases

Using the following radiative forcing data, derive appropriate forcing functions (Eq. 8.25, 8.26, and 8.27) for each of the gases. Assume they have been in the linear, square root, or logarithmic regions from preindustrial times to the current time.

Greenhouse gas	Concentration in 1850 (ppb)	Concentration in 1992 (ppb)	ΔF (W/m²)
CO_2	278,000	356,000	1.56
CH_4	700	1714	0.47
CFC–11	0	0.268	0.06

Solution   For CFC–11, using the linear relationship (8.25) with forcings in W/m² and concentrations in parts per billion by volume (ppb) gives

$$\text{CFC–11:} \quad 0.06 = k_1(0.268 - 0) \quad \text{so } k_1 = \frac{0.06}{0.268} = 0.22\left(\frac{W/m^2}{ppb}\right)$$

so that

$$\Delta F_{CFC-11} = 0.22[(CFC–11) - (CFC–11)_0] \quad (8.28)$$

where the notations $(CFC–11)_0$ and $(CFC–11)$ mean the initial concentration and new concentration of CFC–11, respectively.

For methane, using (8.26) gives

$$CH_4: \quad 0.47 = k_2(\sqrt{1714} - \sqrt{700})$$

$$k_2 = \frac{0.47}{\sqrt{1714} - \sqrt{700}} = 0.031\left(\frac{W/m^2}{ppb^{0.5}}\right)$$

so

$$\Delta F_{CH_4} = 0.031\left[\sqrt{(CH_4)} - \sqrt{(CH_4)_0}\right] \quad (8.29)$$

For CO_2, using the logarithmic relationship of (8.27) gives

$$1.56 = k_3 \ln\left(\frac{356,000}{278,000}\right) = 0.247k_3 \quad \text{so } k_3 = 6.3(W/m^2)$$

and therefore

$$\Delta F_{CO_2} = 6.3 \ln\left[\frac{(CO_2)}{(CO_2)_0}\right] \tag{8.30}$$

For (8.28) and (8.29), the greenhouse gas concentrations must be in parts per billion by volume (ppb) in order to give radiative forcings in W/m². In (8.30) for CO_2 the concentration units cancel, so any volumetric units can be used. ∎

By combining the climate sensitivity factor λ, introduced in (8.17), with the sum of the forcings from each greenhouse gas, we can estimate the global average temperature change. Example 8.6 illustrates the procedure.

EXAMPLE 8.6   Estimating future global temperature change

Suppose an emission scenario estimates the following concentrations for CO_2 and CH_4 in the year 2100. Using a climate sensitivity factor λ equal to 0.57 °C per W/m², estimate the equilibrium global temperature change caused by the forcing of these two gases.

Gas	1992 (ppb)	2100 (ppb)
CO_2	356,000	710,000
CH_4	1714	3616

Solution   Equations (8.29) and (8.30) let us estimate individual forcings:

$$\Delta F_{CO_2} = 6.3 \ln\left(\frac{710,000}{356,000}\right) = 4.35 \text{ W/m}^2$$

$$\Delta F_{CH_4} = 0.031\left(\sqrt{3616} - \sqrt{1714}\right) = 0.58 \text{ W/m}^2$$

The sum of the forcings is

$$\Delta F = 4.35 + 0.58 = 4.93 \text{ W/m}^2$$

From (8.17) we get our estimate of the equilibrium surface temperature change in the year 2100:

$$\Delta T_s = \lambda \Delta F = 0.57 \times 4.93 = 2.8 \text{ °C}$$

∎

Equivalent CO_2 Concentration

Quite often the combination of radiative forcings is expressed as an *equivalent concentration* of carbon dioxide. For example, when comparing climate change models it is common to compare model predictions for radiative forcing that are equivalent to a doubling of CO_2 over the preindustrial period concentration of 278 ppm.
    From (8.30) we can write

$$\Delta F = 6.3 \ln\left[\frac{(CO_2)_{equiv}}{278}\right] \tag{8.31}$$

where $(CO_2)_{equiv}$ = equivalent concentration of CO_2 (ppm) to produce forcing $\Delta F(W/m^2)$.

Solving (8.31) gives

$$(CO_2)_{equiv} = 278 \exp(\Delta F/6.3) \tag{8.32}$$

The following example shows how (8.32) might be used.

EXAMPLE 8.7   Equivalent CO_2

The combined radiative forcing for the principal greenhouse gases (CO_2, CH_4, N_2O, and halocarbons) relative to preindustrial times is estimated to be 2.45 W/m^2. What is the equivalent concentration of CO_2 today? Compare it to the actual value of 356 ppm.

Solution   From (8.32) the equivalent concentration today is

$$(CO_2)_{equiv} = 278 \, (ppm) \times \exp(2.45/6.3) = 410 \text{ ppm}$$

The actual concentration of CO_2 was 356 ppm, but the radiative forcing of all the greenhouse gases combined is equivalent to a concentration of CO_2 of 410 ppm.   ■

Another Climate Sensitivity

The climate sensitivity parameter λ in (8.17) makes the important link between estimates of radiative forcing and the resulting change in surface temperature. There is, however, another way to express climate sensitivity that is even more commonly used. The increase in surface temperature that results from the equivalent of a doubling of CO_2 in the atmosphere is called the climate sensitivity ΔT_{2x}. From that definition, we can write

$$\Delta T_s = \frac{\Delta T_{2x}}{\ln 2} \ln\left[\frac{(CO_2)}{(CO_2)_0}\right] \tag{8.33}$$

where

ΔT_s = the equilibrium global mean surface temperature change

ΔT_{2x} = the equilibrium temperature change for a doubling of atmospheric CO_2

$(CO_2)_0$ = the initial concentration of CO_2

(CO_2) = the concentration of CO_2 at a later time

Notice what happens to (8.33) when (CO_2) is double the initial amount. The change in surface temperature is what it should be. That is,

$$\Delta T_s = \frac{\Delta T_{2x}}{\ln 2} \ln\left[\frac{2(CO_2)_0}{(CO_2)_0}\right] = \frac{\Delta T_{2x}}{\ln 2} \cdot \ln 2 = \Delta T_{2x}$$

If the concentration of CO_2 is quadrupled,

$$\Delta T_s = \frac{\Delta T_{2x}}{\ln 2} \cdot \ln \left[\frac{4(CO_2)_0}{(CO_2)_0} \right] = \frac{\Delta T_{2x}}{\ln 2} \cdot \ln (2^2) = \frac{\Delta T_{2x}}{\ln 2} \cdot 2 \ln 2 = 2\Delta T_{2x}$$

In other words, the logarithmic function suggests that for every doubling of CO_2, the surface temperature goes up by the same amount. For example, if ΔT_{2x} is 2.5 °C, then the first doubling raises the surface temperature by 2.5 °C, doubling it again to four times its initial value raises the temperature by another 2.5 °C, and so on, as illustrated in Figure 8.30.

If we connect (8.32), (8.33) and (8.17), we can establish the relationship between these two climate sensitivities:

$$\Delta T_s = \frac{\Delta T_{2x}}{\ln 2} \ln \left[\frac{(CO_2)}{(CO_2)_0} \right] = \lambda \, \Delta F = \lambda \cdot 6.3 \ln \left[\frac{(CO_2)}{(CO_2)_0} \right]$$

so that

$$\Delta T_{2x} = 6.3\lambda \ln 2 = 4.37\lambda \tag{8.34}$$

For example, we have been using a climate sensitivity parameter λ of 0.57 (°C W^{-1}m^2), which corresponds to

$$\Delta T_{2x} = 4.37 \times 0.57 = 2.5 \, °C$$

Different general circulation models (GCMs) that are run with doubled CO_2 as an input yield different estimates of surface warming. The climate sensitivity ΔT_{2x} is a convenient benchmark that is often used to compare these computer models. It has

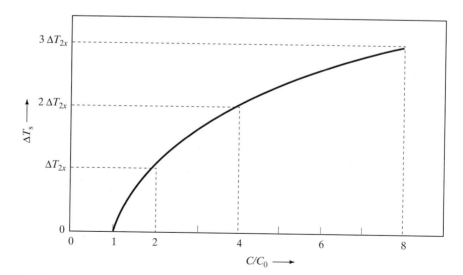

FIGURE 8.30   For every doubling of CO_2, the global equilibrium temperature increases by ΔT_{2x}.

long been estimated to be between 1.5 °C and 4.5 °C, with the best estimate currently being 2.5 °C (IPCC, 1996).

8.9 GLOBAL WARMING POTENTIAL

The 1992 United Nations Framework Convention on Climate Change (FCCC) has established a commitment to "prevent dangerous anthropogenic interference with the climate system." One of the tools being developed to assist policy makers who must try to implement that commitment is the use of an index, called the *Global Warming Potential* (GWP). The GWP is a weighting factor that enables comparisons to be made between the global warming impact of 1 kg of any greenhouse gas and 1 kg of CO_2. It is a dimensionless quantity and includes a time horizon during which the impact will be felt. For example, the 20-year GWP for N_2O is 280, which means 1 kg of N_2O emitted today will exert 280 times as much global warming over the next 20 years as would 1 kg of CO_2 emitted today. Stated slightly differently, 1 kg of N_2O emitted today will have the same impact over the next 20 years as a release today of 280 kg of CO_2.

When the GWP for each greenhouse gas is multiplied by this year's emission rate, a true measure of the impact of each gas can be found. For example, if some process emits 1 kg of N_2O and 1000 kg of CO_2, the 20-year forcing of the pair would be equivalent to $1 \times 280 + 1000 = 1280$ kg of CO_2. The N_2O would be causing 22 percent of the total forcing ($280/1280 = 0.22$).

By including the time horizon, the GWP accounts for the greater impact that a gas with a long atmospheric lifetime will have compared with one that quickly disappears once it has been emitted. For example, 1 kg of CFC–12 in the atmosphere and 1 kg of HCFC–22 in the atmosphere will each have roughly the same immediate impact on radiative forcing. However, the longer lifetime of CFC–12 (102 years vs. 12.1 years for HCFC–22) means the CFC–12 emitted today will have an impact for a much longer period of time, and that is reflected in their GWPs. The 100-year GWP for CFC 12 is about 8000, while the GWP for HCFC–22 is "only" 1700. Notice that comparisons between gases other than CO_2 are easy to make. In the preceding example, 1 kg of CFC–12 emitted today will provide almost five times ($8000/1700 = 4.7$) as much warming in 20 years as 1 kg of HCFC–22.

The choice of time horizon that policy makers might use depends on the issues that are being addressed. For example, shorter time horizons might be used when it is the rate of change of global temperature, rather than the ultimate temperature increase, that is of concern. Short time horizons would also seem appropriate when the most effective greenhouse abatement strategy is needed to head off possibly abrupt climate changes that might be triggered when warming reaches some threshold level. On the other hand, long time horizons might be used to evaluate strategies to avoid long, slow, irreversible impacts such as sea-level changes. The standard time periods that have been incorporated into the tables of GWPs are 20 years, 100 years, and 500 years.

Some examples of how such an index might be used include the following:

1. By combining GWPs with estimates of the cost of curtailing emissions of each greenhouse gas, a country-by-country, least-cost approach to prevention of climate change could be identified.

2. GWPs could facilitate trading of emissions reductions among countries. For example, one country might decide the least expensive way to offset emissions of CO_2 might be to reduce emissions of CH_4 in another. Buying and selling international carbon emission offsets could lead to technology and economic transfer from the developed to the developing countries (Swisher and Masters, 1991).

3. Country-by-country rankings of individual contributions to climate change would be made possible. Quantifiable goals could then be established for future reductions.

4. Combined with other indexes, GWPs could be part of an overall environmental impact assessment for products and industrial process audits. Included in an environmental labeling system, consumer choices could be affected.

Calculating the GWP

There are three primary factors that affect GWPs. The first is the radiative forcing associated with the addition to the atmosphere of a unit mass of each greenhouse gas. The second is based on estimates of how long the gas will remain in the atmosphere. The third is related to the cumulative radiative forcing that a unit addition to the atmosphere will have over some period of time into the future.

The GWP is a ratio of the cumulative radiative forcing for 1 kg of a greenhouse gas over some period of time to the cumulative radiative forcing for 1 kg of a reference gas, chosen to be carbon dioxide, over that same time period. Mathematically, the GWP can be expressed as

$$\text{GWP} = \frac{\displaystyle\int_0^T \Delta F_g \cdot R_g(t)\,dt}{\displaystyle\int_0^T \Delta F_{CO_2} \cdot R_{CO_2}(t)\,dt} \tag{8.35}$$

where

$$\Delta F_g = \text{radiative forcing of greenhouse gas in question per kg (W m}^{-2}\text{ kg}^{-1})$$
$$\Delta F_{CO_2} = \text{radiative forcing of CO}_2 \text{ per kg (W m}^{-2}\text{ kg}^{-1})$$
$$R_g(t) = \text{fraction of the 1 kg of greenhouse gas remaining in the atmosphere at time } t$$
$$R_{CO_2}(t) = \text{fraction of the 1 kg of CO}_2 \text{ remaining at time } t$$
$$T = \text{the time period for cumulative effects (years)}$$

In the usual analysis of GWP, the radiative forcings ΔF_g and ΔF_r are treated as constants for the gases in question, which is to say it is assumed that the composition of the atmosphere does not change during the time horizon. In reality, that is not a valid assumption since the atmosphere does change, and as it changes the radiative forcing changes as well (Figure 8.28). Treating ΔF_g and ΔF_r as constants lets them be taken out of the integrals, which simplifies the calculation.

With one very important exception (CO_2), the rate that a greenhouse gas is removed from the atmosphere is considered to be proportional to the amount of gas present. When that is the case, the amount remaining in the atmosphere after an emission of 1 kg of gas is given by the simple exponential decay function:

$$R_g(t) = e^{-t/\tau} \tag{8.36}$$

where the time constant τ is called the *atmospheric lifetime,* or the *e-folding time.* Notice that at time $t = 0$, $R_g(0) = 1$ kg. The integral in the numerator of (8.35) is

$$\int_0^T R_g(t)dt = \int_0^T e^{-t/\tau}dt = \tau\left(1 - e^{-t/\tau}\right) \tag{8.37}$$

The GWP then becomes the following:

$$\text{GWP} = \left(\frac{\Delta F_g}{\Delta F_{CO_2}}\right) \cdot \frac{\tau\left(1 - e^{-t/\tau}\right)}{\displaystyle\int_0^T R_{CO_2}(t)\,dt} \tag{8.38}$$

The IPCC has assembled tables of time constants τ and radiative forcing ratios, some examples of which are given later in Table 8.6.

The remaining quantity to pin down is the fraction of a 1-kg addition of CO_2 that remains in the atmosphere as a function of time—that is, R_{CO_2} in (8.38). It was mentioned previously that CO_2 is a gas that cannot be characterized with the simple exponential function given in (8.36). Decay is rapid during the first few decades as the biosphere absorbs the carbon; then for next few hundred years it decays at a much slower rate corresponding to the slow uptake of the oceans. To complicate matters further, the decay also a depends on the assumptions made for the background concentration of CO_2. Two curves for R_{CO_2} are shown in Figure 8.31; one corresponds to an atmosphere in which CO_2 concentration is unchanging, while the other has been drawn for a more realistic atmosphere in which CO_2 reaches 650 ppm by the year 2200. Since the GWP is referenced to the behavior of CO_2, different assumptions about the decay rate will produce different estimates of GWP. The decay rate corresponding to a constant concentration of CO_2 is usually used, but there are differences of opinion about which is more appropriate. Working backwards from published GWPs yields the approximate values needed for (8.38) shown in Table 8.5.

TABLE 8.5   Approximate values for the integral of the impulse response of CO_2 based on the Bern model with fixed CO_2 (354 ppm)

Time Horizon, T (yrs)	$\int_0^T R_{CO_2}(t)dt$ (yrs)
20	13.2
100	43.1
500	138.0

Source: Estimated from AGWP data given in IPCC (1995).

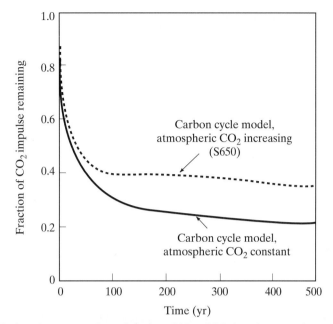

FIGURE 8.31   The impulse response for an injection of 1 kg of CO_2 into the atmosphere; that is, $R_{CO_2}(t)$ in (8.38). The solid line is for an atmosphere in which CO_2 concentration is constant; the dashed line is for an atmosphere that stabilizes at 650 ppm of CO_2 by the year 2200. (*Source*: adapted from IPCC, 1995)

EXAMPLE 8.8   Estimating the GWP for HFC–134a

One kilogram of HFC–134a in the atmosphere causes a radiative forcing that is 4129 times the forcing of 1 kg of CO_2. The decay of HFC–134a is exponential with a time constant of 14.6 years. Estimate the GWP over 20-year and 500-year periods.

Solution   Substituting the data given on HFC–134a into (8.38) along with appropriate time period information for CO_2 given in Table 8.5 gives

$$\text{GWP} = \left(\frac{\Delta F_g}{\Delta F_{CO_2}} \right) \cdot \frac{\tau(1 - e^{-t/\tau})}{\displaystyle\int_0^T R_{CO_2}(t)\, dt}$$

For the 20-year time horizon,

$$\text{GWP}_{20} = (4129) \cdot \frac{14.6(1 - e^{-20/14.6})}{13.2} = 3400$$

and for the 500-year time period,

$$\text{GWP}_{500} = (4129) \cdot \frac{14.6(1 - e^{-500/14.6})}{138.0} = 435$$

Notice how the relatively short time-constant for HFC–134a results in a much higher GWP in the 20-year time period than it does over the much longer 500-year period.   ■

TABLE 8.6   Direct GWPs relative to unchanging future CO_2 concentration.

Species	Chemical formula	Lifetime τ (yrs)	Forcing per unit mass relative to CO_2 $\Delta F_g/\Delta F_{CO_2}$	Global Warming Potential (time horizon in years) 20	100	500
Carbon dioxide	CO_2		1	1	1	1
Methane[a]	CH_4	12.2	58	56	21	6.5
Nitrous oxide	N_2O	120	206	280	310	170
CFC–11	$CFCl_3$	50	3970	5000	4000	1400
CFC–12	CF_2Cl_2	102	5750	7900	8500	4200
CFC–113	$CF_2ClCFCl_2$	85	3692	5000	5000	2300
CFC–114	CF_2ClCF_2Cl	300	4685	6900	9300	8300
HCFC–22	CF_2HCl	13.3	5440	4300	1700	520
HCFC–141b	CH_3CFCl_2	9.4	2898	1800	630	200
HCFC–142b	CH_3CF_2Cl	18.4	4446	4200	2000	630
HFC–134a	CF_3CH_2F	14.6	4129	3400	1300	420
HFC–32	CH_2F_2	5.6	5240	2100	650	200
H–1301	CF_3Br	65	4724	6200	5600	2200
Methyl chloroform	CH_3CCl_3	5.4	913	360	110	35
Carbon tetrachloride	CCl_4	42	1627	2000	1400	500

Typical uncertainty is $\pm 35\%$. Indirect effects of halocarbons offset to a significant degree the direct warmings given here.
[a]Methane includes direct and indirect components.
Source: IPCC (1995 and 1996) and Wuebbles (1995).

Table 8.6 shows GWPs for a number of widely used chemicals. As can be seen, all of these species have very high GWPs relative to CO_2 because they all absorb in the atmospheric window. An important caveat is in order, however. Except for methane, all of the GWPs are only for direct effects of the chemicals in the table. Almost all of these gases are halocarbons, however, and halocarbons have chlorine or bromine atoms that can destroy ozone in the stratosphere. That indirect effect of ozone destruction can offset to a significant degree the direct positive warming of halocarbons. Brominated compounds, including halons and methyl bromide CH_3Br, are so effective at destroying stratospheric ozone that their net forcing effect is negative—that is, they act to cool the atmosphere (IPCC, 1995).

The special case that methane presents needs to be noted as well. The direct positive forcing of methane is *enhanced* by its indirect effect. As previously mentioned, methane adds water vapor and ozone to the stratosphere, leading to additional positive forcing. In Table 8.6 the GWP for methane includes these indirect effects, which means the approach used in Example 8.8, which computes only the direct impacts, will underestimate the GWPs shown in Table 8.6.

The purpose of GWPs is to assess the relative importance of various emissions, which means the GWP for each gas should be multiplied by the emission rate for that gas. The following example illustrates the process.

EXAMPLE 8.9   Using GWPs to rank greenhouse gases

In 1992, anthropogenic emissions of CO_2 were approximately $24{,}000 \times 10^9$ kg per year; emissions of CH_4 were about 375×10^9 kg/yr; and emissions of N_2O were roughly 9×10^9 kg/yr. Compare the impacts of these three gases over a 20-year time horizon.

Solution   The comparison will be based on the products of emission rates and GWPs. GWPs can be taken from Table 8.6, and the emission rates are given. Carbon dioxide has a GWP of 1 (by definition) over all time horizons, so its product does not change:

$$CO_2: \quad GWP \times emissions = 1 \times 24{,}000 \times 10^9 = 2.4 \times 10^{13} \text{ kg } CO_2$$

The 20-year GWP times emission-rate product for methane is

$$CH_4: \quad GWP_{20} \times emissions = 56 \times 375 \times 10^9 = 2.1 \times 10^{13} \text{ kg as } CO_2$$

And, for N_2O, the 20-year impact will be

$$N_2O: \quad GWP_{20} \times emissions = 280 \times 9 \times 10^9 = 0.25 \times 10^{13} \text{ kg as } CO_2 \qquad ■$$

Notice that Example 8.9 suggests that the warming impact over the next 20 years from today's emissions of CH_4 is almost as large as the impact of CO_2. The 100-year and 500-year GWP × emissions products are summarized in Table 8.7 and Figure 8.32. Notice that over longer time horizons CO_2 is by far the dominant greenhouse gas.

8.10 THE CARBON CYCLE

Since carbon dioxide plays such an important role in climate change, it has received much more attention than the other gases and aerosols. Its concentration has increased by 30 percent since the early 1800s, and it is continuing to grow by about 1.53 ppm per year as humankind continues to burn fossil fuels and change the amount of biomass on the earth's surface.

Natural processes continuously transport enormous amounts of carbon back and forth between the atmosphere, biosphere, and oceans. Carbon moves into the food chain during photosynthesis and returns to the atmosphere during respiration. The oceans dissolve carbon dioxide and store almost all of it in the form of bicarbonate ions (HCO_3^-), but some becomes part of the marine food chain. A very small portion of nonliving organic matter each year ends up in sediments. The slow, historical accumulation of that organic carbon is the source of our fossil fuels—oil, natural gas, and

TABLE 8.7   Relative radiative forcings for the three main greenhouse gases.

Time Period (yr)	Gas	GWP	Emissions (10^9 kg/yr)	Product (10^{13} kg CO_2)	Percent of Total
20	CO_2	1	24,000	2.4	51%
20	CH_4	56	375	2.1	44
20	N_2O	280	9	0.25	5
100	CO_2	1	24,000	2.4	69
100	CH_4	21	375	0.8	23
100	N_2O	310	9	0.3	8
500	CO_2	1	24,000	2.4	86
500	CH_4	6.5	375	0.24	9
500	N_2O	170	9	0.15	5

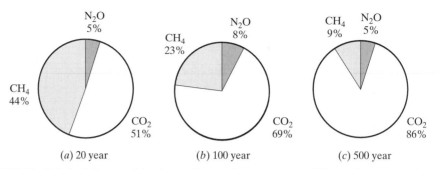

FIGURE 8.32 Relative influence of the three main greenhouse gases over different time horizons based on the products of GWP times one year's emissions.

coal. When these are burned, ancient carbon is returned to the atmosphere. By comparison with the natural fluxes of carbon, the additional amounts added to the atmosphere by combustion and changes in land use are modest, but they are enough to cause a significant response by the climate system.

The flows of carbon from one reservoir to another and the amounts of carbon stored in those reservoirs are summarized in Figure 8.33. As can be seen, the pool of carbon in the atmosphere in the 1980s is estimated at 750 GtC, where 1 GtC means 1 gigatonne of carbon (10^9 tonnes or 10^{12} kg). Since almost all of that carbon is stored in the form of CO_2 (less than 1 percent is in other carbon-containing compounds such as methane and carbon monoxide), in most circumstances it is reasonable to assume that atmospheric carbon is entirely CO_2. The amount of carbon locked up in terrestrial vegetation (610 GtC) is on the same order of magnitude as that in the atmosphere, but both of these amounts are dwarfed by the carbon stored in the oceans. The oceans contain more than 50 times as much carbon as the atmosphere.

During the 1980s, combustion of fossil fuels and production of cement released an average of 5.5 GtC per year into the atmosphere (by 1990 those emissions reached 6.1 GtC/yr). Estimates of the reduction in carbon stored in tropical vegetation released through activities such as biomass burning and harvesting of forests are difficult to make, but indications are that these tropical land-use changes release an average of 1.6 GtC per year. Regrowth of forests in the northern hemisphere is offsetting some of those tropical land-use changes by removing 0.5 GtC/yr from the atmosphere. An additional terrestrial sink is attributed to stimulated plant growth by increased amounts of carbon dioxide in the air and increased nitrogen deposition on the soils from fossil fuel combustion. That nitrogen and CO_2 fertilization is thought to be causing photosynthesis to exceed respiration, yielding a net removal of 1.3 GtC/yr from the atmosphere. The movement of carbon between the oceans and atmosphere is large, on the order of 90 GtC/yr, and the net effect of those fluxes is removal of about 2 Gt/C/yr from the atmosphere.

Table 8.8 summarizes the carbon additions to the atmosphere caused by human activities, and the partitioning of those additions among the various carbon reservoirs. In the 1980s, fossil fuel combustion and land-use changes in the tropics added 7.1 GtC/yr to the atmosphere. Of that 7.1 GtC/yr, 3.8 GtC/yr ended up in the oceans or other terrestrial sinks and the other 3.3 GtC/yr remained in the atmosphere. The ratio of the amount of anthropogenic carbon emitted to the amount that remains in the atmosphere is known as the *airborne fraction*. Using these data, the airborne fraction has been

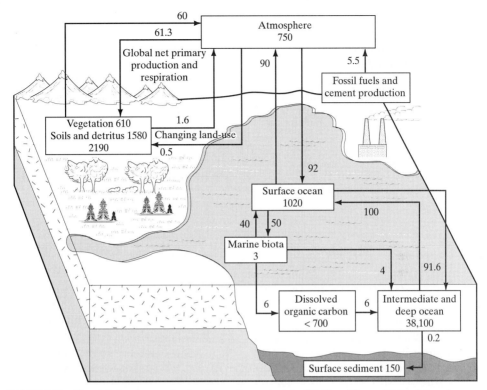

FIGURE 8.33   The global carbon cycle, showing carbon reservoirs (GtC) and fluxes (GtC/yr), based on annual averages over the period 1980 to 1989. (*Source:* IPCC, 1996)

TABLE 8.8   Average annual budget of CO_2 perturbations for 1980–1989

Carbon dioxide sources	
Emissions from fossil fuel combustion and cement production	5.5 ± 0.5
Net emissions from changes in tropical land use	1.6 ± 1.0
Total anthropogenic emissions	7.1 ± 1.1 GtC/yr
Partitioning among reservoirs	
Ocean uptake	2.0 ± 0.8
Uptake by northern hemisphere forest regrowth	0.5 ± 0.5
Other terrestrial sinks (CO_2 and nitrogen fertilization, climatic effects)	1.3 ± 1.5
Net storage in the atmosphere	3.3 ± 0.2 GtC/yr

Source: IPCC, 1996.

$$\text{Airborne fraction} = \frac{3.3 \text{ GtC/yr remaining in atmosphere}}{7.1 \text{ GtC/yr anthropogenic additions}} = 0.46 \qquad (8.39)$$

Thus, roughly speaking, about half of the carbon we emitted in the 1980s stayed in the atmosphere. The airborne fraction is a very useful parameter that helps in the predictions of future CO_2 concentrations for various emission scenarios.

The airborne fraction is not necessarily a fixed quantity. For example, if large areas of land are deforested, the ability of the biosphere to absorb carbon would be reduced and the atmospheric fraction would increase. Likewise, CO_2 fertilization of terrestrial biomass can stimulate plant growth, which increases the rate of removal of atmospheric carbon, so the airborne fraction could get smaller. The airborne fraction also depends on how fast carbon is being added to the atmosphere. For scenarios with little or no growth in emissions or even declining emissions, the oceans and plants have more time to absorb carbon so the atmospheric fraction is lower, perhaps in the vicinity of 35 percent. For rapidly increasing emission rates, carbon sinks cannot keep up and the fraction remaining in the atmosphere may be closer to 55 percent.

The following example develops another useful relationship, this time between the concentration of CO_2 and the tonnes of carbon in the atmosphere.

EXAMPLE 8.10 Carbon content of the atmosphere

Find a relationship between the concentration of carbon dioxide and the total amount of carbon in the atmosphere. The total mass of the atmosphere is estimated to be 5.12×10^{21} g.

Solution We will need to know something about the density of air. That, of course, varies with altitude, but finding it under some particular conditions will work. From Table 8.1 we know the concentration of each gas in air. Recall from Section 1.2 that one mole of each gas occupies 22.4×10^{-3} m^3 at Standard Temperature and Pressure (0 °C and 1 atm), which is 44.6 mol/m^3. The following table organizes the calculation:

Gas	$\dfrac{\text{m}^3 \text{ gas}}{\text{m}^3 \text{ air}}$	×	g/mol	×	$\dfrac{\text{mol}}{\text{m}^3 \text{ gas}}$	=	$\dfrac{\text{g}}{\text{m}^3 \text{ air}}$
N_2	0.7808		28		44.6		975.1
O_2	0.2095		32		44.6		299.0
Ar	0.0093		40		44.6		16.6
CO_2	0.00035		44		44.6		0.7
						Total	1291.4 g/m^3

If all of the atmosphere were at standard temperature and pressure, it would have a density of 1291.4 g/m^3 and its mass would still be 5.12×10^{21} g. Putting these together gives

$$1 \text{ ppm} = \frac{1 \text{ m}^3 \, CO_2}{10^6 \text{ m}^3 \text{ air}} \cdot 44.6 \, \frac{\text{mole}}{\text{m}^3 \, CO_2} \cdot 12 \, \frac{\text{g C}}{\text{mole}} \cdot \frac{5.12 \times 10^{21} \text{ g air}}{1291.4 \, \dfrac{\text{g air}}{\text{m}^3 \text{ air}}} \cdot 10^{-15} \, \frac{\text{GtC}}{\text{g C}} = 2.12 \text{ GtC}$$

Notice that this calculation has taken advantage of the fact that volumetric concentrations (ppm) are independent of temperature or pressure. ∎

From Example 8.10, we have the following useful relationship:

$$1 \text{ ppm } CO_2 = 2.12 \text{ GtC} \tag{8.40}$$

For example, in Figure 8.33 the total amount of carbon in the atmosphere during the 1980s was about 750 GtC. Using (8.40), we can find the concentration of CO_2:

$$\frac{750 \text{ GtC}}{2.12 \text{ GtC/ppm } CO_2} = 354 \text{ ppm of } CO_2$$

Figure 8.33 also indicates that the net increase in carbon stored in the atmosphere in the 1980s was about 3.3 GtC/yr. Using (8.40), we can find the rate of change of CO_2:

$$\frac{3.3 \text{ GtC/yr}}{2.12 \text{ GtC/ppm } CO_2} = 1.6 \text{ ppm/yr}$$

8.11 CARBON EMISSIONS FROM FOSSIL FUELS

Combustion of fossil fuels accounts for over 80 percent of CO_2 emissions, with the rest being mostly the result of deforestation and a small amount is released during cement manufacture. Our use of energy is an important source of other greenhouse gases as well, as Figure 8.34 shows. In addition, fossil fuels are significant sources of the ozone precursors, carbon monoxide, nitrogen oxides, and volatile organic chemicals, as well as being the main source of sulfur aerosols.

Use of energy and the resulting emissions of carbon vary considerably from country to country. The United States, with less than 5 percent of the world's population, emits 22 percent of the energy-related CO_2, which is double the amount emitted by the world's most populous country, China. On a per capita basis, the United States again emits more carbon than any other country. As shown in Figure 8.35, the per capita emission rate is roughly twice as high as the other most developed countries, such as Japan, Germany, and the United Kingdom. While the developed countries dominate carbon emissions today, in the near future the rapid growth in energy demands in the developing countries will reverse that situation.

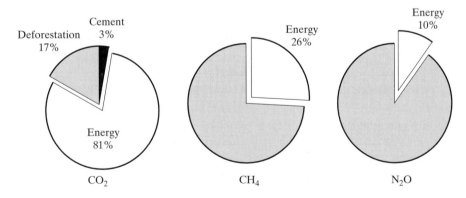

FIGURE 8.34   Use of energy is a major source of greenhouse gases. The percentages represent the fraction of anthropogenic emissions that are associated with energy.

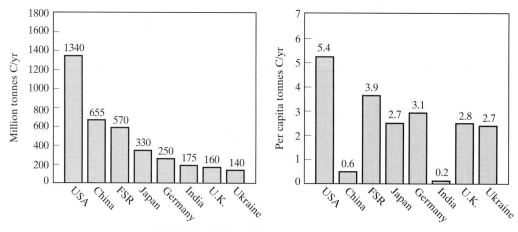

FIGURE 8.35 1992 fossil fuel emissions of carbon.

Carbon Intensity of Fossil Fuels

The amount of carbon released per unit of energy delivered is called the *carbon intensity*. Some fuels have high carbon intensity, such as coal, and some release no carbon at all, such as hydroelectric or nuclear power. Interestingly, biomass fuels can also be used in ways that emit no net amount of carbon. That is, burning biomass fuels simply returns carbon to the atmosphere that those plants had once removed from the atmosphere. As long as the biomass that is burned is replaced with an equal mass of new plant material each year, the net carbon release is zero.

EXAMPLE 8.11   Carbon intensity of methane

The energy released during combustion of methane is 890 kJ/mol. Find the carbon intensity of methane based on its higher heating value (HHV) of 890 kJ/mol (which includes the energy of condensation of the water vapor formed; you might want to review Examples 2.4 and 2.5 in Chapter 2, which describe HHV and LHV). Then find the carbon intensity based on the lower heating value (LHV) of 802 kJ/mol.

Solution   First write a balanced chemical reaction for the oxidation of methane:

$$CH_4 \;+\; 2\,O_2 \;\longrightarrow\; CO_2 \;+\; 2\,H_2O$$

So, burning 1 mol of CH_4 liberates 890 kJ of energy while producing 1 mol of CO_2. Recall that the HHV is based on the assumption that the energy content of the water vapor formed is usable energy, which is not usually the case.

Since 1 mol of CO_2 has 12 g of carbon, the HHV carbon intensity of CH_4 is

$$\text{HHV carbon intensity} = \frac{12 \text{ gC}}{890 \text{ kJ}} = 0.0135 \text{ gC/kJ} = 13.5 \text{ gC/MJ}$$

Similarly, the LHV carbon intensity would be

$$\text{LHV carbon intensity} = \frac{12 \text{ gC}}{802 \text{ kJ}} = 0.015 \text{ gC/kJ} = 15.0 \text{ gC/MJ}$$

■

The LHV and HHV carbon intensities for fossil fuels, combined with estimates of total carbon emissions, are given in Table 8.9. Note the slight difference between the carbon intensity of natural gas (15.3 g C/MJ) and the intensity of methane itself (15.0 g C/MJ) found in Example 8.11. Natural gas is mostly methane, but it also includes other hydrocarbons. Another comment worth making is that energy consumption figures in the United States are often based on higher heating value of a fuel, while most of the rest of the world, including the IPCC, uses LHV values. LHV carbon intensities should be used with energy consumption data that are based on LHVs, and HHV intensities should be used with energy based on HHV. In this chapter, all energy data are based on the LHV to be consistent with the IPCC. This can cause modest discrepancies and confusion when dealing with literature that does not make this distinction clear.

The carbon intensity data given in Table 8.9 suggest that sizable reductions in carbon emissions are possible by switching from coal to natural gas. It is unfortunately the case, however, that most of the world's fossil fuel resources are in the form of coal. It is interesting to note that an estimated 87 percent of the world recoverable resources of coal are in just three countries: the United States, the former USSR, and China.

Table 8.10 presents data on the world's fossil-fuel resources. These resources are shown as a resource base, which consists of already identified reserves plus an estimate at the 50 percent probability level of remaining undiscovered resources. The resources are also described as coming from conventional sources of the type now being exploited, as well as unconventional sources that might be usable in the future. Unconventional sources of oil include oil shale, tar sands, and heavy crude; unconventional natural gas sources include gas in Devonian shales, tight sand formations, geopressured aquifers, and coal seams. An additional column in Table 8.10 is labelled "additional occurrences." These are additional resources with unknown certainty of occurrence and/or with unknown or no economic significance in the forseeable future.

TABLE 8.9   LHV and HHV Carbon Intensities and 1990 World Energy and Carbon Emissions[a]

Fuel	LHV Carbon intensity (gC/MJ)	LHV Energy consumption[b] (10^{12} MJ/yr)	Carbon emissions (GtC/yr)	HHV Carbon intensity (gC/MJ)
Natural gas	15.3	71	1.1	13.8
Oil	20.0	128	2.6	19.7
Coal (bituminous)	25.8	91	2.3	24.2
Nuclear, hydro	0.0	40	0	0
Totals		*330*	*6.0*	

[a]1 GtC = 10^9 tonne C = 10^{15} gC; 1 MJ = 10^6 Joules.
[b]IPCC/OECD use LHV values; United States tends to report energy as HHV.
Source: Nakicenovic, 1996.

TABLE 8.10 Global fossil fuel resources and occurrences, in EJ[a]

Fuel	Conventional resources	Unconventional resources	Total resource base	Additional occurrences
Gas	9,200	26,900	36,100	> 832,000
Oil	8,500	16,100	24,600	> 25,000
Coal	25,200	100,300	125,500	> 130,000
Totals	42,900	143,300	186,200	> 987,000

[a]1 EJ = 10^{18} J.
Source: Nakicenovic, 1996.

Enormous amounts of methane locked in methane hydrates ($CH_4 \cdot 6\, H_2O$) under the oceans (estimated at over 800,000 EJ) are the most important of these.

The following example suggests what might happen to the CO_2 concentration in the atmosphere if we were to burn all of the world's coal resource base.

EXAMPLE 8.12   Burning the world's coal

Estimate the increase in CO_2 in the atmosphere if the 125,500 EJ of coal were to be burned assuming an airborne fraction of 50 percent.

Solution: We can first estimate the carbon content using Table 8.9

$$125,500 \text{ EJ} \times 25.8 \text{ gC/MJ} \times 10^{12} \text{ MJ/EJ} \times 10^{-15} \text{ GtC/gC} = 3238 \text{ GtC}$$

which is roughly four times as much carbon as currently exists in the atmosphere. Converting this to CO_2 and accounting for the estimated amount that would remain in the atmosphere gives

$$\Delta CO_2 = \frac{3238 \text{ GtC} \times 0.50}{2.12 \text{ GtC/ppmCO}_2} = 763 \text{ ppm } CO_2$$

That is enough to more than triple the amount of CO_2 in the atmosphere today.   ∎

It is interesting to note that if we use the carbon emission factors in Table 8.9 for the oil and natural gas in the remaining resource base, they would have the potential to add only about one-third as much CO_2 as the coal resource base would. That calculation is less certain, however, because of the relatively unknown carbon emission factors that would be appropriate for the unconventional oil and gas resources. It is also complicated by the fact that the principal constituent of natural gas, methane, is a much more potent greenhouse gas than CO_2, which means methane leakage can amplify the potential global warming.

The carbon intensity factors given in Table 8.9 suggest that switching from coal to oil or natural gas will reduce emissions significantly. For example, it would appear that switching from coal to natural gas would reduce carbon emissions by about 40 percent while delivering the same amount of energy. There are, however, other factors that complicate that simple estimate. The following two examples illustrate these complications. The first points out the efficiency advantages as well as carbon advantages associated with using natural gas. The combination of effects causes total carbon emission savings to be considerably more than 40 percent. The second illustrates the methane leakage, which can offset some of the advantages of natural gas.

EXAMPLE 8.13   Efficiency and Carbon Intensity Combined

Compare the carbon emissions to heat household water using the following three energy systems: (1) a very good, 37 percent efficient coal-fired power plant delivering electricity to a 100 percent efficient electric water heater; (2) a new, 45 percent efficient natural-gas-fired combined cycle power plant for that same electric water heater; and (3) an 85 percent efficient gas-fired water heater.

Solution   Let us base our comparison on 100 MJ of energy provided to each system. Using the factors given in Table 8.9, burning 100 MJ of coal releases 2580 g of carbon, while 100 MJ of natural gas releases 1530 gC. As suggested in the following diagrams, the coal-plant system delivers 37 MJ to heat water; the more efficient gas-fired power plant delivers 45 MJ to heat water; and using gas directly in the water heater delivers 85 MJ of heat.

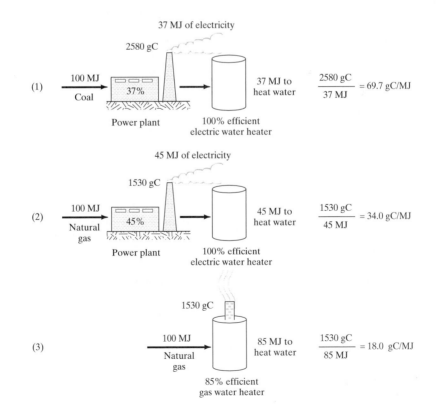

The carbon intensities for each system are shown in the diagrams. Switching from coal to gas in the power plant reduces carbon emissions by more than half, and using gas directly in the water heater reduces carbon emissions by 75 percent.   ■

The next example suggests that switching to natural gas may not cause as much climate mitigation as might have been hoped for. As long as the carbon in methane ends up as CO_2, the 40 percent carbon savings suggested in Table 8.9 will be realized. But if methane is inadvertently released to the atmosphere in the process of mining, transporting, and using natural gas, its greater GWP can offset some of those climate gains.

EXAMPLE 8.14   GWP for a Leaky Gas Pipeline

Consider two fuel systems each delivering 100 MJ of energy to some end use that burns the fuel releasing CO_2. One system delivers that amount of energy in the form of coal. In the other methane is the fuel, but the pipeline leaks 1 MJ of methane for every 100 MJ it delivers to the end user. Compare the 20-year, emissions-weighted GWPs of the two systems.

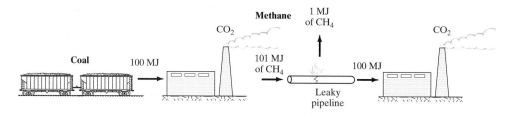

Solution   First let's analyze the coal case. From Table 8.9, coal releases 25.8 gC/MJ. So it emits

$$100 \text{ MJ} \times 25.8 \text{ gC/MJ} = 2580 \text{ gC}$$

To use GWPs we need to know the amount of CO_2, not the amount of C. The atomic weight of carbon is 12 and the molecular weight of CO_2 is 44, so the amount of CO_2 emitted by the coal is

$$CO_2 \text{ emissions from coal} = 2580 \text{ gC} \times \frac{44 \text{ gCO}_2}{12 \text{ gC}} = 9460 \text{ gCO}_2$$

For the methane system, we know from Example 8.11 that 15.0 gC in the form of CO_2 is released per MJ (LHV) of methane energy. Since 100 MJ are burned, the CO_2 emissions are

$$CO_2 \text{ from burning CH}_4 = 100 \text{ MJ} \times 15.0 \text{ gC/MJ} \times \frac{44 \text{ gCO}_2}{12 \text{ gC}} = 5500 \text{ gCO}_2$$

Now we need to determine the amount of methane leakage. From Example 8.11 we know the LHV of CH_4 is 0.802 MJ per mole. The molecular weight of methane is $(12 + 4 \times 1 = 16 \text{ g/mol})$. Thus leaking 1 MJ of methane is equivalent to

$$\text{Methane leaked} = \frac{1 \text{ MJ} \times 16 \text{ gCH}_4/\text{mol}}{0.802 \text{ MJ/mol}} = 19.95 \text{ gCH}_4$$

Now we can find the weighted global warming potential for the two systems. For the coal system all the emissions are in the form of CO_2, which by definition has GWP = 1.

$$\text{Coal: } GWP_{20} \times CO_2 \text{ emissions} = 1 \times 9460 \text{ g} = 9460 \text{ gCO}_2$$

For the methane system, we must add the effect of CO_2 emissions and CH_4 emissions. From Table 8.6 methane has 56 times the warming potential of CO_2 over the 20-year time period, so the total weighted emissions are

$$\text{Methane system: } (CO_2) = 5500 \text{ gCO}_2 \times 1 = 5500 \text{ gCO}_2$$

$$(CH_4) = 19.95 \text{ gCH}_4 \times 56 = 1117 \text{ gCO}_2 \text{ equivalent}$$

$$\text{Total weighted GWP} = 5500 + 1117 = 6617 \text{ gCO}_2 \text{ equivalent}$$

Thus the ratio of the two systems is

$$\frac{\text{20-yr global warming impact of methane system}}{\text{20-yr global warming impact of coal system}} = \frac{6617}{9460} = 0.70$$

Without that leakage, the ratio would have been $5500/9460 = 0.58$. So, without leakage, methane reduces the warming impact by 42 percent; with leakage it only saves 30 percent.   ■

Carbon Emission Rates

Predicting future concentrations of carbon dioxide depends on numerous assumptions about population growth, economic factors, energy technology, and the carbon cycle itself. The usual approach involves developing a range of emission scenarios that depend on those factors and then using those scenarios to drive mathematical models of how the atmosphere and climate system will react to those inputs. At the level of treatment given in this short section we cannot begin to approach the complexity of those models; however, we can make a few simple calculations to at least give a sense of some of the important factors.

A sense of the airborne fraction (8.39) along with the relationship between the carbon content of the atmosphere and the concentration of CO_2 (8.40) are starting points for our simple modeling exercise.

EXAMPLE 8.15   Constant Carbon Emissions

Suppose anthropogenic emissions remain constant at 7.1 GtC/yr for the next century (an optimistic goal of the UN Framework Convention on Climate Change). What equilibrium temperature change would be expected if the climate sensitivity is 2.5 °C for a doubling of CO_2? Use an airborne fraction of 0.46 and assume that the initial CO_2 concentration is 350 ppm.

Solution   The increase in atmospheric carbon over the 100-year period would be

$$7.1 \text{ GtC/yr} \times 100 \text{ yrs} \times 0.46 = 327 \text{ GtC}$$

Using (8.40), the increase in CO_2 would be

$$\Delta(CO_2) = 327 \text{ GtC} \cdot \frac{\text{ppm CO}_2}{2.12 \text{ GtC}} = 154 \text{ ppm CO}_2$$

where $\Delta(CO_2)$ = the change in CO_2 concentration. That means the concentration in 100 years would be

$$350 \text{ ppm} + 154 \text{ ppm} = 504 \text{ ppm CO}_2$$

Notice that even though emissions of carbon are stabilized at current rates, the concentration of CO_2 continues to rise, reaching more than 500 ppm by 2100.

Equation (8.33) can now be used to compute the temperature change:

$$\Delta T_s = \frac{\Delta T_{2x}}{\ln 2} \ln \left[\frac{(CO_2)}{(CO_2)_0} \right] = \frac{2.5}{\ln 2} \ln \left[\frac{504}{350} \right] = 1.3 \,°C$$

■

One way to build more complex models of carbon emissions is to start with the notion that impacts are driven by population, affluence, and technology, as (8.41) suggests:

Environmental impact = (Population) × (Affluence) × (Technology)   (8.41)

An example of the application of (8.41) is the following disaggregation of the key factors that drive carbon emissions associated with our demands for energy:

$$E_{carbon} = \text{Population} \times \frac{GDP}{\text{Person}} \times \frac{\text{Energy}}{GDP} \times \frac{\text{Carbon}}{\text{Energy}} \tag{8.42}$$

where

E_{carbon} = carbon emission rate (GtC/yr)

$\dfrac{GDP}{\text{Person}}$ = per capita gross domestic product ($/person-yr)

$\dfrac{\text{Energy}}{GDP}$ = *energy intensity*, primary energy per unit of GDP (EJ/$)

$\dfrac{\text{Carbon}}{\text{Energy}}$ = *carbon intensity*, carbon emissions per unit of primary energy (GtC/EJ)

Equation (8.42) incorporates the key quantities that drive our energy-related carbon emissions. It includes economic and population scenarios plus two factors that are central to energy: energy intensity and carbon intensity. Carbon intensity has already been introduced. *Energy intensity* is the amount of energy required to create a unit of economic activity as measured by gross domestic product. It is usually thought of as a surrogate for country's energy efficiency. For example, Japan, which only needs half the energy to produce a unit of GDP, is often considered to be roughly twice as energy efficient as the United States. While there is some truth to that assertion, it sometimes masks differences in the standard of living and the climate in each country. For example, the United States has larger houses that are kept warmer in more severe winters, so if more energy is required it may have more to do with those factors than whether or not homes are better insulated in one country or the other.

Figure 8.36 shows the history of energy intensity for a number of industrialized countries, along with a range of projections for the less developed countries (LDCs). The energy intensities of industrialized countries show a common pattern of rising intensity during the early stages of industrialization followed by continuous decreases that are now about 1 percent per year. The LDCs are still in the early stages of industrialization, and their energy intensity may grow unless they can leapfrog over their current inefficient technologies and begin to take advantage of efficiency gains demonstrated by the more developed countries.

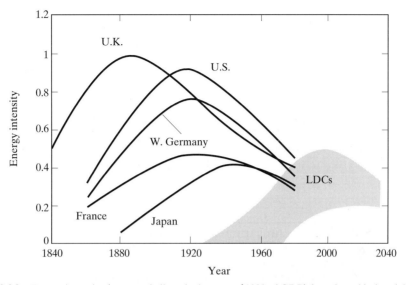

FIGURE 8.36   Energy intensity (tonnes of oil equivalents per $1000 of GDP) for selected industrialized countries, with projections for the less developed countries (LDCs). (*Source:* Reddy and Goldemberg, 1990)

Table 8.11 shows population, economic growth, carbon intensity, and energy intensity values that have been used in one of the most-cited IPCC emission scenarios (IS92a) for energy.[1] For the world as a whole, energy intensity and carbon intensity are both improving over time, which helps offset population and economic growth.

Equation (8.42) expresses the carbon emission rate as the product of four terms: population, GDP, carbon intensity, and energy intensity. Recall from Chapter 3 that if each of the factors in such a product can be expressed as a quantity that is growing (or decreasing) exponentially, then the overall rate of growth is merely the sum of the growth rates of each factor. That is, assuming that each of the factors in (8.42) is growing exponentially, then the overall growth rate of carbon emissions r is given by

$$r = r_{\text{pop}} + r_{\text{GDP/person}} + r_{\text{energy intensity}} + r_{\text{carbon intensity}} \qquad (8.43)$$

By adding the individual growth rates, as has been done in (8.43), an overall growth rate is found that can be used in the following emission equation:

$$E = E_0 e^{rt} \qquad (8.44)$$

where

E = carbon emission rate (GtC/yr)
E_0 = initial emission rate (GtC/yr)
r = overall exponential rate of growth (yr^{-1})

---

[1]The IPCC emission scenarios are described in the next section.

TABLE 8.11   1990 to 2020 average annual growth rates (%/yr) used in the IS92a scenario for energy-related carbon emissions

Region	Population	GDP Capita	Energy GDP	Carbon Energy
China and centrally planned Asia	1.03	3.91	−1.73	−0.32
Eastern Europe and former USSR	0.43	1.49	−0.66	−0.24
Africa	2.63	1.25	0.26	−0.21
United States	0.57	2.33	−1.81	−0.26
World	1.40	1.53	−0.97	−0.24

EXAMPLE 8.16   An Exponential Growth Scenario

(a) The carbon emission rate from fossil fuels in 1990 was 6.0 GtC/yr. Estimate the fossil-fuel carbon emission rate in 2020 using world carbon growth data given in Table 8.11:

$$\text{Population} = 1.40\%/\text{yr} \qquad \frac{\text{Energy}}{\text{GDP}} = -0.97\%/\text{yr}$$

$$\frac{\text{GDP}}{\text{Person}} = 1.53\%/\text{yr} \qquad \frac{\text{Carbon}}{\text{Energy}} = -0.24\%/\text{yr}$$

(b) Suppose carbon emissions associated with land-use changes contribute an average of 1.2 GtC/yr over this period of time. Assuming an airborne fraction of 46 percent, estimate the CO_2 concentration in 2020. Assume the concentration in 1990 was 354 ppm.

Solution   We will use the disaggregation given in (8.42):

$$E_{\text{carbon}} = \text{Population} \times \frac{\text{GDP}}{\text{Person}} \times \frac{\text{Energy}}{\text{GDP}} \times \frac{\text{Carbon}}{\text{Energy}}$$

and the emission equation given by (8.44):

$$E_{\text{carbon}} = E_0 e^{rt}$$

The overall growth rate of carbon emissions is just the sum of the individual growth rates:

$$r = 1.40 + 1.53 - 0.97 - 0.24 = 1.72\%/\text{yr} = 0.0172 \text{ yr}^{-1}$$

(a) The emission rate in 30 years would be

$$E = 6.0 \text{ GtC/yr } e^{0.0172 \times 30} = 10.1 \text{ GtC/yr}$$

(b) As described in Chapter 3, for an emission rate that is growing exponentially, total emissions over a period of time T can be found from the integral of (8.44):

$$E_{\text{tot}} = \int_0^T E_0 e^{rt} dt = \frac{E_0}{r} (e^{rt} - 1) \qquad (8.45)$$

so the total fossil-fuel carbon emissions over these 30 years will be

$$E_{tot} = \frac{6.0 \text{ GtC/yr}}{0.0172/\text{yr}} \left(e^{0.0172 \times 30} - 1\right) = 236 \text{ GtC}$$

Adding that to the deforestation emissions gives a total of

$$236 \text{ GtC(energy)} + 1.2 \text{ Gt/yr} \times 30 \text{ yr (deforestation)} = 272 \text{ GtC}$$

Using the airborne fraction given and the handy link between tonnes of carbon and ppm of CO_2 (8.40) gives the total change in CO_2:

$$\Delta(CO_2) = \frac{272 \text{ GtC}}{2.12 \text{ GtC/ppm}} \times 0.46 = 59 \text{ ppm}$$

The concentration of CO_2 in 2020 would be

$$(CO_2) = 354 \text{ ppm} + 59 \text{ ppm} = 413 \text{ ppm}$$

■

8.12 IPCC EMISSION SCENARIOS

The IPCC has created a set of greenhouse-gas and aerosol emission scenarios that have become standard inputs used by climate modelers. These are scenarios, not forecasts or predictions, that have been designed to cover a wide range of possible emission futures. In fact, no attempt has been made to assess the likelihood of any particular scenario.

A set of six scenarios were developed and reported on in the 1992 report of IPCC (IPCC, 1992). They are referred to as IS92a through IS92f (the "IS" stands for "IPCC Scenarios" and the 92 refers to the year). Figure 8.37 shows the carbon emission portions of these scenarios and the resulting CO_2 concentrations. The full scenarios include the other greenhouse gases (CH_4, N_2O, and halocarbons) plus sulfur oxides, and the NO_x, VOC, and CO precursors of tropospheric ozone. For the full scenarios, worldwide commitments to reduce use of stratospheric ozone-depleting substances are already included.

IS92a is somewhat of an intermediate scenario, using, for example, medium projections of population and economic growth. In fact, recall that Example 8.16 was based on the IS92a scenario. The IS92c scenario assumes the lowest rates of growth and includes severe constraints on fossil fuel supplies, and its emissions actually decrease over time. At the other extreme is IS92e, which assumes intermediate population growth, high economic growth, and abundant fossil fuels.

The IS92 scenarios depicted in Figure 8.37 show total worldwide emissions of just one gas, CO_2. The complete scenarios include emission estimates for all of the greenhouse gases on a regional and worldwide basis. Figure 8.38 presents regional emission rates for the two most important greenhouse gases, CO_2 and CH_4, for 1990 and for 2100, using the middle-ground IS92a scenario. Notice especially how the distribution of carbon emissions shifts over time. In 1990, the 2.8 GtC emitted by OECD countries

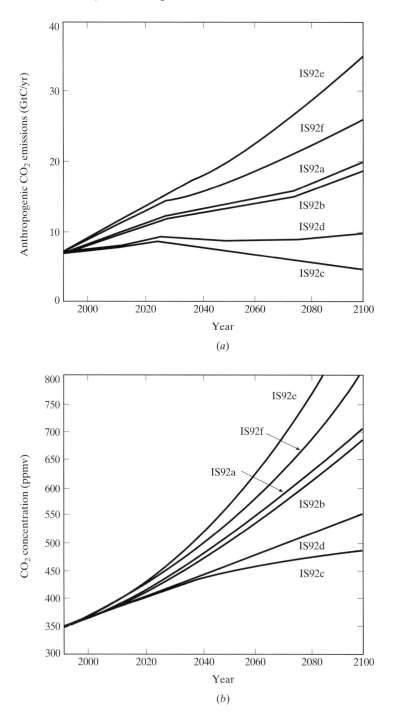

FIGURE 8.37   The six IPCC 1992 scenarios for carbon. (*a*) Annual CO_2 emissions from energy, cement production, and tropical deforestation. (*b*) Resulting CO_2 concentrations. (*Source:* IPCC, 1996)

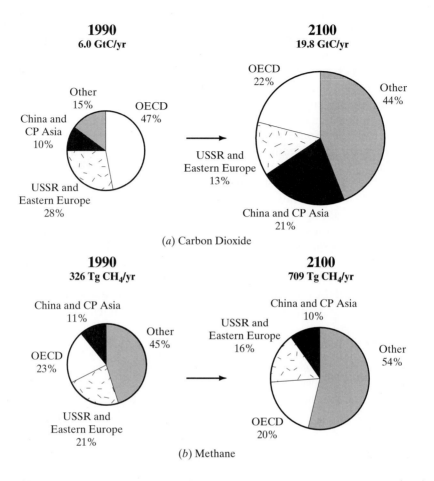

FIGURE 8.38 IS92a emission rates for CO_2 and CH_4 in 1990 and 2100. These rates from IPCC (1992) differ slightly from more recent IPCC reports.

accounted for almost half of the total carbon emissions. By 2100, those emissions increase to 4.3 GtC per year but by then the rest of the world has increased its emission rate so much that the OECD countries would only account for 22 percent of the total. In this scenario, the carbon emission rate for China and other centrally planned economies of Asia, for example, grows from 0.6 GtC/yr in 1990 to 4.2 GtC/yr in 2100. That shift, which is driven primarily by growing energy demands in the developing countries, is much more dramatic than the corresponding percentage shifts in CH_4 emissions shown in Figure 8.38b.

Global Mean Temperature Projections

The IS92 scenarios provide six different estimates for greenhouse gas emissions over the next 100 years or so. Those six emission-rate scenarios are used as inputs for models that estimate future mean global temperature. Those temperature models usually have as a parameter the climate sensitivity, ΔT_{2x}, which is the change in mean global temperature resulting from a doubling of equivalent CO_2 concentration. General circulation models (GCMs) tend to produce estimates of ΔT_{2x} in the range of 1.5 °C to 4.5 °C. It is common practice now for mean global temperature models to use three possible values of that climate sensitivity: 1.5 °C, 2.5 °C, and 4.5 °C. Multiplying 6 scenarios by 3 climate sensitivities yields 18 possible estimates of future global temperature. An additional confounding factor is the very uncertain effect of future aerosol concentrations. Some model runs are made assuming no change in aerosols—that is, the negative forcing caused by aerosols is keep constant at the 1990 levels. Other runs are then made using estimates of increasing aerosols in the future. Six IS92 emission scenarios, three choices of climate sensitivity, and two ways of accounting for aerosols lead to 36 different mean global temperature futures.

The resulting range of mean global temperature change associated with the IS92 emission scenarios, climate sensitivities, and aerosol levels is shown in Figure 8.39. By the year 2100, the range of estimates for realized, global mean temperature change is between 0.8 °C and 4.5 °C. The middle estimate, corresponding to scenario IS92a with constant aerosol emissions and a climate sensitivity of 2.5 °C, suggests that the temperature of earth in 2100 could be about 2.4 °C warmer than it is now. That would make the planet warmer than it has been for hundreds of thousands of years.

Stabilizing Greenhouse Gas Concentrations

We have already seen in Example 8.15 that stabilizing the rate at which carbon is emitted into the atmosphere does not stop the growth of CO_2 in the atmosphere. In fact, with a constant carbon emission rate equal to the level of the 1990s, the CO_2 concen-

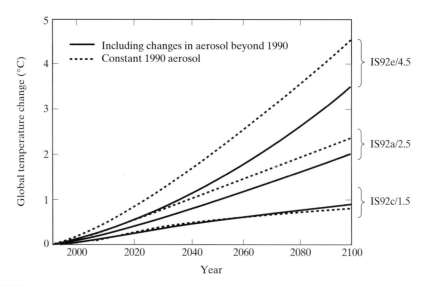

FIGURE 8.39 Range of possible changes in global mean temperature. The upper curve is for IS92e assuming constant aerosol concentrations beyond 1990 and a high climate sensitivity ($\Delta T_{2x} = 4.5$ °C); the lowest curve is for IS92c, also assuming constant aerosol, but with a low climate sensitivity ($\Delta T_{2x} = 1.5$ °C). The middle scenario for IS92a leads to a realized temperature change of 2.0 to 2.4 °C. (*Source:* IPCC, 1996)

tration climbs to around 500 ppm by 2100 (nearly double the preindustrial concentration of 280 ppm). Clearly, the emission rate of carbon must decrease in the future if we ever expect to stabilize the concentration of CO_2.

There are no unique emission scenarios that will lead to CO_2 stabilization since it is primarily the total amount of carbon emitted over a period of time rather than the emission profile that establishes the desired stabilization level. Figure 8.40 shows

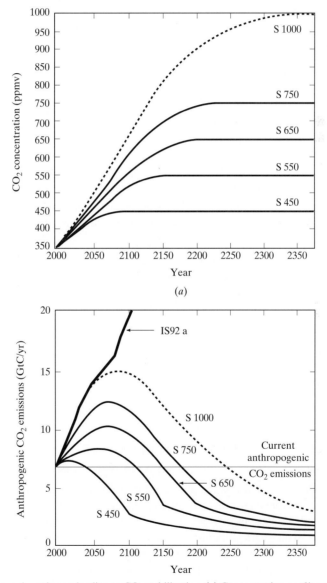

FIGURE 8.40 Example pathways leading to CO_2 stabilization. (*a*) Concentration profiles reaching CO_2 stabilization at 450, 550, 650, and 1000 ppm. (*b*) Carbon emission profiles that lead to those CO_2 stabilization plateaus. (Based on IPCC, 1996)

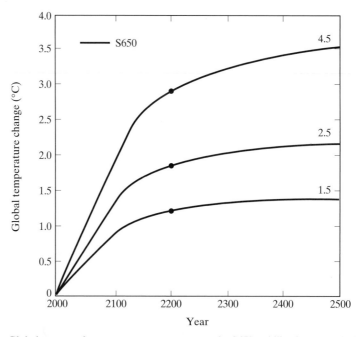

FIGURE 8.41   Global mean surface temperature response to the S650 stabilization scenario for climate sensitivities of 1.5, 2.5, and 4.5 °C. The dots indicate the year in which CO_2 stabilizes at 650 ppm. (Based on IPCC, 1996)

example emission profiles and resulting CO_2 levels. All of the emission scenarios ultimately must have rates far below those of today or else CO_2 will continue to increase.

Even after CO_2 stabilizes at some constant level, the inherent inertia of the climate system means global temperature continues to increase for many years. Figure 8.41 shows temperature profiles for the S650 scenario in which CO_2 stabilizes at 650 ppm in the year 2200. For example, for a midlevel climate sensitivity of 2.5 °C the realized temperature in 2100 is 1.8 °C warmer than it was in 1990. After that time, CO_2 remains constant at 650 ppm yet temperatures continue to coast upward for several hundred years, finally reaching about 2.2 °C by the year 2500.

8.13 REGIONAL IMPACTS OF TEMPERATURE CHANGE

The relatively simple models introduced in this chapter are useful for estimating average global surface temperature, but the regional implications of a global warming are of more importance and are much harder to predict. A starting point in assessing regional changes is provided by three-dimensional general circulation models (GCMs), which can indicate how parameters such as temperature, cloud cover, soil moisture, and ice cover will vary by geographical location and time of year. The most common atmospheric 3-D models are based on a cartesian grid in which the atmosphere is divided into cells such as are shown in Figure 8.42. Each cell, which may be on the order of 250 km on a side and 1 km thick, is characterized by a number of variables such as temperature, humidity, pressure, cloudiness, and so forth. Those cells then com-

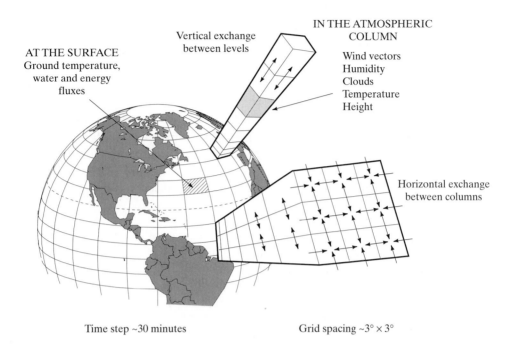

FIGURE 8.42   A cartesian (or rectangular) grid GCM in which horizontal and vertical exchanges are handled between adjacent columns and layers. (*Source:* Henderson-Sellers and McGuffie, 1987)

municate (mathematically) with adjacent atmospheric cells and are updated with some specified time increment. Advanced GCMs couple the atmosphere with oceans, sea ice, and land surfaces.

In spite of their complexity, coupled atmosphere-ocean GCMs are unable to predict the weather in the coming weeks, months, and years (although they can do an adequate job of weather forecasting for a few days at a time), but it is hoped that they can predict the most important statistical properties of future climate. Given their grid parameterization, they are unable to capture climate variations that might occur within each individual cell. There are still enormous uncertainties associated with key factors such as the proper way to represent clouds and their radiative properties, the coupling of ocean and atmospheric circulation models, and biological processes on the surface of the land. Confidence in these models, in spite of these weaknesses, is continually growing as they become increasingly refined.

Coupled atmosphere-ocean models that include the impacts of aerosols as well as greenhouse gases show the following regional climate changes in the future (IPCC, 1996):

- Generally greater surface warming of the land than of the oceans in winter
- A minimum warming around Antarctica and in the northern North Atlantic due to changes in deep oceanic mixing in those areas

- Maximum warming in high northern latitudes in late autumn and winter associated with reduced sea ice and snow cover
- Little warming over the Arctic in summer
- A reduction in diurnal temperature range over land in most seasons and most regions
- An enhanced global hydrological cycle
- Increased precipitation in high latitudes in winter
- Asian summer monsoon rainfall decreases
- Increased soil moisture in high northern latitudes in winter
- An increase in extremely high temperature events and a decrease in extremely low temperatures
- More severe droughts and/or floods in some places but less severe in others.

8.14 THE OCEANS AND CLIMATE CHANGE

The oceans cover 70 percent of the earth's surface and they contain over 97 percent of the earth's water. They have high heat capacity, so they warm slowly, and their circulations distribute heat around the globe, greatly affecting local and regional temperatures on the land. They store 50 times as much carbon as the atmosphere and they exchange carbon dioxide back and forth across the atmosphere-ocean interface at a rate that dwarfs the contribution by humankind. In other words, the oceans significantly affect and are affected by climate change. Greenhouse enhancement can affect the ocean in many ways, but two of the most important changes have to do with rising sea level and changes in the ocean's circulation. In addition, biological processes that result in the removal of carbon from the atmosphere can be affected by climate change and in turn those changes can affect climate.

Rising Sea Level

Water expands as it warms, decreasing its density and increasing its volume, which causes sea level to rise. In addition, melting glaciers and ice caps contribute to increasing ocean volume and rising sea levels. Finally, mining groundwater increases runoff to the oceans, while filling surface water reservoirs tends to transfer ocean water back to the land, lowering sea level. Those changes in surface water and groundwater storage tend to have opposite effects that may be to some extent canceling each other out. Thermal expansion and melting ice, on the other hand, are both significant in size and comparable in magnitude, and both cause rising sea level. Over the past 100 years, the oceans are believed to have risen by about 10 to 25 cm, with about 2 to 7 cm of that thought to have been caused by thermal expansion and another 2 to 5 cm attributed to the observed retreat in glaciers and ice caps. The cause of the remaining rise is as yet unexplained (IPCC, 1996).

Almost all of the world's nonoceanic water is stored in ice caps and glaciers, and almost all of that is contained in just two great ice sheets—the Antarctic ice sheet and the Greenland ice sheet. Moreover, most of the volume of these two ice sheets lies on land above sea level so any net melting contributes to rising sea level. Floating ice that melts would not affect sea level. A portion of Antarctic ice rests on the ocean floor, and there has been some concern that if that West Antarctic Ice Sheet were to break loose and slide into the ocean, it could cause a rapid rise in sea level of perhaps 5 or 6 m. While the likelihood of that occurring in the next 100 years is considered extremely low, it is one of the potential nonlinear "surprises" that we do not know enough about to dismiss. Table 8.12 shows the volume of ice in these ice sheets plus the remaining ice stored in glaciers and ice caps. As shown there, the Antarctic ice sheet stores enough water to raise the level of the oceans by 73 m.

Projections of sea-level rise have been made using the IS92a–f emission scenarios coupled to climate sensitivities of 1.5 °C, 2.5 °C, and 4.5 °C and are shown in Figure 8.43. The midlevel projection (IS92a with $\Delta T_{2x} = 2.5$ °C) suggests a sea-level rise of 49 cm by the year 2100, with a range of 13 to 94 cm for other scenarios. Of that 49 cm, 28 cm would be caused by thermal expansion, and 22 cm would be the result of glacier and ice cap melting, including melt from the Greenland ice sheet. The Antarctic ice sheet, on the other hand, is projected to grow slightly due to increased precipitation coupled with extremely cold temperatures, and it is projected to cause a decrease in sea level of about 1 cm.

TABLE 8.12   Some characteristics of the world's ice

	Area (10^6 km^2)	Volume (10^6 km^3)	Sea Level Equivalent (m)
Antarctic ice sheet	12.1	29	73
Greenland ice sheet	1.71	2.95	7.4
Glaciers and ice caps	0.64	0.1 ± 0.02	0.3 ± 0.05

Source: IPCC, 1996.

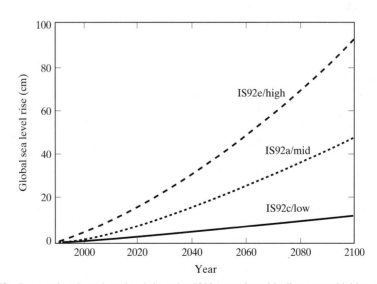

FIGURE 8.43   Range of projected sea-level rise using IS92 scenarios with climate sensitivities of 1.5 °C (low), 2.5 °C (mid), and 4.5 °C (high). (*Source:* IPCC, 1996)

The sea-level rise shown in Figure 8.43 would continue after 2100 as the oceans slowly adjust to the increase in radiative forcing. For example, the IPCC scenario S650, in which CO_2 stabilizes at 650 ppm in the year 2200, projects an eventual rise in sea level of over 1.5 m for the midrange climate sensitivity of 2.5 °C and an increase of over 3 m when ΔT_{2x} is 4.5 °C (IPCC, 1996). The implications of such a rise in sea level are very uncertain. Damage due to storm surges with higher sea level could be devastating to coastal areas that have not been fortified, which means in richer countries the impacts could be much smaller than they would be in countries that do not have the resources to adapt to rising water levels. Tropical cyclones and hurricanes could be more powerful when driven by elevated sea-surface temperatures, which compounds the potential for damaging storms. Other potential impacts include increased shoreline erosion, exacerbated coastal flooding, inundation of coastal wetlands, increased salinity of estuaries and aquifers, and the drowning of coral reefs.

Ocean Circulation and the Biological Pump

The oceans play a critical role in the carbon cycle presented in Figure 8.33. Atmospheric carbon dioxide is absorbed by the oceans and forms inorganic dissolved bicarbonate and carbonate ions. A small fraction of that carbon is taken up by phytoplankton during photosynthesis and becomes part of the food chain in the upper layer of the oceans (the euphotic zone). Photosynthesis, respiration, and decomposition taking place in the surface layers of the ocean remove and replace some 40 to 50 GtC per year. There is a net removal, however, of about 10 GtC per year (nearly twice the carbon emission rate from the combustion of fossil fuels) from the surface waters as dissolved organic carbon, particulate carbon, and $CaCO_3$ in the "hard" parts of marine algae and animals sink into the intermediate and deep ocean. That removal process is known as the *biological pump*. Some studies have noted that phytoplankton productivity is often limited by the availability of dissolved iron, which has led some to propose improving the removal of CO_2 from the atmosphere by artificially fertilizing the oceans with iron. Further studies suggest that a side effect of iron fertilization might be increased emissions of the greenhouse gas N_2O, which would offset any CO_2 gains.

Related to the biological pump is another process that transports carbon-rich surface waters into the deep oceans. As shown in Figure 8.44, there is an enormous flow of ocean water around the globe, equal to the combined flow of all the world's rivers, that transports heat and nutrients from one place to another. Relatively warm seawater near the surface flows into the North Atlantic, where winds encourage evaporation and help sweep the surface waters aside, allowing the warm subsurface water to rise. That rising warm water gives off an enormous amount of heat as it reaches the surface and cools. It is this source of heat (and not the Gulf Stream) that accounts for Western Europe's relatively mild winters. As that water evaporates and cools, the surface becomes more salty and dense, eventually becoming so dense that it sinks down to the ocean bottom. That North Atlantic deep water then moves southward, around the southern tip of Africa. Some emerges off the coast of India, but most upwells in the Pacific Ocean, where it starts its path back to the North Atlantic.

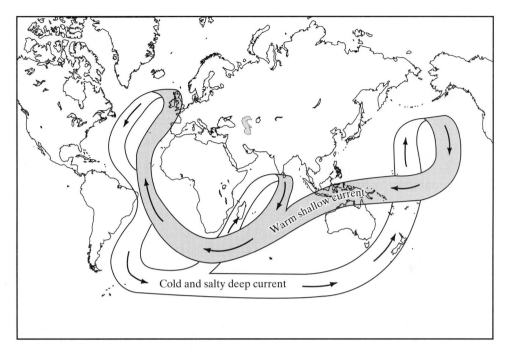

Cold and salty deep current

FIGURE 8.44 The great oceanic conveyor belt. Cold, salty water sinking into the North Atlantic abyss and upwelling near India and in the Pacific creates a climate moderating global oceanic circulation pattern.

This *thermohaline* circulation pattern initiated in the North Atlantic is known as the *conveyor belt circulation,* or the *Atlantic conveyor.* What makes it especially important in studies of climate change is the fact that it appears to have several stable states. Either it runs at a fast rate, an extremely slow rate, or at its present rate, which is somewhere in the middle. The transition between those stable states can occur in a relatively few years. Shifting from one stable state to another is thought to be responsible for a rather dramatic climate shift that occurred in northern Europe just after the last glacial period ended. Some 11,000 years ago, while glaciers were retreating and temperatures were back to interglacial levels, northern Europe and northeastern North America suddenly plunged back into glacial conditions. The temperature in Greenland dropped by 6 °C in less than 100 years, stayed that way for 1000 years, and then jumped back to warm conditions in about 20 years (Broeker and Denton, 1990). The explanation for this event, known as the *Younger Dryas* (named for an Arctic flower that grew in Europe during this period), is based on the sudden stopping, then restarting 1000 years later, of the Atlantic conveyor. Apparently glacial meltwater draining from the North American ice sheet into the North Atlantic diluted the salty surface water enough to halt the density-driven dive into the abyss, which stopped the entire conveyor. Europe lost the warming influence of the current and glacial conditions suddenly returned.

Most coupled ocean-atmosphere models show a future decrease in the strength of the Atlantic conveyor, which reduces the projected warming around the North

Atlantic. The possibility of a complete collapse of the thermohaline circulation if precipitation and runoff patterns change sufficiently is another of the potential "surprises" that worry some climatologists.

8.15 CHANGES IN STRATOSPHERIC OZONE

The changes occurring in the stratosphere's protective layer of ozone are closely linked to the greenhouse problem just discussed. Many of the same gases are involved, including CFCs, halons, methane, and nitrous oxide, as well as aerosols and ozone itself. Some of the gases that enhance the greenhouse effect, such as methane, actually reduce stratospheric ozone depletion, so the two problems really must be considered together.

Ozone has the unusual characteristic of being beneficial when it is in the stratosphere protecting us from exposure to ultraviolet radiation, while ozone formed near the ground in photochemical smog reactions is, as we have seen in Chapter 7, harmful to humans and other living things. About 90 percent of the atmosphere's ozone is contained in the stratosphere between roughly 10 and 50 km in what is commonly referred to as the *ozone layer.* The remaining 10 percent or so is contained in the troposphere, near the ground, above cities and industrialized areas. Figure 8.45 shows this distribution of ozone in the atmosphere along with some comments on its beneficial and harmful attributes.

If all of the atmospheric ozone overhead at any given spot on the earth were to be brought down to ground level so that it would be subjected to 1 atm of pressure, it would form a layer only a few millimeters thick. In fact, one of the common methods of expressing the total amount of ozone overhead is in terms of *Dobson units* (DU),

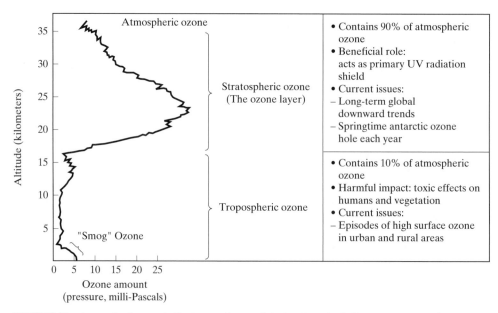

FIGURE 8.45 Amounts of ozone in the troposphere and stratosphere, including comments on the beneficial and harmful roles it plays in each region. (*Source:* UNEP, 1994)

where 1 DU is equivalent to a layer of ozone 0.01 mm thick at 1 atm of pressure at 0 °C. At midlatitudes, the ozone overhead is typically about 350 DU (3.5 mm at 1 atm), while near the equator it is closer to 250 DU. During the annual appearance of the ozone hole over Antarctica, the ozone column drops well below 100 DU.

The Ultraviolet Portion of the Solar Spectrum

Ozone (O_3) is continuously being created in the stratosphere by photochemical reactions powered by short-wavelength ultraviolet (UV) radiation, while at the same time it is continuously being removed by other photochemical reactions that convert it back to diatomic oxygen molecules. The rates of creation and removal at any given time and location dictate the concentrations of ozone present.

The energy required to drive some of these photochemical reactions comes from sunlight penetrating the atmosphere. As you may recall, it is convenient to consider such electromagnetic radiation in some contexts as a wave phenomenon, and in others as a packet of energy called a photon. The energy content of a single photon is the proportional to its frequency, and inversely proportional to its wavelength:

$$E = h\nu = \frac{hc}{\lambda} \tag{8.46}$$

where

E = the energy of one photon (J)
h = Planck's constant (6.626218×10^{-34} J/s
c = the speed of light (2.997925×10^8 m/s)
ν = frequency (s^{-1})
λ = wavelength (m)

Thus, photons with shorter wavelengths have higher energy content.

The entire solar spectrum was introduced in Figure 8.10, where it was used to help derive the effective temperature of the earth. In the context of stratospheric ozone depletion, however, it is only the UV portion of the spectrum that is of interest. Photons associated with those shorter wavelengths can have sufficient energy to break apart molecules, initiating *photochemical reactions*. Molecules that are dissociated by UV light are said to have undergone *photolysis*.

The energy required to drive photochemical reactions is often expressed as an amount of energy per mole of gas under standard conditions. If we assume that the energy required under stratospheric conditions is not much different, we can easily derive the maximum wavelength that a photon must have to cause photolysis. A single photon (with sufficient energy) can cause photolysis in only one molecule. Any extra energy in the photon is dissipated as heat. On the other hand, photons with less energy than needed to cause photolysis also dissipate their energy as heat. Since there are Avogadro's number (6.02×10^{23}) of molecules in one mole of gas, we can convert energy needed per mole into the energy that one photon must have, as follows:

$$E(\text{J/photon}) = \frac{E(\text{J/mol})}{1 \text{ photon/molecule} \times 6.02 \times 10^{23} \text{ molecule/mol}} \tag{8.47}$$

Now we can use (8.46) to solve for the wavelength.

EXAMPLE 8.17   Photon Energy for Photolysis

What is the maximum wavelength that a photon must have to photodissociate diatomic oxygen if 495 kJ/mol are required?

$$O_2 \;+\; h\nu \;\longrightarrow\; O \;+\; O$$

Solution   The energy that a photon must have is given by (8.47):

$$E(\text{J/photon}) = \frac{495 \text{ kJ/mol} \times 10^3 \text{J/kJ}}{1 \text{ photon/molecule} \times 6.02 \times 10^{23} \text{ molecule/mol}} = 8.22 \times 10^{-19}\,\text{J/photon}$$

Rearranging (8.46) and substituting the energy that the photon must have gives us

$$\lambda_{max} = \frac{hc}{E} = \frac{6.626 \times 10^{-34}\,(\text{J s}) \cdot 2.998 \times 10^8 (\text{m/s})}{8.22 \times 10^{-19}\,\text{J}} = 241.6 \times 10^{-9}\,\text{m} = 241.6 \text{ nm}$$

Thus, photons with wavelengths longer than 241.6 nm will not have sufficient energy to break apart the oxygen molecule.   ■

The wavelengths that are most importance in the context of stratospheric ozone depletion are in the ultraviolet portion of the spectrum. As shown in Figure 8.46, UV radiation has wavelengths between about 50 nm and 400 nm ($0.05-0.40\ \mu$m). It is called ultraviolet since wavelengths just above 400 nm are violet in color and visible to the human eye. For analogous reasons, wavelengths above 750 are called infrared, since 750 nm is the edge of the red portion of the visible spectrum. The most important portion of the UV spectrum is divided into three regions, designated as UV-A, UV-B, and UV-C. The UV-A portion extends from 320 to 400 nm; the UV-B portion is from 280 to 320 nm; and the UV-C portion of the spectrum extends from 200 to 280 nm. Wavelengths shorter than 200 nm are sometimes referred to as the *far UV*. As we will see, the UV-B wavelengths are the ones that most affect biological organisms; they are, for example, usually responsible for human sunburn and skin cancer.

The Ozone Layer as a Protective Shield

Ozone formation in the stratosphere can be described by the following pair of reactions. In the first, atomic oxygen (O) is formed by the photolytic decomposition of diatomic oxygen (O_2).

$$O_2 \;+\; h\nu \;\longrightarrow\; O \;+\; O \qquad\qquad (8.48)$$

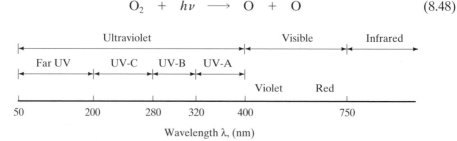

FIGURE 8.46   The ultraviolet (UV), visible, and infrared (IR) portions of the solar spectrum, including the biologically important UV-A, UV-B, and UV-C (not drawn to scale).

where $h\nu$ represents a photon. As shown previously, for this dissociation to take place the photon must have a wavelength of no more than 242 nm. Photons with less energy (longer wavelengths) cannot cause the reaction to take place; all they can do is heat up the molecule that absorbs them. Photons with higher energy than the minimum required use some of their energy to cause photolysis and what is left over is dissipated as heat. Diatomic oxygen (O_2) has its maximum absorption at about 140 nm, and it is a very effective absorber of UV-C radiation between about 130 and 180 nm.

The atomic oxygen formed by (8.48) reacts rapidly with diatomic oxygen to form ozone:

$$O \;+\; O_2 \;+\; M \;\longrightarrow\; O_3 \;+\; M \tag{8.49}$$

where M represents a third body (usually a nearby N_2 molecule) needed to carry away the heat generated in the reaction.

Opposing the preceding ozone formation process is ozone removal by photodissociation:

$$O_3 \;+\; h\nu \;\longrightarrow\; O_2 \;+\; O \tag{8.50}$$

The absorptance of ozone extends from about 200 to 320 nm and reaches its peak at 255 nm. The reaction shown in (8.50) is very effective in removing UV-C radiation and some of the UV-B before it reaches the earth's surface.

The preceding combination of reactions (8.48 to 8.50) forms a chain in which oxygen atoms are constantly being shuttled back and forth between the various molecular forms. A principal effect is the absorption of most of the short-wavelength, potentially damaging, UV radiation as it tries to pass through the stratosphere. In addition, that absorption also heats the stratosphere, causing the temperature inversion shown in Figure 8.1. That temperature inversion, which is what defines the stratosphere, produces stable atmospheric conditions that lead to long residence times for stratospheric pollutants.

The effectiveness of these reactions in removing short-wavelength UV is demonstrated in Figure 8.47, in which the extraterrestrial solar flux and the flux actually reaching the earth's surface are shown. The radiation reaching the surface has been drawn for a clear day with the sun assumed to be 60° from the zenith (overhead), corresponding roughly to a typical midlatitude site in the afternoon. As can be seen, the radiation reaching the earth's surface is rapidly reduced for wavelengths less than about 320 nm. In fact, an intact ozone layer shields us from almost all of the UV having wavelengths shorter than about 290 nm.

Catalytic Destruction of Stratospheric Ozone

As is well known by now, the natural formation and destruction of ozone in the stratosphere is being affected by gases that we release into the atmosphere. In particular, ozone destruction is enhanced by catalytic reactions that can be represented as follows:

$$X \;+\; O_3 \;\longrightarrow\; XO \;+\; O_2$$

and

$$\underline{XO \;+\; O \;\longrightarrow\; X \;+\; O_2}$$

for a net

$$O \;+\; O_3 \;\longrightarrow\; 2\,O_2 \tag{8.51}$$

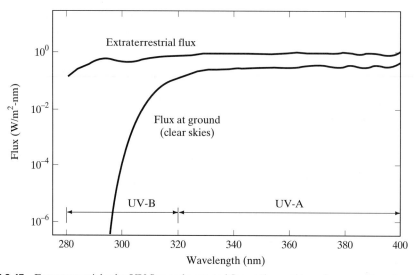

FIGURE 8.47   Extraterrestrial solar UV flux and expected flux at the earth's surface on a clear day with the sun 60° from the zenith (*Source:* Frederick, 1986)

where X is a free radical such as Cl, Br, H, OH, or NO (recall that free radicals are highly reactive atoms or molecules that have an odd number of electrons, which means one electron is not paired with another atom). Notice in (8.51) how the X radical that enters the first reaction is released in the second, freeing it to go on and participate in another catalytic cycle. The net result of the preceding pair of reactions is the destruction of one ozone molecule. The original catalyst that started the reactions, however, may go on to destroy thousands more ozone molecules before it eventually leaves the stratosphere.

Chlorofluorocarbons (CFCs) are a major source of ozone destroying chlorine. Recall that CFCs have long atmospheric lifetimes since they are inert and non-water soluble, so they are not removed in the troposphere by chemical reactions or rainfall. Eventually CFCs drift into the stratosphere, where they are exposed to UV radiation, which breaks apart the molecules, freeing the chlorine. CFCs typically require photons in the UV-C range with wavelengths shorter than 220 nm for photolysis. For example, CFC–12, which has an atmospheric lifetime of 102 years (Table 8.6), undergoes photolysis as follows:

$$CF_2Cl_2 \ + \ UV\text{-}C \ \longrightarrow \ CF_2Cl \ + \ Cl \tag{8.52}$$

Eventually the second chlorine atom is also released. The freed chlorine then acts as a catalyst, as described by (8.51).

$$\begin{aligned}
Cl \ + \ O_3 &\longrightarrow ClO \ + \ O_2 \\
ClO \ + \ O &\longrightarrow Cl \ + \ O_2 \\
\hline
\text{Net:} \qquad O \ + \ O_3 &\longrightarrow 2\,O_2
\end{aligned} \tag{8.53}$$

The chlorine atom makes this loop tens of thousands of times, but eventually it reacts with methane to become water-soluble HCl that can diffuse into the troposphere, where it can be washed out of the atmosphere by rainfall.

$$Cl \ + \ CH_4 \ \longrightarrow \ HCl \ + \ CH_3 \qquad\qquad (8.54)$$

Chlorine that has been incorporated into HCl, as in (8.54), is *inactive*—that is, it does not participate in the catalytic destruction of ozone represented by (8.53). Notice that methane, which is a potent greenhouse gas, helps remove the ozone-destroying, active form of chlorine from the stratosphere. In the context of climate change, methane is part of the problem; in the context of ozone depletion, it is part of the cure.

Another way that chlorine is removed from the catalytic chain reaction is by its incorporation into chlorine nitrate, $ClONO_2$:

$$ClO \ + \ NO_2 \ \longrightarrow \ ClONO_2 \qquad\qquad (8.55)$$

At any given time, as much as 99 percent of the chlorine in the stratosphere is tied up in these inactive molecules, HCl and $ClONO_2$. Unfortunately, those inactive molecules are subject to chemical reactions and photolysis that can restore the chlorine to its active forms, Cl and ClO, especially when there are nearby particles available that provide platforms on which the reactions can take place. Sulfate aerosols, especially from volcanic eruptions, and clouds of nitric acid ice over the Antarctic provide such surfaces. When gases and particles are both involved, the reactions are referred to as *heterogeneous* chemistry, as opposed to *homogeneous* reactions involving gases alone.

Bromine in the stratosphere is another potent ozone-depleting gas, as the following reactions suggest:

$$Br \ + \ O_3 \ \longrightarrow \ BrO \ + \ O_2$$
$$BrO \ + \ ClO \ \longrightarrow \ Br \ + \ ClO_2 \qquad\qquad (8.56)$$

As is the case with chlorine, bromine acts as a catalyst, destroying ozone until it eventually combines with methane to form hydrogen bromide (HBr). Hydrogen bromide deactivates the bromine, but not for long. Photolysis rather quickly decomposes HBr, sending bromine back into its catalytic loop with ozone. In this regard, bromine is much more potent than chlorine since stratospheric chlorine exists mostly in its inactive forms, HCl and $ClONO_2$.

The main source of bromine in the stratosphere is methyl bromide (CH_3Br), about half of which comes from the oceans and about half of which is due to anthropogenic sources. Methyl bromide is used extensively in agriculture to sterilize soil and fumigate some crops after they are harvested. Other anthropogenic sources include biomass burning and automobile exhaust emissions when leaded fuels are used. Bromine has also been used as a flame retardant in fire extinguishers.

The Antarctic Ozone Hole

Chlorine that has been incorporated into hydrogen chloride (HCl) or chlorine nitrate ($ClONO_2$) is inactive, so it does not contribute to the destruction of ozone. In the Antarctic winter, however, a unique atmospheric condition known as the *polar vortex* traps air above the pole and creates conditions that eventually allow the chlorine to be activated. The polar vortex consists of a whirling mass of extremely cold air that forms over the South Pole during the period of total darkness in the Antarctic winter. The vortex effectively isolates the air above the pole from the rest of the atmosphere until the Antarctic spring arrives in September. Stratospheric temperatures in the vortex may drop to below $-90\,°C$, which is cold enough to form polar stratospheric clouds (PSCs) even though the air is very dry. The ice crystals that make up polar clouds play a key role in the Antarctic phenomenon by providing reaction surfaces that allow chemical species to stay together long enough to react with each other.

A number of reactions take place on the surfaces of polar stratospheric cloud particles that result in the formation of chlorine gas (Cl_2). For example,

$$ClONO_2 \ + \ H_2O \ \longrightarrow \ HOCl \ + \ HNO_3 \tag{8.57}$$

$$HOCl \ + \ HCl \ \longrightarrow \ Cl_2 \ + \ H_2O \tag{8.58}$$

$$ClONO_2 \ + \ HCl \ \longrightarrow \ Cl_2 \ + \ HNO_3 \tag{8.59}$$

Once the sun rises in the Antarctic spring in September, the chlorine gas (Cl_2) formed during the darkness of winter photolytically decomposes into atomic chlorine:

$$Cl_2 \ + \ h\nu \ \longrightarrow \ 2\,Cl \tag{8.60}$$

which then destroys ozone, produces active chlorine monoxide (ClO), and reinitiates the catalytic destruction of ozone described in (8.53):

$$Cl \ + \ O_3 \ \longrightarrow \ ClO \ + \ O_2 \tag{8.61}$$

The destruction of ozone as the sun first appears in the Antarctic spring proceeds as described until the nitric acid (HNO_3) formed in (8.57) and (8.59) photolyzes and forms the inactive chlorine nitrate ($ClONO_2$), which stops the ozone destruction process. As the vortex breaks down in the spring, ozone from nearby areas rushes in and replenishes the ozone above Antarctica. Conversely, the transport of ozone-depleted air from polar regions is thought to be contributing to decreases in ozone at middle latitudes.

Figure 8.48 shows how drastic the depletion of ozone is over Antarctica in the middle of spring (October) compared with the concentrations in winter (August). In August of 1993, the total ozone column measured in at 272 ± 26 DU. In mid-October in 1992, the column dropped to 105 DU, while in 1993 it reached a record low of 91 DU. During that record low in 1993, there was total destruction of ozone between 14 and 19 km of altitude and the hole covered an area of 23 million km^2, which is nearly three times the size of the United States. Some of this loss has been attributed to the eruption of Mount Pinatubo in 1991, which greatly increased the amount of sulfate aerosol in the stratosphere. That sulfate aerosol acts as tiny platforms on which the heterogeneous reactions take place. In 1994, 1995, and 1996, the ozone hole was not as deep as it was in 1993, which tends to confirm the transient nature of the impact from Mount Pinatubo in 1991.

Measurements of springtime total ozone over Antarctica have been made since 1957. As can be seen in Figure 8.49, the decline in springtime total ozone over Antarctica accelerated during the 1970s. Concern over the possible destruction of

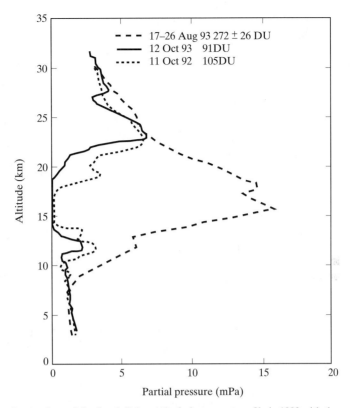

FIGURE 8.48   Comparison of the South Pole predepletion ozone profile in 1993 with the profile observed when total ozone reached a minimum in 1992 and 1993. (*Source:* UNEP, 1994)

stratospheric ozone was first expressed by two scientists, F. Sherwood Rowland and Mario J. Molina from the University of California, Irvine (Molina and Rowland, 1974). However, it was not until 1985, with the dramatic announcement of the discovery of a "hole" in the ozone over Antarctica, that the world began to recognize the seriousness of the problem. In 1995, Rowland, Molina, and Paul Crutzen from the Max-Planck-Institut fur Chemie were awarded the Nobel Prize in Chemistry for their pioneering work in explaining the chemical processes that lead to the destruction of stratospheric ozone.

The very low stratospheric temperatures over Antarctica, which form those polar stratospheric clouds, are not duplicated in the Arctic. The combination of land and ocean areas in the Arctic results in warmer temperatures and much less of a polar vortex, factors that do not encourage the formation of polar clouds. The result is a much less dramatic thinning of the ozone layer over the Arctic in the spring, although it is still significant. There is concern, however, that greenhouse effect enhancement could lead to stratospheric cooling, which would make an Arctic ozone hole more likely. Should an extensive ozone hole develop in the northern hemisphere, far more people would be exposed to elevated UV radiation.

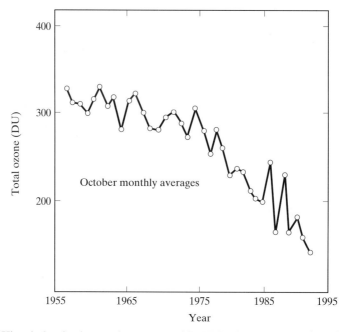

FIGURE 8.49 Historical springtime total ozone record for Halley Bay, Antarctica (76 °S). (*Source:* UNEP, 1994)

Ozone Depletion Potential

The concept of ozone depletion potentials (ODPs) for ozone-depleting substances (ODSs) is analogous to the GWPs for greenhouse gases already described. In the case of ODPs, the reference gas is CFC–11 ($CFCl_3$). The ODP of a gas is defined as the change in total ozone per unit mass emission of the gas, relative to the change in total ozone per unit mass emission of CFC–11. Notice that the definition of ODPs is independent of the length of time required for the depletion to take place, so in that sense they are somewhat different from time-dependent GWPs.

ODPs have become an integral part of the regulatory approach to controlling emissions of ozone-depleting substances. The Montreal Protocol on Substances that Deplete the Ozone Layer and its subsequent Amendments, which governs international production and use of halocarbons, and the U.S. Clean Air Act Amendments of 1990, which controls domestic uses, both dictate the phase-out of ozone-depleting substances based on their ODPs.

Table 8.13 lists ODPs for a number of CFCs, halons, and possible replacements. By definition, CFC–11 has an ODP of 1.0, which is the highest ODP of any of the CFCs. Halons–1211 and –1301 have much higher ODPs than CFC–11 due to the higher reactivity of bromine and the reduced effectiveness of deactivation traps compared with chlorine. Notice that the replacement HCFCs have low ODPs, typically less than 0.1, which is a reflection of their short atmospheric lifetimes.

TABLE 8.13   Steady-state ODPs and 100-year GWPs measured relative to CFC–11

Species	Chemical formula	Ozone depletion potential relative to CFC–11	Global warming potential relative to CFC–11
CFC–11	$CFCl_3$	1.0	1.0
CFC–12	CF_2Cl_2	0.9	2.1
CFC–113	$C_2F_3Cl_3$	0.9	1.3
HCFC–22	CF_2HCl	0.04	0.4
HCFC–123	$C_2F_3HCl_2$	0.014	0.02
HCFC–124	C_2F_4HCl	0.03	0.12
HCFC–141b	$C_2FH_3Cl_2$	0.10	0.16
HFC–134a	$C_2H_2F_4$	$< 5 \times 10^{-4}$	0.33
Halon–1211	CF_2ClBr	5.1	—
Halon–1301	CF_3Br	13	1.4
Methyl chloroform	CH_3CCl_3	0.12	0.03
Methyl bromide	CH_3Br	0.6	—

Source: Based on Wuebbles (1995).

For comparison, 100-year GWPs relative to CFC–11 rather than the usual CO_2 reference molecule have been added to Table 8.13. These GWPs tell us that the HCFC and HFC replacements for CFCs are not only much better in terms of ozone depletion, but they also cause considerably less global warming. The halons, on the other hand, are worse than CFCs in both categories.

Effect of Ozone Depletion on Surface UV Radiation

As Figure 8.47 indicates, the ozone layer greatly reduces the amount of UV-B reaching the earth's surface; indeed, without it, life as we know it could not exist. Thinning of the ozone layer increases exposure to UV-B radiation, which can cause adverse effects on human health, terrestrial and aquatic plants, and durability of materials.

The amount of ozone overhead (total column ozone) has been monitored since the 1920s using ground-based instruments and more recently with satellite measurements. Although there have been difficulties with unreliable instrumentation, those problems have been addressed and the overall record is now considered to be quite good. Figure 8.50 shows the record of total ozone over a 60°S to 60°N latitude range for the period 1979 to 1994. The graph has been adjusted to eliminate the effect of variations in solar intensity. A trend line fitted to the data before Mount Pinatubo in 1991 shows a decrease in ozone of 2.9 percent per decade. The impacts of increased aerosols on ozone concentrations after the eruption of El Chichon in 1982 and Mount Pinatubo in 1991 are clearly visible. Total column ozone in northern hemisphere midlatitudes (30 to 60°), over the period 1979 to 1994, has been dropping at an even faster rate, averaging about 6 percent per decade in winter and spring, and about 3 percent per decade in summer and fall months.

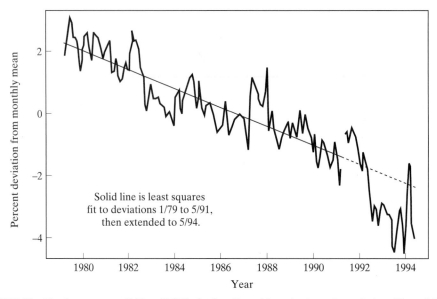

FIGURE 8.50   Total ozone over 60 °S to 60 °N latitude, adjusted for solar intensity variations. The solid line fits the data up to the eruption of Mount Pinatubo in 1991. (*Source:* UNEP, 1994)

Surface measurements of UV-B radiation show increases when the amount of ozone overhead decreases. For example, Figure 8.51 shows wavelength-dependent, clear-sky irradiance in Germany compared with similar measurements made in New Zealand with 25 percent less ozone. The difference in UV-A radiation is negligible, but in the UV-B region the difference in irradiance increases rapidly as wavelengths decrease. It is unfortunately the shorter wavelengths, which are more damaging to living things, that increase most dramatically as the ozone layer thins.

Impacts of Increased Exposure to UV

The biological response to increased ultraviolet radiation is often represented by an *action spectrum*. The action spectrum provides a quantitative way to express the relative ability of various wavelengths to cause biological harm. For example, the action spectra for plant damage, DNA damage, and erythema (sunburn) are shown in Figure 8.52*a*. All of these spectra show increasing damage as wavelength decreases and they all show significant damage in the UV-B (280 to 320 nm) region. There are some differences, however. Plants, for example, show no damage for wavelengths above about 315 nm, but erythema damage occurs well into the UV-A (320 to 400 nm) portion of the solar spectrum. Sunscreens that block UV-B, but not UV-A, can give a false sense of security by allowing people to spend more time exposed to the sun without burning, but skin damage that may lead to skin cancer is occurring nonetheless.

By combining wavelength-dependent action spectra, such as are shown in Figure 8.52*a*, with solar irradiances, such as are shown in Figure 8.51, a biological

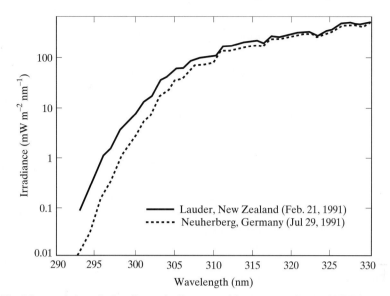

FIGURE 8.51   Measured clear-sky irradiances in Germany with an ozone column of 352 DU compared with New Zealand with 266 DU. As wavelengths decrease, the differences become more dramatic. (*Source:* UNEP, 1994)

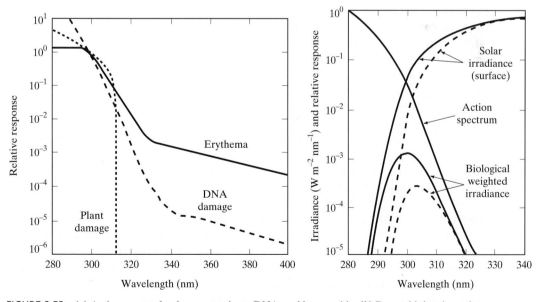

FIGURE 8.52   (*a*) Action spectra for damage to plants, DNA, and human skin. (*b*) By combining the action spectra with solar irradiance, a biological weighted irradiance is obtained. The dotted lines represent irradiance under normal ozone conditions, while the solid lines correspond to a reduction in ozone. (*Source:* Simon, 1993, and Madronich, 1993)

weighted effectiveness of radiation is obtained. Figure 8.52*b* shows a representative action spectrum superimposed on two solar irradiance curves—one showing irradiance with an intact ozone layer, the other with an ozone layer that has been thinned somewhat. As the ozone column is depleted, the irradiance in those critical short UV-B wavelengths, where organisms are most sensitive, rises rapidly, as does the potential biological damage. The biologically weighted irradiance shows that the critical wavelengths in terms of biological damage are between about 290 and 310 nm. Shorter wavelengths are effectively blocked by the remaining ozone layer, while longer wavelengths do little biological damage.

Most of our concern for increasing exposure to UV radiation has focused on the increased probability of developing skin cancer. Numerous studies have shown that skin cancer is readily induced in laboratory animals exposed to broad-spectrum UV radiation, and The International Agency for Research on Cancer (1992) states unequivocally that "there is sufficient evidence in humans for the carcinogenicity of solar radiation. Solar radiation causes cutaneous malignant melanoma and non-melanocytic skin cancer." In fact, exposure to solar radiation is probably the main cause of all forms of skin cancer and an important, if not the main cause, of cancer of the lip (Armstrong, 1993).

Skin cancer is usually designated as being either the more common non-melanoma (basal cell or squamous cell carcinomas) or the much more life-threatening malignant melanoma. Epidemiological studies directly link skin cancer rates with exposure to UV-B. At lower latitudes there is more surface-level UV-B, and skin cancer rates increase proportionately. Using such data, it has been estimated that a 1 percent increase in UV-B is estimated to cause a 0.5 percent increase in the incidence rate of melanoma and a 2.5 percent increase in nonmelanoma (Armstrong, 1993). Both types of skin cancer are more prevalent among people with fair complexions than among individuals who have more of the protective pigment, melanin.

Nonmelanoma is most likely to occur on areas of the skin habitually receiving greatest exposure to the sun, such as the head and neck. While it is rarely fatal, it often causes disfigurement. Melanomas, on the other hand, are as common on intermittently exposed parts of the body, such as the trunk and legs, as they are on the head and neck. Melanomas seem to occur more frequently on individuals who have had severe and repeated sunburns in childhood. This observation has led to a hypothesis that melanoma is less related to the accumulated dose of radiation, as seems to be the case for nonmelanoma, than it is to the pattern of exposure.

The incidence rate of melanoma among fair-skinned individuals in the United States is increasing at 2 to 3 percent annually, which is faster than any other form of cancer. Approximately 40,000 melanomas are detected each year in the United States, and that may be only half of the actual rate of incidence. While it is tempting to relate that increase to the thinning of the ozone layer, it is much more likely to be the result of lifestyle changes rather than increased surface-level UV irradiation. More people are simply spending more time outdoors in the sun than in the past. Moreover, the decrease in stratospheric ozone is a relatively recent phenomenon and evidence suggests that skin cancer may have a long latency period. There may well may be a 20-year lag time between changes in solar irradiance and measurable increases in skin cancer caused by that exposure.

Other human health problems associated with UV exposure include ocular damage (cataracts and retinal degeneration) and immune system suppression. The effects of increasing UV-B radiation would not be felt by humans alone, of course. There is considerable evidence that UV-B reduces photosynthesis and other plant processes in terrestrial and aquatic plants. UV-B radiation is known to penetrate to ecologically significant depths in the oceans, and besides impairing photosynthesis it threatens marine organisms during their critical developmental stages. For example, amphibian eggs have little protection when exposed to UV-B irradiation, and is thought that the worldwide decline in frogs and toads, especially at high altitudes where solar exposure is greatest, may be in part caused by our thinning ozone layer.

The UV Index

Early in 1992 satellite and aircraft measurements found unusually high levels of reactive chlorine in the Arctic, which led to warnings that large ozone reductions at high latitudes in the northern hemisphere might occur in the coming months. Worried about citizen response to the warning, Canada initiated two public information programs in the spring of 1992: the Ozone Watch Program and the UV Index Advisory Program. The Ozone Watch Program compares the previous two-week's total ozone column over its major cities with pre-1980 reference ozone levels for the same period. It then issues weekly values of the percentage difference between the two, which are reported to the public.

Canada's UV Index Advisory Program is somewhat different in that it tries to forecast the intensity of UV radiation for the next day. The index is a nondimensional, single number in the range of 0 to 12 that represents an erythemal-weighted irradiance expected at solar noon (when the sun is at its highest). Using a procedure analogous to that shown in Figure 8.51b for finding biologically weighted irradiances, Canada uses a standardized erythemal action spectrum (called the Diffey curve) multiplied by the expected wavelength-dependent irradiance to produce the index.

In Canada, the UV Index would not be expected to rise above 10, but numbers in the 10 to 12 range are possible at lower latitudes. To make it more appealing and understandable for the public, the Index is linked to descriptive words ranging from "Low" to "Extreme," and it also provides an estimate of the length of time required for a fair-skinned person to get sunburned, as shown in Table 8.14.

TABLE 8.14 Canada's UV Index and Descriptors

Index range	Category descriptor	Sunburn context
0–3.9	Low	More than 1 hour
4.0–6.9	Moderate	About 30 minutes
7.0–8.9	High	About 20 minutes
> 9.0	Extreme	Less than 15 minutes

In 1994, the United States initiated its own UV Index system. While the procedure is modeled after that used in Canada, the measurement techniques differ somewhat but the end result is a single number in the same range as those issued by the Canadian Atmospheric Environment Service.

Political Response to the Ozone Depletion Problem

The political response to warnings of ozone depletion by the scientific community has been unusually effective and timely. Table 8.15 provides a thumbnail sketch of the historical development of the scientific and political milestones that began in 1930 when the Frigidaire division of General Motors first began to produce CFCs for use in refrigerators, up to 1995, when Rowland, Molina, and Crutzen were awarded the Nobel Prize in Chemistry for their pioneering work on explaining the chemistry of ozone depletion by CFCs.

When CFCs were first created, they were hailed as wonder chemicals. They replaced toxic, explosive, and unstable refrigerants and made possible the refrigeration and air conditioning luxuries that we now take for granted. A British scientist, James E. Lovelock, began experimenting with CFCs in the 1970s and concluded in 1973 that essentially all of the CFCs ever produced were still in the atmosphere. The first hint of potential problems came with the publication of Molina and Rowland's paper in 1974, which warned of the danger to the stratospheric ozone layer that CFCs presented. The EPA responded in a remarkably short period of time and banned nonessential spray-can uses of CFCs by 1979. A few countries followed the U.S. lead, including Canada and some of the Scandinavian countries, but most continued to manufacture and use them without con-

---

TABLE 8.15 Thumbnail sketch of milestones in ozone depletion

1930	First production of CFCs for refrigeration by Frigidaire.
1974	Rowland and Molina publish their ozone-depletion hypothesis.
1978	EPA acts to ban the use of CFCs in nonessential, aerosol spray-can propellants by 1979. Canada and several Scandinavian countries follow.
1985	Ozone hole over Antarctica discovered by researchers from the British Antarctic Survey.
1987	Montreal Protocol on Substances that Deplete the Ozone Layer: 23 nations, including the United States, agree to cut use of key CFCs by 50 percent by 1999.
1988	DuPont Corporation acknowledges the role of CFCs in ozone depletion and agrees to cease production.
1990	London Amendment: 93 nations agree to phase out the production of CFCs and most halons by the year 2000. U.S. Clean Air Act Amendments of 1990 ban production of most Class I ozone-depleting substances by 2000 and Class II substances by 2030.
1991	Eruption of Mount Pinatubo in the Philippines resulting in greatly increased sulfuric aerosols in stratosphere. Reduced global warming and increased ozone depletion predicted (and realized).
1992	Copenhagen Amendment: CFC and halon ban moved up to 1996 (with a 10-year grace period for developing countries); phase-out of HCFCs by 2030. Canada creates its UV Index.
1993	Record drop in Antarctic ozone. EPA speeds up phase-out schedule: production of CFCs to cease in 1996, and most HCFCs by 2003.
1995	Rowland, Molina, and Crutzen receive Nobel Prize in Chemistry for their work on stratospheric ozone depletion.
1996	Production and importation of CFCs, halons, carbon tetrachloride, and methylchloroform in the United States ceases.

cern for the consequences. It was not until the discovery of the ozone hole over Antarctica in 1985 that the rest of the world awakened to the lurking danger. At that time, worldwide CFC use as an aerosol propellant was still the largest single source of CFC emissions.

In 1987 an international meeting convened in Montreal that led to the signing of the *Montreal Protocol on Substances that Deplete the Ozone Layer*. That Protocol called for a 50 percent reduction in use of CFCs by 1999. It was soon realized, however, that the called-for reductions were inadequate, and in 1990, 93 nations convened in London and created a new timetable calling for a complete phase-out of CFCs, most halons, and carbon tetrachloride by the year 2000. The plight of the developing countries was recognized and their use of CFCs was extended to 2010. A fund was established to help pay for technology transfer to enable the less developed countries to take advantage of proposed replacement chemicals.

The next year, 1991, Mount Pinatubo erupted and the ozone hole grew to record proportions, which led to the Copenhagen Amendments to the Protocol in 1992. The phase-out of CFCs was moved up to 1996.

Meanwhile, in the United States, Title VI of the Clean Air Act Amendments of 1990 brought the United States into compliance with the London Amendment. Major provisions included the following:

- Two categories of ozone-depleting substances were created: Class I chemicals are those with the highest ozone depletion potentials and include CFCs, halons, and carbon tetrachloride, while Class II substances have lower ODPs and include the HCFCs. Production of Class I chemicals was to cease by the year 2000, while Class II chemicals were to be phased out by 2030. Exceptions for necessary medical, aviation, safety, and national security purposes and for export to developing countries were allowed. These phase-out dates were later accelerated.

- Motor vehicle air conditioners receive special attention since they have been the largest single source of CFCs. CFC recycling equipment for air conditioner servicing is required and only certified repair persons are allowed to do the work. CFCs must be removed from car air conditioners before the cars are crushed for disposal.

- Nonessential products that use ozone-depleting substances, such as noise horns, "silly string," and some commercial photographic equipment, are banned.

- Warning labels are required for products that contain ODSs, such as refrigerators and foam insulation.

- Modifications to the Montreal Protocol that accelerate the phasing out of ODSs must be adhered to.

Two years later, in 1992, the Copenhagen Amendments to the Montreal Protocol called for participating nations to move the ban of many ODSs to 1996. The EPA responded in 1993 with an accelerated phase-out schedule. Production and imports of halons ceased in 1994 and CFCs, methyl chloroform (1,1,1-trichloroethane), and carbon tetrachloride ceased in 1996. Most HCFCs are to be eliminated by 2003, but some will still be allowed until 2030. An interesting side-effect of the banning of CFCs is that in 1996 they were the second most lucrative substance smuggled into the United States, after cocaine.

Has this flurry of regulatory activity been effective? Stratospheric ozone depletion is one of those problems that cannot be reversed in a short period of time.

Most of the CFCs ever produced are still drifting around in the atmosphere, and they will be with us for decades into the future. The production cuts dictated by the Montreal Protocol and subsequent Amendments have already stopped the steady, historical climb in CFC atmospheric concentration (Figure 8.22). The good news is that stratospheric ozone losses are expected to peak near the year 2000 and decline thereafter. Countless lives have been saved. Perhaps just as important is the model of effective international cooperation on global environmental problems that the Protocol provides—a model that needs to be replicated.

PROBLEMS

8.1. Suppose the ratio of $(^{18}O/^{16}O)$ in standard ocean water is 0.00200. If another sample of ocean water has an $(^{18}O/^{16}O)$ ratio of 0.00199, what would be the value of $\delta^{18}O$? Would the sample correspond to a warmer or colder climate?

8.2. A relationship between the mean annual surface temperature of Greenland and the value of $\delta^{18}O$ of the snow pack on the Greenland ice sheet is given by

$$T(°C) = 1.5\delta^{18}O(^0/_{00}) + 20.4$$

An ice core sample dating back to the last glacation has a value of $\delta^{18}O$ equal to -35. What would the estimated surface temperature have been at that time?

8.3. Suppose the earth is really flat (Figure P8.3). Imagine an earth that is shaped like a penny, with one side that faces the sun at all times. Also suppose that this flat earth is the same temperature everywhere (including the side that faces away from the sun). Neglect any radiation losses off of the rim of the earth and assume there is no albedo or greenhouse effect. Treating it as a perfect blackbody, estimate the temperature of this new, flat earth.

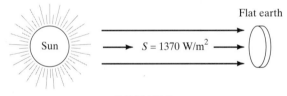

FIGURE P8.3

8.4. The solar flux S striking a planet will be inversely proportional to the square of the distance from the planet to the sun. That is, $S = k/d^2$, where k is some constant and d is the distance. Using data from Table 8.2 to find k,

  (a) Estimate the solar constant for the planet Mercury.

  (b) If Mercury is 58×10^6 km from the sun and has an albedo of 0.06, find its effective temperature.

  (c) At what wavelength would the radiation spectrum from Mercury reach its peak?

8.5. The solar flux S arriving at the outer edge of the atmosphere varies by ± 3.3 percent as the earth moves in its orbit (reaching its greatest value in early January). By how many degrees would the effective temperature of the earth vary as a result?

8.6. In the article "The Climatic Effects of Nuclear War" (*Scientific American,* August 1984), the authors calculate a global energy balance corresponding to the first few months following a 5000-megaton nuclear exchange. The resulting smoke and dust in the atmosphere absorb 75 percent of the incoming sunlight (257 W/m^2) while the albedo is reduced to 20 percent. Convective and evaporative heating of the atmosphere from the earth's surface is negligible, as is the energy reflected from the earth's surface. The earth's surface radiates 240 W/m^2, all of which is absorbed by the atmosphere. Assuming that the earth can be modeled as a blackbody emitter as shown in Figure P8.6, find the following (equilibrium) quantities:

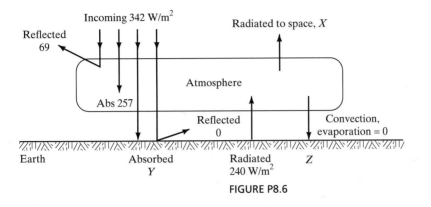

FIGURE P8.6

(a) The temperature of the surface of the earth (this is the infamous "nuclear winter")

(b) X, the rate at which radiation is emitted from the atmosphere to space

(c) Y, the rate of absorption of shortwavelength solar radiation at the earth's surface

(d) Z, the rate at which the atmosphere radiates energy to the earth's surface

8.7. Consider the two-layer atmospheric model shown in Figure P8.7. An advantage of this model is it allows us to model radiation so that each layer radiates the same amount off its top and bottom. Find the unknown quantities $W, X, Y,$ and Z to make this model balance. What values of T_1 and T_2 would radiate W and Z?

8.8. In Figure 8.13, the average rate at which energy is used to evaporate water is given as 78 W/m^2. Using 2465 kJ/kg as the latent heat of vaporization of water, along with the surface

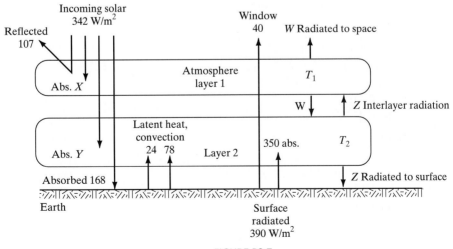

FIGURE P8.7

area of the earth, which is about 5.1×10^{14} m^2, estimate the total world annual precipitation in m^3/yr (which is equal to the total water evaporated). Averaged over the globe, what is the average annual precipitation in meters of water?

8.9. Suppose a greenhouse effect enhancement raises the surface temperature of the earth to 291 K, melting enough snow and ice to reduce the albedo to the point where only 100 W/m^2 are reflected (see Figure P8.9). The atmospheric window is closed somewhat so that only 30 W/m^2 now pass directly from the earth's surface to space. If the latent and sensible heat transfer to the atmosphere do not change, and if incoming solar energy absorbed by the atmosphere does not change, find the solar radiation absorbed at the surface W, the surface radiation X absorbed by the atmosphere, the back radiation from atmosphere to the surface Y, and the outgoing radiation from the atmosphere Z so that energy balance is maintained in each of the three regions (space, atmosphere, surface).

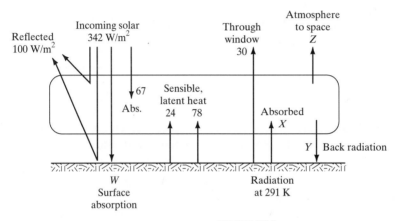

FIGURE P8.9

8.10. Suppose sunspots cause the solar constant to increase by 0.1 percent. Assuming that the albedo stays at 31 percent, what change in radiative forcing (in W per m^2 of surface) would result? Using a climate sensitivity parameter λ of 0.55, estimate the equilibrium change in the earth's surface temperature.

8.11. Suppose a change in greenhouse gas concentration causes a radiative forcing of 4 W/m^2, and suppose that this change, in turn, causes the albedo to change from 0.31 to 0.30. Calculate the total change in radiative forcing. If the climate sensitivity parameter λ is 0.55, estimate the change in earth's surface temperature.

8.12. Indicate whether the following halocarbons are CFCs, HCFCs, HFCs, or halons, and give their designation numbers:
  (a) C_3HF_7
  (b) $C_2FH_3Cl_2$
  (c) $C_2F_4Cl_2$
  (d) CF_3Br

8.13. Write chemical formulas for the following:
  (a) HCFC–225
  (b) HFC–32
  (c) H–1301
  (d) CFC–114

8.14. The radiative forcing as a function of concentration for N_2O is modeled as a square root dependence:

$$\Delta F = k_2(\sqrt{C} - \sqrt{C_0})$$

Assuming it has been in that region since preindustrial times when the concentration was 275 ppb, find an appropriate k_2 if the current concentration is 311 ppb and the forcing is estimated to be 0.14 W/m². Estimate the added radiative forcing in 2100 if it reaches a concentration of 417 ppb.

8.15. The following is an estimate for radiative forcing caused by the principal greenhouse gases:

$$\Delta F = 6.3 \ln \frac{[(CO_2)]}{[(CO_2)_0]} + 0.031(\sqrt{CH_4} - \sqrt{(CH_4)_0}) + 0.133(\sqrt{(N_2O)} - \sqrt{(N_2O)_0})$$
$$+ 0.22[(CFC–11) - (CFC–11)_0] + 0.28[(CFC–12) - (CFC–12)_0]$$

where concentrations are expressed in ppb and ΔF is in W/m². Using the following data for atmospheric concentrations:

Year	CO$_2$ (ppm)	CH$_4$ (ppb)	N$_2$O (ppb)	CFC–11 (ppb)	CFC–12 (ppb)
1850	278	700	275	0	0
1992	356	1714	311	0.268	0.503
2100	710	3616	417	0.040	0.207

(a) What would be the combined radiative forcing caused by these gases from 1850 to 1992?

(b) What would be the forcing from 1992 to 2100?

(c) What would be the forcing from 1850 to 2100?

8.16. What would be the equilibrium temperature change from 1850 to 2100 for the combination of greenhouse gases described in Problem 8.15 if the climate sensitivity parameter λ is 0.57 °C W^{-1} m²?

8.17. What would be the equivalent CO_2 concentration in 2100 for the combination of greenhouse gases in Problem 8.15?

8.18. Suppose the climate sensitivity to a doubling of CO_2 is $\Delta T_{2x} = 2.0\,°C$. Also suppose the combination of greenhouse gases in 2050 is equivalent to 600 ppm, while the actual concentration of CO_2 is 500 ppm.

(a) What would be the value of the other climate sensitivity factor λ °C W^{-1} m²?

(b) What would be the radiative forcing in 2050 caused by CO_2 alone (compared with 278 ppm CO_2 in 1850)?

(c) What would be the radiative forcing caused by the non-CO_2 greenhouse gases in 2050?

(d) What would be the equilibrium temperature change in 2050 (relative to 1850)?

(e) What fraction of the equilibrium warming in 2050 would be caused by CO_2?

8.19. Equations (8.30) and (8.31) assume that the radiative forcings of methane and nitrous oxide are independent. An adjustment to those equations can be made to account for their overlapping absorption bands, as follows (IPCC, 1990, p. 52):

$$CH_4: \quad \Delta F_{CH_4} = 0.036\left[\sqrt{(CH_4)} - \sqrt{(CH_4)_0}\right] - f(M, N_0) + f(M_0, N_0)$$

$$\text{N}_2\text{O:} \quad \Delta F_{\text{N}_2\text{O}} = 0.14 \left[\sqrt{(\text{N}_2\text{O})} - \sqrt{(\text{N}_2\text{O})_0} \right] - f(M_0, N) + f(M_0, N_0)$$

where

$$f(M, N) = 0.47 \ln \left[1 + 2.01 \times 10^{-5}(MN)^{0.75} + 5.31 \times 10^{-15}M(MN)^{1.52} \right]$$

and

$$M = (\text{CH}_4), M_0 = (\text{CH}_4)_0, N = (\text{N}_2\text{O}), N_0 = (\text{N}_2\text{O})_0 \text{ all in ppb}$$

In Example 8.6 with $(\text{CH}_4)_0 = 1714$ ppb in 1992 and $(\text{CH}_4) = 3616$ ppb in 2100, the forcing calculated using Equation (8.29) was found to be $\Delta F_{\text{CH}_4} = 0.58$ W/m². Suppose nitrous oxide concentrations increase from 311 ppb to 417 ppb during the same time. Compute the forcing for CH_4 using the preceding corrected equations that account for their overlapping absorption bands and compare it to the 0.58 W/m² found before.

8.20. Compute the global warming potential for a greenhouse gas having atmospheric lifetime $\tau = 42$ years and relative forcing per unit mass that is 1630 times that of CO_2 over the following time periods:

(a) 20 years

(b) 100 years

(c) 500 years

8.21. The IS92a emission scenario suggests anthropogenic emissions in 2025 equal to the following: $44{,}700 \times 10^9$ kg CO_2; 320×10^9 kg CH_4; and 22×10^9 kg N_2O. Weighting these by their global warming potentials, what fraction of the total impact of that year's emissions of these three gases would be caused by each gas over the following time periods?

(a) 20 years

(b) 100 years

(c) 500 years

8.22. Direct greenhouse gas radiative forcing in the mid-1990s is estimated to be about 2.45 W/m². If realized temperature is 75 percent of equilibrium temperature, what negative forcing by aerosols, etc., would be needed to match the 0.6 °C realized global temperature increase actually observed? Assume that an appropriate climate sensitivity factor λ is 0.57.

8.23. The United States derives about 25 percent of its energy from coal, 45 percent from oil, 20 percent from gas, and 10 percent from non-carbon-emitting sources.

(a) Using HHV values for carbon intensity for each fuel, estimate the overall carbon intensity for the aforementioned combination of sources.

(b) Suppose all of that coal consumption is replaced with nuclear and renewable energy systems, but everything else remains constant (total energy and distribution of sources). Estimate the new ratio of carbon to energy.

(c) Suppose that transition takes 100 years. Modeled as an exponential change, what rate of growth in carbon intensity (carbon/energy) would have prevailed?

8.24. Suppose that in 100 years the world is virtually out of natural deposits of oil and gas and energy demand is double the current rate of 330 EJ/yr. At that time, imagine that 28 percent of world energy comes from coal, 60 percent is synthetic gas and oil from coal with a LHV carbon intensity of 44 gC/MJ, and the remainder is derived from non-carbon-emitting sources. Using LHV values of carbon intensity,

(a) What would be the carbon emission rate then (Gt/yr)?

(b) If carbon emissions grow exponentially from the current rate of 6.0 GtC/yr to the rate found in (a), what rate of growth r would have prevailed during that 100 years?

(c) If the airborne fraction is 50 percent, how much of the carbon emitted in the next 100 years would remain in the atmosphere?

(d) What would be the atmospheric concentration of CO_2 in 100 years if there is no net contribution from biomass (and there are 750 Gt of carbon in the atmosphere now)?

(e) If the equilibrium temperature increase for a doubling of CO_2 is 3 °C, what would be the equilibrium temperature increase in 100 years if the initial concentration is 356 ppm?

8.25. Suppose in 100 years energy consumption is still 330 EJ/yr, but coal supplies only 20 percent of total demand, natural gas supplies 30 percent, and oil 10 percent. The remaining 40 percent of demand is met by non-carbon-emitting sources such as nuclear and solar energy. Using this conservation scenario, repeat parts (a) through (e) of Problem 8.24.

8.26. Disaggregated growth rates for three scenarios are given in the following table:

	Population (%/yr)	GDP/person (%/yr)	Energy/GDP (%/yr)	C/Energy (%/yr)	Airborne fraction	Sensitivity ΔT_{2x} (°C)
(A)	1.0	0.3	−2.0	−0.7	0.4	3
(B)	1.5	1.5	−0.2	0.4	0.5	2
(C)	1.4	1.0	−1.0	−0.2	0.5	3

For each scenario,

(a) Predict CO_2-induced global equilibrium temperature increase in 70 years, if the initial CO_2 concentration is 356 ppm, the initial emission rate is 6.0 Gt/Cyr, and the initial atmospheric carbon content is 750 GtC.

(b) If those growth rates continue, in what year would CO_2 concentrations be equal to double the initial 356 ppm?

8.27. Consider the following four ways to heat a house. The first uses a new high-efficiency pulse-combustion gas furnace; the second uses a conventional gas furnace; the third uses an electric heat pump that delivers 3 units of energy to the house for each unit of electrical energy that it consumes (the other 2 units are heat taken from the ambient air); and the fourth uses conventional electric heating. Using HHV values of carbon intensity (since these efficiencies are based on American definitions) and powerplants fueled with coal, compute the carbon emissions per unit of heat delivered to the house (gC/MJ).

Option	Description	Furnace efficiency (%)	Power plant efficiency (%)
1	Pulse-gas	95	—
2	Conventional gas	70	—
3	Heat pump	300	35
4	Electric	100	35

8.28. Supplement Example 8.13 with a water heater that burns propane (C_3H_8) with 85 percent efficiency.

(a) Propane releases 2200 kJ of energy per g-mol; find its carbon intensity (gC/MJ).

(b) Find the ratio of carbon released to energy that actually heats the water (gC/MJ).

(c) What percentage of carbon emissions are saved using propane compared with electric heating from a natural-gas-fired power plant as computed in Example 8.13?

8.29. Based on a 1990 world fossil-fuel carbon emission rate of 6.0 Gt/yr and total fossil fuel carbon ever emitted equivalent to the consumption of 200,000 EJ of coal (LHV), for each of the following values of maximum emission rate, plot a graph of carbon emissions vs. time using a Gaussian emission function described in Chapter 3 (Equations 3.17 to 3.20). In what year would maximum emissions occur?

(a) Maximum emission rate of 22 GtC/yr

(b) Maximum emission rate of 34 GtC/yr

(c) Maximum emission rate of 58 GtC/yr

8.30. For total carbon emissions of fossil fuels equivalent to the consumption of 200,000 EJ of coal, using the LHV carbon intensity, and assuming an airborne fraction of 73 percent (this may seem high, but this is the result of the diminishing effectiveness of the oceans to absorb excess carbon), estimate the ultimate concentration of CO_2 added to the "natural" 280 ppm. Find the resulting equilibrium global temperature increase assuming $\Delta T_{2x} = 3\,°C$ compared to preindustrial times when there were 280 ppm CO_2.

8.31. For each of the following fuels, find the carbon intensity based on the higher heating values (HHV) given:

(a) Ethane, C_2H_6, HHV = 1542 kJ/mol

(b) Propane, C_3H_8, HHV = 2220 kJ/mol

(c) n-Butane, C_4H_{10}, HHV = 2878 kJ/mol

8.32. Many economists favor a carbon tax as a way to discourage CO_2 emissions. Suppose such a tax were to be set at $20 per tonne of carbon emissions as (CO_2). Consider a small, 50-MW, 35 percent efficient, coal-fired power plant. Using a carbon intensity for coal of 24 gC/MJ,

(a) What would the annual carbon tax be for this power plant assuming capacity factor of 100%?

(b) Suppose a tree plantation sequesters (removes from the atmosphere and stores in biomass) on the order of 5000 kg of carbon per year per acre over the 40 years that the trees are growing (after which time the forest is mature and no further accumulation of carbon occurs). What area of forest would have to be planted to "offset" the power plant's emissions over the next 40 years (roughly the lifetime of the power plant)?

(c) How much could the owners of the plant pay for the forestry project ($/acre per year) and still have it be cheaper than paying the $20/tonne carbon tax?

8.33. Suppose a landfill leaks 10 tonnes of methane CH_4 into the atmosphere each year.

(a) Using methane's 20-year GWP, what would the warming (radiative forcing) be equivalent to in terms of tonne/yr of CO_2 emissions?

(b) Suppose a soil-vapor extraction system is installed at the landfill to suck up the methane before it leaks into the atmosphere. If that methane is burned, the methane is converted to CO_2. What would be the CO_2 emissions per year now?

(c) What is the equivalent CO_2 savings by burning the methane? How many tonnes per year of C emissions (as CO_2) would be saved?

(d) If a carbon tax of $20/tonne of C (as CO_2) is enacted, how much tax could be saved per year by burning the methane instead of letting it leak out of the landfill?

(e) A carbon tax of $20/tonne of C is the same as a tax of $5.45 per tonne of CO_2 (carbon is 12/44ths of the mass of CO_2) Using the tons of CO_2 equivalents found in (a), how much tax could be saved per year if the methane is burned instead of letting it leak out of the landfill?

8.34. Gasoline is approximately C_7H_{15}, and one gallon of it weighs about 6.15 pounds. Assume that all of the carbon in gasoline is emitted as CO_2 when it is burned.

(a) Suppose an old car that only gets 12 miles per gallon (mpg) will be driven 40,000 more miles before it ends up in the junk yard. How much carbon will it emit during that time?

(b) If the car weighs 4000 pounds and it is driven 10,000 miles per year, what is the ratio of the weight of carbon emitted per year to the weight of the car?

(c) Suppose there is a carbon tax of $15/ton (1 ton = 2000 lb). What would the carbon tax be on 1 gallon of gasoline?

(d) How much would carbon emissions be reduced over those 40,000 miles if that old car is replaced with a new one that gets 40 mpg?

(e) If an electric utility is trying to reduce its carbon tax by offering incentives to get older cars off the road, how much should the utility be willing to pay to encourage the owner of the old clunker to trade it in on a 40 mpg car?

8.35. Consider the potential carbon emissions from a gasoline-powered car compared with the carbon emissions to make the electricity for an electric car.

(a) Suppose the conventional car gets 40 miles per gallon (mpg). What are the carbon emissions (gC/mi) assuming that gasoline contains 5.22 pounds of carbon per gallon?

(b) An efficient combined-cycle, natural-gas-fired power plant needs about 8000 kJ of heat to generate 1 kWh of electricity. At 13.8 gC/MJ for natural gas, what would be the carbon emissions (gC/mi) for an electric vehicle that gets 5 miles/kWh?

(c) Suppose an older, 30 percent efficient coal-fired power plant generates the electricity that powers that electric car. At a carbon intensity of 24 gC/MJ for coal, what would the carbon emissions be (gC/mi) for that 5 mi/kWh electric car?

8.36. The photon energy required to cause the following reaction to occur is 306 kJ/mol. what is the maximum wavelength that the photon can have (you might want to refer back to Example 2.6)?

$$NO_2 \ + \ h\nu \ \longrightarrow \ NO \ + \ O$$

8.37. Photodissociation of oxygen requires 495 kJ/mol. What maximum wavelength can the photon have (refer to Example 2.6)?

$$O_2 \ + \ h\nu \ \longrightarrow \ O \ + \ O$$

REFERENCES

Armstrong, B. K., 1993, Implications of increased solar UVB for cancer incidence, *The Role of the Stratosphere in Global Change*, M. Chanin (ed.), Springer-Verlag, Berlin.

Arrhenius, S., 1896, On the influence of carbonic acid in the air upon the temperature of the ground, *Philosophical Magazine and Journal of Science*, S. 5, 41(251):237–276.

Baird, C., 1995, *Environmental Chemistry*, W. H. Freeman, New York.

Broecker, W. S., and G. H. Denton, 1990, What drives glacial cycles?, *Scientific American,* January, 49–56.

Dansgaard, W., and H. Oeschger, 1989, Past environmental long-term records from the arctic, *The Environmental Record in Glaciers and Ice Sheets,* H. Oeschger and C. C. Langway (eds.) John Wiley & Sons, New York.

Edmonds, J., and J. Reilly, 1983, Global energy production and use to the year 2050, *Energy,* 8:419.

Frederick, J. E., 1986, The ultraviolet radiation environment of the biosphere, *Effects of Changes in Stratospheric Ozone and Global Climate,* Vol. 1, USEPA and UNEP, Washington, DC.

Henderson-Sellers, A., and K. McGuffie, 1987, *A Climate Modelling Primer,* John Wiley & Sons, New York.

Imbrie, J., J. D. Hays, D. G. Martinson, A. McIntyre, A. C. Mix, J. J. Morley, N. G. Pisias, W. I. Pell, and N. J. Shackleton, 1984, *Orbital theory of the Pleistocene climate: Support from a revised chronology of the marine $\delta^{18}O$ record,* in Milankovitch and Climate, Part I, A. Berger, J. Imbrie, J. Hays, G. Kukla, and B. Saltzman (eds), D. Reidel Publishing Co., Dordrecht. pp 269–305.

Intergovernmental Panel on Climate Change (IPCC), 1990, *Climate Change, The IPCC Scientific Assessment,* Cambridge University Press, Cambridge, U.K.

Intergovernmental Panel on Climate Change (IPCC), 1992, *Climate Change 1992,* Cambridge University Press, Cambridge, UK.

Intergovernmnental Panel on Climate Change, 1995, *Climate Change 1994, Radiative Forcing of Climate Change,* Cambridge University Press, Cambridge, UK.

Integovernmental Panel on Climate Change, 1996, *Climate Change 1995, The Science of Climate Change,* Cambridge University Press, Cambridge, UK.

International Agency for Research on Cancer, 1992, *Solar and ultraviolet radiation,* IARC monographs on the evaluation of carcinogenic risks to humans, Vol. 55, IARC, Lyon.

Jouzel, J., N. I. Barkov, J. M. Barnola, M. Bender, J. Chappellaz, C. Genthon, V.M. Kotlyakov, V. Lipenkov, C. Lorius, J. R. Petit, D. Raynaud, G. Raisbeck, C. Ritz, T. Sowers, M. Stievenard, F. Yiou and P. Yiou, 1993, Extending the Vostok ice-core record of paleoclimate to the penultimate glacial period, *Nature,* 364:407–412.

Kiehl, J. T., and K. E. Trenberth, 1996, Earth's annual global mean energy budget, *Bulletin of the American Meteorological Society* (submitted).

Lean, J., and D. Rind, 1996, The Sun and climate, *Consequences,* Vol. 2, No 1, 27–36.

Madronich, S., 1993, Trends in surface radiation, *The Role of the Stratosphere in Global Change,* M. Chanin (ed.), Springer-Verlag, Berlin.

Molina, M. J., and F. S. Rowland, 1974, Stratospheric sink for chlorofluormethanes: Chlorine atom catalyzed destruction of ozone, *Nature,* 249:810–812.

Nakicenovic, N., 1996, Energy Primer, in *Climate Change 1995, Impacts, Adaptations and Mitigation of Climate Change: Scentific-Technical Analyses, Intergovernmental Panel on Climate Change,* Cambridge University Press, Cambridge, UK.

Reddy, A. K. N., and J. Goldemberg, 1990, Energy for the developing world, *Scientific American,* September, 263(3):110–118.

Simon, P. C., 1993, Atmospheric changes and UV-B monitoring, *The Role of the Stratosphere in Global Change,* M. Chanin (ed.), Springer-Verlag, Berlin.

Swisher, J. N., and G. M. Masters, 1991, Buying environmental insurance: Prospects for trading of global climate-protection services, *Climatic Change,* 19:233–240.

United Nations Environment Programme (UNEP), 1994, *Montreal Protocol on Substances that Deplete the Ozone Layer, Scientific Assessment of Ozone Depletion,* World Meteorological Organization Global Ozone Research and Monitoring Project—Report No. 37, Geneva, Switzerland.

Warren, S. G., and S. H. Schneider, 1979, Seasonal simulation as a test for uncertainties in the parameterization of a Budkyo-Zellers zonal climate model, *Journal of the Atmospheric Sciences,* 36:1377–1391.

Wuebbles, D. J., 1995, Weighing functions for ozone depletion and greenhouse gas effects on climate, *Annual Review of Energy and Environment,* 20:45–70.

Solid Waste Management and Resource Recovery

> *The Congress hereby declares it to be the national policy of the United States that pollution should be prevented or reduced at the source whenever feasible; pollution that cannot be prevented should be recycled in an environmentally safe manner whenever feasible; pollution that cannot be prevented or recycled should be treated in an environmentally safe manner whenever feasible; and disposal or other release into the environment should be employed only as a last resort and should be conducted in an environmentally safe manner.*   —*Pollution Prevention Act of 1990*

9.1 INTRODUCTION

On March 22, 1987, the garbage barge *Mobro 4000* and its tug, the *Break of Dawn,* left the port of New York loaded with solid waste that could no longer be disposed of in the local Blydenburgh Landfill on Long Island. In a futile search for a friendly port, the barge traveled on a six-month odyssey of over 6000 miles; it was refused entry in six states and three foreign countries before it returned to New York Harbor, a symbol of a nation producing too much trash with no convenient place to put it. Hoping our trash will just "go away," when there is no convenient "away " anymore, has become a serious problem that encompasses a range of social, political, and technical issues. Some of those issues are perhaps best illustrated by three acronyms (White et al., 1995): NIMBY (Not In My Back Yard), NIMET (Not In My Elected Term), and BANANA (Build Absolutely Nothing Anytime Near Anyone).

Historically (certainly well into the twentieth century), solid waste management was of the most rudimentary sort even in the most developed countries of the world. Garbage, if it was collected at all, usually ended up at the local dump, where open burning was commonly used to control both the volume and public health dangers associated with the waste. It was not until roughly 1950 that most urban areas in the United States ceased operating unsightly, rat-infested open dumps. More recently, and for a number of reasons, the technologies used to manage our wastes have become increasingly sophisticated. As the quote at the beginning of this chapter illustrates, present policies emphasize waste reduction and recycling to minimize the volume and toxicity of the materials that must be disposed of. The "away" that the remaining wastes are heading toward may include incineration, perhaps with energy recovery, and/or disposal in carefully engineered and managed sanitary landfills.

9.2 BACKGROUND

While most of the focus in waste management is on municipal solid waste (MSW) management, it is important to note that MSW is but a small fraction of the total amount of waste generated in the United States. The Resource Conservation and Recovery Act (RCRA), which is the major federal statute that governs solid waste, delineates two categories of waste: *hazardous* and *nonhazardous* waste. Subtitle C of RCRA defines and regulates hazardous waste, and Subtitle D addresses nonhazardous waste.

In the United States, we generate some 11.7 billion tons of wastes each year, including about 0.7 billion tons of hazardous waste and 11 billion tons of nonhazardous waste.[1] As shown in Figure 9.1, most of that 11 billion tons is generated by industry during raw material extraction, material processing, and product manufacturing, while less than 2 percent is municipal solid waste. In this chapter, we will be primarily looking at just that 2 percent.

The magnitude of the earth- and ore-moving task needed to provide the minerals society demands is truly monumental. Digging holes, removing the ores, and piling up the leftover residues creates enormous aesthetic, environmental, economic, and energy problems. Table 9.1 provides an estimate of typical amounts of materials that must be moved and processed to produce some of our most important minerals.

Even though MSW is dwarfed in size (and environmental impact) by the other sectors, it should be remembered that all of that industrial waste is created in the process of providing us with the material things that ultimately end up in our trash. Consuming less therefore saves not only the wastes that would have ended up in the municipal waste stream, but it also reduces the energy, materials, and waste associated with providing those items that we may be able to live without. Clearly, attention needs

---

[1]In this chapter most of the data are from American sources that use short tons (2000 lb) rather than metric tons (1000 kg). To distinguish between the two, metric tons will always be expressed as tonnes (1 tonne = 1000 kg = 2200 lb = 1.1 ton).

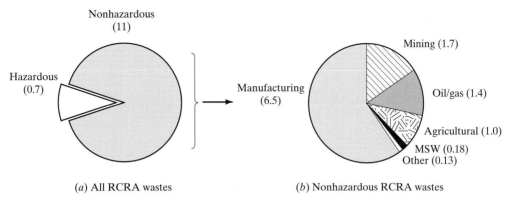

FIGURE 9.1   Solid wastes as defined under the Resource Conservation and Recovery Act (RCRA). Units are billions of tons per year. (*Source:* Office of Technology Assessment [OTA], 1992)

to be directed not just at the management of consumer wastes, but at the complete set of processes that result in the products our society seems to demand.

Definitions

Before we can more carefully consider the material flows and environmental impacts of our economy, some definitions need to be provided. Figure 9.2 may help keep some of these terms straight.

> *Solid wastes* are wastes that are not liquid or gaseous, such as durable goods, non-durable goods, containers and packaging, food scraps, yard trimmings, and miscellaneous inorganic wastes. Solid waste is more or less synonymous with the term *refuse,* but *solid waste* is the preferred term.

> *Municipal solid waste* (MSW) is solid waste from residential, commercial, institutional, and industrial sources, but it does not include such things as construction waste, automobile bodies, municipal sludges, combustion ash, and industrial process wastes even though those wastes might also be disposed of in municipal waste landfills or incinerators.

> *Garbage,* or *food waste,* is the animal and vegetable residue resulting from the preparation, cooking, and serving of food. This waste is largely putrescible

TABLE 9.1   Ore Required to Yield 100 kg of Product

Mineral	Average Grade (%)	Ore (kg)	Product (kg)	Residue (kg)
Aluminum	23.00	435	100	335
Copper	0.91	10,990	100	10,890
Iron	40.00	250	100	150
Lead	2.50	4000	100	3900
Nickel	2.50	4000	100	3900
Others (avg.)	8.10	1234	100	1134

Source: Based on data from Young (1992).

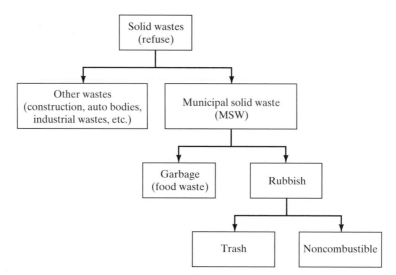

FIGURE 9.2   Illustrating some terms used in solid waste management.

organic matter and moisture. Home kitchens, restaurants, and markets are sources of garbage, but the term usually does not include wastes from large food processing facilities such as canneries, packing plants, and slaughterhouses.

Rubbish consists of old tin cans, newspaper, tires, packaging materials, bottles, yard trimmings, plastics, and so forth. Both combustible and noncombustible solid wastes are included, but rubbish does not include garbage.

Trash is the combustible portion of rubbish.

Generation refers to the amounts of materials and products that enter the waste stream. Activities that reduce the amount or toxicity of wastes before they enter the municipal waste system, such as reusing refillable glass bottles or reusing plastic bags, are not included.

Materials recovery is the term used to cover the removal of materials from the waste stream for purposes of recycling or composting.

Discards are the solid waste remaining after materials are removed for recycling or composting. These are the materials that are burned or buried. In other words,

$$\text{Waste generation} = \text{Materials recovered} + \text{Discards} \qquad (9.1)$$

Municipal Solid Waste

The Environmental Protection Agency estimates that the United States generated 207 million tons (188 million tonnes) of municipal solid waste in 1993, which is about 4.4 pounds (2 kg) per person per day (U.S. EPA, 1994a). A materials breakdown for that 207 million tons is shown in Figure 9.3. Paper and paperboard products are the largest single component of MSW (37.6 percent by weight) and yard trimmings are the second largest (15.9 percent). Figure 9.4 shows the historical growth in generation of MSW, by source. Over the years the single largest component—paper and paperboard

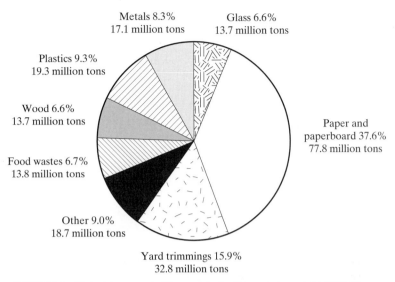

FIGURE 9.3   Materials generated in municipal solid waste by weight, 1993. (*Source:* U.S. EPA, 1994a).

products—has grown fairly steadily. In recent years the second largest component, yard trimmings, has been declining in importance as backyard composting and mulching lawnmowers have become more common. Generation of metals and food wastes has been fairly constant.

Of the 207 million tons of MSW generated in 1993, almost 22 percent was recovered for recycling or composting, up from 17 percent only three years earlier. Of the waste stream that was not recycled or composted, three-fourths went to landfills and the remainder was incinerated (Figure 9.5).

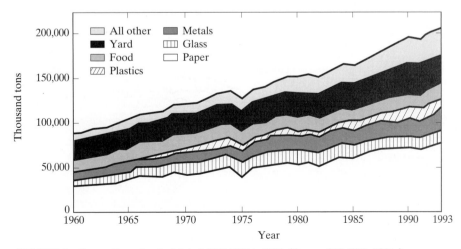

FIGURE 9.4   Generation of materials in MSW, 1960 to 1993. (*Source:* U.S. EPA, 1994a)

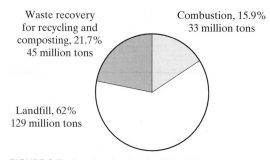

Waste recovery for recycling and composting, 21.7% 45 million tons

Combustion, 15.9% 33 million tons

Landfill, 62% 129 million tons

FIGURE 9.5 Destination for the 207 million tons of MSW generated in the United States in 1993. (*Source:* U.S. EPA, 1994a)

Trends in U.S. solid waste generation, recovery, and discards, from 1960 to 1993, with EPA projections to the year 2000, are shown in Figure 9.6. Between 1960 and 1993, the generation of wastes more than doubled, from 88 million tons to 207 million tons. On a per capita basis, MSW generation increased from 2.7 lb per person per day to 4.4 lb during that period of time, but EPA's projections for the year 2000 show a slight decline to 4.3 lb per person. Waste recovery for recycling and composting increased dramatically, from 7 percent of MSW generated in 1960 to 22 percent by 1993. The total amount of refuse sent to landfills or incinerators is projected to decline from its peak of 162 million tons in 1990 to 152 million tons in the year 2000—even though total waste generation is projected to increase to 218 million tons in 2000. The fraction of municipal solid waste that ends up in landfills is dropping rapidly. In 1985, 83 percent of all MSW was buried in landfills, but by 1993 that figure had dropped to 62 percent.

Figure 9.6 also illustrates the interesting history of combustion of municipal solid waste. In 1960, about 30 percent of MSW was burned in incinerators, typically with no

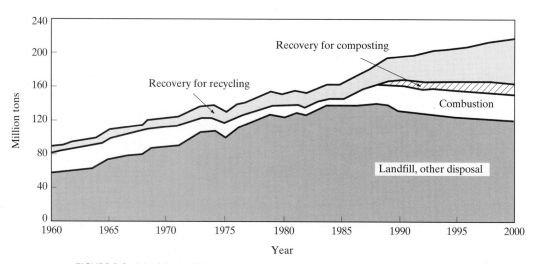

FIGURE 9.6 Municipal solid waste management, 1960–2000 (*Source:* U.S. EPA, 1994a)

air pollution controls and no energy recovery. As concern for air quality grew in the 1960s and 1970s, those old incinerators were closed and the fraction of our wastes that was burned dropped to a low of 10 percent of MSW in the early 1980s. As energy-recovery incinerators with emission controls have begun to come back on line, the fraction of MSW burned has held fairly steady at about 16 percent in the 1990s.

The per capita generation of municipal solid waste in the United States is typically at least twice as high as it is for most developed countries, as shown in Table 9.2. For comparison, the percentages of waste generated that is landfilled and the population densities are also shown in the table. While there is some correlation between population density and landfilling, there are notable exceptions. Sparsely populated countries, such as Australia and Canada, tend to landfill very high fractions of their wastes, as might be expected, while densely populated countries such as Japan and Italy landfill less and incinerate more. The population density of Japan, for example, is 12 times that of the United States and it landfills only 33 percent of its municipal solid waste while incinerating 64 percent. The United Kingdom, on the other hand, is also much more densely populated than the United States, but it sends a higher fraction of its waste to landfills than we do.

9.3 LIFE-CYCLE ASSESSMENT

Solid waste management all too often focuses almost entirely on what to do with a given waste stream, with the key decision being whether to incinerate the waste or bury it. As landfills filled up or were closed because they could not meet new environmental regulations, as incinerators were shut down due to poor performance, and as communities became more agitated by the environmental effects of living near either type of facility, a new, broader approach to the problem has emerged. Using an energy and materials balance approach at every stage in the life cycle of a product can yield new insights into not only the solid waste problem, but also the problems of air and water pollution already described in this book.

TABLE 9.2   Municipal Solid Waste In Selected Countries

Country	Year of Estimate	Per Capita Generation (kg/yr)	Percent Landfilled (%)	Percent Incinerated (%)	Relative Population Density
Australia	1980	681	98	2	0.1
Austria	1988	355	68	8	3.4
Canada	1989	625	84	9	0.1
France	1989	303	45	41	3.8
Germany	1987	318	66	30	9.2
Italy	1989	301	31	16	7.1
Japan	1988	394	33	64	12.1
Spain	1988	322	77	5	2.9
United Kingdom	1989	357	78	14	8.7
United States	1993	730	62	16	1.0

Note: Some data may not be directly comparable given differences in reporting methods.
Source: World Resources Institute (1992) and U.S. EPA (1994a).

PRODUCT LIFE CYCLE

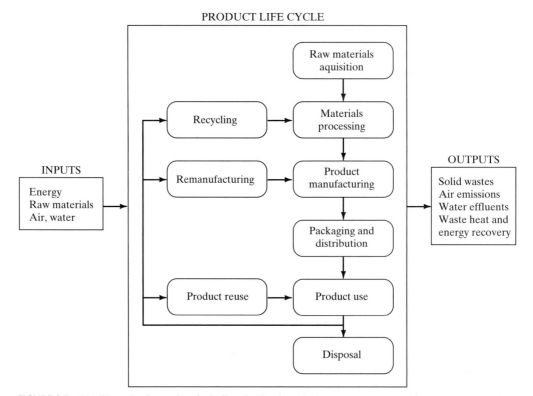

FIGURE 9.7 The life cycle of a product, including clarification of the terms *reuse, remanufacturing,* and *recycling.*

A simple conceptual diagram for the life-cycle assessment of a product is presented in Figure 9.7. Inputs include energy and raw materials utilized in each stage of the production, use, and disposal of the product. The central box in the figure suggests the various stages in the product life cycle, which includes the acquisition and processing of materials (mining, smelting, etc.), the actual manufacturing of the product itself, the packaging and distribution of the product, the use of the product, and, finally, the ultimate disposal of the product. Outputs are the air, water, and solid waste effluents associated with each stage, along with the waste heat dissipated in the environment plus energy that may be recovered during disposal.

Figure 9.7 also helps define some of the terms that are used to describe the potential recovery of materials in the product life cycle.

> *Reusing* a product in the same application for which it was originally intended saves energy and resources. For example, a plastic bag can carry groceries home from the market over and over again, and a polystyrene cup might be used several times before disposal. Returnable glass bottles for soft drinks are another example. A product can also be reused for some other purpose, such as occurs when glass jars are reused in a workshop to hold small objects such as screws or nails.
>
> *Remanufacturing* refers to the process of restoring a product to like-new condition. The restoration begins by completely disassembling the product, cleaning

and refurbishing the reusable parts, and then stocking those parts in inventory. That inventory, along with new parts, is used to remanufacture products that are equal in quality to new units. Some distinguish between remanufacturing and repair. Repair means that only those parts that have failed are replaced. For example, malfunctioning components on a faulty electronic circuit board might be replaced rather than throwing out the entire product.

Recycling is the term used to describe the act of recovering materials from the waste stream and reprocessing them so they become raw materials for new applications.

An Example Life-cycle Assessment: Polystyrene Cups

As an example of the value (and difficulty) of performing a complete life-cycle assessment, consider a comparison of the environmental impacts of single-use, 8-ounce, hot-drink containers made from polystyrene foam with similar cups made from uncoated paper (Hocking, 1991). As shown in Table 9.3, the raw materials inputs for the two types of cups are very different. A paper cup requires about 21 g of wood and bark plus 1.2 g of other chemicals to produce an 8.3-g cup, while a 1.9-g polystyrene cup requires 2.4 g of petroleum feedstock plus 0.08 g of other chemicals. From the materials-input perspective, it is very difficult to compare the environmental impacts associated with logging with the impacts associated with drilling, transporting, and processing petroleum.

In terms of energy requirements, processing of raw wood, to pulp, to fully bleached kraft paper for the paper cup uses large amounts of process heat (steam) and electricity. If we assume that the steam and electricity are produced by burning residual fuel oil, a total of about 3.8 g of fuel per cup are required. Converting petroleum feedstock into polystyrene also uses significant quantities of heat and electricity. Crude oil or natural gas are first cracked in the presence of steam to yield ethylene or benzene, which are in turn catalyzed to produce ethylbenzene. Ethylbenzene is then thermally decomposed to styrene, which is converted to polystyrene. Hocking's analysis indicates that roughly the same amount of process heat per cup is required to make polystyrene or bleached kraft paper, but about 13 times as much electricity is required for the paper cup. Assuming that fuel oil is burned in a 33 percent efficient power plant to produce the electricity needed, the paper cup uses roughly 80 percent more energy than the styrofoam cup (3.8 g of oil equivalents per cup vs. 2.1 g). However, if we add in the petrochemical feedstock needed for the polystyrene cup, the paper cup uses roughly 15 percent less petroleum (3.8 g of oil vs. 4.5 g).

Most of the air and water effluents are considerably higher for the paper cup, although the comparison is complicated by the fact that the manufacturing processes for paper and polystyrene are so different. Production of polystyrene, for example, emits significant quantities of pentane (the foam blowing agent that has replaced chlorofluorocarbons), while paper production does not emit any. While pentane is not an ozone-depleting substance, it can contribute to photochemical smog. Again, it would be difficult to decide whether, for example, pentane emissions from the manufacture of polystyrene are more important than the higher-criteria air pollutants emitted during paper production.

TABLE 9.3 Life-Cycle Assessment for Single-Use, Hot-Drink Cups (per 1000 cups)

	8.3 g Paper Cup	1.9 g Polyfoam Cup
Raw Materials		
Wood and bark (kg)	21	0
Petroleum feedstock (kg oil)	0	2.4
Other chemicals (kg)	1.2	0.08
Purchased Energy		
Process heat (kg oil)	1.8	1.9
Electricity (kg oil)[a]	2	0.15
Water Effluent		
Volume (m^3)	1	0.05
Suspended solids (g)	80	1
BOD (g)	90	0.4
Organochlorines (g)	20	0
Inorganic salts (g)	500	30
Fiber (g)	10	0
Air Emissions		
Chlorine (g)	2	0
Chlorine dioxide (g)	2	0
Reduced sulfides (g)	10	0
Particulates (g)	20	0.8
Carbon monoxide (g)	30	0.2
Nitrogen oxides (g)	50	0.8
Sulfur dioxide (g)	100	7
Pentane (g)	0	80
Ethylbenzene, styrene (g)	0	5
Recycle/Reuse Potential		
Reuse	Possible	Easy
Recycle	Acceptable	Good
Ultimate Disposal		
Proper incineration	Clean	Clean
Heat recovery (MJ)	170	80
Mass to landfill (kg)	8.3	1.9
Volume in landfill (m^3)	0.0175	0.0178
Biodegradability (landfill)	Yes	No

[a] Calculated using 33 percent efficient power plant burning residual fuel oil.
Source: Based on data in Hocking (1991).

Consider the use, reuse, and recyclability of the two cups. The polystyrene cups are stiffer and stronger, especially when holding hot liquids, and their natural insulation helps keep the drinks hotter and the outside surface cooler. It is easy to imagine reusing the same polyfoam cup several times while at a party or conference, for example, but the poor structural integrity of a paper cup—especially when hot—makes it less likely to be reused. Both types of cups can be recycled, but the hot melt adhesive used in paper cups makes them somewhat less attractive.

At the disposal end of the life cycle, if the cups are incinerated properly they both burn cleanly and they both produce about the same amount of ash. If the incineration includes energy recovery, the paper cups yield roughly twice as much heat recovery per cup. However, since paper cups have four times the mass, paper cups yield about half as much energy per kilogram. If they are buried in a landfill rather than being burned in an incinerator, the polystyrene cup is compressed to roughly the same volume as the paper cup. The paper cups, however, are biodegradable (but very slowly), which means they can eventually take up less space, but their degradation produces a potent greenhouse gas, methane, and the liquids that filter through the landfill will contribute to the BOD of the leachate. If the landfill is well managed, the methane and leachate are easily controlled, but if it is poorly managed it might be better to have the nonbiodegradable polystyrene to deal with.

As should be obvious by now, the paper vs. polystyrene cup analysis illustrates how difficult it is to reach solid conclusions when performing a life-cycle assessment. Besides the many assumptions that must be made to generate any of the numerical comparisons, what conclusions can be drawn when the two approaches produce different sorts of environmental impacts? Perhaps Hocking's conclusion that "it would appear that polystyrene foam cups should be given a much more even-handed assessment as regards their environmental impact relative to paper cups than they have received during the past few years" (Hocking, 1991, p. 745) is about as strong a statement as can be made in this controversial debate.

Soft-drink Containers

As another example of life-cycle analysis, consider the comparison of environmental impacts associated with a number of common soft-drink containers. The energy requirements, air and water emissions, and volume of solid waste for each container system per liter of soft drink are presented in Table 9.4. For almost every measure, refillable glass bottles cause the least environmental impact. Unfortunately, although they were common in the past, supermarkets and convenience stores have been switching to throwaways to avoid having to accept and store returnables (although they are still popular in Europe). By comparison, nonrefillable glass bottles use several

TABLE 9.4   Environmental Impacts Associated with the Delivery of 1 Liter of Soft Drink in Various Containers[a]

Type of container	Energy (kJ/L)	Air emissions (g/L)	Water emissions (g/L)	Solid waste ($10^{-4}\,m^3$)
0.47 L (16 oz) refillable glass bottle used 8 times	4290	6.5	1.0	2.2
0.47 L (16 oz) nonrefillable glass bottle	9700	18.2	2.0	6.7
1 L (34 oz) nonrefillable glass bottle	10,200	19.8	2.1	7.5
0.29 L (10 oz) nonrefillable glass bottle	11,600	22.0	2.4	8.1
0.47 L (16 oz) PET bottle	8810	11.1	1.9	3.4
1 L (34 oz) PET bottle	7110	8.9	1.6	2.6
2 L (68 oz) PET bottle	5270	6.7	1.2	1.8
3 L (102 oz) PET bottle	5180	6.5	1.2	1.7
0.35 L (12 oz) aluminum can	9170	11.0	3.2	1.6

[a]At 1987 recycling rates.
Source: Rhyner et al. (1995) based on data from Sellers (1989).

times as much energy and their emissions, and solid waste requirements are correspondingly higher as well.

In the United States, plastic bottles and aluminum cans are becoming the containers of choice. Although not as benign as returnable bottles, from an energy standpoint both would seem to be better than one-way glass bottles. Unfortunately, making comparisons for air and water emissions for glass, polyethylene terephthalate (PET), and aluminum containers is more difficult than the table would suggest since the types of pollution for each will be considerably different.

Integrated Solid Waste Management

The life-cycle assessment approach is meant to provide a useful framework for manufacturers, who must choose materials and technologies, as well as consumers who want to include environmental considerations in their choice of products. An analogous process, called *integrated solid waste management,* is intended to help guide decisions about the generation of wastes, recycling of materials, and ultimate disposal of waste residues. An outline of some of the highlights of integrated waste management is suggested in Table 9.5. Source reduction and recycling in almost all circumstances are given the highest priority, and it is easy to understand why:

- To reduce the amount of solid waste that has to be burned or buried
- To reduce pollution associated with mining, use, and disposal of resources
- To reduce the rate of consumption of scarce resources.

While the order in which the entries in Table 9.5 have been listed is purposely intended to suggest a hierarchy of priorities, that hierarchy is not rigid. For example, is lightweight packaging that uses less material always better than more substantial packaging? If the heavier packaging can more easily be recycled, it might be better than the lightweight materials that go straight to the dump. Perhaps the heavyweight materials

---

TABLE 9.5   Integrated Solid Waste Management

Source Reduction
- Reduce toxicity
- Less packaging
- Product reuse
- More durable products
- On-site mulching and composting

Recycling
- Collection
- Processing
- Use of recycled materials in products
- Composting

Disposal
- Combustion with energy recovery
- Landfill
- Incineration without energy recovery

---

would burn cleanly in waste-to-energy incinerator, while toxics in the lightweight packaging might make it inappropriate for combustion. The disposal priorities suggested in Table 9.5, which would appear to rank combustion with energy recovery above landfilling, are controversial. Not only would combustion characteristics of the waste affect the decision, but also the relative suitability of local topography and meteorology for landfills versus incinerators would need to be evaluated.

9.4 SOURCE REDUCTION

"Garbage that is not produced does not have to be collected" is a simple enough concept, yet, at least in the United States, it has not received the attention that it deserves That is not the case in many other advanced countries in the world—especially those with modest domestic resources and limited land space for disposal, such as Germany and Japan.

Green Product Design Strategies

Design that concerns itself with reducing the environmental impacts associated with the manufacture, use, and disposal of products is an important part of any pollution prevention strategy. Companies that engage in such green product design are finding that products that combine environmental advantages with good performance and price have additional market appeal.

The Office of Technology Assessment identifies two complementary goals of green design: waste prevention and better materials management (OTA, 1992). As suggested in Figure 9.8, waste prevention can be achieved by reducing the weight, toxicity, and energy use of products along with increasing the useful life of products. Better materials management facilitates remanufacturing, recycling, and composting along with enhanced energy recovery opportunities.

A number of strategies have been identified that contribute to good green design practices, including the following (Keoleian et al., 1994).

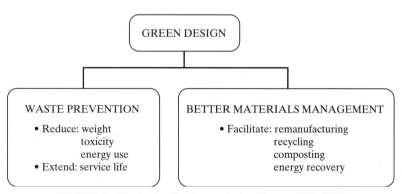

FIGURE 9.8   The dual goals of green design (*Source:* OTA, 1992)

Product System Life Extension.   Products that do not wear out as quickly do not have to be replaced as often, which usually means that resources are saved and less waste is generated. Sometimes products are discarded for reasons that have nothing to do with their potential lifetime, such as computers that become obsolete and clothing fashions that change, but many products can continue to remain in service for extended periods of time if they are designed to be durable, reliable, reusable, remanufacturable, and repairable.

Extending product life, of course, means that consumers replace their products less often, which translates into decreased sales volume for manufacturers. It is hoped that the temptation toward planned obsolescence will be tempered by the realization that market share in the future may be driven to some extent by consumer demand for greener products. Consumers, of course, must make that happen.

Material Life Extension.   Once a product has reached the end of its useful life, the materials from which it was made may still have economic value, and additional savings can result from avoiding disposal. The key design parameter for extending the life of materials is the ease with which they can be recycled.

Products that have been designed to be recycled easily are becoming especially common in Germany, where tough recycling requirements are in place. For example, Germany's requirement that automakers take back and recycle old automobiles has stimulated green design in companies such as BMW and Volkswagen, both of which have built pilot plants to study disassembly and recycling of recovered materials. BMW, which already sells cars with plastic body panels that have been designed for disassembly and that are labeled as to resin type so they may more easily be recycled, has set a goal of learning how to make an automobile entirely out of reusable and recyclable parts by the year 2000 (OTA, 1992).

Material Selection.   A critical stage in product development is selection of appropriate materials to be used. In green design, attempts are made to evaluate the environmental impacts associated with the acquisition, processing, use, and retirement of the materials under consideration. In some cases substituting one material for another can have a modest impact on the quality and price of the resulting product but can have a considerable impact on the environmental consequences.

Designers today have a much greater range of materials to choose from in developing their products. Ceramics and composites offer superior strength with lighter weight than traditional materials such as steel and aluminum. High-strength alloys and plastics are quickly displacing the metals used in the past. For example, the telecommunications cables that AT&T used in the 1950s consisted mostly of steel, lead, and copper, with small percentages of aluminum and about 1 percent plastics. By the mid-1980s, polyethylene had replaced virtually all of the lead that had been used in the cables' sheathing so that the fraction of plastics is now more than 35 percent. If replacing lead sheathing with polyethylene had not occurred, AT&T's lead requirements might now be close to a billion pounds annually (Ausubel, 1989). Such progress continues with fiber-optic cables that weigh only 3 percent as much as traditional copper cables and that use only 5 percent of the energy to produce (OTA, 1992).

An especially important aspect of materials selection is the need to reduce the toxicity of materials whenever possible. Toxic substances used in products can create

TABLE 9.6   Examples of Toxic Metal Sources in MSW

Metal	Applications
Arsenic	Wood preservatives, household pesticides
Cadmium	Rechargable Ni-Cad batteries; pigments in plastics (kitchenware, etc.), colored printing inks (magazines, catalogs), and enamels (pots, pans, mugs, colored glassware, porcelain, ceramics)
Lead	Automobile batteries, solder (plumbing, gutters, electronics, food cans); pigments or stabilizers in plastics, glass, ceramics
Mercury	Batteries; fluorescent lamps; fungicides; paints
Zinc	Flashlight batteries; rubber products (including tires)

serious environmental risks when those products end up in landfills or incinerators. Heavy metals, such as lead, cadmium, chromium, mercury, arsenic, copper, tin, and zinc, are commonly used in consumer products and are especially dangerous when they enter the waste stream. Examples of sources of heavy metals in the municipal waste stream are shown in Table 9.6.

Reduced Material Intensiveness.   Green design strategies include reducing the amount, and/or toxicity, of materials required to make a given product while maintaining the product's usefulness and value. Changes in battery technology provide a good example. In 1974, a typical car battery contained about 30 lb of lead, but modern batteries use less than 20 lb. Common household batteries have also been redesigned to use only 15 percent as much mercury those produced a decade or so ago (OTA, 1992).

The history of beverage containers provides another example of sizable reductions in materials intensiveness. Nonrefillable (one-way) 0.47 L (16 oz) glass bottles, for example, are 37 percent lighter than they were in the 1970s. Similar "lightweighting" for steel cans, aluminum cans, and PET one-piece plastic bottles is shown in Table 9.7.

Process Management.   Manufacturing products requires raw materials and energy inputs, both of which often can be managed more efficiently.

The energy required to manufacture a product is an important component of a life-cycle assessment. Process improvements that take advantage of waste-heat recovery, more efficient motors and motor controls, and high-efficiency lighting are almost always very cost effective. Since electric motors account for two-thirds of the electricity used by industry (and half of the electricity used in the United States for all purposes),

TABLE 9.7   Mass Reductions in Soft-Drink Containers

Type of Container	1972 (g)	1992 (g)	Percent change
One-way glass bottle (0.47 L, 16 oz)	344	218	−37%
Steel can (0.36 L, 12 oz)	48	33	−31%
Aluminum can (0.36 L, 12 oz)	20	16	−22%
PET bottle (2 L, 68 oz, one piece)	66	54	−18%

Note: Does not include weight of labels and caps. PET data for 1977, not 1972.
Source: Franklin Associates (1994).

a good place to look for efficiency improvements is there. It is not uncommon for motors in use to be oversized and to run at constant speed, with both factors contributing to low energy efficiency. When motors are used to pump fluids, the pumping rate is usually controlled by adjusting valves or dampers rather than by adjusting the speed of the motor itself, which wastes electricity. New electronic adjustable-speed drives, coupled with more efficient motors, can often save half of the energy normally used.

Better materials management can also lead to lower environmental impacts. Wastes can be minimized by more carefully estimating and ordering needed inputs, especially when they are hazardous materials, and by more careful inventory control.

Efficient Distribution.   Methods of packaging and transporting products greatly affect the overall energy and environmental impacts associated with those products.

Transportation costs are affected by the quantity and type of material shipped, which is in turn affected by packaging, trip distance, and type of carrier. The type of carrier is constrained by the terrain to be covered as well as the speed required for timely arrival. In general, shipping by boat is the least energy-intensive option, followed by rail, truck, and then air, which is fastest but requires the most energy per ton-mile transported. If pipelines are an option (as, for example, transport of petroleum), they can be even less energy intensive per ton-mile than transport by ship.

Packaging can be considered a component of distribution. Reducing the amount of packaging can reduce the environmental costs of distribution, but may increase deterioration of the product if it is damaged or loses freshness. Using recycled materials for packaging along with strategies that reduce the amount of packaging can contribute to reduced environmental life-cycle costs.

Policy Options.   Germany has shifted the burden of packaging disposal from the consumer back to manufacturers and distributors. Germany's *Packaging Waste Law* requires manufacturers and distributors to recover and recycle their own packaging wastes. The concept may even be extended to large durable goods, such as household appliances and automobiles. Germany's take-back policy is one of many examples of approaches that governments can take to encourage reduction in the environmental costs of products. Table 9.8 gives an extensive list of regulatory instruments and market-based incentives that could affect material flows in the American economy.

Labeling

Surveys have consistently shown that American consumers, if given a choice, would purchase products that are environmentally superior to competing products, even if they cost a bit more (U.S. EPA 1991). Attempts by manufacturers to capture that environmental advantage have led to an overuse of poorly defined terms such as *recyclable, recycled, eco-safe, ozone-friendly,* and *biodegradable* on product labels. Unfortunately, without a uniform and consistent standard for such terms, these labels are all too often meaningless or even misleading. For example, all soaps and detergents have been "biodegradable" since the 1960s, and "CFC-free, ozone-friendly" aerosols have been the norm in the United States since the banning of CFCs for such applications in 1978. (Such aerosols may, however, use HCFCs, which still have some ozone depletion potential, or pentane, which can contribute to photochemical smog.) Labels

TABLE 9.8   Policy Options That Could Affect Materials Flows

Life-cycle stage	Regulatory Instruments	Economic Instruments
Raw material extraction and processing	Regulate mining, oil, and gas non-hazardous solid wastes under the Resource Conservation and Recovery ACT (RCRA). Establish depletion quotas on extraction and import of virgin materials.	Eliminate special tax treatment for extraction of virgin materials, and subsidies for agriculture. Tax the production of virgin materials.
Manufacturing	Tighten regulations under Clean Air Act, Clean Water Act, and RCRA. Regulate nonhazardous industrial waste under RCRA. Mandate disclosure of toxic materials use. Raise Corporate Average Fuel Economy Standards for automobiles. Mandate recycled content in products. Regulate product composition (e.g., volatile organic compounds or heavy metals). Establish requirements for product reuse, recyclability, or biodegradability. Ban or phase out hazardous chemicals. Mandate toxic use reduction.	Tax industrial emissions, effluents, and hazardous wastes. Establish tradable emission permits. Tax the carbon content of fuels. Establish tradable recycling credits. Tax the use of virgin toxic materials. Create tax credits for use of recycled materials. Establish a grant fund for clean technology research.
Purchase, use, and disposal.	Mandate consumer separation of materials for recycling.	Establish weight/volume-based waste disposal fees. Tax hazardous or hard-to-dispose products. Establish a deposit-refund system for packaging or hazardous products. Establish a fee/rebate system based on a product's energy efficiency. Tax gasoline.
Waste management	Tighten regulation of waste management under RCRA. Ban disposal of hazardous products in landfills and incinerators. Mandate recycling diversion rates for various materials. Exempt recyclers of hazardous wastes from RCRA Subtitle C. Establish a moratorium on construction of new landfills and incinerators.	Tax emissions or effluents from waste management facilities. Establish surcharges on wastes delivered to landfills or incinerators.

claiming a product is "eco-safe" or "environmentally safe" have little meaning since the terms are largely undefined.

Clearly, some sort of credible labeling system certifying that the products and packaging bearing such labels have been independently certified to meet certain environmental standards would be a powerful motivator in the marketplace. To that end there are now several competing eco-labels being promulgated by private organizations in the United States. *Green Seal* labels provide a simple, overall stamp of approval, analogous to the Underwriter Laboratories (UL) label on electrical appliances or the Good Housekeeping Seal of Approval, while the *Scientific Certification*

Systems label attempts to use life-cycle analysis to compare products based on resource inputs and waste outputs.

An early example of the Scientific Certification Systems approach, in which a hypothetical product is compared against a "benchmark" product, is shown in Figure 9.9. The advantage of this label is that specific data are provided for such inputs as water and energy and for such outputs as air emissions and solid wastes. The simple bar chart then attempts to convert those data into an easily understandable pictorial representation. Unfortunately, such a label would not really enable a consumer to decide whether one product is "greener" than another unless some measures of the relative environmental impacts of the various categories of data are provided. For example, how would a consumer know whether a better than average rating on the nonrenewable energy input scale offsets a worse than average rating on the sulfur dioxide scale,

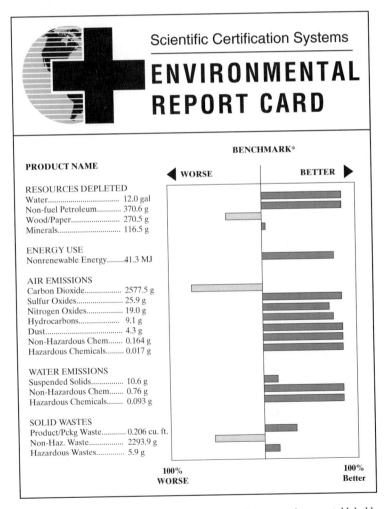

FIGURE 9.9   An early Scientific Certification Systems approach to an environmental label based on life-cycle analysis. A newer version with a simpler set of environmental indicators is being developed. (*Source:* OTA, 1992)

or whether a pound of nitrogen oxides added to the atmosphere is better or worse than a pound of solid wastes going to the landfill? Still, it can be argued that the risk of causing confusion is more than offset by the value of providing more information.

It is interesting to note that the EPA once proposed the establishment of a national eco-label, but has since abandoned the idea, leaving such labels in the hands of the private sector. In other parts of the world, however, national labels are prevalent. Examples of such labels shown in Figure 9.10 include the "Blue Angel" (*umweltfreundlich*)

FIGURE 9.10   Eco-labels from around the world. (*Source:* OTA, 1992).

in Germany, the "White Swan" label used in Sweden, Norway, and Denmark, Japan's "EcoMark," and the "Environmental Choice" label used in Canada.

Germany's Source Reduction Program

The former West Germany has long been recognized as a leader in solid waste reduction efforts. Its Blue Angel environmental labeling scheme was initiated in 1978, which makes it the world's most well-established system. The Blue Angel designation is given to categories of products that meet certain environmental criteria. By 1991, 3600 products in 64 product categories were so designated. Labels are found on categories of goods such as paints, lacquers, and varnishes when they have low volatility, batteries when they have sufficiently low amounts of hazardous substances, copiers that reduce waste and have low emissions, water-efficient plumbing systems, energy-efficient windows, and biodegradable lubricating oils (Newman, 1991). The rating system is controlled by an Environmental Label Jury made up of individuals from a variety of organizations such as environmental and consumer groups, scientific organizations, labor unions, industry representatives, and federal states. Most of the funding comes from publications and product certifications, supplemented with small amounts of federal money.

While the original goal of the Blue Angel labeling system was to have product evaluations be based on a wide range of environmental criteria, in practice that has usually not been the case since such evaluations would be so complex. The program has, however, been credited with a number of successes in helping to shift the market toward environmentally superior products.

A much more radical component of Germany's solid waste reduction effort has been the enactment of a strict new law called the *German Federal Ordinance Concerning Avoidance of Packaging Waste,* which was passed in April 1991. With landfills rapidly filling to capacity, packaging wastes, which constitute roughly one-third of all municipal solid wastes, were targeted for drastic reductions. The law places the burden for packaging waste reductions directly on industry by requiring that companies take back and recycle used packaging.

Three types of packaging are defined by Germany's packaging law and their regulation went into effect in three stages. The first stage, which went into effect in 1991, deals with packaging that protects products during transport from manufacturers to distributors, such as large, corrugated shipping containers and wooden pallets. Transport packaging, which accounts for approximately one-third of all packaging, must be taken back by the manufacturers and distributors to be reused or recycled outside of the public waste system. German industry is responding by creating more durable, versatile, and reusable transport container systems designed to handle a wide range of products.

The second phase, initiated in April 1992, requires that secondary packaging, such as boxes around toothpaste tubes, cellophane wraps around cardboard boxes, and plastic blister packs, be collected by distributors. Distributors (including retailers) must either remove secondary packaging before products reach the consumer, or they must provide clearly marked bins near the check-out counter that enable consumers to deposit excess packaging before they leave the store. Retailers, not wanting to become local garbage dumps, have responded by putting pressure on suppliers to reduce or eliminate secondary packaging. The results are striking, especially to visitors from the United States. Blister packs have almost disappeared, products such as toothpaste tubes no longer have boxes, and plastic foam packing materials have all but been abandoned completely. While various studies suggest dramatic decreases in secondary packaging because of this law, the impact is somewhat symbolic since such packaging constituted less than 1 percent of municipal solid waste before the law went into effect (Steuteville, 1994).

Finally, phase three, which went into effect in 1993, covers sales packaging. Sales packaging comes in direct contact with the product, such as beverage containers, food wrappers, and cans, plastic containers and boxes holding products like laundry detergent, fabric softener, peanut butter, and spray paints. German industry is given a choice—either achieve strict recycling quotas or the government will require large deposit fees and stores will be required to take back these packaging wastes. The amount of sales waste is enormous—constituting roughly two-thirds of the packaging waste stream—so industry has been strongly motivated to set up an effective source reduction and recycling system to avoid the threat of their retail outlets having to take back such large quantities of materials. What has resulted is a not-for-profit consortium of hundreds of companies that have banded together to create a labeling, collection, sorting and recycling system called the *Duales System Deutschland* (DSD). DSD is a private system separate from the existing municipal solid waste system—hence the word *dual*.

A key element of the Duales system is the Green Dot (*der grüne punkt*) label, which identifies packages that can be discarded into special DSD recycling bins. An example of the green dot recycling label (which may not be green in color) is shown in Figure 9.11. DSD charges a licensing fee for use of the Green Dot, which depends on the material and weight of the packaging. Fees, which run as high as 20 pfennig (U.S. $0.12) per package, are expected to raise 4 billion Deutsch marks per year ($2.4 billion) (Steuteville, 1994).

The Green Dot label appears only on one-way packaging, so it is not really an environment label per se. It simply indicates that the packaging will be accepted by the DSD system. Reusable containers, for example, which are usually considered to be the best for the environment, do not use the Green Dot. In spite of that potential confusion, almost all beverages in Germany now come in refillable containers, and hefty deposit fees help assure that they will be returned for reuse.

FIGURE 9.11   The *Duales System Deutschland* Green Dot label, which identifies products that can be discarded in special recycling bins.

The DSD has to collect and sort packaging waste but is not in the recycling business. The DSD collection system includes bins conveniently located near residences and in stores. And the collection system must harmonize with other local waste management systems, which means the *der grüne punkt* bins sit side by side with recycling bins that accept glass and paper and other recyclables.

The net effect of the Duales system is that management of roughly one-third of all of Germany's solid waste has been shifted from the taxpayers to the private sector, with the resulting costs being built right into the price of products purchased by consumers. There is concern for international implications of the law since it applies to all products sold in Germany, regardless of their country of origin. Many European countries, including France, Netherlands, and Austria, have put similar programs in place, and the European Community is attempting to implement an EC eco-label.

9.5 COLLECTION AND TRANSFER OPERATIONS

The collection of solid wastes and their transport to incinerators and landfills accounts for roughly three-fourths of the total cost of refuse service. In the past, collection and transport decisions by professionals involved in solid waste management were focused on selecting the proper number and size of trucks, choosing the most efficient collection routes and schedules, locating transfer stations if they were to be used, and administering the whole system. With the growing importance of recycling and composting, those basic operations have become more complicated. Now a municipality may have separate trucks, routes, schedules, and destinations for recyclables and compostable materials—all of which need to be coordinated with already existing refuse collection systems.

A number of collection options are available for communities with recycling programs. In some areas drop-off centers, typically located in convenient places such as supermarket parking lots, accept recyclables that the consumer delivers in person. These may provide the only recycling service in a community, which is often the case for rural areas, or they may supplement curbside collection. Curbside collection systems vary considerably, and the collection vehicles that are used depend on the degree of sorting that residents are expected to perform. It has been common for a separate truck for recyclables to make its own collection run, but co-collection systems, in which a single truck picks up both recyclables and residential solid waste, are beginning to appear more often.

Collection of MSW

Conventional collection vehicles for municipal solid waste are distinguished by their size, how they are loaded, and how much compaction of refuse they can accomplish. Trucks designed for residential refuse service are usually rear loaded or side loaded. Rear-loaded trucks more readily accommodate large, bulky items, and they tend to be able to provide greater refuse compaction, while side-loaded trucks tend to cost less and are often more appropriate in densely populated areas where collection takes place on one side of the street at a time. A range of sizes for either type of collection truck is available, from roughly 5 to 30 m³ (6 to 40 yd³). Obviously, larger trucks cost more, but they do not have to make as many trips back and forth to the disposal site, which can more than offset the higher capital costs. Larger trucks, however, are also less maneuverable in crowded urban areas. Examples of typical collection vehicles are shown in Figure 9.12.

Typical municipal solid waste at the curbside has a density of roughly 120 kg/m³ (200 lb/yd³). At that low density, collection vehicles fill too fast, which means multiple, time-wasting trips to the disposal site would be needed. Modern trucks, called packers, have hydraulic compactors that can compress that waste to as much as 900 kg/m³ (1500 lb/yd³). Most compactor trucks, however, have compaction densities closer to 450 kg/m³ (750 lb/yd³). The ratio of the density of waste in the truck to the density at curbside is known as the compaction ratio. Ratios of 3 to 4 were the norm in the recent past, but newer generations of trucks are being offered with 8-to-1 compaction.

FIGURE 9.12   Various trucks for municipal solid waste and recyclables collection. (*a*) Recycling trailer, (*b*) low-profile closed-body recycling truck, (*c*) crane truck for "igloo" containers, (*d*) rear-loading packer truck, (*e*) hook-lift truck for dumpsters, (*f*) dual-compartment co-collection truck. (*Source:* Graham, 1993)

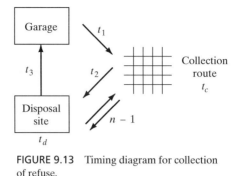

FIGURE 9.13   Timing diagram for collection of refuse.

Direct Haul Systems

We can use some simple estimates to help gain a sense of how truck sizes and collection patterns might be planned. Consider Figure 9.13, which shows a basic collection system consisting of the garage, where collection vehicles are parked overnight, the collection route, and the disposal site, where refuse is deposited. Later, the possibility of incorporating a transfer station between the collection route and the disposal site to reduce lost time by the collection vehicles will be introduced.

To work up a basic plan for refuse collection, let us introduce the following notation:

t_1 = time to drive from garage to beginning of collection route
t_2 = time to drive between collection route and disposal site
t_3 = time to drive from disposal site back to garage at end of day
t_c = total time spent collecting refuse
t_d = time spent at the disposal site dropping one truck load
t_b = time per workday spent on breaks, etc.
T_t = total time for one day of refuse collection
n = number of runs from the collection route to the disposal site

No matter what, each day's collection requires at least a drive from the garage to the collection route, time spent on the route, a run to the disposal site, time spent there unloading, and a drive back from the disposal site to the garage. More than one run may be made between the collection route and the disposal site, depending on how big the truck is, how many pick-ups are to be made, and how much refuse is picked up at each stop. The following describes a single day of collection:

$$T_t = t_1 + t_c + t_2 + t_d + (n - 1)(2t_2 + t_d) + t_3 + t_b$$
$$T_t = t_1 + (2n - 1)t_2 + nt_d + t_3 + t_b + t_c \tag{9.2}$$

EXAMPLE 9.1   Time Spent Collecting Refuse

Suppose it takes 0.4 hour (24 minutes) to drive from the garage to the beginning of the route, 0.4 hour (24 minutes) to drive between the route and disposal site, and 0.25 hour (15 minutes) to return from the disposal site to the garage. It takes 0.2 hour (12 minutes) to offload a truck at the disposal site. The crew is given two 15-minute breaks per day, and another 30 minutes is allowed for unexpected delays (total 1 hour). If two runs are made to the disposal site each day, how much time is left in an 8-hour shift for actual refuse collection?

Solution   Rearranging (9.2) and substituting appropriate values,

$$t_c = T_t - t_1 - (2n - 1)t_2 - nt_d - t_3 - t_b$$
$$= 8 - 0.4 - 3 \times 0.4 - 2 \times 0.2 - 0.25 - 1 = 4.75 \text{ hours left for actual collection}$$ ■

To decide how many customers can be served per day by one collection truck, we need the average length of time needed to service each stop. A simple approach suggests that the time per stop is equal to the average time that it takes to drive from one stop to another, plus the time taken at each stop to empty the containers. For example, if we assume that the truck speed averages about 2 m/s (about 5 mph) between stops, and if it takes roughly 10 seconds to empty a container, we might start with an expression such as the following:

$$t_s = 0.5D + 10N_c \tag{9.3}$$

where

t_s = average time for one stop (s)
D = average distance between stops (m)
N_c = number of containers to empty per stop

To find the volume of an appropriate collection vehicle, we need to know the number of stops, the volume of refuse per stop, and the vehicle compaction ratio:

$$V = \frac{vN}{r} \tag{9.4}$$

where

V = collection vehicle volume (m^3)
v = average volume of refuse at each collection point $(m^3/stop)$
r = compaction ratio (curbside volume/compacted volume in truck)
N = number of stops per truck load

EXAMPLE 9.2   Vehicle sizing

Continuing with the situation described in Example 9.1, suppose the average distance between stops along the route is 80 m, and there are typically two containers at each stop. Each stop is for an individual household that puts 0.25 m^3 of refuse at the curb each week, with density 120 kg/m^3. How many stops would be made per truckload, and how large should the truck be if it has a compaction ratio of 3.5?

Solution   Using (9.3), the time required to service each stop would be

$$t_s = 0.5D + 10N_c$$

$$= 0.5 \text{ (s/m)} \times 80 \text{ m} + 10 \text{ (s/container)} \times 2 \text{ containers/stop} = 60 \text{ s/stop}$$

In Example 9.1, we found that there are 4.75 hours of collection time per day. At 60 seconds per stop and two truckloads per day being collected, the number of stops that could be made per truckload is

$$N = \frac{4.75 \text{ h/d} \times 3600 \text{ s/h}}{60 \text{ s/stop} \times 2 \text{ truckloads/d}} = 142.5 \text{ stops/truckload}$$

The truck volume needed would be

$$V = \frac{vN}{r} = \frac{0.25 \text{ m}^3/\text{stop} \times 142.5 \text{ stops/load}}{3.5(\text{m}^3 \text{ at curb/m}^3 \text{ in truck})} = 10.2 \text{ m}^3$$

So you would choose a truck size that provides at least 10.2 m^3 of hauling capacity.   ■

The size of the collection vehicle in Example 9.2 was based on an assumption that each truck would make two runs to the disposal site per day. If the truck is large enough, perhaps only one run per day to the disposal site would be needed. In fact, if we rework Examples 9.1 and 9.2 using only one run, we find that a 24.6-m^3 truck would be needed. With three runs per day, there is so much time taken up making trips back and forth to the disposal site that the time remaining to fill a truck is so low that a 5.4-m^3 truck is all that would be needed. So the natural question to ask is, Which choice is best? Should we design the collection system around large, expensive trucks that make fewer trips, or go with smaller trucks, which are cheaper, but since they make more back-and-forth runs to the disposal site each day they serve fewer customers per truck?

Economics of Collection

To decide on the truck size that would provide the cheapest waste transport, we need to know the annual cost of owning and operating trucks, including the cost of the crew that makes the pick-ups. Finding the annualized cost of each truck involves using an engineering economy calculation in which the capital cost, amortized over the lifetime of the vehicle, is added to the estimated annual maintenance and fuel costs. The relationship between the purchase price of capital equipment, such as trucks, and the amortized yearly costs is given by

$$A = P\left[\frac{i(1 + i)^n}{(1 + i)^n - 1} \right] \tag{9.5}$$

where

$$A = \text{annual cost (\$/yr)}$$
$$P = \text{purchase price (\$)}$$
$$i = \text{discount factor (yr}^{-1})$$
$$n = \text{amortization period (yr)}$$

The quantity in brackets in (9.5) is known as the *capital recovery factor*. One way to interpret (9.5) is to think of A as being the annual payments that would pay off a loan of P dollars in n years if the loan interest rate is i (the decimal fraction representation of i should be used; e.g., 10 percent should be entered as 0.10). For example, a mid-size collection vehicle costing \$125,000, if amortized over a five-year period using a 10 percent discount factor, would have an annualized cost of

$$A = 125{,}000 \left[\frac{0.10(1 + 0.10)^5}{(1 + 0.10)^5 - 1} \right] = 125{,}000 \times 0.2638 = \$32{,}975/\text{yr}$$

These vehicles are heavy, and their start-and-stop driving pattern leads to very low fuel efficiency—1 or 2 miles per gallon is typical. An average truck driven 30 miles per day, 260 days per year, at 2 mpg and \$1.25 per gallon would cost almost \$5000 per year in fuel alone. If we add maintenance, insurance, and other costs, this typical packer truck can easily cost on the order of \$50,000 to \$80,000 per year—not counting the labor cost of the crew.

The annualized cost will depend on the size of the vehicle, and one approach is to model it as follows:

$$\text{Annualized cost (\$/yr)} = \alpha + \beta V \tag{9.6}$$

where α and β are empirically determined estimates based on a survey of available vehicles, and V is the volume of the truck.

The following example illustrates how such an economic decision could be made.

EXAMPLE 9.3   An Economic Analysis

Suppose the annualized cost of purchasing, fueling, and maintaining a compactor truck is given by the following expression:

$$\text{Annualized truck cost (\$/yr)} = 10{,}000 + 4000V$$

where V is the truck volume in m^3. Suppose these trucks require two-person crews, with labor charged at \$15 per hour each (including benefits). Do an economic analysis of the collection system in Example 9.2 in which a 10.2-m^3 truck collects refuse from 142.5 households on each of its two trips per day to the disposal site. The trucks and crew work five days per week, and curbside pick-up is provided once a week for each house. What is the cost per tonne of refuse collected if each household puts out 0.25 m^3 per week of waste with curbside density 120 kg/m^3? Also, what is the cost per year per household?

Solution   If we assume 8-hour days, 5 days/week, 52 weeks/yr, the annualized cost of labor per truck would be

Labor cost = 2 person $\times$ \$15/hr $\times$ 8 hr/day $\times$ 5 days/week $\times$ 52 week/yr = \$62,400/yr

The annualized cost of the truck would be

Annualized cost of the truck (\$/yr) = 10,000 + 4000 $\times$ 10.2 = \$50,700/yr

Since the truck is full after collecting waste from 142.5 houses and two trips per day are made to the disposal site, a total of 285 houses are served each day. Over a five-day week, 1425 houses ($5 \times 285 = 1425$) are served by each truck. The total amount collected by one truck in one year is

$$\text{Refuse/truck} = \frac{0.25 \text{ m}^3/\text{house} \cdot \text{week} \times 120 \text{ kg/m}^3 \times 1425 \text{ homes} \times 52 \text{ weeks/yr}}{1000 \text{ kg/tonne}}$$

$$= 2223 \text{ tonne/yr}$$

The annual cost per tonne of waste collected is therefore

$$\text{Annual cost} = \frac{(62{,}400 + 50{,}700)\$/\text{yr}}{2223 \text{ tonne/yr}} = \$50.88/\text{tonne}$$

The annual cost per household for this collection service is

$$\text{Annual cost} = \frac{\$50.88/\text{tonne} \times 2223 \text{tonne/yr}}{1425 \text{ households}} = \$79.38/\text{yr}$$

■

Table 9.9 extends the analysis of Examples 9.2 and 9.3 to trucks making one run or three runs per day to the disposal site. Under the assumptions made in this analysis, midsize trucks are optimum. Larger trucks make fewer runs to the disposal site, but their capital cost is too high. Use of smaller trucks holds their capital cost down, but the extra labor costs more than offset that advantage.

The solution represented by Table 9.9 is quite sensitive to the distance between collection area and disposal site. As that distance increases, having fewer trucks driving that long distance only once per day begins to become the most cost effective design, as is shown in Figure 9.14.

Transfer Stations

As convenient local landfills close, it is often the case that the replacement site is located many miles away from the community that is being served. As that distance from the collection area increases, it takes more and more time to haul the refuse to the disposal site, which leaves less and less time for each truck to collect the wastes. At some point, it is better to construct a facility close to town, called a *transfer station,* that acts as a temporary repository for wastes dropped off by local garbage trucks. Larger, long-haul trucks are then used to transport wastes from the transfer station to the disposal site.

TABLE 9.9   Comparison of Costs for Trucks Making One, Two, or Three Trips Per Day to the Disposal Site Under Conditions Given in Examples 9.2 and 9.3

Number of trips per day	Houses served per truck	Minimum truck size (m³)	Annualized Costs Trucks ($/yr)	Annualized Costs Labor ($/yr)	Cost per tonne ($/tonne)	Cost per household ($/yr)
1	1725	24.6	108,570	62,400	63.53	99.11
2	1425	10.2	50,710	62,400	50.88	79.38
3	1125	5.4	31,430	62,400	53.46	83.40

Note: The one-way trip to the site is assumed to take 0.4 hour.

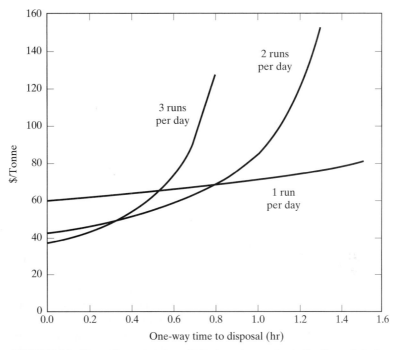

FIGURE 9.14   The optimum number of trips to make per day to the disposal site is very sensitive to the time needed to make the drive. For short distances, smaller trucks making more trips is optimum. For long distances, large trucks that can collect all day long before disposal become more cost effective. This figure is based on data in Examples 9.2 and 9.3.

Again, a design decision must be made: At what point is the disposal site so far away that the added costs of a transfer station are justified? The problem we need to analyze is shown in Figure 9.15. The procedure for deriving the cost versus distance graph in Figure 9.15*a*, corresponding to a direct haul by the collection vehicles, has already been demonstrated. Figure 9.15*c* is an example of what that might look like. What needs to be done now is an analysis of the transfer station and its associated fleet of long-haul trucks.

The cost of a transfer station itself depends on many factors, including its size, local construction costs, and the price of land. In addition, transfer stations can employ a variety of technological features that also contribute to costs. Simple transfer stations may consist of just a lightweight building shell with a thick concrete slab called a tipping floor. Collection vehicles drop their refuse onto the tipping floor while a front loader scoops it up and loads the transfer vehicles. More complex facilities may include hoppers for direct deposit of refuse from collection vehicles into transfer vehicles. They may also have compaction equipment to compress the wastes before loading. Some transfer stations compress wastes into dense bales ($600 \, \text{kg/m}^3$) that can be loaded with forklifts onto flatbed trucks. Baled refuse is especially convenient for rail or barge transport.

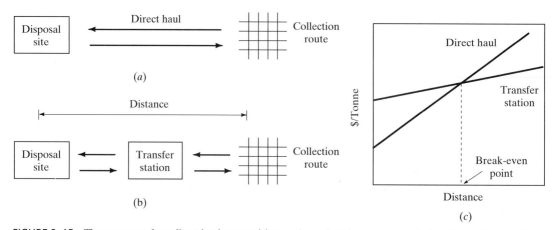

FIGURE 9. 15 The cost curve for a direct haul system (*a*) starts lower but rises more steeply than the cost curve when a transfer station is included (*b*). At some distance, where the curves cross, the transfer station option becomes more cost effective (*c*).

Trucks that haul wastes from transfer stations to the disposal site are usually large tractor-trailer vehicles. The trailers may be open topped or closed, but closed trailers are preferred since they are less likely to spread debris along the highway and the wastes can be more easily compacted in the trailer. Open trailers carry on the order of 30 to 100 m^3 of refuse with densities of 120 to 180 kg/m^3. Since the density of wastes in closed trailers may be higher (300 to 500 kg/m^3), closed trailers are usually smaller (30 to 60 m^3) to keep the total vehicle weight below highway load limits.

EXAMPLE 9.4 Costs of a Transfer Station and Vehicles

A transfer station handling 300 tonnes/day, five days per week, costs $4 million to build and $100,000 per year to operate.

An individual tractor-trailer costs $100,000 and carries 15 tonnes per trip. Operation and maintenance costs (including fuel) of the truck are $30,000/yr; the driver makes $40,000 per year (including benefits). The capital costs of the building and transfer trucks are to be amortized over a 10-year period using a 12 percent discount factor.

Suppose it takes 1-hour to make a round trip from the transfer station to the disposal site, and seven trips per day are made. Find the transfer/hauling cost in dollars per tonne.

Solution Using (9.5), the capital recovery factor for a 10-year amortization period and 12 percent discount factor is

$$\text{CRF} = \frac{i(1 + i)^n}{(1 + i)^n - 1} = \frac{0.12(1 + 0.12)^{10}}{(1 + 0.12)^{10} - 1} = 0.177/\text{yr}$$

The annualized cost of the transfer station is therefore

$$\$4,000,000 \times 0.177/\text{yr} + \$100,000/\text{yr} = \$808,000/\text{yr}$$

Per tonne of waste, that works out to be

$$\text{Transfer station} = \frac{\$808,000/\text{yr}}{300 \text{ tonnes/day} \times 260 \text{ days/yr}} = \$10.36/\text{tonne}$$

The annualized cost of a single truck and its driver is

$$\$100,000 \times 0.177/\text{yr} + \$30,000/\text{yr} + \$40,000/\text{yr} = \$87,700/\text{yr}$$

Per tonne of waste hauled, the truck and driver cost

$$\text{Truck and driver} = \frac{\$87,700/\text{yr}}{15 \text{ tonne/trip} \times 7 \text{ trips/day} \times 260 \text{ days/yr}} = \$3.21/\text{tonne}$$

Total cost of transfer station and trucks is therefore

$$\$10.36 + \$3.21 = \$13.57/\text{tonne}$$ ∎

Example 9.4 was worked out for a round trip from transfer station to disposal site of one hour. Figure 9.16 extends this calculation to show how costs depend on trip length. For the one-hour round trip (including unloading time), the fixed cost of the transfer station of $10.36/tonne and the variable cost for the trucks (0.5 hour each way) of $3.21/tonne are shown.

Suppose an ideal site for a transfer station is located 0.3 hours from town. Continuing Example 9.4, Figure 9.14 shows that the cheapest trash collection system could use either two or three runs per day from town to the transfer station at a cost of $48/tonne. Adding the cost of a transfer system to carry the waste the rest of the way to the disposal site is shown in Figure 9.17. From Figure 9.17 it can be seen that

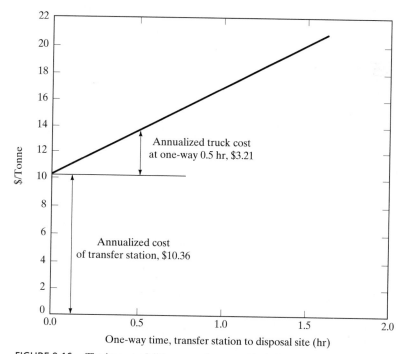

FIGURE 9.16   The impact of distance on the cost of including a transfer station in the waste collection system. Drawn for the data given in Example 9.4.

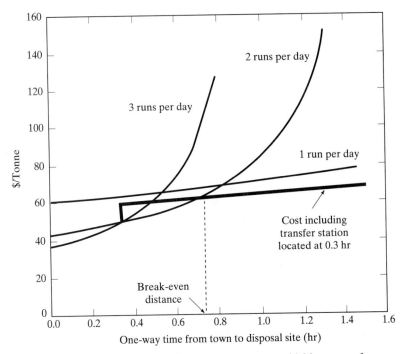

FIGURE 9.17 The implications of a transfer station located 0.3 hour away from the collection route, using data given in Examples 9.3 and 9.4.

for one-way distances between the collection route and the disposal site of less than about 0.75 hr, no transfer station is warranted—it is cheaper to make two runs per day from town all the way out to the disposal site. If a one-way drive to the disposal site takes longer than 0.75 hour, a transfer station should be used since at that point it offers the lowest total cost. These results, of course, are specific to the example presented. They should, however, give a sense of the tradeoffs that need to be evaluated in decision making.

9.6 RECYCLING

After source reduction, which is given the highest priority in the solid waste management hierarchy, the recovery of materials for recycling and composting is generally thought to be the next most important component of integrated solid waste management programs. Resource recovery means that the materials have not only been removed from the municipal waste stream, but they also must be, in essence, purchased by an end user. That distinction, for example, means yard trimmings composted at home are considered to be source reduction, not recycling. On the other hand, yard trimmings that are delivered to an off-site composting facility and then sold are considered to be recovered or recycled materials.

The term *recycling* is often misconstrued to include activities such as refilling bottles for reuse and remanufacturing products for resale to consumers, but it is better to use the term *recycling* only when materials are collected and used as raw materials for new products. The process of recycling includes collecting recyclables, separating them by type, processing them into new forms that are sold to manufacturers, and, finally, purchasing and using goods made with reprocessed materials.

Another distinction that can be made is between *preconsumer* and *postconsumer* recyclable materials. Preconsumer materials are associated with the manufacture of products and consist of scrap that is recycled back into the original manufacturing process without ever having been turned into a useful product. Postconsumer recyclables are products that have already been used by consumers for their originally intended purposes and that are now part of the waste stream. Preconsumer recycled materials are not included in EPA data on generation, recovery, and discards, and they generally are not counted in procurement laws that mandate government purchases of products with significant recycled-materials content.

The recycling symbol most often used in the United States also has subtle features. The three arrows chasing each other in a triangle can mean the product is made from recycled materials, or they can mean the materials are recyclable, as shown in Figure 9.18.

Fine distinctions between terms such as *resource recovery, reuse, remanufacturing, recyclable, recycled, preconsumer* and *postconsumer* recycled materials, and so forth may seem annoying or even unnecessary, but they are important when systematic collection, reporting, and unambiguous interpretation of recycling data are needed.

The rate of recovery of materials from the waste stream, as reported by the Environmental Protection Agency, has increased dramatically in the last decade or so. Resource recovery grew from less than 10 percent of all municipal solid waste in 1980 to 22 percent in 1993. Most of that increase is attributable to greater rates of recovery of paper and paperboard and increased composting of yard trimmings. Table 9.10 shows U.S. recovery rates for a number of categories of MSW. Notice how dominant the paper and paperboard category is in terms of total tonnes of material recovered (almost 60 percent), which tends to mask the importance of recovery rates for other materials. For example, recovered aluminum is only about 3 percent of the total mass of recovered materials, but in terms of its total economic value it far exceeds the paper products category.

Examples of products that can be recycled, and typical applications of the raw materials that are so produced, are illustrated in Figure 9.19. Details of the products,

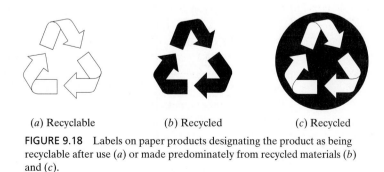

(*a*) Recyclable   (*b*) Recycled   (*c*) Recycled

FIGURE 9.18 Labels on paper products designating the product as being recyclable after use (*a*) or made predominately from recycled materials (*b*) and (*c*).

TABLE 9.10 Generation and Recovery of Materials in MSW in the United States, 1993 (millions of tonnes)

	Amount Generated	Amount Recovered	Percent Recovered
Paper and paperboard	70.6	24.0	34.1
Glass	12.4	2.7	22.0
Metals			
Ferrous metals	11.7	3.1	26.1
Aluminum	2.7	1.0	35.4
Other nonferrous metals	1.1	0.7	62.9
Total metals	15.5	4.7	30.4
Plastics	17.5	0.6	3.5
Rubber and leather	5.6	0.4	5.9
Textiles	5.5	0.6	11.7
Wood	12.4	1.2	9.6
Other materials	3.0	0.6	22.1
Total materials in Products	142.7	34.9	24.5
Other wastes			
Food wastes	12.5	Negligible	Negligible
Yard trimmings	29.8	5.9	19.8
Miscellaneous inorganic wastes	2.8	Negligible	Negligible
Total other wastes	45.1	5.9	13.1
Total municipal solid waste	187.7	40.8	21.7

Source: U.S. EPA, 1994a.

CONSTRUCTION WASTE, TIRES	PLASTICS, DRINK BOTTLES
Reprocessed for pressed board, roads, and other construction projects	Reprocessed for auto parts, fiberfill, strapping

ALUMINUM CANS	YARD WASTE
Reprocessed for can sheet and castings	Composted for landscaping, gardens

OTHER METALS	GLASS
Cleaned and reprocessed as scrap and structural products	Cullet for jars, bottles, construction material

PAPER	ANIMAL WASTE
Reprocessed as newsprint, paperboard, insulation	Used as fertilizer

FIGURE 9.19 Recyclables in the waste stream. (*Source:* U.S. EPA, 1989)

recycling potentials, processes, and applications of recycled materials will be considered in the following sections.

Paper and Paperboard Recycling

The largest single category of waste generation and waste recovery is paper and paperboard products. In 1993, 70,600 tonnes of paper products were generated and 34 percent (24,000 tonnes) of that was recovered. Two types of paper products, corrugated cardboard boxes and newspapers, together account for half of the paper products generated and three-fourths of paper recovered. The other major categories of paper products, along with their generation and recovery rates, are shown in Figure 9.20.

The Chinese invented paper 2000 years ago when they discovered that vegetable fibers, such as bamboo or mulberry, when soaked in water formed matts of fibrous material that bonded itself together and when dried could be used as a writing medium. Soon after, their process shifted to use of wood as the principal fiber source. The U.S. papermaking industry began in 1860 using recycled fiber from cotton, linen rags, and waste paper, but now virgin paper production is based almost entirely on pulps made from wood chips. Other sources of fiber can include straw, jute, flax, bagasse from sugar cane, and hemp—in fact, there is a surge of interest these days in legalizing hemp as a source of fiber to reduce the need for forest products.

Papermaking begins with pulping. For quality paper, the cellulose fibers in wood chips must be separated from the lignins that hold the fibers together. The process

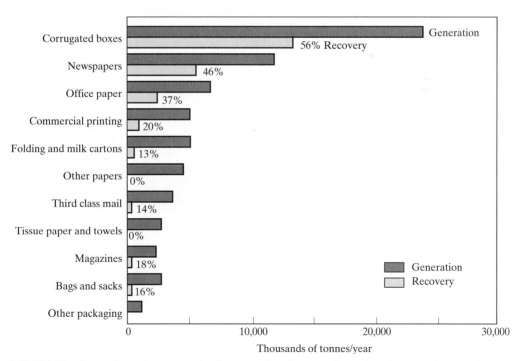

FIGURE 9.20   Generation and recovery of major categories of paper and paperboard products in the United States in 1993. (*Source:* U.S. EPA, 1994a)

involves cooking, to soften the wood, and chemical or mechanical treatment, to help separate the fibers. Most wood pulp is prepared using a chemical process. For low-quality paper, such as newsprint, whole chips of wood can be pulped mechanically, without removing the lignins, creating a product called *groundwood.* Pulp is naturally tan in color, so chemical bleaching may be required if the end product needs to be white. If the finished pulp needs to be transported some distance to the papermill, it will be dried and baled; otherwise it is pumped as a thick slurry to the mill.

Recycled wastepaper has long been used as a source of fiber to augment pulp made from virgin materials. Depending on the application, various fractions of recycled and virgin materials can be incorporated—all the way up to 100 percent recycled. Each recycling of wastepaper, however, results in losses so it is not possible to use the same fiber over and over again. Each time paper is soaked in water, heated, and processed, some of the cellulose breaks down into starches and is lost; also, in each processing some of the smaller fibers do not bond together and are lost as well. Not only is less fiber made available after each recycling, but the fibers are shorter and the resulting product is not as strong as product made from virgin materials. Nonetheless, there are abundant uses for recycled wastepaper.

As Figure 9.21 shows, roughly half of U.S. recycled paper products are used to make paperboard (cereal boxes, tablet backs, shoe boxes, etc.) and containerboard (corrugated cardboard). For such applications, product whiteness is not important. When paper products are to be recycled into newsprint or other papers, an additional de-inking stage must be added to the processing, which increases costs. Recycled wastepaper is also used to make various construction products such as cellulose insulation, fiberboard, and some flooring materials. Other applications include animal bedding, nursery pots, garden mulch, and trays.

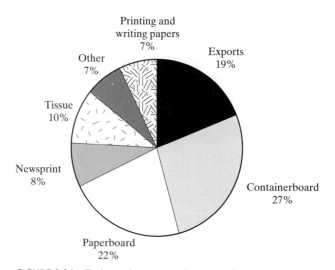

FIGURE 9.21   End uses for recovered paper and paperboard.
(Based on data in American Forest and Paper Association, 1993.)

The economic value of recycled paper very much depends on the characteristics of the paper collected. In fact, the Paper Stock Institute has defined 51 grades of recycled paper and paper products, and each category has its own pricing structure when sold. In the early 1990s, old newspapers (ONP) were worth roughly $15 to $20 per tonne, while old corrugated containers (OCC) were selling for approximately $25 to $35 per tonne. Among office papers, prices for grade 40 "Sorted White Ledger" were roughly $120 to $140 per tonne, while grade 42 "Computer Printout" was selling for close to $200 per tonne (Rabasco, 1995). The paper recycling business received a boost in October 1993, when President Clinton signed an executive order directing every federal agency to purchase printing and writing paper containing at least 20 percent postconsumer material by the end of 1994, and 30 percent by the end of 1998. That, along with other factors such as better recycling technologies and relatively scarce supplies with respect to demand, led to a sudden increase of prices in 1994, with most recycled paper more than doubling in value in a single year. That surge lasted for only a year or so, and by 1996 prices had declined back to their more normal levels.

Recycling Plastics

The rate at which plastics enter the municipal solid waste stream in the United States has grown phenomenally from 0.36 million tonnes per year in 1960 to almost 18 million tonnes in 1993 (U.S. EPA, 1994a). Plastics now comprise almost 9.3 percent of the weight of all municipal solid waste generated, but since they are so bulky they account for nearly one-fourth of the total volume in landfills. Although all plastics are theoretically recyclable, only 3.5 percent are actually recovered.

The term *plastics* encompasses a wide variety of resins or polymers with different characteristics and uses. Plastics are produced by converting basic hydrocarbon building blocks such as methane and ethane into long chains of repeating molecules called polymers. While there are natural polymers such as cellulose, starch, and natural rubbers, it is only the synthetic materials that are referred to as plastics. One way to categorize plastics is based on their behavior when heated. Roughly 80 percent are *thermoplastics,* which can be remelted and remolded into new products. The other 20 percent are *thermoset* plastics, which decompose when they are heated so they are more difficult to recycle.

The Plastic Bottle Institute of the Society of the Plastics Industry has developed a voluntary coding system for plastic bottles and other rigid containers that uses the familiar three arrows in a triangle, this time with a numerical designation that indicates the type of resin used in the plastic. Table 9.11 describes the labeling system and highlights the main uses of the designated plastics.

Although the three-arrow symbol is usually thought to mean the product is recyclable, in the case of plastics it is really only polyethylene terephthalate (PETE or PET) and high-density polyethylene (HDPE) that are recycled to any significant degree. Since most plastics are not recycled, the symbol has been severely criticized for potentially giving consumers the wrong impression, which has led to the possibility that it may be changed in the future.

PET and HDPE together account for 79 percent of all plastics recovered from the waste stream (U.S. EPA, 1994a), with most of that being milk jugs and clear plastic

TABLE 9.11 Types of Resins and 1993 Generation and Recycling Rates

Designation	Polymer	Applications	Generation (1000 tonnes/yr)	Recovery (%)
1 PETE	Polyethylene terepthalate (PET)	Soft-drink bottles, peanut butter jars; frequently recycled	890	26
2 HDPE	High-density polyethylene	Milk, water, juice, detergent, motor oil bottles; often recycled	3750	6
3 V	Vinyl/polyvinyl chloride (PVC)	Cooking oil bottles, credit cards, household food wrap, building materials; rarely recycled, toxic if burned	1120	0
4 LDPE	Low-density polyethylene	Shrink wrap, garbage and shopping bags, lids, cups; rarely recycled	5660	0.1
5 PP	Polypropylene	Snack food wrap, margarine tubs, straws, car batteries; occasionally recycled	1540	6
6 PS	Polystyrene	Pharmaceutical bottles, Styrofoam cups, clear salad bowl lids, packing peanuts; rarely recycled	2250	0.1
7 Other resins		Multilayered, mixed-material containers, squeezable bottles; rarely recycled	2340	0.1
TOTALS			17,500	3.5

Source: Generation and recycling data from U.S. EPA, 1994a.

soda bottles. PET provides a better barrier to transmission of gases than HDPE, which makes it a better container for carbonated beverages and other foods whose aromas need to be contained. PET also fetches a much higher price in the recycling market.

When plastics are recycled, it is often important to separate them carefully by resin type and color. Different resins have different melting points, so if a batch of mixed plastics is being heated and reformed into new products some resins may not melt at all and some may end up burning. A single PVC bottle in a batch of 20,000 PET bottles, for example, can ruin the whole batch and potentially damage the manufacturing equipment (Moore, 1994). Similarly, if polypropylene, which is difficult to distinguish from polyethylene, contaminates a batch of recycled polyethylene, the resulting blend would be useless. Plastic recycling is also complicated by the potential for contamination by the products that they once contained. Plastics do absorb, to varying degrees, materials that they come in contact with. Substances such as pesticides and various oils will migrate slowly into their plastic containers, and some will survive the melting and reforming processes when recycled. For that reason, the Food and Drug Administration keeps a careful eye on the use of recycled plastics for direct-food-contact applications.

Recycled plastics find application in a number of products. Recycled PET bottles, for example, are being transformed into quality fibers for jackets and other outdoor clothing, as well as fiberfill material for cushions. Recycled PET is also used to make carpeting, surfboards, sailboat hulls, as well as new PET soft-drink bottles. To avoid the contamination problem mentioned previously, recovered PET used in new soft-drink bottles is usually sandwiched between an inner and outer layer of virgin plastic. HDPE and LDPE are often recycled into new detergent bottles, trash cans, and drainage pipes. For some of the most popular uses for recycled HDPE and LDPE, minor color variations in the final product are not particularly important. For other products, such as colored detergent bottles, variations in color would affect consumer perceptions of the product, so presorting of recyclables by color can be important.

While it is usually important for recovered plastics to be sorted carefully, some applications are emerging that can use co-mingled plastics. Mixtures of plastics can be shredded, melted, and extruded into useful forms. Plastic lumber, which can be used for fence posts, siding, timbers, park benches, docks, and so forth, is formed in this way. Plastic lumber is more expensive than wood lumber, but it resists insect and weather damage better than wood. In countries such as Japan, which have little timber, plastic lumber is enjoying a growing niche in the construction market.

PVC in the waste stream poses particular problems. As already mentioned, small amounts can cause major contamination problems for recycled PET. It is also a source of potentially serious air pollution problems if it ends up being burned. At elevated temperatures, the chlorine in PVC combines with hydrogen to form toxic hydrogen chloride, which can be emitted into the atmosphere. To help avoid these problems, a new mechanical separation system has been devised that automatically separates PVC from other plastics at the recycling facility. Whole or crushed plastic bottles are moved on a conveyor belt past a sensor that detects the chlorine atoms in PVC. When PVC is detected, air jets are triggered that kick the PVC bottles away from the remaining containers (Moore, 1994).

Glass Container Recycling

The amount of glass in the municipal waste stream is one of the few materials that has shown a decrease over time. In 1980, 13.6 million tonnes of glass entered the waste stream, while in 1993 that amount had dropped to 12.4 million tonnes (U.S. EPA, 1994a). The impact of much lighter-weight aluminum cans and plastic drink bottles is evident. Even at those lower 1993 rates, glass entering the waste stream is still a rather large amount, 48 kg (105 lb) per person per year. The glass recovery rate in the United States is about 22 percent, most of which is made into new glass containers.

Figure 9.22 shows the rate at which the Unites States generates glass for various applications and the recovery percentage for each category. Ninety percent of the glass used in the United States is for bottles and jars of one sort or another. The remaining 10 percent, referred to as durable goods, consists of items such as window glass, fiberglass, mirrors, ceramic dishes, porcelain, glassware, ovenware, and light bulbs. The recycling rate for those durable goods is essentially zero, and in fact they are considered to

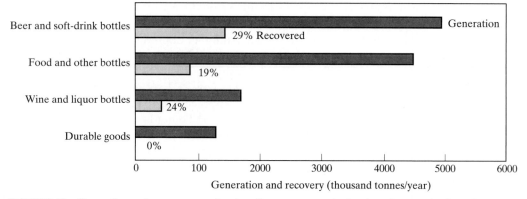

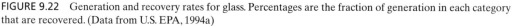

FIGURE 9.22   Generation and recovery rates for glass. Percentages are the fraction of generation in each category that are recovered. (Data from U.S. EPA, 1994a)

be contaminants in container glass recycling. Ceramics and heat-resistant cookware melt at much higher temperatures than container glass so they form solid or slightly melted chunks of impurities when the two are mixed together in a glass furnace. Metal caps and rings left on bottles also cause problems when they are mixed in with glass. Metals melt at lower temperatures than glass, and that molten metal can cause expensive corrosion problems in the glass furnace itself.

Recovered glass that will be remade into new container glass needs to be separated by color: clear (flint), green (emerald), or brown (amber). Those colors have traditionally been created with permanent dyes, but there are now colored polymer coatings that can be applied to the outside of clear glass containers. If these coatings would allow all glass used in containers to be clear, the need to separate by color as well as the need to use care to prevent bottle breakage during collection (which mixes bits and pieces of different colored glass together in the residue) would be eliminated. An additional advantage of the polymer coatings is that they make the container considerably stronger, which could allow bottles to be made even lighter than they are now (Rhyner, et al., 1995).

Once recovered glass has been separated by color and the metal rings and caps removed, it needs to be crushed into smaller pieces, called cullet. When significant quantities of color-separated cullet are accumulated, they are usually shipped to container manufacturers to be remade into new glass bottles and jars. Glass has the unusual property of being 100 percent recyclable; that is, the same bottle can be melted down and remade over and over again without any product degradation. Moreover, cullet melts at lower temperature than the raw materials from which new glass is made (silica sand, soda ash, and limestone), which helps save energy.

Not all cullet is remade into containers. Some is used in a mix with asphalt to form a new road-paving material called *glasphalt.* Cullet can also be used as a road base material in areas where suitable supplies of gravel may be limited. Other products that can be made using cullet include fiberglass, abrasives, reflective paint for road signs (made from small glass beads), lightweight aggregate for concrete, glass polymer composites, and glass wool insulation.

Aluminum Recycling

By most measures, the most valuable collectable in a municipal recycling program is aluminum cans. A tonne of aluminum is typically worth 10 times as much as a tonne of PET or HDPE, and it generates on the order of 20 times as much revenue per tonne as glass, steel cans, or newspapers. Even though it may comprise only 2 or 3 percent of the total tonnage of collected recyclables, it often generates on the order of 20 to 40 percent of the total revenues. This economic incentive is directly attributable to the fact that recycled aluminum uses only 2 to 3 percent of the energy required to make new aluminum from bauxite ore.

In 1993, aluminum beer and soft-drink cans made up 54 percent (1.46 million tonnes/yr) of the total 2.45 million tonnes of aluminum in municipal solid waste. Almost two-thirds of those cans (63 percent), roughly 60 billion cans per year, are recovered from the waste stream and recycled back into the production of aluminum (U.S. EPA, 1994a). It is estimated that 95 percent of recovered aluminum goes back into the production of the next generation of cans, with a turnaround time of as little as 90 days (Buckholz, 1993). The next biggest source of aluminum in MSW is from durable and nondurable goods such as home appliances and furniture, with almost none of that being recovered for recycling. These data are displayed in Figure 9.23.

Recycling aluminum not only helps reduce the demands placed on landfills, but doing so saves large amounts of energy as well. The production of new aluminum begins with the mining of bauxite ore. Then, using a procedure known as the Bayer process, impurities are removed and the aluminum hydrates and hydroxides in bauxite are converted to aluminum oxide, or alumina. Alumina is then electrolytically reduced to aluminum metal. That electrolysis, known as the Hall-Heroult process, consumes enormous amounts of electricity. On the average, the two-step conversion of bauxite to aluminum metal requires 15 kWh of electricity plus 60,000 kJ of fossil fuel energy per kilogram of metal produced (Atkins, 1991).

To recycle aluminum, the scrap is melted down in a furnace, mixed with various alloys and primary aluminum as needed, and then cast into ingots. The process requires

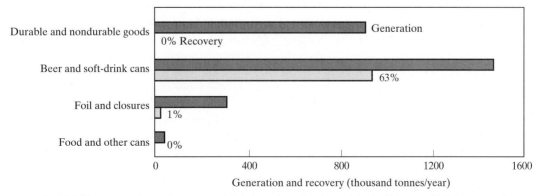

FIGURE 9.23   Generation and recovery of aluminum in 1993 U.S. municipal solid waste. (U.S. EPA, 1994a)

0.08 kWh of electricity and 4200 kilojoules of fossil fuel per kilogram of aluminum produced (Atkins, et al., 1991). Thus, recycling saves roughly 99 percent of the electricity and 93 percent of the fossil fuel energy compared with primary aluminum production.

It is awkward to keep expressing these energies both in terms of kilowatt-hours of electricity and kilojoules of fossil fuel. Combining them into a single, total amount of primary energy requires that we convert a kilowatt-hour of electrical output into the equivalent amount of heat that an average thermal power plant would need to generate that kilowatt-hours. This is known as the *heatrate,* which for the United States is close to 11,700 kJ/kWh (10,600 Btu/kWh). This conversion has been used in Table 9.12. Also shown there are estimates of the amount of the greenhouse gas, carbon dioxide, that would be expected to be released using virgin ores versus recycled aluminum. The carbon emissions are based on the assumption that half the electricity is obtained from hydroelectric sources and half from coal-fired power plants for primary aluminum production, and the fossil fuel for remelting scrap is in the form of oil and gas. About 96 percent of the carbon emissions are saved by recycling.

TABLE 9.12 Energy Requirements and CO_2 Emissions per Kilogram of Aluminum Produced From Bauxite and From Recycled Aluminum

Source	Electricity (kWh/kg)	Fossil Fuel (kJ/kg)	Primary Energy[a] (kJ/kg)	CO_2 Emissions[b] (kg/kg)
Bauxite	15	60,000	235,000	13.1
Recycled	0.08	4,200	5,150	0.48

Source: Based on data in Atkins et al. (1991).
[a] at 10,600 Btu/kWh
[b] half hydroelectric, half coal-fired power plants

EXAMPLE 9.5 Energy for the Aluminum in a Can

a. If the current recycling rate for aluminum cans is 63 percent, find the primary energy required to produce the aluminum in a 355 mL (12 oz) aluminum can having a mass of 16 g.

b. How much energy is saved when a can is recycled instead of being thrown away?

c. What is the equivalent amount of gasoline wasted when one can is thrown away? (gasoline has an energy content of about 35,000 kJ/L.)

Solution

a. The primary energy needed for the 37 percent of the can made from bauxite is

$$\text{Primary energy} = \frac{16 \text{ g/can} \times 0.37 \times 235,000 \text{ kJ/kg}}{1000 \text{ g/kg}} = 1391 \text{ kJ/can}$$

and the primary energy needed to produce the rest from recycled aluminum will be

$$\text{Primary energy} = \frac{16 \text{ g/can} \times 0.63 \times 5150 \text{ kJ/kg}}{1000 \text{ g/kg}} = 52 \text{ kJ/can}$$

So the total for an average can with 63 percent recycled aluminum is

$$\text{Primary energy} = 1391 + 52 = 1443 \text{ kJ/can}$$

b. Every time an aluminum can is not recycled, a new can must be made from bauxite to replace it. The energy to produce the aluminum for a new can from ore is

$$\text{Primary energy} = \frac{16 \text{ g/can} \times 235,000 \text{ kJ/kg}}{1000 \text{ g/kg}} = 3760 \text{ kJ/can}$$

If the can is recycled, it can be melted down and made into the next new can, which requires

$$\text{Primary energy} = \frac{16 \text{ g/can} \times 5150 \text{ kJ/kg}}{1000 \text{ g/kg}} = 82 \text{ kJ/can}$$

The energy saved by recycling is therefore

$$\text{Energy saved} = 3760 - 82 = 3678 \text{ kJ/can}$$

c. The equivalent amount of gasoline thrown away when a can is not recycled is

$$\text{Gasoline equivalent} = \frac{3678 \text{ kJ/can}}{35,000 \text{ kJ/L}} = 0.105 \text{ L} = 105 \text{ mL}$$

Since the can holds 355 mL of soda, that is like throwing away almost one-third of a can of gasoline. ∎

The analysis in Example 9.5 tells us the amount of energy needed to produce the aluminum for one can given the current recycling rate. It does not tell us the total amount of energy required to make an aluminum can, however. As shown in Figure 9.24, once an ingot of aluminum is cast, more energy is required to roll it into sheets, to transport those sheets to the can manufacturing plant, to manufacture the cans, and to transport used cans back to the plant that remelts the aluminum. Table 9.13 provides an estimate of the energy needed to fabricate aluminum cans. The total energy invested in a single 0.355-L (12-oz), 16-g can is about 2826 kJ. At the current 63 percent recycling rate, that is about double the 1443 kJ of energy required to produce the aluminum itself from bauxite and recycled cans.

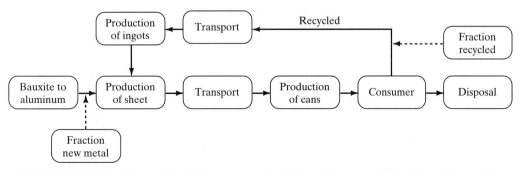

FIGURE 9.24 A flow diagram for aluminum cans produced from primary materials and recycled materials.

TABLE 9.13 Energy Required to Make a Single 16-g, 12-oz aluminum can[a]

Process	Energy (kJ)
Bauxite to make 6 g of aluminum metal	1391
Recycled cans to make 10 g of aluminum	52
Production of 16 g of sheet from ingot	421
Transport 16 g sheet to can manufacturer	4
Production of single 16 g can	955
Transport 10 g of recycled cans to foundry	3
TOTAL	2826

[a]Assumes 63 percent recycling rate.
Source: Analysis based on data in Atkins et al. (1991) and International Petroleum Industry Environmental Conservation Association (1992).

Other Metals

Metals can be conveniently classified into two categories: ferrous and nonferrous. Nonferrous metals (e.g., aluminum, copper, lead, zinc) have little or no iron in them, while ferrous metals do. In keeping with the EPA definitions, the only metals we will consider here are those that end up in the municipal solid waste stream as discarded refrigerators, stoves, water heaters, and so forth. That excludes, for example, preconsumer scrap that is recovered during metal manufacturing, and it also excludes old automobile and truck bodies that are crushed and recycled.

In 1993, 15.6 million tonnes of metals entered the municipal solid waste system. Ferrous metals account for the largest chunk of that total, 76 percent. Aluminum is the second largest at 17 percent, and the remaining 7 percent are other nonferrous metals. Total generation and recovery of these three categories of metals are shown in Figure 9.25. Overall, 30 percent of metals generated were recovered from the waste stream.

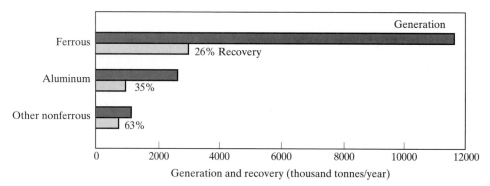

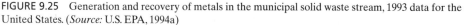

FIGURE 9.25 Generation and recovery of metals in the municipal solid waste stream, 1993 data for the United States. (*Source:* U.S. EPA, 1994a)

Recovery of ferrous metals in white goods (appliances) was estimated to be 68 percent; 48 percent of the steel food and beverage cans were recycled; and 35 percent of the aluminum was recovered. The recovery rate of "other nonferrous" metals is quite high, averaging 63 percent. That category is dominated by lead, most of which is in car batteries, with 95 percent of that battery lead being recovered (U.S. EPA, 1994a).

Materials Recovery Facilities

Recyclable materials need to be sorted to separate the glass, plastics, newspapers, cans, and so forth. If a community has a recycling system in place, some of this separation is usually done by consumers, some may be done by the crew of the recycling truck as they make their pick-ups, and some is typically done at a *materials recovery facility* (MRF). If there is no recycling system geared toward consumers (that is, materials that could be recycled are mixed in with all of the other refuse collected), there still may be some attempt at recovery. A facility that tries to recover recyclable materials that are mixed in with all the usual municipal solid waste is referred to as a *waste processing facility* (WPF) or sometimes as a *front-end processing system.* These definitions are not precise; for example, a not uncommon collection system has consumers put their commingled recyclables into separate (usually blue) bags, which are tossed into the truck with the rest of the mixed waste. The blue color of those bags helps them stand out visually from dark green, plastic garbage bags, which facilitates their removal at a processing facility.

The primary function of a materials recovery facility is to separate bottles by color, plastics by resin, cans by their metal content, as well as old newspapers (ONP), old corrugated containers (OCC), and compostable organics. The second function is to densify those separated materials so that they can be more easily shipped to end users. Densification includes crushing bottles, flattening metal cans, granulating and baling plastics, and baling waste paper.

Since different communities have differing degrees of preseparation by consumers and truck crews, materials recovery facilities also have a range of equipment and processes. Almost all will include a significant amount of hand separation supplemented by varying amounts of automated equipment. Smaller facilities tend to use more labor since they may not process enough waste to justify the cost of automated equipment, but even large facilities use workers for some amount of handpicking and quality control. An example MRF is illustrated in Figure 9.26.

The usual MRF is designed to receive old newspapers and corrugated cardboard that have already been separated from the bottles, cans, plastics, and glass. This not only simplifies processing, but it helps reduce contamination of paper products from food and chemical wastes that may accompany the other recyclables. A separate area of the tipping floor, with its own conveyor belt, receives newspapers and cardboard. Sorters stand by the conveyor belt, removing cardboard, which will be baled later, along with paper and plastic bags, string, and other foreign materials, which are discarded. Newspapers remain on the belt and are fed into a bailer that compacts them and wraps them with wire to form bales. A typical bale is approximately 0.75 m by 1.2 m by 1.5 m (30 × 48 × 60 inches). The same baler can be used to make bales of corrugated cardboard, PET, HDPE, aluminum, or ferrous metals. The removal of specific

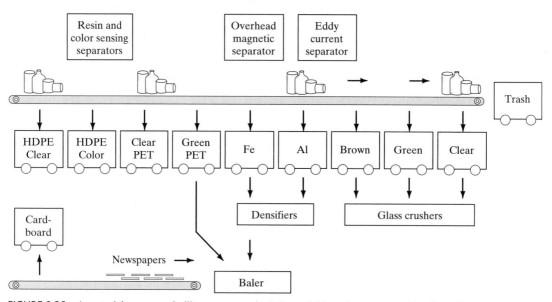

FIGURE 9.26 A materials recovery facility sorts commingled recyclables using some combination of hand and machine separation.

items, such as cardboard, from the conveyor is referred to as *positive sorting*, to distinguish it from *negative sorting*, in which the product in question (say, newspaper) is left on the conveyor belt.

Another conveyor belt has sorters, who push bottles, cans, and other objects into individual hoppers. Since plastics are voluminous and lightweight, they tend to stay on top so they are usually pushed into the appropriate hoppers first. To avoid mixing resins, at least three hoppers are needed: HDPE, PET, and mixed plastics. More careful separation would have hoppers for HDPE, green PET, clear PET, PVC, and mixed residue. The plastics may be fed into a granulator, or knife shredder, which cuts them into pieces roughly 1 cm in size. Or they may be crushed and baled. As was mentioned earlier, PVC is a particularly critical contaminant of PET, and it can also be a source of chlorine emissions if burned, so sorters must keep a careful watch for it. Manual separation is difficult since PVC and PET are visually very similar. Mechanical sorters, which remove PVC, are now available, however. One, marketed by National Recovery Technologies, Inc., is called the VinylCycle® system. In it, plastic bottles pass over a detector array that can sense the presence of chlorine atoms in PVC bottles. When chlorine is detected, air jets blow the PVC bottles over a partition away from the remaining materials (Moore, 1994). Another sorting technology, called Autosort®, analyzes light that has been transmitted through plastic bottles to identify the type of resin and color.

Ferrous metals can be removed from the conveyor belt by hand, or magnets can pull them off. One version of a magnetic separator uses a stationary magnet that is suspended above the main conveyor belt. The magnet draws the ferrous metals to another belt running perpendicular to the main conveyor and then drops them into their hopper. Some aluminum, either from bimetal cans having ferrous metal sides and aluminum lids or aluminum cans that get wedged in steel cans, will end up in the ferrous metal hopper. This aluminum contamination reduces the value the recycled metal when the product is sold to detinning plants, but the aluminum can actually be somewhat beneficial in steelmaking.

Aluminum cans and aluminum food trays can be picked off the conveyor by hand, as is usually the case in a small MRF, but it is also possible to do this using an *eddy-current separator.* When good electrical conductors, such as aluminum, are placed in a changing magnetic field, small eddy currents are created in the conductor. Those eddy currents, in turn, create their own magnetic field, which interacts with the field created by the magnets. That interaction generates a force that can be used to nudge aluminum out of the waste stream. Other metals, such as copper, will also experience forces induced by eddy currents, but they are much heavier than aluminum and are not moved as easily. One version of an eddy-current separator has the materials to be separated sliding down a ramp under a series of alternating polarity permanent magnets. As the aluminum slides under these magnets it experiences a changing magnetic field that results in eddy-current forces that push the aluminum to one side of the ramp. Nonconductors slide straight down the ramp, so separation is achieved.

Bottles need to be separated by color, into clear, green, and brown. Prior to color separation it is important to avoid breakage, but afterward the glass is purposely crushed. Vibrating screens, which allow the crushed glass to fall through the openings, can be used to help remove larger, unwanted contaminants such as bottle caps, labels, and other debris.

Processing Equipment for Commingled Wastes

Many communities collect commingled recyclables and mixed wastes. Such wastes could be landfilled as collected, but they contain valuable resources that should be recovered. Not only can recyclables be removed, but combustible products can be prepared for waste-to-energy incineration, and materials to be landfilled can be densified to take up less space. Processing facilities that handle such mixtures may utilize a combination of large, heavy-duty machines to screen, shred, separate, and classify the waste, as is suggested in Figure 9.27.

Raw solid waste consists of a mix of boxes, bags, newspapers, chunks of wood, bottles, cans, garden clippings, tires, and so forth, with varying sizes and shapes. To facilitate recovery of various components of the waste, it must be broken down into

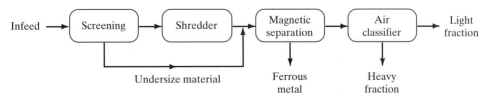

FIGURE 9.27   An example resource recovery system for mixed solid waste.

smaller pieces. A combination of machines that screen and shred the waste accomplishes this task.

Screening. Figure 9.27 suggests a simple screening of the waste as a first step, though this is not always the case. The most common screen is a large-diameter, tilted, rotating drum with appropriately sized holes, called a trommel. These drums have a diameter of roughly 1 to 3 m, and the holes are often about 10 cm across—big enough to let most of the metal and glass fall through, but too small to pass much of the paper and larger objects. The rate of rotation is such that particles move at least halfway up the side of the drum before falling back. If the speed is too high, centrifugal forces can hold the wastes against the drum so they never fall back to the bottom. Typical rotation speeds are on the order of 20 to 30 rpm. A sketch of a trommel is shown in Figure 9.28a.

Shredding. Shredders are brute-force machines that pound, crush, pulverize, and shred the wastes. A number of types of shredders are available, but the one most commonly used in recycling operations is the horizontal axis hammermill shown in Figure 9.28b. Wastes dropped into the top are pounded by the hammers until they are reduced to a small enough size to fall through the bottom grate. As a way to help prolong the relatively short life of the hammers, wastes are often prescreened before the hammermill to remove objects that are already sufficiently small. Other types of shredders include flail mills, which are similar to hammermills except that the hammers are

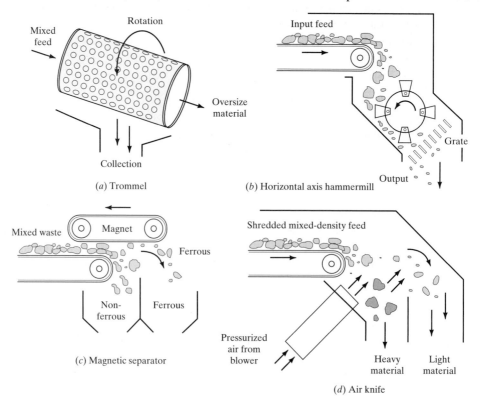

FIGURE 9.28 Examples of typical equipment used to process commingled recyclables and mixed wastes.

more like knives and there are no grates so materials make only one pass by the hammers. There are also knife shredders, which slice up objects like rags and bottles; shear shredders, with counterrotating blades that cut materials into strips; and pulverizers, which crush glass.

Magnetic separators.   After materials have been broken into small enough particles, they can be passed by magnets to remove ferrous metals. A number of design variations are possible, but they are all similar in nature. Ferrous metals are picked up by the magnet and moved away from the remaining materials on the main conveyor belt. Figure 9.28c shows one version of a magnetic separator.

Air classification.   Once the waste stream has been reduced to relatively small particle sizes, a blast of air, along with gravity, can be used to separate heavy particles from light ones. In one version of an air classifier, called an air knife, light particles such as plastic and paper are blown over barriers that separate a number of collection bins. Heavy particles are not affected much by the air stream, so they fall into the first hopper. The air knife shown in Figure 9.28d has two receiving bins, but more are possible if more sensitive separation is needed.

Environmental Impacts of Recycling

A recent study by the Environmental Defense Fund (EDF) (Denison, 1996) attempts to compare the environmental impacts of recycling with impacts associated with waste incineration and landfilling (see Figure 9.29). The EDF's analysis includes the direct impacts associated with each of the three practices—recycling, incineration, and landfilling—as well as the indirect impacts. The indirect impacts for incineration with energy recovery, for example, include the reduction in emissions from conventional power plants that would result because new electricity is generated in waste-to-energy facilities. For the recycling option, indirect impacts include the reduction in energy and emissions associated with mining and manufacturing based on virgin materials.

Economics of Recycling

While the environmental and resource value of recycling is clear, the economic viability under current market and regulatory conditions is less certain. To make an economic assessment, we need to explore the collection costs, the cost of materials processing, the market value of recycled materials, and the tipping fees avoided by reducing the amount of waste that ends up at the local landfill or incinerator.

To a significant extent, there is a tradeoff between the cost of collection and the cost of processing. For example, separate collection of source-separated recyclables is expensive, but the processing that follows collection is relatively cheap. Moreover, overall resource recovery rates may be lower if consumers balk at being asked to do most of the source separation. At the other extreme, using the same packer to pick up recyclables and mixed waste cuts collection costs, but raises the cost of processing. In this case, everyone participates whether they like it or not, so potential recovery rates are high. Also complicating the economics is the tradeoff between extensive hand sorting at the materials recovery facility, common when volumes are low, and mechanized

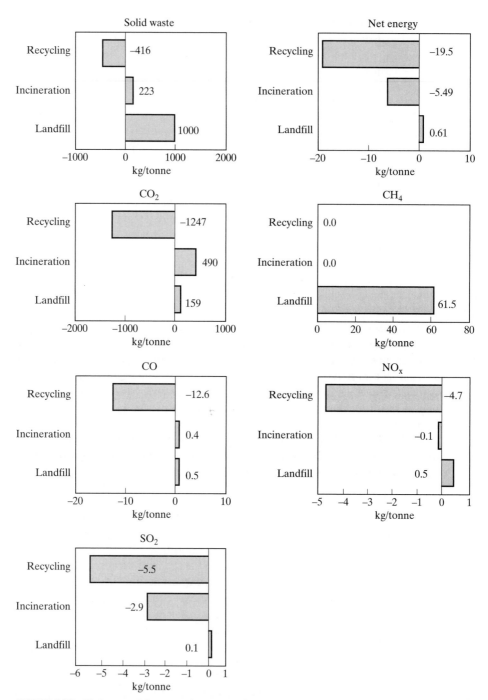

FIGURE 9.29 Estimates of the net environmental impacts of recycling, incineration, and landfilling of MSW, per tonne of waste. Negative values result from the displacement of energy and processing of raw materials when incineration or recycling are used. (*Source:* Denison 1996)

sorting, which predominates in larger operations. Finally, data on costs are hard to come by, difficult to interpret, and often based on widely varying assumptions and accounting practices.

A study by the National Solid Wastes Management Association (Miller, 1993) attempts to evaluate collection costs for a hypothetical "typical" curbside recycling program. Households are assumed to set out newspapers in a bundle along with a single bin containing commingled glass, aluminum, steel, and plastic containers. Most of the households are single-family, with an assumed distance of 46 m (150 ft) between stops. The start-up to on-route time is 30 minutes, with 60 minutes assumed for the round-trip time between route and processing site. Crews work $8\frac{1}{2}$ hours per day. Variables include truck size, crew size (one or two persons per truck), and fraction of households actually participating in the recycling program (the "set-out" rate). In the study, truck size had almost no effect on the collection cost per ton, and crew size was only modestly important.

As can be seen in Table 9.14, as the set-out rate increases, costs per tonne decrease. This seems reasonable since greater participation means trucks do not have to drive so far to accumulate a full load. It is important to note that the costs to collect recyclables are considerably higher than the collection costs for a packer collecting ordinary MSW. Since the density of recyclables is much lower than the density of ordinary wastes, and since recyclables are compacted very little if at all during pick-up (to help prevent glass breakage), a full truck holds far fewer tonnes of recyclables than a packer full of regular MSW. In addition, a recycling truck picks up far less material at each stop and drives further between stops since not every household participates, which increases the driving time per tonne collected. The combination of factors results in a cost per tonne to collect recyclables of typically at least twice the cost of ordinary mixed waste collection.

The National Solid Wastes Management Association also sponsored a study on the cost of materials recovery facilities (Miller, 1993). The analysis is based on a study of 10 actual MRFs, with throughputs ranging from 90 to 275 tonnes/day (100 to 300 tons/day), that process papers and commingled, recyclable containers. The study looked only at processing costs—revenues from the sale of recycled materials, avoided costs of disposal, and collection costs were not included. According to the study, the cost of processing a tonne of recyclable materials at an MRF ranges from $31 to $80

TABLE 9.14   Collection Costs, $/Tonne, For a Hypothetical, "Typical" Curbside Recycling Program[a]

Crew size	Set-out rate		
	25%	50%	75%
1 person	$175	$134	$123
2 person	164	126	115

[a]23.7 m^3 dual side-loading truck.
Source: Miller (1993).

per tonne ($28 to $72/ton), with an average cost of $55/tonne ($50/ton). If the recyclables are not commingled, then the recovery facility merely consolidates incoming materials that have already been separated and costs are closer to $5 to $15 per tonne (Kreith, 1994).

The big economic question is, of course, How does the cost of collection and processing recyclables, which seems to be roughly around $170 per tonne ($115 for collection, $55 for MRF), compare with the revenues received when the recovered materials are sold? Unfortunately, that comparison is not a simple one for the general case. Prices vary considerably over time, both within a given year and from year-to-year. For example, prices for recycled aluminum cans tend to rise in the spring as can-makers prepare for the heavy summer demand for beverages. In the fall, prices drop as the supply of returned cans increases and demand drops. Prices also vary from year to year, sometimes by large amounts. For example, mills were overloaded with supplies of recycled paper in 1992 and paper processors often had to pay mills to take it off their hands, but in 1994 it was not uncommon for mills to be paying over $100 per tonne. Corrugated cardboard that sold for $30/tonne in 1993 jumped to $70 per tonne in 1994 (Rabasca, 1995).

Table 9.15 shows examples of prices paid for various processed recycled materials in 1997. These are national average prices for a selected set of U.S. cities for truckload quantities of baled, high-quality, well separated materials.

Prices paid for recycled materials, such as those shown in Table 9.15, coupled with estimates of the amounts of material that are recovered, provide a basis for evaluating the potential economic value of a community's recycling programs. An example of such an estimate is presented in Table 9.16. The quantities given there, which are meant to be somewhat representative of nationally averaged resource recovery rates, have been normalized to 100 tons of materials. Notice corrugated cardboard accounts for just about half of the total mass and one-third of the total revenue received, while aluminum, with only 3 percent of the mass, provides nearly 40 percent of the revenues.

TABLE 9.15 Examples of National Average Prices for Truckload Quantities of Baled, High-Quality Recycled Materials in May, 1996 and 1997

Material	May, 1997	May, 1996
Steel cans ($/ton)	91	88
Aluminum cans (¢/lb)	59	58
Clear PET (flaked, ¢/lb)	21	37
Green PET (flaked, ¢/lb)	13	28
Natural HDPE (flaked, ¢/lb)	28	23
Corrugated cardboard ($/ton)	59	48
Newspaper #6 ($/ton)	16	24
High-grade office paper ($/ton)	76	81
Clear glass ($/ton)	42	40
Green glass ($/ton)	13	14
Brown glass ($/ton)	23	26

Source: Recycling Manager, May 1997 (http://grn.com/prices/rm-prices.htm)

TABLE 9.16   An Example Estimate of the Recycling Rates and Resulting Revenues for 100 Tons of Recovered Materials

Material	Mass tons	Price $/ton	Revenue $	Percent of revenue
Aluminum cans	3	1180	3540	39
Corrugated cardboard	49	59	2891	32
High-grade office paper	9	76	684	8
Natural HDPE	1	560	560	6
Steel cans	5	91	455	5
Newspaper #6	22	16	352	4
Glass	10	30	300	3
Clear PET	0.5	420	210	2
Green PET	0.5	260	130	1
TOTAL	100		$9122	100

Newspapers account for a significant fraction of the mass of materials recycled (22 percent), but generate a modest 4 percent of revenues.

The data presented in Table 9.16 suggest the revenues generated by a recycling program might be on the order of $100 per tonne of materials ($91/ton). If we use an estimate of about $170 per tonne to collect and process recyclables, our estimate for the average net cost of recycling is about $70 per tonne. Is this good or bad? If the alternative is to simply collect MSW and send it to the local landfill, we need to know the cost of collection and the tipping fees at the landfill—both of which vary considerably from community to community. As a rough estimate, with a typical cost of MSW collection of around $50/tonne and tipping charges of say $30/tonne, the total cost to dispose of ordinary refuse is around $80/tonne versus our estimate of $70/tonne to recycle. Recycling would appear to be a bit less expensive, so money would be saved. This is not a robust conclusion, however, since there are simply too many uncertainties and variations in the numbers used here to justify a strong general statement. In some communities and circumstances, recycling will lose money; in others it will make money.

9.7 COMPOSTING

Yard trimmings and food waste account for almost one-fourth of the mass of all municipal solid waste generated in the United States. Prior to the 1990s, essentially all of that ended up as discards sent to the local landfill or incinerator. As the limited remaining life of landfills becomes more critical, it is apparent that source reduction and recycling programs should be implemented for these wastes as well as all the others. In fact, by the mid-1990s, over half of the U.S. population lived in states having regulations restricting disposal of yard trimmings (U.S. EPA, 1994b). The impact of those programs is beginning to be felt, as Figure 9.30 demonstrates. The generation rate of yard trimmings is dropping as more and more households implement backyard composting, and the recovery rate for yard trimmings is increasing rapidly as municipalities implement their own composting programs. The number of facilities in the United States that handle yard trimmings has grown rapidly, from around 800 in 1988 to over 3200 in 1994

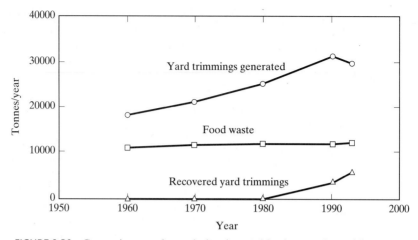

FIGURE 9.30 Generation rates for yard trimmings and food waste, along with recovery rates for yard trimmings. (Based on data in U.S. EPA, 1994b)

(Steuteville, 1995). Notice the distinction between backyard composting, which is a form of source reduction since it reduces the amount of waste that has to be collected, and municipal composting, which is considered a form of recycling since it creates a marketable product out of collected wastes.

Composting is the term used to describe the aerobic degradation of organic materials under controlled conditions, yielding a marketable soil amendment or mulch. It is a natural process that can be carried out with modest human intervention, or it can be carefully controlled to shorten the composting time and space required and to minimize offensive odors. The stabilized end product of composting is rich in organic matter, which makes it a fine soil conditioner, but the concentrations of key nutrients such as nitrogen, phosphorus, and potassium are typically too low for it to compete with commercial fertilizers.

In the simplest composting systems, yard wastes (sometimes mixed with sewage sludge) are stacked in long, outdoor piles called windrows (Figure 9.31). A typical windrow might be 2 m high, 3 or 4 m wide, and tens of meters long. Their length is

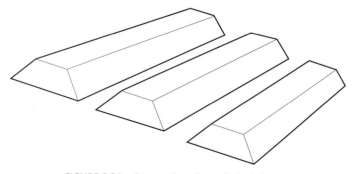

FIGURE 9.31 Composting piles, called windrows.

determined by the rate of input of new materials, the length of time that materials need for decomposition, and the cross-sectional area of the pile. More complex systems use biological reactor vessels. Some systems involve both technologies: Reactor vessels are used to start the decomposition process, and windrows are used to for the finishing stages.

The composting process is affected by temperature, moisture, pH, nutrient supply, and the availability of oxygen. Bacteria and fungi are the principal players in the decomposition process, but macroorganisms such as rotifers, nematodes, mites, sowbugs, earthworms, and beetles also play a role by physically breaking down the materials into smaller bits that are easier for microorganisms to attack.

Temperature

In the early stages of decomposition, mesophilic microorganisms (bacteria and fungi that grow best at temperatures between 25 and 45 °C) generate heat while they metabolize the waste, which raises the temperature of the pile. When temperatures reach 45 °C, the activity of mesophiles stops and thermophilic microorganisms, which prefer temperatures between 45 and 70 °C, take over the decomposition. Inside the compost pile, temperatures continue to increase as long as nutrient and energy sources are plentiful for the thermophiles. If the pile temperature stays above 55 °C for more than 72 hours, most pathogens and weed seeds will be destroyed, making a more marketable end product. Eventually, the nutrient supply drops, thermophiles die off, the temperature falls, and mesophilic microorganism once again dominate the decomposition process until stable end products are formed. Figure 9.32 shows an example of this rise and fall of pile temperature.

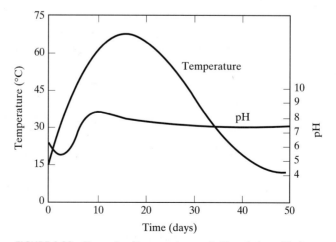

FIGURE 9.32   Example of temperature and pH variation with time in a compost pile. (After U.S. EPA, 1994b)

pH

In the early stages of decomposition, organic acids are formed, which cause the pH of the pile to drop to about 5. When the pH is this low, acid-tolerating fungi dominate. Eventually microorganisms break down the organic acids and the pile pH rises, as suggested in Figure 9.32.

Nutrient levels

A number of nutrients must be available to the microorganisms that are attacking and degrading the compost pile. The most important nutrients needed are carbon for energy, nitrogen for protein synthesis, and phosphorus and potassium for cell reproduction and metabolism. In addition, a number of nutrients are needed in trace amounts, including calcium, cobalt, copper, magnesium, and manganese.

One of the best indicators of the likely health of microorganisms is the ratio of carbon to nitrogen available to them. An ideal C:N ratio is roughly 25 to 35 parts of available carbon to available nitrogen. High C:N ratios inhibit the growth of microorganisms, slowing the decomposition. Low C:N ratios accelerate the rate of decomposition, but may cause loss of nitrogen as ammonia gas and may cause rapid depletion of the available oxygen supply, leading to foul-smelling anaerobic conditions. Leaves, cornstalks, rice hulls, and paper are examples of materials that have high C:N ratios, while grass clippings and sewage sludge have low C:N ratios. Examples of high carbon content and high nitrogen content feedstocks are given in Table 9.17.

By properly combining different kinds of solid wastes, the C:N ratio can be brought into the desired range. Mixing dry leaves with grass clippings, for example, helps balance the ratio, leading to rapid decomposition without foul odors. Sewage sludge, which is high in water as well as nitrogen, nicely complements municipal solid waste, which tends to be low in moisture and nitrogen. When composted properly, high

TABLE 9.17 Carbon-To-Nitrogen Ratios of Various Materials

Type of Feedstock	C:N Ratio
High Carbon Content	
Corn stalks	60:1
Foliage	40–80:1
Dry leaves and weeds	90:1
Mixed MSW	50–60:1
Sawdust	500:1
High Nitrogen Content	
Cow manure	18:1
Food scraps	15:1
Grass clippings	12–20:1
Humus	10:1
Fresh leaves	30–40:1
Nonlegume vegetable scraps	12:1

Source: U.S. EPA, 1994b.

compost temperatures not only kill pathogens but promote the drying of sewage sludges, which helps reduce the cost of sludge dewatering.

Oxygen

As aerobic microbes degrade waste, they remove oxygen from the pile. If the supply of oxygen is insufficient to meet their needs, anaerobic microorganisms will take their place, slowing the degradation process and producing undesirable odors. Oxygen can be supplied by simply mixing or turning the pile every so often, or piles can be aerated with forced ventilation.

Compost piles that are aerated by mixing need to be turned anywhere from once or twice a week to once per year. More frequent turning speeds the composting process and helps prevent anaerobic conditions, but it may cause the pile to dry out or cool down too much. Such turned windrows can complete the composting of yard trimmings in roughly three months to one year. Composting can be speeded up (which also reduces the land area required), by incorporating a forced aeration system. Wastes are stacked on top of a grid of perforated pipes and a blower forces air through the pipes and composting materials. Composting can be completed in as little as three to six months using this method (U.S. EPA, 1994b).

9.8 DISCARDED MATERIALS

Recall the definitions used to describe the flow of solid waste materials through our society. Source reduction (e.g., lightening containers) and reuse (e.g., refilling glass bottles) are activities that reduce the amount of materials that enter the municipal waste stream. What remains are referred to as MSW materials generated. Some of those materials are recovered from the waste stream and recycled or composted, leaving materials that are discarded. Those remaining discards are burned or buried.

Table 9.18 shows historical amounts of materials discarded in the United States, and Figure 9.33 shows the percentage distribution of those materials in 1993. The impacts of recycling and composting in the 1990s are evident. Paper and paperboard discards are beginning to drop, as are yard trimmings and metals. The ever-increasing use of plastics in our society, along with only modest efforts to recycle, is reflected in the continued rapid rise in plastics discards.

9.9 WASTE-TO-ENERGY COMBUSTION

The history of combustion of solid wastes in the United States clearly demonstrates the controversial nature of this method of waste disposal. As Figure 9.34 shows, in 1960 a little over 30 percent of MSW was burned. That percentage dropped rapidly during the 1960s and 1970s, reaching a low of less than 10 percent in 1980. With increased emphasis on avoiding landfilling, with better emission control systems and with better incinerators that were designed to allow the recovery of energy along with volume reduction, combustion of MSW increased to about 16 percent of MSW generation by 1990. EPA projections suggest that incineration will remain in the 15 percent range

TABLE 9.18   Materials Discarded in the Municipal Waste Stream, 1960 to 1993 (millions of tonnes per year)

Materials	1960	1970	1980	1990	1993
Paper and paperboard	22.2	33.3	38.8	47.4	46.5
Yard trimmings	18.1	21.0	24.9	27.9	23.8
Plastics	0.4	2.8	7.1	14.9	16.9
Food wastes	11.0	11.6	11.9	11.9	12.5
Wood	2.7	3.6	6.1	10.8	11.2
Metals	9.4	12.3	12.0	11.7	10.8
Glass	6.0	11.3	12.9	9.5	9.6
Rubber and leather	1.5	2.7	3.8	5.1	5.3
Textiles	1.6	1.8	2.3	5.3	4.9
Other	1.2	2.1	4.2	4.9	5.1
Totals	74.1	102.5	124.0	149.5	146.6

Source: U.S. EPA, 1994a.

through the year 2000, with almost all of that (93 percent) utilizing energy recovery. As Table 9.2 illustrated, a number of developed countries around the world rely much more heavily on incineration than does the United States—especially if they have high population densities. Japan, for instance, burns almost two-thirds of its wastes, while Germany and France burn 30 and 41 percent, respectively.

In the early 1990s, the United States had 190 combustion facilities in operation that burned just under 30 million tonnes of waste. Most of these plants are located in the eastern part of the United States, as Figure 9.35 indicates. Of these, 158 facilities

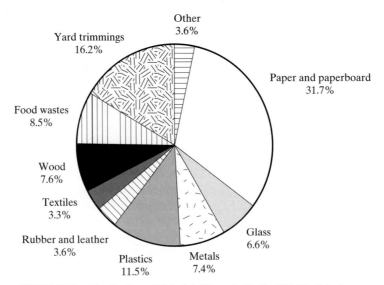

FIGURE 9.33   Distribution of Materials Discarded in the U.S. Municipal Waste Stream, by weight, 1993. (*Source:* U.S. EPA, 1994a)

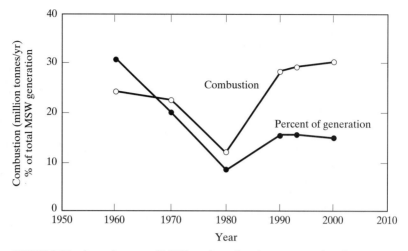

FIGURE 9.34   Annual tonnes of MSW combusted, and percentage of total generation for the United States. Data for the year 2000 are an EPA projection. (*Source:* U.S. EPA, 1994a)

were waste-to-energy plants, and all of those proposed for the future take advantage of energy recovery (Steuteville, 1995).

Incineration of MSW as a method of waste disposal has a number of favorable attributes, including volume reduction, immediate disposal without waiting for slow biological processes to do the job, much less land area requirements, destruction of hazardous materials, and the possibility of recovering useful energy. There are trade-offs. For example, it is reasonable to question whether burning paper for its energy value or recycling it to reduce the environmental impacts of virgin pulping is the higher use for the material. On the negative side of combustion, poorly operated incinerators release toxic substances such as dioxins into the air, the ash recovered may be classified as hazardous materials that require special handling, and the public has generally been reluctant to accept the technology—especially if a facility is being proposed in their own area. There is also a concern for whether incineration competes with recycling. Once an incinerator has been built, a reliable stream of refuse to burn must be maintained to pay it off. There can be reluctance then to expand a recycling program if it may reduce the fuel supply, and hence cash flow, for the incineration system. In the general hierarchy of solid waste management, recycling is considered a greater good than incineration, so there are advocates who feel it is important to write into an incineration contract the freedom to reduce the amount of garbage provided to the facility at any time, without penalty or other economic risk, when that reduction is due to an expansion in recycling (Denison and Ruston, 1990).

Energy Content of MSW

The energy content of municipal solid waste depends on the mix of materials that it contains as well as its moisture content. The standard test used to determine the heating value of a material involves completely burning a sample in a bomb calorimeter

FIGURE 9.35 Distribution of operating and planned municipal waste combustion facilities. The numerator is number in operation; the denominator is number planned, under construction, or inactive. (*Sources:* Pirages and Johnston, 1993)

and then measuring the rise in temperature of a surrounding water bath. The result obtained is known as the higher heat value (HHV), or gross heat value. Included in the HHV is energy contained in the vaporized water that is produced. Since this vapor is not usually condensed, that energy is lost and a more realistic estimate of the energy that can be recovered during combustion results, which is known as the lower heat value (LHV), or net energy. Table 9.19 gives example higher heating values of various components of municipal solid waste "as received"—that is, without any particular effort to remove the moisture. For comparison, the energy content of fossil fuels is included.

EXAMPLE 9.6   Energy Content of MSW Discards

Estimate the energy contained in a kilogram of "as received" discarded solid waste in the United States.

Solution Using data given in Figure 9.33 for discarded materials and data from Table 9.19 for energy content, we can set up a table based on 100 kg of discards. We need to make a few assumptions since some data are missing. For wood, we will use an average HHV of hardwood and softwood $(17,000 + 15,000)/2 = 16,000$ kJ/kg. Since the mix of rubber and leather is not given in the figure, we will use an average value of $(26,100 + 18,500)/2 = 22,300$ kJ/kg. For "miscellaneous other" wastes, we will assign an energy value of zero. Also, metals and glass will be assumed to have zero heating value even though coatings on cans do provide modest amounts of energy.

Material	kg	kJ/kg	kJ
Paper and paperboard	31.7	15,800	500,860
Yard trimmings	16.2	6,300	102,060
Plastics	11.5	32,800	377,200
Food wastes	8.5	5,500	46,750
Wood	7.6	16,000	121,600
Metals	7.4	0	
Glass	6.6	0	0
Rubber and leather	3.6	22,300	80,280
Textiles	3.3	18,700	61,710
Miscellaneous other	3.6	0	0
Total	100.0		1,290,460

Thus, typical discarded solid waste has an energy content of approximately 12,900 kJ/kg (5,550 Btu/lb), which is a little less than half the energy content of coal and roughly 30 percent of the energy content of fuel oil. ∎

The composition of MSW varies considerably around the world, and so does its energy content. In developing countries, a relatively small fraction of the waste consists of manufactured materials such as paper, metals, plastic and glass, which means

TABLE 9.19 Examples of "as received" HHV of various components of MSW, as well as fossil fuels

Material	kJ/kg	Btu/lb
Mixed paper	15,800	6,800
Mixed food waste	5,500	2,400
Mixed green yard waste	6,300	2,700
Mixed plastics	32,800	14,000
Rubber	26,100	11,200
Leather	18,500	8,000
Textiles	18,700	8,100
Demolition softwood	17,000	7,300
Waste hardwood	15,000	6,500
Coal, bituminous	28,500	12,300
Fuel oil, no. 6	42,500	18,300
Natural gas	55,000	23,700

Source: Solid waste data from Niessen (1977).

that the percentage that is food waste is high—usually over 50 percent. Since food waste has a relatively low energy content, the heating value of MSW in the developing world tends to be lower than that of more industrialized countries. Khan and Abu-Ghararah (1991) have developed an equation that attempts to predict the heating value of MSW based on the paper and food fractions plus a term that accounts for plastic, leather, and rubber.

$$\text{HHV (kJ/kg)} = 53.5(\text{F} + 3.6\,\text{CP}) + 372\,\text{PLR} \tag{9.7}$$

where F is the mass percent of food, CP is the mass percent of cardboard and paper, and PLR is the mass percent of plastic, rubber, and leather in the waste mixture (Table 9.20).

Only in unusual circumstances can the HHV of a fuel be captured. Usually the latent heat contained in water vapor is lost to the atmosphere rather than being captured by the power plant. There are two possible sources of that water vapor loss: moisture in the wastes, and hydrogen in the waste that can react with oxygen to form water. Using 2440 kJ/kg as the latent heat of vaporization for water (at 25 °C), each kg of moisture that is vaporized and lost contains 2440 kJ of energy. In addition, each kg of hydrogen in the waste itself can produce another 9 kg of water vapor. The total energy lost in vaporized water is therefore

$$Q_L = 2440(\text{W} + 9\text{H}) \tag{9.8}$$

where

$$Q_L = \text{latent heat of water vapor released (kJ)}$$
$$\text{W} = \text{kg of moisture in the waste}$$
$$\text{H} = \text{kg of hydrogen in dry waste}$$

TABLE 9.20 Estimated High Heating Value of MSW for Various Countries

Country	Paper (%)	Metals (%)	Glass (%)	Food (%)	PLR[a] (%)	HHV (MJ/kg)
Australia	38	11	18	13	0.1	8.1
Colombia	22	1	2	56	5	9.1
Czechoslovakia	13.4	6.2	6.6	41.8	4.2	6.4
England	37	8	8	28	2	9.4
France	30	4	4	30	1	7.8
Germany	20	5	10	21	2	5.7
India	3	1	8	36	1	2.9
Iran	17.2	1.8	2.1	69.8	3.8	8.5
Japan	21	5.7	3.9	50.0	6.2	9.0
Kenya	12.2	2.7	1.3	42.6	1.0	5.0
Sweden	50	7.0	8.0	15	8.0	13.4
U.S.A.	31.7	7.4	6.6	8.5	15.1	12.9

[a]PLR means plastic, leather, and rubber.
Source: based on Khan and Abu-Ghararah (1991), except for United States, which uses MSW discard data from U.S. EPA (1994a).

EXAMPLE 9.7   Lower heating value of MSW

Typical MSW has a moisture content of around 20 percent, and roughly 6 percent of the dry mass of MSW is hydrogen. Using the high heating value of MSW found in Example 9.6, estimate the lower heat value.

Solution   In 1 kg of waste, there will be 0.2 kg of moisture and 0.8 kg of dry waste. In that dry waste, there will be $0.8 \times 0.06 = 0.048$ kg of hydrogen.

Using (9.8) the energy content of the water vapor released when 1 kg of waste is burned is

$$Q_L = 2440 \times (0.2 + 9 \times 0.048) = 1540 \text{ kJ}$$

If that heat is not recovered, the 12,900 kJ/kg found in Example 9.6 is reduced to

$$\text{LHV} = 12,900 - 1540 = 11,360 \text{ kJ/kg}$$

Using the conversion 2.326 kJ/kg = 1 Btu/lb gives 4880 Btu/lb.   ∎

Table 9.21 shows typical moisture and hydrogen contents of various components of solid waste. The values given for the low heating value are based on (9.8).

Mass Burn and Refuse-Derived Fuel

Combustion for energy recovery is typically done in one of two ways: Either MSW is sent directly to a mass-burn incinerator, or it is preprocessed to produce a more homogeneous product called refuse-derived fuel (RDF) that has much better combustion characteristics. As of 1988, 77 percent of the capacity of municipal waste combustors in the United States was mass burn, while the remaining 23 percent utilized RDFs. That ratio is shifting, however, with one-third of the capacity additions being RDF facilities (Denison and Ruston, 1990).

TABLE 9.21   High and low heating values of combustible components of MSW.

Material	Moisture[a] (%)	H[b] (%)	HHV (kJ/kg)	LHV[c] (kJ/kg)
Mixed paper	10.2	5.8	15,800	14,400
Mixed food waste	72.0	6.4	5,500	3,400
Mixed green yard waste	62.0	5.6	6,300	4,300
Mixed plastics	2.0	7.2	32,800	31,200
Rubber	1.2	10.4	26,100	23,800
Leather	10.0	8.0	18,500	16,700
Textiles	22.0	6.4	18,700	17,100
Demolition softwood	7.7	6.2	17,000	15,500
Waste hardwood	12.0	6.1	15,000	13,500

[a]Percent moisture is as received, by weight
[b]H is percent hydrogen in dry mass
[c]LHV is computed from (9.8)
Source: Based on data given in Niessen (1977)

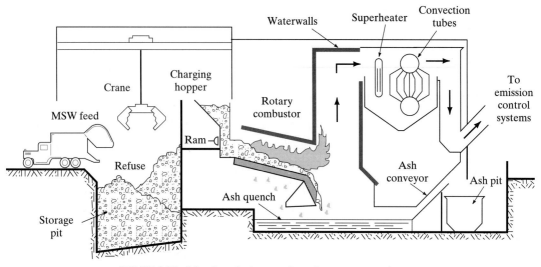

FIGURE 9.36   Mass burn incinerator featuring a rotary combustor.

Figure 9.36 shows a sketch of a mass-burn, rotary-kiln furnace receiving refuse directly from a packer truck. Heat recovery from the waterwalls, convection tubes, and superheater provides steam that can be used for process heat or space heating in nearby industrial facilities, or the steam can be used to generate electricity, which can be transported more easily over long distances.

EXAMPLE 9.8   U.S. Electricity Potential from MSW

The United States discards annually roughly 147 million tonnes of MSW (Table 9.18). The low heating value of those discards is about 11.4 MJ/kg (Example 9.7). A mass-burn waste-to-energy facility can convert those wastes to electricity at a net efficiency of about 20 percent. Estimate the electrical energy that could be produced per year if all of our discards were used in this type of WTE system. Compare it with the total that is now generated, which is about 2.9×10^{12} kWh/yr.

Solution   At 20 percent efficiency, to produce 1 kWh_e of electricity requires 5 kWh_t of heat input to the boiler (to avoid confusion, the subscripts t and e refer to thermal and electrical forms of energy). Using the conversion 1 kWh_t of heat equals 3.6 MJ gives

$$\text{Heat input} = (5 \text{ kWh}_t/1 \text{ kWh}_e) \times 3.6 \text{ MJ/kWh}_t = 18 \text{ MJ/kWh}_e$$

$$\frac{147 \times 10^6 \text{ tonne/yr} \times 1000 \text{ kg/tonne} \times 11.4 \text{ MJ/kg}}{18 \text{ MJ/kWh}_e} = 93 \times 10^9 \text{ kWh}_e/\text{yr}$$

That is about 3 percent of all electricity generated in the United States.   ■

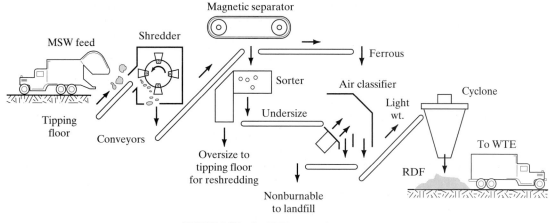

FIGURE 9.37   An RDF processing system.

Refuse-derived fuels have had most of the noncombustible portion of MSW removed, and after shredding and sorting a relatively uniform product is produced. An example flow diagram for a resource recovery facility that yields RDF is shown in Figure 9.37. With the noncombustibles removed, the energy content of RDF is obviously higher than that of unprocessed MSW. The typical range of heat values for MSW is roughly 8000 kJ/kg to 15,000 kJ/kg (3500 to 6500 Btu/lb), while the RDF range is typically 14,000 kJ/kg to 18,000 kJ/kg (6000 to 7800 Btu/lb). Even with these greater heating values, RDF is often used as a supplemental fuel in coal-fired facilities that have been suitably modified. Such co-firing facilities tend to use up to 20 percent RDF, with the remaining portion being higher energy-content fossil fuel. Fluidized-bed combustors, described in Chapter 7, are well suited to refuse-derived fuels since they are able to burn a wide variety of fuels with combustion efficiencies that approach 99 percent. Japan has more than 100 such incinerators, but only recently have they been introduced in the United States.

Environmental Impacts of WTE

There can be significant environmental impacts associated with the solids and gases that are produced during combustion. Solids are in the form of ash, which is made up of bits of glass and metal, unburned carbon particles, and other inert substances such as silica and alumina. Bottom ash is the residue that falls through grates in the combustion chamber or that drops off the walls of the boiler. Fly ash is made up of solid parti-... the boiler by combustion gases. Fly ash should be collected by particulate control systems such as scrubbers, electrostatic precipitators, and/or baghouses. Since combustion is involved, there are all of the usual criteria air pollutants, such as nitrogen oxides, sulfur oxides, and carbon monoxide, to worry about, and in addition there can be emissions of various acid gases, such as hydrochloric and hydro-fluoric acid. Of special concern are emissions of especially toxic substances such as dioxins, furans, polychlorinated biphenyls (PCBs), metals, and polycyclic aromatic hydrocarbons (PAHs).

Dioxins and Furans   Dioxins and furans are chlorinated compounds that have somewhat similar chemical structures. They both consist of two benzene rings that are linked together with oxygen bridges. Dioxins have two oxygen atoms joining the benzene rings, while furans have only one oxygen link, the other link being a direct carbon bond. Figure 9.38 introduces the generic structures of dioxins and furans, showing the conventional numbering system for the chlorine bonding sites.

With chlorine added to the basic dioxin and furan structures shown in Figure 9.38, two families of chemicals are created. There are 76 possible isomers of dioxin, two of which are shown in Figure 9.39, and 136 possible isomers of furan. The most common and most toxic of the dioxins is 2,3,7,8-tetrachlorodibenzo-*p*-dioxin, or 2,3,7,8-TCDD, shown in Figure 9.39.

Dioxins are highly toxic substances that are produced as byproducts of incineration, chemical processing, chlorine bleaching of paper and pulp, the burning of diesel fuel, and the use of some herbicides. They are not manufactured for any commercial purpose and appear to be entirely the result of human activities. Dioxins have become rather well known to the public as a result of several incidents in the 1970s. They were

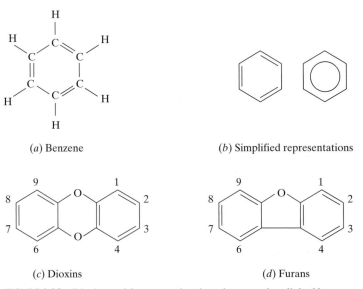

(*a*) Benzene   (*b*) Simplified representations

(*c*) Dioxins   (*d*) Furans

FIGURE 9.38   Dioxins and furans consist of two benzene rings linked by oxygen bridges. (*a*) The benzene ring., (*b*) Simplified representations of the ring. (*c*) The two oxygen bridges that characterize dioxins, along with the conventional site numbering system. (*d*) The single oxygen bridge of furans.

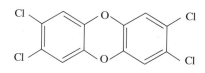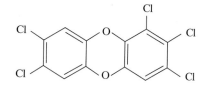

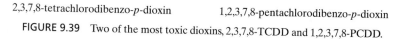

2,3,7,8-tetrachlorodibenzo-*p*-dioxin   1,2,3,7,8-pentachlorodibenzo-*p*-dioxin

FIGURE 9.39   Two of the most toxic dioxins, 2,3,7,8-TCDD and 1,2,3,7,8-PCDD.

identified as the chemical that caused the death of numerous birds, cats, and horses when dioxin-laced waste oil was used as a spray to control dust in Times Beach and other locations in Missouri in 1971. TCDD was a contaminant of the herbicides 2,4-D and 2,4,5-T, which were commonly used for weed control in the United States and which were mixed together and used as the defoliant Agent Orange in Vietnam. In 1976, an explosion at a chemical plant in Seveso, Italy released a cloud of dioxin that exposed thousands of workers and residents. These episodes, tragic as they were, did help to provide data on carcinogenicity and other health effects of exposure.

Dioxins do cause cancer in laboratory animals, but there is enough uncertainty about whether they cause cancers in people that they are still listed as "probable" causes of human cancer. The EPA finished a major study of dioxin in 1994 that emphasized that cancer was not the only disease of concern. Dioxins cause reproductive, developmental, and immune system problems in fish, birds, and mammals, at concentrations that are close to those humans experience. The ability of dioxins to affect cell growth and confuse biological signals early in fetal development suggests that the human embryo may be susceptible to long-term impairment of immune function from in utero exposure. Most of human exposure to dioxin is the result of airborne emissions that settle on plants and grasses and that make their way up the food chain into the beef, dairy, pork, and poultry products that we consume.

Dioxins and furons are created during combustion, so incineration of municipal solid waste (as well as hazardous and medical wastes) has to be done carefully if emissions are to be minimized. When combustion occurs at temperatures over 900 °C, with residence times of over one to two seconds, dioxin and furan emissions in the furnace itself can be reduced to nondetectable levels. However, if dioxin-furan precursors, such as hydrogen chloride, phenols, chlorophenols, and aromatic hydrocarbons, have not been completely destroyed during combustion, they can react in the presence of fly ash to form new dioxins and furans as the exiting flue gases cool. Complete combustion, combined with downstream emission control systems, are _____ ntial to achieve acceptable emission limits.

It is encouraging to note that dioxin levels in the fat ____ American have dropped from 18 parts per trillion in 1976 to ____ lion in 1994 . These gains are thought to have been achieved ____ PCBs in 1978, the ban on the herbicide 2,4,5-T in 1979, an ____ production, use, and disposal practices for a number of other important chlorinated hydrocarbons.

Heavy Metals.   Metals such as lead, zinc, cadmium, arsenic, and mercury are part of the municipal waste stream and, when burned, become part of the gases and particles that leave the combustion chamber or end up as part of the ash residue. In either case, their potential to contaminate the environment requires that they be carefully controlled. Removal prior to combustion as well as after combustion are both necessary parts of the control strategy.

The temperature at which these metals vaporize and condense is an important characteristic that affects the ease with which they can be removed from the combustion gases. During combustion, most metals vaporize into gases, which are difficult to control. When those flue gases are cooled, however, metals with high volatilization temperatures condense and adsorb onto fly ash particles, which simplifies their removal with conventional particulate control systems. Metals condense and adsorb more readily onto small particles, typically below 2 μm in size. Most metals are thus best controlled by a combination of flue gas cooling to below the condensation temperature, followed by an emission control system that effectively removes small particles. A heat exchanger, called an *economizer,* recovers heat from flue gases after they leave the boiler, which not only increases the overall energy recovery rate but also cools and causes condensation of most metals. The emission control system of choice for control of metals is the baghouse since it outperforms electrostatic precipitators at removing very small particles.

The most problematic metal to control in flue gases is mercury, which has the lowest boiling point of all the metals. It is hard to cool flue gases enough to cause mercury vapor to condense, so it may not be effectively removed by particulate control systems. This makes it even more important to focus attention on removing mercury from the waste stream before incineration. By far the biggest source of mercury in municipal solid waste is dry-cell batteries, which contribute 88 percent of the total (U.S. EPA, 1992). Fluorescent lamps contribute 5 percent and mercury thermometers contribute another 2 percent. Not only are dry-cell batteries the major source of mercury in municipal solid waste, but they also contribute 52 percent of another important metal, cadmium. Attempts at community battery collection programs have met with limited success, but are certainly to be encouraged.

MSW Incinerator Ash.   The ash from incineration of municipal solid waste is collected as fly ash and bottom ash, both of which are contaminated with heavy metals and dioxins/furans. There are two major environmental concerns with respect to disposing of this ash. For one, it is the nature of ash that it can easily become airborne, which increases the risk of exposure due to ingestion or inhalation. Second, when ash is disposed of in landfills it becomes exposed to acidic conditions that enhance the ease with which metals dissolve into the leachate. Of special concern are the heavy metals, lead and cadmium, which are highly toxic and are especially common in fly ash.

Until recently, fly ash and bottom ash were typically combined and disposed of in conventional clay or plastic-lined landfills, which is a much cheaper option than disposal in landfills designed for hazardous materials. This practice was based on Subtitle C,

Section 3001(i) of the Resource Conservation and Recovery Act (RCRA), which exempts municipal incinerators from federal hazardous waste regulation as long as they do not burn hazardous wastes. In 1994, however, that section was interpreted by the U.S. Supreme Court to apply only to the waste itself, not the resulting ash. Under their ruling, ash must now be tested and if it fails the toxicity test it must be managed as a hazardous waste. Roughly 90 percent of an incinerator's ash is bottom ash, which may not require disposal as a hazardous waste, but the remaining 10 percent, which is fly ash, is highly toxic and will almost always require special disposal. Mixing them together means the entire ash production may require expensive handling and disposal as hazardous wastes, so that practice is being discontinued.

Of course, if metals are not in the waste stream, they will not end up in incinerator ash. The need to keep toxic metals out of municipal solid waste, especially in areas that utilize incineration, has become especially important.

9.10 LANDFILLS

The image of the local dump, complete with flies, rats, gulls, odors, airborne bits of paper and garbage, and a black cloud of smoke, is well engrained in the consciousness of most of the public. The NIMBY reaction to a proposal for a new site, or expansion of an existing site, comes as no surprise then, even though modern, engineered facilities bear little resemblance to the image just described. In fact, one of the major impacts of RCRA on solid waste management has been the prohibition of such open dumps as of 1984.

There are three classifications for landfills. Class I landfills, or *secure* landfills, are those designed to handle hazardous wastes; Class II landfills, or *monofills,* handle so-called designated wastes, which are particular types of waste, such as incinerator ash or sewage sludge, that are relatively uniform in characteristics and require special handling; and Class III landfills, or *sanitary* landfills, are engineered facilities designed to handle municipal solid waste. Comments were made in Chapter 6 on hazardous waste landfills, and this chapter will deal with Class III, MSW landfills.

While landfilling continues to be the primary means of MSW disposal in the United States, the "3Rs" of *reduction, reuse, recycling* are beginning to have some impact and the amounts landfilled have been on the decline since the late 1980s (see Figure 9.6). The number of MSW landfills in the United States has been declining rapidly of late. In 1988 there were 8000 landfill sites around the country, but by 1994 there were only 3558—a decline of 55 percent in just 6 years (Steuteville, 1995). Most of the sites closed were small, publicly owned facilities that were unable to meet new RCRA Subtitle D landfill regulations that went into effect in 1994. What those statistics mask is the fact that newer facilities tend to be much larger than the smaller ones they replace, so the overall capacity of landfills has not declined appreciably.

With closure of so many local, substandard landfills, coupled with the difficulty in siting new ones, it is not surprising that there are parts of the country that face disposal capacity problems. Additional maintenance requirements and tight capacity constraints are helping to raise the tipping fees charged when a truck drops its load at the landfill site. Landfill tipping fees vary considerably from one part of the country to

TABLE 9.22   Regional MSW Landfill Tipping Fees in the United States (1994)

Region	Tipping Fee ($/tonne)	Tipping Fee ($/ton)
Mid-Atlantic	$62/tonne	$56/ton
New England	53	48
Great Lakes	36	33
Western States	32	29
Southern States	29	26
Rocky Mountain States	18	16
U.S. weighted average	$34	$31

Source: Steuteville (1995).

another, ranging from a low of $9/tonne in New Mexico in 1994 to $83/tonne in New Jersey and Vermont (Steuteville, 1995). Table 9.22 shows tipping fees in various regions of the country.

Federal Regulation of MSW Landfills

As required by Subtitle D of RCRA, the U.S. Environmental Protection Agency has promulgated regulations on the siting of all municipal solid waste landfills, both new and existing, that receive waste on or after October 1993. As shown in Table 9.23, these regulations cover six basic areas: location, operation, design, monitoring, closure, and financial assurance requirements. In addition to federal regulations, individual states have their own design, operation, and closure requirements.

Basic Landfill Construction and Operation

Moisture in a landfill is critically important if the wastes are to decompose properly, so it is an aspect of landfill design that receives considerable attention. The initial moisture contained in the wastes themselves is rather quickly dissipated, so it is water that percolates through the surface, sides, and bottom that eventually dominates the water balance in the landfill. Water percolating through the wastes is called leachate, and its collection and treatment is essential to the protection of the local groundwater.

Subtitle D regulations require MSW landfills to have composite liners and leachate collection systems to prevent groundwater contamination. The composite liner consists of a flexible membrane liner (FML) above a layer of compacted clay soil. Leachate is collected with perforated pipes that are situated above the FML. A final cover over the completed landfill must be designed to minimize infiltration of water. During waste decomposition, methane gas is formed so completed landfills need collection and venting systems. The essential features of such a landfill are shown in Figure 9.40.

TABLE 9.23   Summary of U.S. EPA Regulations for MSW Landfills

Area	Aspect	Requirement
Location	Airport safety	Landfills attract birds, so the site must be far enough from airports to prevent any bird hazards to aircraft.
	Floodplains	Landfills may not be located in areas that are prone to flooding.
	Wetlands	Siting in wetlands is allowed only under restricted conditions.
	Fault zones	New landfills must be designed to withstand local seismic activity.
	Unstable areas	Landfills cannot be located in areas that are subject to landslides, mudslides, or sinkholes.
Operation	Hazardous wastes	MSW landfills must have programs to keep regulated hazardous wastes out of their sites.
	Covering	Each day's waste must be covered to prevent the spread of disease.
	Explosive gases	Explosive methane gas must be monitored and controlled.
	Restricted access	Access to the landfill must be restricted to prevent illegal dumping.
	Stormwater	Levees and ditches must prevent stormwater from carrying pollutants offsite.
	Liquids	To reduce leachate, landfills cannot accept liquid waste from tank trucks or in 55-gallon drums.
	Air emissions	Clean Air regulations must not be violated. Among other things, this prohibits open burning.
Design	Groundwater protection	Landfill liners and leachate collection systems must prevent contaminant levels that exceed federal drinking-water limits.
Monitoring	Groundwater protection	Monitoring systems to detect groundwater contamination are needed, with twice-a-year sampling generally required.
Closure	Environmental protection	For 30 years after closure, landfill cover maintenance, groundwater, and gas monitoring required.
Financial	Assurance	Landfill owners/operators must show they have the financial means to cover closure, postclosure maintenance, and cleanups.

The operational phase of a sanitary landfill is organized around the concepts of *cells, daily covers,* and *lifts.* Each day's wastes are received and compacted into cells, which are then covered at the end of the day with a thin layer of soil or other material. The size of a cell depends on the daily volume of refuse to be buried, but typically they are on the order of 3 m (10 ft) thick (including daily cover) and their individual area is determined by the amount and density of the compacted refuse. Cells are covered each day, or more often if necessary, to prevent windblown spread of refuse, reduce odor, control the amount of water entering the cell, and control rodent, bird, and fly access to

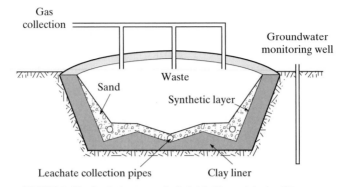

FIGURE 9.40   Basic features of a Subtitle D municipal solid waste landfill.

the garbage. When a given, active area of the landfill is filled with cells, other layers, called lifts, can be added on top. Figure 9.41 illustrates an active landfill with a second lift being added to the first.

Sizing a landfill requires estimates of the rate at which wastes are discarded and the density of those wastes when they are compacted in the fill. The density of waste depends on how well it is spread and compacted by heavy machinery making multiple passes over the refuse. For municipal solid waste, the density in a landfill will usually be somewhere between 325 kg/m³ and 700 kg/m³ (550 to 1200 lb/yd³).

Up until the time that wastes reach the actual landfill, waste quantities and charges are all based on easily measured mass or weights of refuse. At the landfill, however, it is not weight that matters, it is volume. An attempt to disaggregate refuse volumes in landfills into the proportions taken by paper, cans, bottles, etc., was done as part of the Garbage Project administered by the University of Arizona, Bureau of Applied Research in Anthropology (Franklin Associates, 1990). The EPA applied

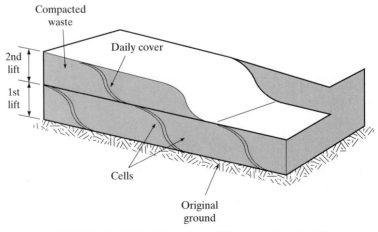

FIGURE 9.41   Cells, daily cover, and lifts in a sanitary landfill.

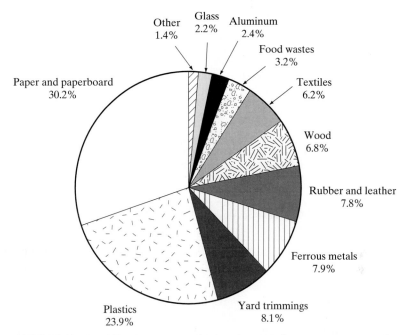

FIGURE 9.42   Landfill volume of materials in MSW, 1993. (*Source:* U.S. EPA, 1994)

those results to the national flow of discards, with results for 1993 shown in Figure 9.42. Notice that the combination of paper products and plastics accounts for over half of the total landfill volume.

Some examples of volume taken up by individual categories of products are shown in Table 9.24. Notice those controversial disposable diapers account for only 1.1 percent, junk mail (third class) is another 2.0 percent, while plastic cups and plates use up 0.4 percent of landfill volume. The overall density of waste in a landfill, derived by combining the densities found for each component, yields an estimate of 440 kg/m^3 (745 lb/yd^3). That estimate is likely to be on the low side, since it does not account for small items such as bits of paper and glass that fill gaps left by bigger items such as plastic containers.

EXAMPLE 9.9   Estimating Landfill Requirements

Estimate the landfill area needed to handle one year's MSW for a town of 100,000 people. Assume national average discards, no combustion, a landfill density of 600 kg/m^3, and a single 3-m lift. Assume that 20 percent of the cell volume is soil used for cover.

Solution   From Table 9.24, the United States discards 146.6 million tonnes of MSW per year. If we assume a population of roughly 260 million, the landfill volume of refuse for 100,000 people would be

$$V_{\text{MSW}} = \frac{146.6 \times 10^6 \text{ tonne} \times 10^3 \text{ kg/tonne} \times 100,000 \text{ people}}{260 \times 10^6 \text{ people} \times 600 \text{ kg/m}^3} = 93,975 \text{ m}^3$$

Since only 80 percent of a cell is landfill, the volume of cell needed is

$$V_{cell} = \frac{93,975 \text{ m}^3}{0.8} = 117,468 \text{ m}^3$$

The area of lift, at 3 m cell depth, is

$$A_{lift} = \frac{117,468 \text{ m}^3}{3 \text{ m}} = 39,155 \text{ m}^2 \quad (9.7 \text{ acre})$$

The actual sizing of a landfill would include a number of additional factors, such as additional area requirements for access roads and auxiliary facilities, reduction in landfill volume as biological decomposition takes place, and increases in compaction as additional lifts are added.   ∎

TABLE 9.24   Estimated Amounts, Density, and Landfill Volume of Selected Discards (Per Year, 1993)

Items	Discards (10^3 tonne/yr)	Landfill Density (kg/m^3)	Landfill Volume[a] (% of total)
Newspapers	6,350	475	4.0
Office papers	4,090	475	2.6
Third-class mail	3,140	475	2.0
Plastic plates & cups	300	210	0.4
Disposable diapers	2,440	650	1.1
Glass bottles	8,340	1,660	1.5
Aluminum cans	570	150	1.2
Steel cans	1,310	330	1.2
Plastic bottles	680	210	0.9
Paper bags and sacks	1,670	440	1.2
Plastic bags and sacks	930	400	0.7
Yard trimmings	23,800	890	8.1
All discards	146,600	440[b]	100%

[a]Assumes all discards are landfilled, but some are combusted.
[b]Derived from individual densities; actual should be higher.
Source: U.S. EPA (1994a).

Decomposition in Landfills

The decomposition of landfill materials can be thought of as a four-stage process. As shown in Figure 9.43, these stages are as follows:

I. *Aerobic phase:* When wastes are placed in a landfill there is enough entrained oxygen to allow aerobic decomposition to take place for the first few days. Oxygen levels drop, and at the end of this phase anaerobic conditions begin.

II. *Acid phase:* During this phase anaerobic conditions prevail and a two-step process begins. First, hydrolyzing-fermentative organisms produce enzymes that break down complex organics such as cellulose and starch into simpler products, which can be fermented into hydrogen, carbon dioxide, fatty acids, and alcohols. In the second step, those products are converted by bacteria, called acetogens,

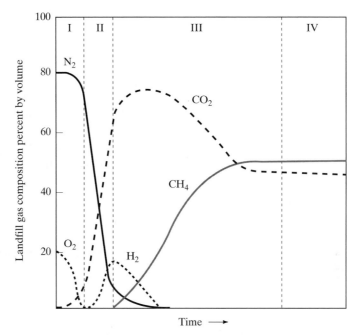

FIGURE 9.43   Changes in the composition of landfill gases help define a four-phase decomposition process. Phase I—aerobic; Phase II—acidogenesis; Phase III—unsteady methanogenesis; Phase IV—steady methanogenesis. (*Source: Solid Wast Landfill Engineering and Design* by McBean & Rovers, © 1994. Reprinted by permission of Prentice-Hall, Inc., Upper Saddle River, NJ.)

into simpler organic acids, typified by acetic acid (CH_3COOH). As these acids form, the pH of the leachate drops, which can allow heavy metals to be solubilized. The CO_2 concentration in the waste rises and small amounts of hydrogen gas (H_2) are produced.

III. *Methanogenesis, unsteady:* Another group of microorganisms, called *methane formers* or *methanogens,* convert the organic acids into CH_4 and CO_2. The pH begins to return toward more neutral conditions and the release of heavy metals into the leachate declines. This phase can last for months.

IV. *Methanogenesis, steady:* The duration of each phase depends on the availability of moisture and nutrients, but typically on the order of a year or so after a landfill cell is completed the generation rate of CH_4 and CO_2 settles down into nearly equal percentages, which is the characteristic of Phase IV. After many years, perhaps several decades, the decomposition process and the rate of production of methane decline significantly.

Under the assumption that CO_2, CH_4, and NH_3 are the principal gases liberated during decomposition, an equation that describes the complete decomposition of organic materials under anaerobic conditions is given by the following:

$$C_aH_bO_cN_d \ + \ n\,H_2O \ \longrightarrow \ m\,CH_4 \ + \ s\,CO_2 \ + \ d\,NH_3 \qquad (9.9)$$

where

$$n = (4a - b - 2c + 3d)/4$$
$$m = (4a + b - 2c - 3d)/8$$
$$s = (4a - b + 2c + 3d)/8$$

To use (9.9) to determine methane production from waste requires that we first determine a chemical formula for that waste. Niessen (1977) has developed estimates of the mass percentage of carbon, hydrogen, oxygen, and nitrogen that make up fully dried components of solid waste. Such a breakdown by elements is called an *ultimate analysis,* or *elemental analysis.* The following example illustrates the process.

EXAMPLE 9.10   Estimating the methane potential of discards

Based on data given in Example 9.6, 67.3 percent of typical "as delivered" MSW discards is paper, yard trimmings, food waste, textiles, and wood, which are decomposable. Of that, 32.3 percent is moisture. An elemental analysis of the dried decomposable components yields the following mass percentages:

Element:	C	H	O	N	Other	Total
Dry mass percentage:	44.17	5.91	42.50	0.73	6.69	100%

Find the chemical formula for the C, H, O, N portion of the decomposables. Then find the energy content of the methane that would be generated, per kilogram of discards. The HHV of methane is 890 kJ/mol.

Solution   Starting with 1000 g of as received discards, 673 g are decomposable. Of that, 32.3 percent is moisture, leaving a dry mass of

Dry mass of decomposables $= (1 - 0.323) \times 673 = 456.1$ g/kg discards

which gives the following for each element:

Element	% of dry mass	Mass (g)
C	44.17	$0.4417 \times 456.1 = 201.50$
H	5.91	$0.0591 \times 456.1 = 27.00$
O	42.50	$0.4250 \times 456.1 = 193.90$
N	0.73	$0.0073 \times 456.1 = 3.34$
		$C_aH_bO_cN_d = 425.7$ g/mol

Thus, for example, 1 g-mol of $C_aH_bO_cN_d$ has 201.5 g of carbon, so using an atomic weight of 12 for carbon, we can compute a as follows:

$$12 \text{ g/mol} \times a \text{ mol} = 201.5 \text{ g}, \quad \text{so } a = 201.5/12 = 16.8 \text{ mol}$$

Similarly, using atomic weights of H, O, and N of 1, 16, and 14, respectively, gives us

$$\text{H: } 1 \text{ g/mol} \times b \text{ mol} = 27.0 \text{ g} \quad b = 27.0 \text{ mol}$$
$$\text{O: } 16 \text{ g/mol} \times c \text{ mol} = 193.9 \text{ g} \quad c = 12.1 \text{ mol}$$

$$N: \ 14 \text{ g/mol} \times d \text{ mol} = 3.34\text{g} \qquad d = 0.24 \text{ mol}$$

So the chemical formula is $C_{16.8}H_{27.0}O_{12.1}N_{0.24}$.

We can now write the equation that describes the complete decomposition of this material using (9.9):

$$C_aH_bO_cN_d \ + \ n\,H_2O \ \longrightarrow \ m\,CH_4 \ + \ s\,CO_2 \ + \ d\,NH_3 \qquad (9.10)$$

where

$$n = (4a - b - 2c + 3d)/4 = (4 \times 16.8 - 27.0 - 2 \times 12.1 + 3 \times 0.24)/4 = 4.17$$
$$m = (4a + b - 2c - 3d)/8 = (4 \times 16.8 + 27.0 - 2 \times 12.1 - 3 \times 0.24)/8 = 8.65$$
$$s = (4a - b + 2c + 3d)/8 = (4 \times 16.8 - 27.0 + 2 \times 12.1 + 3 \times 0.24)/8 = 8.14$$

So complete decomposition can be described by

$$C_{16.8}H_{27.0}O_{12.1}N_{0.24} \ + \ 4.17\,H_2O \ \longrightarrow \ 8.65\,CH_4 \ + \ 8.14\,CO_2 \ + \ 0.24\,NH_3$$

Notice that 8.65 mol out of 17.03 mol of gas produced $(8.65 + 8.14 + 0.24 = 17.03)$ is methane. That is, 51 percent by volume of the gas produced is CH_4, 48 percent is CO_2, and about 1 percent is NH_3. That is, landfill gas is about half methane and half CO_2.

One mole of dried discards yields 8.65 mol of CH_4. At 890 kJ/mol, that equals

$$\text{HHV of } CH_4 \text{ released} = 8.65 \text{ mol} \times 890 \text{ kJ/mol} = 7698 \text{ kJ}$$

To summarize, 1 kg of as received moist discards has 425.6 g of C, H, O, and N, which makes up 1 mol of $C_{16.8}H_{27.0}O_{12.1}N_{0.24}$, and which produces 8.65 mol of CH_4 having an HHV of about 7700 kJ. ∎

The calculation performed in Example 9.10 assumes that everything that can decompose, does decompose. This is an overly optimistic assumption. Excavations of old dumps routinely find newspapers that can still be read a decade or two after disposal. One reason for the slow decomposition is that landfills dry out, and without the moisture assumed in (9.9) the process proceeds slowly, if at all. Also, the calculation does not provide any insight into the rate at which methane is generated. Some models suggest that one-half of the total methane production occurs over the first 10 years, while 30 years are required to generate 90 percent of the ultimate volume (McBean et al., 1995).

Methane is explosive when it is in the presence of air in concentrations between 5 and 15 percent, so some attention must be paid to its management. Within the landfill itself, there is not enough oxygen to cause such an explosion, but when it outgases from the landfill there is some potential for it to cause problems if it is allowed to accumulate under buildings and other enclosed spaces that are located nearby. Methane is lighter than air, so it tends to rise. If there is an impermeable cover over the landfill, the rising gas creates pressure, which can cause it to move horizontally hundreds of meters before emerging from the soil. Pressure relief can be obtained passively, with simple perforated vent pipes that penetrate the cover and extend into a layer of gravel above the wastes; or actively, with collection systems that suck the gases out of the landfill. Passive venting is generally no longer practiced in large landfills because of strict air emission regulations.

Collected landfill gases are usually flared, which wastes the energy that could be recovered but does help control odors and converts the potent greenhouse gas, methane, into carbon dioxide, which is less capable of causing global warming.

Only a few percent of MSW landfills currently take advantage of the energy in landfill gases, but that fraction is growing. One drawback to the use of landfill gases is the relatively low heating value (about 18,300 kJ/m^3, or 500 Btu/ft^3), which is about half the heating value of natural gas. This results, of course, from the fact that only about half of landfill gas is methane, while natural gas is almost entirely methane. Raw landfill gas can be used as is, but it is highly corrosive, so some treatment is usually required. It can be upgraded to a medium-energy-content gas by removing moisture and some of the contaminants while leaving the CO_2 intact. In that form, it can be used in steam or gas turbines to produce electricity. Further upgrading is possible by removing the carbon dioxide, resulting in pipeline-quality gas that can be sold or used to generate electricity on site.

PROBLEMS

9.1 How long would it take to fill a 30-yd^3 packer truck that compresses waste to 750 lb/yd^3 if it travels 100 ft between stops at an average of 5 mph and it takes 1 minute to load 200 lb of waste at each stop? If each stop services four homes and two collection runs are made per day, how many customers could be provided with once-per-week service by this single truck (assuming a five-day week)?

9.2 Consider the following data for a municipal waste collection system:

Travel time, garage to route:	20 min
Travel time, route to disposal site:	20 min
Time to unload at disposal site:	15 min
Time from disposal site to garage:	15 min
Time spent on worker breaks:	40 min/day
Packer truck volume:	25 yd^3
Compaction ratio:	4
Curb volume per service:	0.2 yd^3/customer
Travel time between stops:	30 sec
Customers served per stop:	4
Time loading per stop:	1 min

(a) How many hours per day would the crew have to work if it fills the truck twice per day?

(b) Making two runs per day, how many customers would be served per truck if each home has once-per-week service and the truck is used five days per week?

(c) Suppose the cost of a crew for one truck is $40 per hour for the first eight hours per day, plus $60 per hour for any hours over that amount. Assume that the crew works 52 weeks per year. Furthermore, suppose a packer truck has an annualized cost of $10,000 + $3500/yd^3. What would be the annual cost of service (crew plus truck) per customer?

9.3 To avoid overtime pay, the crew in Problem 9.2 are to work only eight hours per day. They still make two runs to the disposal site each day, but their truck is not always full.

How many customers would the truck operated 5 days/week provide once-per-week service to now?

9.4 To avoid overtime pay, the crew in Problem 9.2 are to work only eight hours per day, which means the truck can be smaller. What minimum-size truck would be needed to serve this route? If the cost of a crew for one truck is $40 per hour for 52 weeks per year, and if packer trucks have an annualized cost of $10,000 + $3500/yd^3, what would be the annual cost of service (crew plus truck) per customer? You might want to compare this with the results found in Problem 9.2.

9.5 Suppose each customer puts out 0.25 m^3 of waste each week. Packer trucks with a compaction ratio of 4 take 0.4 minutes per customer to collect the waste. Two trucks are being considered: one that makes two trips per day to the disposal site, and another that makes three.

Trips per day to disposal site	2	3
Truck volume (m^3)	27	15
Annualized truck cost ($/yr)	120,000	70,000
Time driving, unloading, breaks (min/day)	160	215

(a) Operating five days per week, with once-per-week pick-up, how many customers would each truck service?

(b) How many hours per day would each truck and crew have to operate to fill the trucks each day?

(c) With the cost of crew being $40 per hour, what is the total annual cost of trucks and crew for each system? From that, find the annual cost per customer. Which system is less expensive?

9.6 Suppose a packer truck costs $150,000 to purchase; it uses 2 gallons of fuel per mile at $1.50 per gallon and it is driven 10,000 miles per year; and annual maintenance costs are estimated at $20,000.

(a) If it is amortized over an eight-year period at 12 percent, what is the annualized cost to own and operate this truck?

(b) If labor costs $25 per hour each and the truck has a crew of two who work 40-hour weeks all year, what is the annualized cost of labor?

(c) If this truck hauls 10 tonnes per day, 260 days per year, what is the cost per tonne?

9.7 Rework Examples 9.1–9.3 to confirm the costs per tonne of waste given in Table 9.9 for

(a) One trip to the disposal site per day

(b) Three trips to the disposal site per day

9.8 Suppose a transfer station that handles 200 tonnes per day, five days per week, costs $3 million to build and $100,000 per year to operate. Transfer trucks cost $120,000, carry 20 tonnes per trip, and have an annual cost for maintenance and driver of $80,000 per year. If individual trucks make four trips per day, five days per week from transfer station to disposal site and if trucks and station are amortized at 10 percent interest over 10 years, what is the cost of this operation ($/tonne)?

9.9 Suppose the cost of a direct haul system is given by

$$\text{Direct haul cost}(\$/\text{tonne}) = 40 + 30t_1$$

where t_1 is the time (hrs) required to make a one-way run to the disposal site (or transfer station). Furthermore, suppose the cost of owning and operating a transfer station with its associated long-haul trucks is given by

$$\text{Transfer station cost }(\$/\text{tonne}) = 10 + 10t_2$$

where t_2 is the time (hrs) to make a one-way run from the transfer station to the disposal site. Suppose the disposal site is 1.5 hr from the collection route so that $t_1 + t_2 = 1.5$ hr (see Figure P9.9).

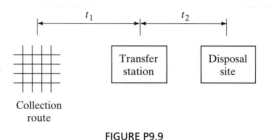

FIGURE P9.9

(a) What is the cost ($/tonne) of a direct haul to the disposal site?

(b) What would be the total cost of the collection system if a transfer station is located 0.3 hr from the collection route?

(c) If the transfer station location is under consideration, what minimum separation (one-way hours) must there be between the transfer station and disposal site to justify having a transfer station?

9.10 A 0.355-L (12-oz) aluminum can has a mass of about 16 g. Assume a 50 percent recycling rate and, using data for the energy intensity of various can production processes given in Table 9.13, find the energy required to produce cans per liter of drink.

9.11 Aluminum cans in the 1970s were considerably heavier than they are now. For a 0.355-L (12-oz) aluminum can, with mass 0.0205 kg, and using a 25 percent recycling rate, find the primary energy required to produce the aluminum in one such can. What percent reduction in primary energy has resulted from the lightweighting of cans and the higher recycling rate illustrated in Example 9.5?

9.12 In the United States approximately 1.6×10^6 tonnes/yr of aluminum is used in beverage cans, 63 percent of which is recovered and recycled. Using estimates from Table 9.12, determine the following:

(a) The total primary energy used to make the aluminum for those cans

(b) The total primary energy that would have been needed if no recycling had occurred

(c) The reduction in CO_2 emissions that result from that recycling

9.13 About 3 million tonnes/yr of aluminum are used in the United States, 35 percent of which is recovered and recycled. Using estimates from Table 9.12, determine the following:

(a) The total CO_2 emissions resulting from production of aluminum

(b) The total primary energy used to produce that aluminum

9.14 "As received" newsprint is 5.97 percent moisture and has an HHV of 18,540 kJ/kg. When dried, 6.1 percent is hydrogen. Find the lower heating value of this waste.

9.15 "As received" corrugated boxes have 5.2 percent moisture and an HHV of 16,380 kJ/kg. When dried, 5.7 percent is hydrogen. Find the LHV.

9.16 A 2-L PET bottle has a mass of 54 g. Polyethylene is roughly 14 percent hydrogen by mass and has an HHV of 43,500 kJ/kg. Find the net energy that might be derived during combustion of one such bottle.

9.17 One way to estimate the energy content of wastes is based on an empirical equation described in Rhyner et al. (1995), which utilizes an elemental analysis (also called an ultimate analysis) of the material in question:

$$HHV (kJ/kg) = 339(C) + 1440(H) - 139(O) + 105(S)$$

where (C), (H), (O), and (S) are the mass percentages of carbon, hydrogen, oxygen, and sulfur in dry material (e.g., cardboard is 43.73 percent carbon so C = 43.73). For the following materials, find the higher heating values (HHV) of the dry waste and the HHV of "as received" waste accounting for the moisture content.

Material	Moisture (% as received)	Dry weight percentages			
		C	H	O	S
(a) Corrugated boxes	5.20	43.73	5.70	44.93	0.21
(b) Junk mail	4.56	37.87	5.41	42.74	0.09
(c) Mixed garbage	72.0	44.99	6.43	28.76	0.52
(d) Lawn grass	75.24	46.18	5.96	36.43	0.42
(e) Demolition softwood	7.7	51.0	6.2	41.8	0.1
(f) Tires	1.02	79.1	6.8	5.9	1.5
(g) Polystyrene	0.20	87.10	8.45	3.96	0.02

Source: Niessen (1977).

9.18 Draw the chemical structures of the following dioxins and furans:

  (a) 1,2,3,4,7,8-hexachlorodibenzo-*p*-dioxin

  (b) 1,2,3,4,6,7,8-heptochlorodibenzo-*p*-dioxin

  (c) Octachlorodibenzo-*p*-dioxin (8 chlorines)

  (d) 2,3,4,7,8-pentachlorodibenzofuran

  (e) 1,2,3,6,7,8-hexachlorodibenzofuran

9.19 The United States sends about 129 million tons of municipal solid waste to landfills. What landfill area would be required for one year's worth of MSW if the landfill density is 800 lb/yd^3, cell depth is 10 ft with one lift per year, and 80 percent of the cell is MSW. What is the area in acres/yr per 1000 people (1 acre = 43,560 ft^2)?

9.20 Suppose a city of 50,000 people generates 40,000 tons of MSW per year. At current recovery and recycling rates, 22 percent of that is recovered or recycled and the rest goes to a landfill. Suppose also that the landfill density is 1000 lb/yd^3, cell depth is 10 ft, and 80 percent of the cell is MSW.

  (a) What lift area would be required per year?

  (b) If the current landfill site covers 50 acres, including 10 acres needed for access roads and auxiliary facilities, and two more lifts are envisioned, how long would it take to complete this landfill?

9.21 If the city described in Problem 9.20 increases its recovery and recycling rate to 40 percent, how many years would it take to complete the landfill?

9.22 One kilogram of as received yard trimmings is made up of approximately 620 g of moisture, 330 g of decomposible organics represented by $C_{12.76}H_{21.28}O_{9.26}N_{0.54}$, and 50 g of other constituents.

  (a) At 0.0224 m^3/mol of CH$_4$, what volume of methane gas would be produced per kilogram of as received yard trimmings?

  (b) At 890 kJ/mol, what is the energy content of that methane?

9.23 Food wastes are estimated to be 72 percent moisture, with the remaining portion containing 45 percent C, 6.4 percent H, 28.8 percent O, 3.3 percent N, and 16.5 percent other constituents.

(a) Write a chemical formula for the C,H,O,N portion of the waste.

(b) Write a balanced chemical reaction showing the production of methane.

(c) What fraction of the volume of gas produced is methane?

(d) At STP (1 atm and 0 °C), 1 mol of gas occupies 22.4×10^{-3} m³. What volume of methane is produced per kilogram of food waste?

(e) Find the HHV value of the methane in kilojoules per kilogram of food waste.

REFERENCES

ARGARWAL, J. C., 1991, Minerals, energy, and the environment, *Energy and Environment in the 21st Century,* J. W. Testor (ed.), MIT Press, Cambridge, MA.

ALTER, H., 1991, The future course of solid waste management in the U.S., *Waste Management and Research,* 9:3–10.

AMERICAN FOREST AND PAPER ASSOCIATION, 1993, *Recovered Paper Statistical Highlights 1992,* Washington, DC.

ATKINS, P. R., H. J., HITTNER, and D. WILLOUGHBY, 1991, Some energy and environmental impacts of aluminum usage, *Energy and Environment in the 21st Century,* J. W. Testor (ed.), MIT Press, Cambridge, MA.

AUSUBEL, J. H., 1989, Regularities in technological development: An environmental view, *Technology and the Environment,* J. H. Ausubel and H. E. Sladovich (eds.), National Academy Press, Washington, DC.

BUKHOLZ, D. M., 1993, Aluminum cans, *The McGraw Hill Recycling Book,* H. F. Lund (ed.), McGraw-Hill, New York.

CHAPMAN, P. F., and F. ROBERTS, 1983, *Metal Resources and Energy,* Butterworths, Boston.

CORBITT, R. A., 1990, *Handbook of Environmental Engineering,* McGraw-Hill, New York.

DENISON, R. A., 1996, Environmental lifecycle comparisons of recycling, landfilling and incineration: A review of recent studies, *Annual Review of Energy and Environment,* Vol. 21, Annual Reviews, Inc., Palo Alto.

DENISON, R. A., and J. RUSTON (eds.), 1990, *Recycling and Incineration,* Environmental Defense Fund, Island Press, Washington, DC.

FRANKLIN ASSOCIATES, LTD., 1990, *Estimates of the Volume of MSW and Selected Components in Trash Cans and Landfills,* for the Council for Solid Waste Solutions, Washington, DC.

FRANKLIN ASSOCIATES, LTD., 1994, *The Role of Recycling in Integrated Solid Waste Management to the Year 2000,* Keep America Beautiful, Inc.

GRAEDEL, T. E., and B. R. ALLENBY, 1995, *Industrial Ecology,* Prentice Hall, Englewood Cliffs, NJ.

GRAHAM, B., 1993, Collection Equipment and Vehicles, *The McGraw-Hill Recycling Handbook,* H. F. Lund (ed.), McGraw-Hill, New York.

HANNON, B. M., 1972, Bottles cans energy, *Environment,* March, 11–21.

HOCKING, M. B., 1991, Relative merits of polystyrene foam and paper in hot drink cups: Implications for packaging, *Environmental Management,* 15(6):731–747.

INTERNATIONAL PETROLEUM INDUSTRY ENVIRONMENTAL CONSERVATION ASSOCIATION, 1992, *Climate Change and Energy Efficiency in Industry,* London.

KHAN, M.Z.A. and Z.H. ABU-GHARAH, 1991, New approach for estimating energy content of municipal solid waste, *Journal of Environmental Engineering,* 117(3):376–380.

KEOLEIAN, G. A., D. MENEREY, B. VIGON, D. TOLLE, B. CORNABY, H. LATHAM, C. HARRISON, T. BOGUSKI, R. HUNT, and J. SELLERS, 1994, *Product Life Cycle Assessment to Reduce Health Risks and Environmental Impacts,* Noyes Publications NJ.

KREITH, F., 1994, *Handbook of Solid Waste Management,* McGraw-Hill, New York.

MCBEAN, E. A., F. A. ROVERS, and G. J. FARQUHAR, 1995, *Solid Waste Landfill Engineering and Design,* Prentice Hall, Englewood Cliffs, NJ.

MILLER, C., 1993, The cost of recycling at the curb, *Waste Age,* October, 46–54.

MOORE, B., 1994, Breakthrough in plastics recovery, *EPA Journal,* Fall 1994, U.S. EPA, Washington, DC.

NEWMAN, A., 1991, The greening of environmental labeling, *Environmental Science and Technology,* 25 (12):–.

NIESSEN, W. R., 1977, *Handbook of Solid Waste Management,* D. G. Wilson (ed.), Van Nostrand Reinhold Co., New York.

OFFICE OF TECHNOLOGY ASSESSMENT, 1992, *Green Products by Design: Choices for a Cleaner Environment,* OTA-E-541, U.S. Government Printing Office, Washington, DC.

PIRAGES, S. W., and J. E. JOHNSTON, 1993, Municipal waste combustion and new source performance standards: Use of scientific and technical information, *Keeping Pace with Science and Engineering, Case Studies in Environmental Regulation,* M.F. Uman (ed.), National Academy of Engineering, Washington, DC. 1993.

RABASCA, L., 1995, Recycling came of age in 1994, *Waste Age,* April, 213–222.

RHYNER, C. R., L. J. SCHWARTZ, R. B. WENGER, and M. G. KOHRELL, 1995, *Waste Management and Resource Recovery,* Lewis Publishers, New York.

SELLERS, V. R. and J. D. SELLERS, 1989, Comparative Energy and Environmental Impacts for Soft Drink Delivery Systems, Franklin Associates, Ltd., Prairie Village, KS.

STEARNS, R. A., 1982, Measuring productivity in residential solid waste collection systems, *Residential Solid Waste Collection,* GRCSA: 3-1/3-19.

STEUTEVILLE, R. A., 1994, Duales System on Firmer Ground in Germany, *BioCycle,* June, 61–63.

STEUTEVILLE, R., 1995, The state of garbage in America, *BioCycle,* April, 54–63.

U.S. EPA, 1989, *Recycling Works! State and Local Solutions to Solid Waste Management Problems,* Office of Solid Waste, Washington, DC.

U.S. EPA, 1991, *Assessing the Environmental Consumer Market,* Office of Policy Planning and Evaluation, 21P-1003, Washington, DC.

U.S. EPA, 1992, *Used Dry Cell Batteries, Is a Collection Program Right for Your Community?,* Solid Waste and Emergency Response, OS-305, Washington, DC.

U.S. EPA, 1994a, *Characterization of Municipal Solid Waste in the United States: 1994 Update,* Office of Solid Waste, Washington, DC.

U.S. EPA, 1994b, *Composting Yard Trimmings and Municipal Solid Waste,* Solid Waste and Emergency Response, EPA530-R-94-003, Washington, DC.

WHITE, P. R., M. FRANKE, and P. HINDLE, 1995, *Integrated Solid Waste Management: A Lifecycle Inventory,* Blackie Academic & Professional, New York.

WORLD RESOURCES INSTITUTE, 1992, *World Resources 1992–93,* Oxford University Press, New York.

YOUNG, J. E., 1992, *Mining the Earth,* Worldwatch Technical Paper 109, Worldwatch Institute, Washington, DC.

Useful Conversion Factors

LENGTH

1 inch	= 2.540 cm
1 foot	= 0.3048 m
1 yard	= 0.9144 m
1 mile	= 1.6093 km
1 meter	= 3.2808 ft
	= 39.37 in.
1 kilometer	= 0.6214 mi

AREA

1 square inch	= 6.452 cm^2
	= 0.0006452 m^2
1 square foot	= 0.0929 m^2
1 acre	= 43,560 ft^2
	= 0.0015625 sq mi
	= 4046.85 m^2
	= 0.404685 ha
1 square mile	= 640 acre
	= 2.604 km^2
	= 259 ha
1 square meter	= 10.764 ft^2
1 hectare	= 2.471 acre
	= 0.00386 sq mi
	= 10,000 m^2

VOLUME

1 cubic foot	= 0.03704 cu yd
	= 7.4805 gal (U.S.)
	= 0.02832 m^3
	= 28.32 L
1 acre foot	= 43,560 ft^3
	= 1233.49 m^3
	= 325,851 gal (U.S.)
1gallon (U.S.)	= 0.134 ft^3
	= 0.003785 m^3
	= 3.785 L
1 cubic meter	= 8.11 × 10^{-4} Ac ft
	= 35.3147 ft^3
	= 264.172 gal (U.S.)
	= 1000 L
	= 10^6 cm^3

LINEAR VELOCITY

1 foot per second	= 0.6818 mph
	= 0.3048 m/s
1 mile per hour	= 1.467 ft/s
	= 0.4470 m/s
	= 1.609 km/hr
1 meter per second	= 3.280 ft/s
	= 2.237 mph

MASS

1 pound (avdp)	= 0.453592 kg
1 kilogram	= 2.205 lb (avdp)
	= 35.27396 oz (avdp)
1 ton (short)	= 2000 lb (avdp)
	= 907.2 kg
	= 0.9072 ton (metric)
1 ton (metric)	= 1000 kg
	= 2204.622 lb (avdp)
	= 1.1023 ton (short)

FLOWRATE

1 cubic foot per second	$= 0.028316 \text{ m}^3/\text{s}$
	$= 448.8 \text{ gal (U.S.)}/\text{min (gpm)}$
1 cubic foot per minute	$= 4.72 \times 10^{-4} \text{ m}^3/\text{s}$
	$= 7.4805 \text{ gpm}$
1 gallon (U.S.) per minute	$= 6.31 \times 10^{-5} \text{ m}^3/\text{s}$
1 million gallons per day	$= 0.0438 \text{ m}^3/\text{s}$
1 million acre feet per year	$= 39.107 \text{ m}^3/\text{s}$
1 cubic meter per second	$= 35.315 \text{ ft}^3/\text{s (cfs)}$
	$= 2118.9 \text{ ft}^3/\text{min (cfm)}$
	$= 22.83 \times 10^6 \text{ gal/d}$
	$= 70.07 \text{ Ac-ft/d}$

DENSITY

1 pound per cubic foot	$= 16.018 \text{ kg/m}^3$
1 pound per gallon	$= 1.2 \times 10^5 \text{ mg/L}$
1 kilogram per cubic meter	$= 0.062428 \text{ lb/ft}^3$
1 gram per cubic centimeter	$= 62.427961 \text{ lb/ft}^3$

CONCENTRATION

1 milligram per liter in water (specific gravity $= 1.0$)	$= 1.0 \text{ ppm}$
	$= 1000 \text{ ppb}$
	$= 1.0 \text{ g/m}^3$
	$= 8.34 \text{ lb per million gal}$

PRESSURE

1 atmosphere	$= 76.0 \text{ cm of Hg}$
	$= 14.696 \text{ lb/in.}^2 \text{ (psia)}$
	$= 29.921 \text{ in. of Hg (32 °F)}$
	$= 33.8995 \text{ ft of H}_2\text{O (32 °F)}$
	$= 101.325 \text{ kPa}$
1 pound per square inch	$= 2.307 \text{ ft of H}_2\text{O}$
	$= 2.036 \text{ in. of Hg}$
	$= 0.06805 \text{ atm}$
1 Pascal (Pa)	$= 1 \text{ N/m}^2$
	$= 1.45 \times 10^{-4} \text{ psia}$
1 inch of mercury (32 °F)	$= 3386.4 \text{ Pa}$
(60 °F)	$= 3376.9 \text{ Pa}$

ENERGY

1 British Thermal Unit	= 778 ft-lb
	= 252 cal
	= 1055 J
	= 0.2930 Whr
1 quadrillion Btu	= 10^{15} Btu
	= 1055×10^{15} J
	= 2.93×10^{11} kWhr
	= 172×10^6 barrels (42-gal) of oil equivalent
	= 36.0×10^6 metric tons of coal equivalent
	= 0.93×10^{12} cubic feet of natural gas equivalent
1 Joule	= 1 N-m
	= 9.48×10^{-4} Btu
	= 0.73756 ft-lb
1 kilowatt-hour	= 3600 kJ
	= 3412 Btu
	= 860 kcal
1 kilocalorie	= 4.185 kJ

POWER

1 kilowatt	= 1000 J/s
	= 3412 Btu/hr
	= 1.340 hp
1 horsepower	= 746 W
	= 550 ft-lb/s
1 quadrillion Btu per year	= 0.471 million barrels of oil per day
	= 0.03345 TW

Index